Nutritional Neurosciences

Series Editor
Mohamed Essa, Sultan Qaboos University
Qaboos, Oman

This book series aims to publish volumes focusing on both basic and clinical research in the field of nutritional neuroscience with a focus on delineating the effect of nutrition on brain function and behavior. The books will examine the role of different nutrients, food agents and supplements (both macro and micro) on brain health, neurodevelopment, neurochemistry, and behaviour. The books will examine the influence of diet, including phytochemicals, antioxidants, dietary supplements, food additives, and other nutrients on the physiology and metabolism of neurons, neurotransmitters and their receptors, cognition, behavior, and hormonal regulations.

The books will also cover the influence of nutrients and dietary supplements on the management of neurological disorders. It details the mechanism of action of phytonutrients on signaling pathways linked with protein folding, aggregation, and neuroinflammation. The books published in the series will be useful for neuroscientists, nutritionists, neurologists, psychiatrists, and those interested in preventive medicine.

Abhishek Kumar Singh
Sachchida Nand Rai
Editors

Rejuvenating the Brain: Nutraceuticals, Autophagy, and Longevity

Springer

Editors
Abhishek Kumar Singh
Manipal Centre for Biotherapeutics Research
Manipal Academy of Higher Education
Manipal, Karnataka, India

Sachchida Nand Rai
Centre of Experimental Medicine and Surgery (CEMS), Institute of Medical Sciences (IMS)
Banaras Hindu University
Varanasi, Uttar Pradesh, India

ISSN 2730-6712 ISSN 2730-6720 (electronic)
Nutritional Neurosciences
ISBN 978-981-95-2789-2 ISBN 978-981-95-2790-8 (eBook)
https://doi.org/10.1007/978-981-95-2790-8

This Springer imprint is published by the registered company Springer Nature Singapore Pte Ltd.
The registered company address is: 152 Beach Road, #21-01/04 Gateway East, Singapore 189721, Singapore

Foreword

Aging is an inevitable biological process, yet its impact on the human brain remains one of the most profound challenges. As we age, the brain undergoes structural and functional changes that can lead to memory loss, cognitive impairment, and, ultimately, neurodegenerative diseases such as Alzheimer's disease and Parkinson's disease. While modern medicine has made great strides in prolonging life, the quality of cognitive health in later years still lags. This disparity highlights the urgent need for innovative strategies that not only treat neurological symptoms but also actively promote brain rejuvenation and resilience against aging.

This book, *Rejuvenating the Brain: Nutraceuticals, Autophagy, and Longevity*, emerges from a growing recognition that the convergence of nutrition, cellular self-renewal mechanisms, and longevity science holds untapped potential for maintaining and even restoring brain health. Nutraceuticals, bioactive compounds derived from foods, have gained scientific traction for their neuroprotective properties, particularly through mechanisms such as anti-inflammatory action, mitochondrial support, and antioxidant defense. Among the most compelling cellular mechanisms discussed herein is autophagy, a self-cleaning process through which cells degrade and recycle damaged organelles and proteins, contributing to cellular renewal and metabolic homeostasis.

The integration of nutraceuticals and autophagy modulation represents a paradigm shift in how we think about brain aging: not as a one-way trajectory toward decline but as a dynamic process that can be influenced by lifestyle, diet, and molecular interventions. This book provides a comprehensive exploration of these concepts through evidence-based chapters covering the biochemical underpinnings of aging, key nutraceutical agents, and the molecular pathways that govern autophagic flux and longevity. Additionally, the book discusses emerging therapeutic strategies such as caloric restriction mimetics, gut–brain axis modulation, and nanoformulation technologies that increase bioavailability and targeted delivery to the brain.

This book is intended for a broad scientific audience, including neuroscientists, gerontologists, pharmacologists, nutrition researchers, and advanced students, as well as clinicians and health policy experts seeking integrative approaches to promote healthy cognitive aging. By synthesizing research from the fields of cellular biology, neurodegeneration, and nutraceutical science, the book aims to offer not only a foundational resource but also a forward-looking vision for interventions that could redefine brain health in the twenty-first century.

I hope that this book will spark curiosity, encourage interdisciplinary collaboration, and inspire new generations of researchers to pursue strategies that enhance the vitality of the aging brain. The path to longevity is no longer just about adding years to life but about adding life to years, especially for the mind.

N. S. Raviraja
Manipal Centre for Biotherapeutics Research,
Academy of Higher Education (MAHE)
Manipal, Karnataka, India

Acknowledgments

Abhishek Kumar Singh gratefully acknowledges the support of the Manipal Centre for Biotherapeutics Research, Manipal Academy of Higher Education (MAHE), Manipal. He also thanks the Anusandhan National Research Foundation (ANRF), Government of India, for financial assistance through the Core Research Grant (CRG/2022/006612), and MAHE for intramural funding. Sachchida Nand Rai sincerely acknowledges the Indian Council of Medical Research (ICMR), Government of India, for supporting his work through the Research Associate Award (No.: 3/1/3/196Neuro/2021-NCD-I).

We extend our heartfelt gratitude to all contributors whose valuable work has enriched this book. Our deepest thanks go to our families for their unwavering support and understanding during the many hours dedicated to this endeavor. We are also grateful to our friends and colleagues for their continuous encouragement and support throughout the writing process.

Lastly, we thank the publishing team at Springer for their professionalism, enthusiasm, and enduring faith in this project. Their expertise in design, production, and dissemination has played a crucial role in bringing this book to life.

Abhishek Kumar Singh
Sachchida Nand Rai

Contents

Editors and Contributors

About the Editors

Abhishek Kumar Singh is an Associate Professor at the Manipal Centre for Biotherapeutics Research (MCBR), Manipal Academy of Higher Education (MAHE), Manipal. With over 12 years of experience in neuroscience, neurobiology of aging, and stem cell research, his interdisciplinary work focuses on brain aging, autophagy, neurodegeneration, nutraceutical product development, and biotherapeutic interventions using advanced stem cell and tissue engineering models. He earned his PhD in Biotechnology from CSIR-Indian Institute of Toxicology Research, Lucknow, where his research focused on xenobiotic metabolism in differentiating human stem cell-derived neurons. He is a Gold Medalist in MSc Biotechnology from Guru Nanak Dev University, Amritsar, and has held various academic and research positions at esteemed institutions, including Washington State University, Guru Ghasidas Central University, Amity University, NIT Rourkela, and the University of Allahabad. He has successfully secured competitive national funding from SERB, MHRD, and MAHE-Intramural schemes for projects on neurogenesis, neurotoxicology, and 3D brain aging models. He has published over 70 peer-reviewed articles in high-impact journals and has contributed several book chapters and edited volumes on aging, neurodegeneration, and regenerative medicine. As a mentor, he has guided PhD and postgraduate research scholars working on neural tissue engineering, nano-senolytics, and neuroinflammatory models applicable for aging and anti-aging research. His contributions extend to editorial responsibilities with leading journals such as *PLOS One*, *Expert Opinion on Drug Metabolism & Toxicology*, and *Frontiers in Molecular Biosciences*. An invited speaker at international conferences, including those hosted by ISN, JNS, APSN, ISN, and SNCI, he has received several accolades, including multiple international travel fellowships, best presentation awards, UGC-Dr. D.S. Kothari Postdoctoral Fellowship, National Postdoctoral Fellowship, and the highest cumulative impact factor award at MCBR in 2024. His current research delves into nanotherapeutic strategies, caloric restriction mimetics, and cellular rejuvenation approaches aimed at combating age-associated neurological decline.

Sachchida Nand Rai is currently working as a Research Associate (ICMR-RA) at the Centre of Experimental Medicine and Surgery, Institute of Medical Sciences, Banaras Hindu University (BHU), Varanasi, a role he has held since January 27, 2023. He earned his PhD in Biochemistry from BHU in 2018, where he began his research journey focusing on neurodegenerative disorders, particularly Parkinson's disease. His current research centers on investigating the neuroprotective and anti-Parkinsonian properties of several Indian medicinal plants, including *Mucuna pruriens*, *Tinospora cordifolia*, and *Withania somnifera*, along with their bioactive constituents such as ursolic acid, betulinic acid, and chlorogenic acid. These studies are primarily conducted in toxin-induced Parkinsonian mouse models to uncover potential mechanisms of action and therapeutic efficacy. Before his current position, he served as a Post-Doctoral Fellow at the Centre of Biotechnology, University of Allahabad, Prayagraj, from January 27, 2020, to January 26, 2023, where he gained extensive experience in neurobiology, experimental pharmacology, and autophagy regulation. He has significantly contributed to the scientific community through his prolific publication record in high-impact journals and his dedicated editorial and peer-review work. He is an Academic Editor for *BMC Research Notes* and serves as Associate Editor for several prestigious journals including *3 Biotech* (Springer Nature), *Molecular Neurobiology*, *Oxidative Medicine and Cellular Longevity*, *Journal of Inflammation*

Research, and *Degenerative Neurological and Neuromuscular Disease*. He is also on the advisory board of *Heliyon Neuroscience*. Throughout his career, he has reviewed more than 2000 manuscripts and edited over 100 papers for reputed international publishers. His publication record is further supported by his authorship of 2 books with Bentham Science, 3 books with CRC Press, and 11 additional books in progress, 7 with Elsevier and 4 with Springer, covering diverse aspects of neuropharmacology, nutraceuticals, and chronic neurological diseases. His research impact is underscored by over 5204 citations on Google Scholar and an h-index of 38. He is an active member of the Indian Academy of Neurosciences and is also engaged in multiple book projects as series editor and volume editor, including contributions to Elsevier, Springer, CRC Press, and Bentham Science. With a strong translational focus, his current research aims to bridge traditional medicinal knowledge with modern neuroscience to explore novel therapeutic avenues for neurodegenerative disorders.

Contributors

Md. Abubakar Department of Pharmacology and Toxicology, National Institute of Pharmaceutical Education and Research, Hajipur, Vaishali, Bihar, India

Krishnendu Adhikary Department of Medical Laboratory Technology, Paramedical College Durgapur, Durgapur, West Bengal, India

Sana Ahmed Department of Pharmacology, Deccan College of Pharmacy Nampally, Hyderabad, Himachal Pradesh, India

Abhishek Ambawatiya Department of Pharmacology and Toxicology, National Institution of Pharmaceutical Education and Research, Hajipur, Vaishali, Bihar, India

Syed Mohammad Nasar Ata Department of Neurology, Henry Ford Health System, Detroit, MI, USA

Jamal Basha Department of Pharmacognosy, Sri Padmavathi School of Pharmacy, Tirupati, Andhra Pradesh, India

Shradha Bisht Uttaranchal Institute of Pharmaceutical Sciences, Uttaranchal University, Dehradun, Uttarakhand, India

Nandini C. Devaru Department of Pharmacology, JSS College of Pharmacy, JSSAHER, Mysuru, Karnataka, India

Krishnendu Ganguly Department of Medical Laboratory Technology, Paramedical College Durgapur, Durgapur, West Bengal, India

Geetika Garg Department of Zoology, Savitribai Phule Pune University, Pune, Maharashtra, India

Vipin Kumar Garg Department of Pharmaceutical Technology, Meerut Institute of Engineering & Technology (MIET), Meerut, Uttar Pradesh, India

Priyanka Gautam Department of Neurology, Institute of Medical Sciences, Banaras Hindu University, Varanasi, India

Shampa Ghosh GloNeuro, Sector 107, Noida, Uttar Pradesh, India

Aditi Giri Department of Pharmaceutical Technology, Meerut Institute of Engineering & Technology (MIET), Meerut, Uttar Pradesh, India

Falguni Goel Department of Pharmaceutical Technology, Meerut Institute of Engineering & Technology (MIET), Meerut, Uttar Pradesh, India

Adrija Gupta Department of Microbiology, University of Calcutta, Kolkata, West Bengal, India

Vishal Jain Center of Neural Basis of Cognition, Carnegie Mellon University, Pittsburgh, PA, USA

Vishakha Jaiswal Amity Institute of Pharmacy, Amity University, Lucknow, Uttar Pradesh, India

Monika Kaushik Amity Institute of Pharmacy, Amity University Madhya Pradesh, Gwalior, Madhya Pradesh, India

Afrin Khan Sri Aurobindo Institute of Pharmacy, Indore, MP, India

Vishi Khattri Department of Pharmaceutical Technology, Meerut Institute of Engineering & Technology (MIET), Meerut, Uttar Pradesh, India

Parameshwar Kondapuram School of Pharmacy, Guru Nanak Institutions Technical Campus, Ibrahimpatnam, Telangana, India

Daksh Kumar Department of Pharmaceutical Technology, Meerut Institute of Engineering & Technology (MIET), Meerut, Uttar Pradesh, India

Nitesh Kumar Department of Pharmacology and Toxicology, National Institution of Pharmaceutical Education and Research, Hajipur, Vaishali, Bihar, India

Parimala Mohanty Department of Community Medicine & Family Medicine, All India Institute of Medical Sciences (AIIMS), Bhubaneswar, India

Baisakhi Mukherjee Department of Medical Laboratory Technology, Paramedical College Durgapur, Durgapur, West Bengal, India

Smruti Navthare Sri Aurobindo Institute of Pharmacy, Indore, MP, India

Sagar Pamu Amity Institute of Pharmacy, Amity University Madhya Pradesh, Gwalior, Madhya Pradesh, India

Department of Pharmacy Practice, Institute of Pharmacy, Nirma University, Gujarat, Ahmedabad, India

Ashish Kumar Parashar Lloyd Institute of Management and Technology, Noida, UP, India

Abhishek Pathak Department of Neurology, Institute of Medical Sciences, Banaras Hindu University, Varanasi, India

Mrugendra Potdar Shri Vile Parle Kelawani Mandal's Institute of Pharmacy, Dhule, Maharashtra, India

Pranamya Prabhu Manipal Centre for Biotherapeutics Research, Manipal Academy of Higher Education, Manipal, Karnataka, India

Swati Prakash Amity Institute of Pharmacy, Amity University, Lucknow, Uttar Pradesh, India

Thumpati Prasanth Department of Chemistry, Virginia Commonwealth University, Richmond, VA, USA

Eswara Rao Puppala Department of Surgery, School of Medicine, St. Louis, MO, USA

Manchala Raghunath ICMR—National Institute of Nutrition, Tarnaka, Hyderabad, Telangana, India

Sachchida N. Rai Centre of Experimental Medicine & Surgery, Institute of Medical Sciences, Banaras Hindu University, Varanasi, Uttar Pradesh, India

Ruchika Rajani Sri Aurobindo Institute of Pharmacy, Indore, MP, India

Devyani Rajput Amity Institute of Pharmacy, Amity University Madhya Pradesh, Gwalior, Madhya Pradesh, India

Ishani Sadhukhan Department of Food Processing, Indian Institute of Engineering Science and Technology, Shibpur, Howrah, West Bengal, India

Sarika School of Biological and Environmental Sciences, Shoolini University of Biotechnology and Management Sciences, Solan, Himachal Pradesh, India

Riya Sarkar Department of Medical Lab Technology, Dr. B. C. Roy Academy of Professional Courses, Durgapur, West Bengal, India

Ishika Singh Manipal Centre for Biotherapeutics Research, Manipal Academy of Higher Education, Manipal, Karnataka, India

Krishna Kumar Singh Symbiosis Centre for Information Technology, Symbiosis International (Deemed University), Rajiv Gandhi InfoTech Park, Hinjawadi, Pune, Maharashtra, India

Payal Singh Department of Zoology, MMV, Banaras Hindu University, Varanasi, Uttar Pradesh, India

Department of Zoology, Banaras Hindu University, Varanasi, Uttar Pradesh, India

Jitendra Kumar Sinha GloNeuro, Sector 107, Noida, Uttar Pradesh, India

Mohammad Talha Department of Pharmacology & Pharmaceutical Chemistry, Anwarul Uloom College of Pharmacy, New Mallepally, Hyderabad, Andhra Pradesh, India

Sahil Thakre NMT Gujarati College of Pharmacy, Indore, MP, India

Prabhash Nath Tripathi Department of Pharmaceutical Technology, Meerut Institute of Engineering & Technology (MIET), Meerut, Uttar Pradesh, India

Department of Pharmaceutical Sciences, College of Pharmacy, University of Arkansas for Medical Sciences, Little Rock, AR, USA

Ritu Varshney P. D. Patel Institute of Applied Sciences, Charotar University of Science and Technology, Anand, Gujarat, India

Akhil Vijay P. D. Patel Institute of Applied Sciences, Charotar University of Science and Technology, Anand, Gujarat, India

Ranjeet Kumar Vishwakarma Department of Physiology, Institute of Medical Sciences, Banaras Hindu University, Varanasi, India

Understanding Brain Aging and Cognitive Decline

1

Daksh Kumar, Falguni Goel, Payal Singh,
Sachchida Nand Rai, and Thumpati Prasanth

Abstract

Brain aging is a multifaceted, multifactorial process involving genetic, environmental, and lifestyle determinants. Cognitive deterioration, the signature of aging, appears as incremental impairments of memory, executive function, and speed of processing, with wide variation in degrees among individuals. This chapter delves into neurobiological processes of brain aging such as synaptic impairment, neuroinflammation, mitochondrial failure, and protein aggregation. It also explores the influence of vascular health, hormonal shift, and oxidative stress on the trajectory of cognition. In addition, we touch on the differentiation between healthy cognitive aging and pathology like Alzheimer's disease and other dementias. We then review interventions promoting cognitive resilience through neuroplasticity-stimulating interventions, dietary change, exercise, and cognitive training. By incorporating knowledge from neuroscience, epidemiology, and clinical studies, the chapter presents an integrated understanding of brain aging and points toward promising avenues for intervention and healthy cognitive longevity.

D. Kumar · F. Goel
Department of Pharmaceutical Technology, Meerut Institute of Engineering & Technology (MIET), Meerut, Uttar Pradesh, India

P. Singh
Department of Zoology, MMV, Banaras Hindu University, Varanasi, Uttar Pradesh, India

S. Nand Rai
Centre of Experimental Medicine & Surgery, Institute of Medical Sciences, Banaras Hindu University, Varanasi, Uttar Pradesh, India

T. Prasanth (✉)
Department of Chemistry, Virginia Commonwealth University, Richmond, VA, USA
e-mail: thumpatip@vcu.edu

A. Kumar Singh, S. Nand Rai (eds.), *Rejuvenating the Brain: Nutraceuticals, Autophagy, and Longevity*, Nutritional Neurosciences,
https://doi.org/10.1007/978-981-95-2790-8_1

Keywords

Cognitive impairment · Neurogenesis · Oxidative stress · Biomarkers · Therapeutic implications

Abbreviations

Aβ	Amyloid beta
AD	Alzheimer's disease
APP	Amyloid precursor protein
BDNF	Brain-derived neurotrophic factor
CNS	Central nervous system
CSF	Cerebrospinal fluid
fMRI	Functional magnetic resonance imaging
HPA	Hypothalamic pituitary adrenal (axis)
MCI	Mild cognitive impairment
NFTs	Neurofibrillary tangles
NMDA	N-methyl D-aspartate (receptor)
ROS	Reactive oxygen species
SIRT	Sirtuins
Tau	Microtubule-associated protein tau
IL-6	Interleukin-6
TNFα	Tumor necrosis factor alpha
PET	Positron emission tomography
MRI	Magnetic resonance imaging
GABA	Gamma-aminobutyric acid
CRP	C-reactive protein
EEG	Electroencephalogram
DG	Dentate gyrus
PFC	Prefrontal cortex
HDAC	Histone *deacetylase*
BBB	Blood-brain barrier
DHA	Docosahexaenoic acid

1.1 Introduction

Aging is a physiological and irreversible process that impacts all organ systems, and the brain is especially sensitive to its manifestations. With increasing age, there are structural, functional, and biochemical alterations in the brain, resulting in cognitive impairments to varying extents. Although mild forgetfulness and processing slowing are generally regarded as normal aging (Lee and Kim 2022), more severe cognitive impairment can be predictive of pathologic conditions like mild cognitive impairment (MCI) or dementia. Understanding the pathophysiologic mechanisms of brain aging and its effects on cognitive function will allow for the formulation of plans for preserving cognitive function and preventing neurodegenerative disease (Yadav et al. 2024).

Brain aging is a multifactorial process that is influenced by intrinsic and extrinsic determinants. Cell-level brain aging is a chronic decline in synaptic vitality, mitochondrial pathology, oxidative damage, and inflammation. These pathways lead to abnormal neuronal signaling and lower plasticity and eventually impair cognitive performance. Structural changes, including cortical thinning, hippocampal loss, and lower integrity of white matter, underlie impaired cognitive function. In addition, neurochemical changes, especially in neurotransmitter systems of the cholinergic, dopaminergic, and glutamatergic systems, also lead to impairments in attention, memory, and executive function (Tripathi et al. 2024).

One of the defining characteristics of pathological brain aging is the accumulation of misfolded proteins amyloid beta and tau into plaques and tangles, respectively, characteristic of Alzheimer's disease. These pathological processes initiate a cascade of neurotoxicity, driving cognitive decline and neurodegeneration (Canepa and Fossati 2021). Not all brains age in this manner. Recent discoveries placed in the foreground the theory of cognitive reserve, according to which people with higher intellectual stimulation, education, and physical fitness exhibit better resistance to age-related cognitive decline. This helped to instill interest in the discovery of modifiable risk factors and interventions that can ensure healthy brain aging (Goel et al. 2025).

Recent developments in neuroimaging, genomics, and biomarker research have deepened knowledge about brain aging and created new possibilities for early detection and intervention. Lifestyle interventions like exercise, nutritional health, cognitive stimulation, and stress reduction have also demonstrated potential to retard age-related decline in cognition. Pharmacologic interventions and new therapeutic strategies to preserve synaptic integrity and decrease neuroinflammation are under investigation (Chauhan et al. 2024).

This chapter discusses the complex brain aging and cognitive decline processes in relation to genetic, environmental, and lifestyle determinants and the interactions between these. It spotlights the latest research efforts towards the development of biomarkers, deciphering aging brain neuroplasticity, and devising personalized interventions aimed at retarding or halting cognitive deterioration. By solving the mystery of brain aging, clinicians and researchers are able to support cognitive longevity and quality of life in elderly adults (Polidori 2024).

1.2 Fundamentals of Brain Aging

Brain aging is a complex process involving continuous structural, cellular, and molecular alterations throughout life, influencing cognitive function and making the brain vulnerable to neurodegenerative diseases. Gray matter is gradually lost with increasing age, especially from areas like the prefrontal cortex, hippocampus, and temporal lobes (Shankar 2010). Losses in all these brain areas are essential for memory, decision-making, and learning, so it is no wonder that age-related cognitive impairment becomes more significant. Sincerely, at the same time, white matter integrity is disrupted, slowing down signal transmission between the various brain areas. Disconnection is the cause of decreased cognitive flexibility and slowed processing speed (Pettit 2014).

At the cellular level, mitochondrial dysfunction is critical in brain aging through decreased ATP production and enhanced oxidative stress. A buildup of ROS results in oxidative damage to DNA, proteins, and lipids, which impairs neuronal function and viability (Misrani et al. 2021). Neuroinflammation, through chronically activated microglia and astrocytes, adds to neuronal injury and plays a role in synaptic dysfunction. Proteostasis, the equilibrium of protein synthesis, folding, and breakdown, is disrupted with aging and leads to the accumulation of misfolded proteins, β-amyloid, and hyperphosphorylated tau, hallmarks of AD (Thapa et al. 2024).

Neurochemical alterations also follow brain aging, including reduced neurotransmitters like dopamine, serotonin, acetylcholine, and GABA. These changes influence mood, memory, and motor regulation, hence the emotional and behavioral alterations experienced by most aging people (Gasiorowska et al. 2021). Synaptic plasticity, the capacity of the brain to change and create new synapses, declines over time, affecting learning and memory. Homeostasis of calcium is also affected, with elevated intracellular calcium levels potentially causing excitotoxicity and hastening neuronal demise (Ekundayo et al. 2024).

Genetic and epigenetic mechanisms determine the course of brain aging. The APOEε4 allele, for example, contributes to a higher risk for Alzheimer's disease, while epigenetic changes, for example, DNA methylation and histone modifications, have the ability to adjust gene expression associated with aging and cognitive impairment (Paniri et al. 2024). Lifestyle and habits, such as nutrition, exercise, and mental activity, also affect brain aging significantly. Caloric restriction and intermittent fasting have been demonstrated to improve mitochondrial function and induce autophagy, which serves to remove damaged cellular components. Exercise enhances neurogenesis and facilitates synaptic plasticity, whereas cognitive training and lifelong learning help to sustain cognitive resilience (Bhushan et al. 2025).

Neurogenesis, the production of new neurons, still takes place in the adult brain, but mostly in the hippocampus, and decreases with age. Brain plasticity, or the ability of the brain to adapt and reorganize, is still a strong force for compensating for age changes, though less efficient with age (Diniz and Crestani 2023). Although some cognitive impairment is unavoidable with age, proactive measures like healthy living, regular physical and mental exercises, and controlling chronic diseases can retard the rate of brain aging and decrease the risk of neurodegenerative disorders. Knowing the mechanisms behind brain aging gives useful insights into designing interventions for healthy aging and maintaining cognitive abilities (Turrini et al. 2023).

1.3 Cognitive Decline: From Normal Aging to Neurodegeneration

Cognitive impairment is a process that is progressive in nature and varies from the minor changes of normal aging to the severe impairment of neurodegenerative disease. Although cognitive function does vary with age, the progression from normal aging to neurodegeneration is characterized by mounting dysfunction in memory, attention, executive function, and other cognitive functions (Leng et al. 2024).

In healthy aging, cognitive impairment is minimal and will usually impact speed of processing, attention, and episodic memory. Older people can process information more slowly and forget things now and then but without substantial enough impairment to greatly affect daily activity (Ivanova et al. 2024). Semantic memory, procedural memory, and lexicon remain somewhat preserved, which enables older individuals to hold firm a good grasp of knowledge and skills. Minor multitasking or name recall difficulties are not uncommon, though they are usually made up by greater dependence upon experience and hard-earned knowledge (Fig. 1.1) (Bhaskar 2021).

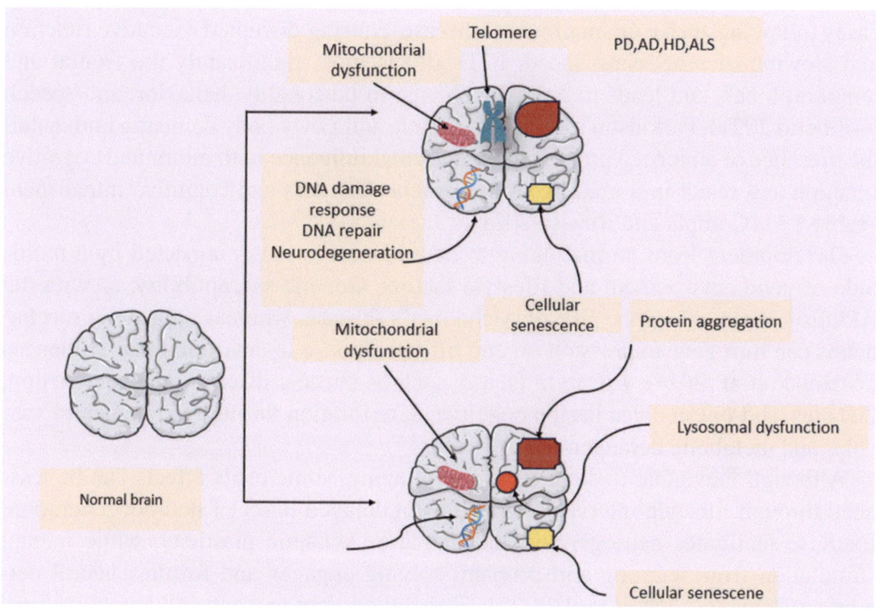

Fig. 1.1 Evolution of cognitive impairment from normal aging to neurodegeneration. The graph presents the continuum of cognitive change that starts with normal age-related decrement in terms of mild impairments in memory, attention, and processing speed. As neurodegenerative pathologies progress, mild cognitive impairment (MCI) occurs in the form of detectable but not disabling cognitive impairments. In the later stages, pathological states like Alzheimer's disease and other dementias arise, with prominent neuronal loss, amyloid beta plaque accumulation, tau tangles, and irreversible cognitive impairment. The main determinants of this pathway are genetic susceptibility, environmental influences, and lifestyle

As the brain grows older, a few people get MCI, an in between phase of normal aging and dementia. MCI is defined by apparent cognitive impairment that surpasses what would be expected for someone's age but does not significantly affect everyday activities. These patients will notice difficulties with memory, word finding, or thinking, but they are still capable of carrying out activities of daily living on their own. MCI, especially amnestic MCI, is a condition that is at high risk of developing into AD or some other dementing illness (Bailey et al. 2022).

Neurodegeneration is defined as the loss of brain structure and function leading to severe, irreversible cognitive damage. AD, the prevalent cause of dementia, is known to be marked by the aggregation of β-amyloid plaques and tau neurofibrillary tangles, which destroys synaptic communication and induces neuronal apoptosis (Mangalmurti and Lukens 2022). The hippocampus, an area important for the creation of new memories, is one of the first regions damaged, resulting in gradual loss of memory, loss of spatial orientation, and problems with language and reasoning. When AD progresses, patients lose even the capacity for simple activities of daily living, and behavioral abnormalities, such as agitation and apathy, predominate (Migliaccio and Cacciamani 2022).

Other neurodegenerative illnesses also account for cognitive impairment. Vascular dementia is caused by chronic diminished perfusion of the brain, in many cases following stroke or small vessel disease, causing disrupted executive function and slowing of processing speed. FTD affects most significantly the frontal and temporal lobes, and leads to a drastic change in personality, behavior, and speech (Giliberto 2021). Parkinson's disease dementia and Lewy body dementia both entail the presence of abnormal protein aggregates that influence both motor and cognitive function and result in a mixture of movement disorders and cognitive impairment (Table 1.1) (Compta and Revesz 2021).

Development from normal aging to neurodegeneration is impacted by a multitude of gene environment and lifestyle factors. Genetic susceptibility, as with the APOEε4 allele, raises the risk of Alzheimer's disease, whereas epigenetic mechanisms can turn gene expression on and off in response to environmental influences (Woźniak et al. 2024). Lifestyle factors such as chronic diseases of hypertension, diabetes, and obesity also hasten cognitive deterioration through the impact of vascular and metabolic derangements.

Although inevitable to some extent with aging, some of its effects can be lessened through lifestyle interventions and even delayed onset of neurodegeneration. Exercise facilitates neurogenesis and increases synaptic plasticity, while mental stimulation from learning and problem solving engages and fortifies neural networks (Clemente-Suárez et al. 2025). High antioxidant and anti-inflammatory content in the diet can minimize oxidative stress, while cardiovascular risk factor control maintains vascular integrity and mental function. Early identification and intervention in those at high risk for cognitive decline could provide the chance to retard progression and enhance quality of life. Familiarity with the spectrum from normal aging to neurodegeneration is key to the establishment of effective means to preserve cognitive function and enhance successful aging (Zhao et al. 2024).

Table 1.1 Stages of cognitive decline—from normal aging to neurodegeneration

Stages	Cognitive features	Neurological changes	Functional impacts	Potential interventions	References
Normal aging	Mild memory lapses	Synaptic slowing	Minimal impact on daily life	Cognitive stimulation	Vejandla et al. (2024)
	Slower processing speed	Reduced neurotransmitter efficiency		Physical activity and diet	Chromiec et al. (2021)
	Difficulty in multitasking	Mild brain atrophy		Social engagement	Stine-Morrow and Manavbasi (2022)
Mild cognitive impairment (MCI)	Noticeable memory difficulties	Hippocampal atrophy	Mild disruption in complex tasks	Cognitive training	Won et al. (2021)
	Difficulty with concentration	Early amyloid beta accumulation	Increased reliance on support	Control of vascular risk factors	Dake (2021)
	Preserved independence	Mild tau pathology		Lifestyle modification	Katayama et al. (2021)
Early Alzheimer's disease (AD)	Significant memory loss	Extensive amyloid beta deposition	Impaired decision-making	Pharmacological treatments (e.g., *cholinesterase* inhibitors)	Khan et al. (2023)
	Impaired learning and recall	Synaptic and neuronal loss	Difficulty in managing finances	Behavioral and cognitive therapy	Reiss et al. (2022)
	Mild language and spatial issues	Increased neuroinflammation	Reduced independence	Caregiver support	Tegiacchi (2023)
Moderate Alzheimer's disease	Severe cognitive decline	Widespread neuronal death	Major assistance required	Structured care environment	Liss et al. (2021)

(continued)

Table 1.1 (continued)

Stages	Cognitive features	Neurological changes	Functional impacts	Potential interventions	References
	Confusion and disorientation	Extensive tau tangles	Impaired basic ADLs (Activities of Daily Living)	Behavioral management strategies	Karunarathna et al. (2024)
	Personality and mood changes	Reduced brain volume		Symptom management	Mirakhori et al. (2022)
Severe Alzheimer's disease	Profound cognitive impairment	Advanced neurodegeneration	Total dependence on caregivers	Palliative care	Culberson et al. (2023)
	Loss of communication abilities	Extensive cortical atrophy	Loss of mobility	Comfort and end of life care	Gupta et al. (2024)
	Inability to perform basic functions	Complete loss of synaptic function			Peng et al. (2022)

1.4 Mechanisms Underlying Cognitive Decline

Cognitive impairment, whether a normal part of aging or the pathway to neurodegenerative diseases, is caused by intricate and highly interlinked biological processes. These mechanisms influence the structure, function, and signaling of the brain, which all, in turn, contribute to memory, attention, executive function, and other cognitive impairments (Haripriyaa and Suthindhiran 2024). Mild cognitive impairment is an inevitable aspect of aging, but pathological processes drive the decline faster and mount the risk of developing Alzheimer's disease and other dementias (Fig. 1.2).

At the center of cognitive aging lies continuous synaptic integrity loss, which impinges on the brain's effective signal transmission. Synaptic plasticity, the capacity of neurons to strengthen or weaken connections based on experience, weakens with age. The learning and memory mechanism, LTP, loses effectiveness, reducing the brain's capacity for information encoding and storage (Kelling 2025). Lessened also are dendritic arborization and spine density, in hippocampus and prefrontal cortex, those areas that subserve executive functions and memory consolidation.

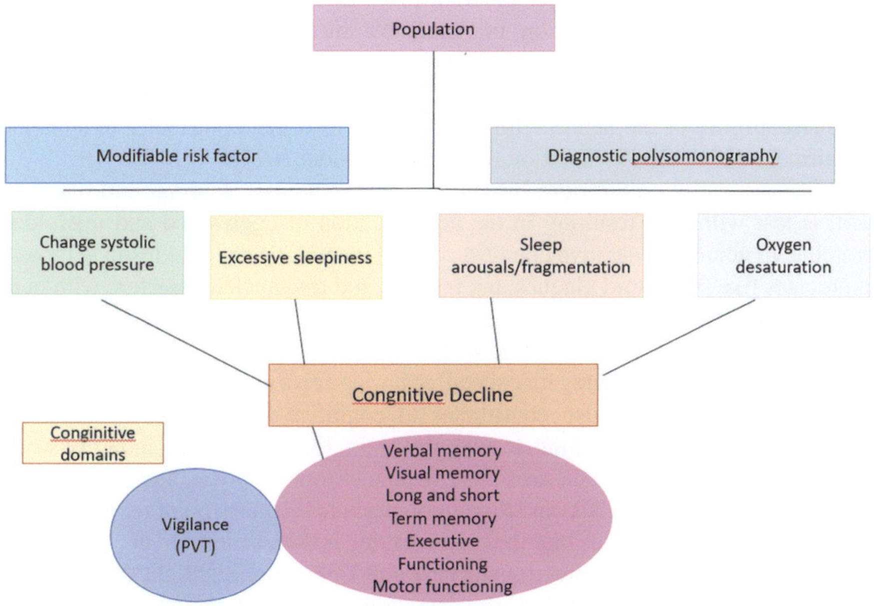

Fig. 1.2 Mechanisms of cognitive decline. This illustration depicts the multifactorial processes that lead to cognitive decline with aging. The primary mechanisms are synaptic dysfunction, neuroinflammation, oxidative stress, mitochondrial dysfunction, and amyloid beta plaque and tau tangle accumulation. Vascular dysfunction and decreased neurotrophic support also contribute to neuronal loss and cognitive impairment. Hormonal changes and dysregulation of the hypothalamicpituitaryadrenal (HPA) axis also play a role in brain aging. These interrelated processes together result in progressive cognitive impairment, from mild cognitive impairment (MCI) to neurodegenerative diseases like Alzheimer's disease

When weaker synapses disappear, the neuronal circuits grow less efficient at message processing and gradually slower, to produce less agile thinking (Ruyle 2021).

Mitochondrial dysfunction is a key mechanism of cognitive impairment. Mitochondria, the cell's energy furnaces, also become less functional with age and produce less ATP and more ROS. Accumulating oxidative damage to DNA, proteins, and lipids in neurons leads to breakdown of cellular function and to apoptosis. Because neurons are highly energy-dependent, any disruption in mitochondrial function disproportionately damages brain function and leads to cognitive dysfunction (Nelson 2023).

Neuroinflammation is also a key baddie in cognitive decline. The resident brain immune cells, microglia, get dysregulated with aging, shifting from a protective to a proinflammatory phenotype. Low grade chronic inflammation, commonly known as "inflammaging," leads to the secretion of proinflammatory cytokines like IL1β, TNFα, and IL6. Repeated inflammation dysregulates synaptic function, inhibits neurogenesis, and promotes neuronal degeneration, leading to cognitive impairment. Astrocytes, as metabolic supporters for neurons, become reactive and make a contribution to the inflammatory environment (Komleva et al. 2021).

The other critical mechanism is the dysregulation of calcium homeostasis that becomes more intense with age. Calcium plays a crucial role in neurotransmitter release and synaptic transmission; but excessive intracellular concentration of calcium can induce excitotoxicity and result in neuronal death. Aging-induced alterations of the regulation of calcium channels and defective buffering capacity enhance the susceptibility of the neurons for calcium overload that contributes to synaptic impairment and neurodegeneration (Rahi and Kaundal 2024).

Proteostasis, the equilibrium between protein synthesis, folding, and degradation, is lost with age, resulting in the accumulation of aggregated and misfolded proteins. In neurodegenerative conditions like Alzheimer's and Parkinson's, protein aggregates like β-amyloid plaques, tau tangles, and α-synuclein interfere with neuronal communication and induce massive synaptic dysfunction and cell death. Ubiquitinproteasome system failure and autophagy pathways failure worsen the buildup of toxic protein aggregates, speeding up cognitive decline (Singh et al. 2024).

Vascular dysfunction also contributes to cognitive impairment by impairing blood supply and delivery of nutrients to the brain. Cerebrovascular integrity worsens with aging, resulting in decreased perfusion, chronic hypoxia, and disruption of the blood-brain barrier (BBB) and disruption provides an entry point for pathogenic molecules and immune cells into the brain, driving inflammation and neuronal damage. VCID (Vascular Cognitive Impairment and Dementia) highlight the necessity of vascular integrity maintenance in sustaining cognitive function (Ramakrishna et al. 2023).

Epigenetic alterations also affect cognitive decline by altering gene expression patterns in reaction to aging and environmental factors. DNA methylation, histone modification, and microRNA imbalance hit genes that are essential for synaptic plasticity, mitochondrial integrity, and inflammation. These changes contribute to worsening cognitive decline and susceptibility to neurodegenerative disease, a fact that underlines the importance of epigenetic regulation in the brain (Tripathi et al. 2019).

Neurogenesis, the production of new neurons in the adult brain, occurs primarily in the hippocampus but declines with increasing age. Dysfunctional neurogenesis reduces the capacity of the brain to learn and recover, limiting cognitive resilience. While neurogenesis continues to some extent throughout life, aging modifies the microenvironment, for example, inflammation and oxidative stress, to render it less permissive for neuronal growth and integration (Srivastava et al. 2019).

Cumulatively, these processes initiate a cascade of events that increasingly compromise cognitive function. The interaction between synaptic loss, mitochondrial dysfunction, chronic inflammation, calcium dysregulation, failure of proteostasis, and vascular compromise underlies the gradual breakdown of cognitive capacities. Knowledge of these underlying mechanisms offers important insight into possible therapeutic targets and lifestyle interventions to forestall cognitive decline and facilitate healthy brain aging (Rai et al. 2020).

1.5 Risk Factors and Prevention Strategies

Cognitive impairment is driven by both nonmodifiable and modifiable risk factors that shape the tempo and magnitude of brain function change over time. Age is the most important nonmodifiable risk factor, inasmuch as the risk for cognitive impairment is directly related to increasing age (Ramakrishna et al. 2024). The structural and functional alterations within the brain, including synaptic loss, decreased neurogenesis, and the development of oxidative stress, all contribute to the normal diminution. Genetic influences also have a significant role, and the APOEε4 allele is among the most powerful genetic risk factors for AD. People who carry this allele are more likely to get Alzheimer's, though not every carrier becomes mentally impaired. In addition, there is also a family history of dementia, implying a genetic factor, while there could also be sex-related variations in susceptibility with women at increased risk due to hormonal modifications after menopause (Hogervorst et al. 2023).

On the other hand, there are many modifiable risk factors that can be treated by specific interventions to decrease the risk of cognitive impairment and neurodegenerative diseases. Reduced cardiovascular health, such as hypertension, diabetes, hyperlipidaemia, and obesity, increases greatly the risk of cognitive impairment through damage to blood vessels and decreased cerebral perfusion (Patel and Edison 2024). Vascular dysfunction over time causes hypoxia, inflammation, and disruption of the blood-brain barrier, thus speeding up neuronal damage. Control of cardiovascular risk by frequent monitoring and management of blood pressure, glucose, and cholesterol, coupled with healthy weight and abstinence from smoking, can prevent deterioration of cognitive functions (Krivanek et al. 2021).

Physical exercise is one of the best methods to avoid cognitive deterioration. Exercise, on a regular basis, improves cerebral blood flow, encourages neurogenesis, and enhances synaptic plasticity while diminishing inflammation and oxidative stress (Bhattacharya et al. 2023). Aerobic exercise, resistance training, and mind–body exercise such as yoga and tai chi have all been shown to have beneficial influences on cognitive functioning in older individuals. Participation in at least 150 minutes per week of moderate intensity aerobic activity, supplemented with strength exercises and balance training, enhances brain health and decreases the likelihood of neurodegeneration (Grzywniak 2025).

Cognitive stimulation and lifelong learning also play a role in cognitive resilience. Brain- stimulating activities like reading, acquiring new skills, and working on problem solving activities stimulate the growth of synapses and consolidate neural networks. Acquiring new languages, playing a musical instrument, or engaging in cognitively stimulating hobbies offers ongoing cognitive stimulation, which improves neural plasticity and develops cognitive reserve, making the brain more resistant to age-related decline (Knight 2023).

Diet and nutrition are also critical in promoting brain health. The Mediterranean diet that consists of high fruit, vegetable, whole grain, lean protein, and healthy fat intake has been linked with reduced risk for AD and other neurodegenerative disorders (Bhuiyan et al. 2023). Antioxidants, omega-3 fatty acids, and anti-inflammatory medications guard against oxidative stress and inflammation, and key vitamins and minerals like B vitamins, vitamin D, and magnesium maintain cognitive capacity. Reduced intake of processed food, added sugar, and saturated fat will also guard against metabolic maladaptation and vascular damage and is a reason for increased long-term brain capacity (Fathima et al. 2023).

Stresslessness and sleep are also required in preventing cognitive impairment. Disturbances in sleep, such as insomnia and sleep apnea, impair brain function to eliminate toxic waste products such as β-amyloid that had built up while awake. Stress produces chronic activation of the HPA axis (Sharan and Vellapandian 2024), which leads to excessive levels of cortisol that damage the hippocampus and memory. Prioritizing regular sleep habits, creating a soothing bedtime routine, and coping with stress using mindfulness, meditation, and relaxation methods can counteract these adverse effects and enhance cognitive resilience (Abbo Bacia 2024).

Social interaction and emotional health also contribute importantly to staying cognitively healthy. A good social relationship is a cognitive stimulus, buffers stress, and is good for emotional health, while social isolation and loneliness lead to a high risk of decline and depression (Guarnera et al. 2023). Being involved in group activities, staying connected to family and friends, and active involvement in local community events induce emotional support and cognitive stimulation that enhance brain resilience over time (Table 1.2).

Environmental toxins, such as long-term exposure to heavy metals, pesticides, and air pollutants, may augment the risk for cognitive impairment. Neurotoxins interfere with neuronal function and enhance oxidative stress, leading to neuronal injury. Reducing environmental toxin exposure, avoiding excessive alcohol use, and abstaining from illicit drug use can minimize the cumulative load of toxins on the brain, and thereby promote sustained cognitive function (Soyka et al. 2017).

Implementation of multiple prevention measures provides the strongest protection against cognitive impairment. Interventions based on multidomain approaches

Table 1.2 Risk factors and prevention strategies for cognitive decline

Risk factor	Mechanism/impact	Prevention/ intervention strategy	References
Age	Natural decline in neuronal function	Early cognitive stimulation	Connell et al. (2022)
	Increased vulnerability to neurodegeneration	Lifelong learning and mental exercises	Lisko et al. (2021), Bieri et al. (2023)
Genetic predisposition	The presence of APOEε4 allele	Genetic counseling and risk assessment	Saragea (2024), Choudhury et al. (2021)
	Higher risk of Alzheimer's disease	Personalized preventive strategies	Azar et al. (2021), Lista et al. (2022)
Cardiovascular health	Hypertension, diabetes, and obesity	Blood pressure and glucose control	Azar et al. (2021)
	Impaired blood flow to the brain	Healthy diet (Mediterranean, DASH)	Chen et al. (2024)
Physical inactivity	Reduced blood flow and oxygen to the brain	Regular physical activity (aerobic and strength training)	Daniele et al. (2022)
Poor diet	Deficiency in essential nutrients	Antioxidant-rich diet (vitamins, omega-3)	Barta et al. (2022)
	Increased oxidative stress	Limited saturated fats and processed foods	Onyango (2021)
Smoking and alcohol use	Vascular damage and inflammation	Smoking cessation programs	Jalali et al. (2021), Attard et al. (2021)
	Increased oxidative stress	Moderate or no alcohol consumption	Viola et al. (2023)
Sleep disorders	Disrupted brain clearance processes (e.g., glymphatic system)	Sleep hygiene and management of sleep apnea	Yi et al. (2022), Nepozitek et al. (2025)
Chronic stress and depression	Dysregulation of HPA axis	Stress management techniques (yoga, meditation)	Ningthoujam et al. (2021), Nesin and Sathyamoorthy (2025)
	Increased neuroinflammation	Early treatment of depression	Xiao et al. (2021), Troubat et al. (2021)
Social isolation	Reduced cognitive stimulation	Active social engagement	Cardona and Andrés (2023), Xiong et al. (2023)
	Increased risk of depression	Participation in group activities	Motillon-Toudic et al. (2022), Cené et al. (2022)
Traumatic brain injury (TBI)	Chronic neuroinflammation,	Use of protective headgear	Sone et al. (2017)
	Increased risk of neurodegeneration	Fall prevention strategies	DeKosky and Asken (2017)
Environmental toxins	Oxidative stress and neuronal damage	Avoidance of heavy metals and pollutants	Münzel and Daiber (2018)
Hormonal changes	Estrogen decline in postmenopausal women	Hormone replacement therapy (under medical supervision)	Pinkerton (2020)

that deal with lifestyle interventions like diet, physical exercise, cognitive training, and vascular risk management have proved to significantly enhance cognitive outcomes and lower the risk of dementia (Meng et al. 2022). Modifiable risk factors can be targeted and, through a proactive strategy for brain health, enhanced cognitive resilience, delayed onset of neurodegenerative diseases, and healthy aging can be supported, ultimately ensuring a better quality of life (Pappalettera et al. 2024).

1.6 Diagnostic Tools and Biomarkers

Diagnosis of cognitive decline, especially the differentiation between normal aging and neurodegenerative diseases, involves a combination of clinical evaluation, neuroimaging methods, and the determination of valid biomarkers (Koníčková et al. 2022). Early and precise diagnosis is necessary for timely treatment and management, and the progress in diagnostic technologies and biomarker studies has considerably enhanced the capacity to identify cognitive impairment and follow disease progression (Kale et al. 2024).

Clinical assessment is the basis of cognitive testing and usually consists of a comprehensive medical history, neuropsychological examination, and functional evaluation. Doctors take a history of the onset, duration, and course of cognitive symptoms, and also inquire about comorbid illnesses, medications, and family history of dementia (Reiss et al. 2022). Standardized cognitive examinations, for example, MMSE, MoCA, and Clock Drawing Test, yield quantitative measures of cognition, memory, executive function, and visuospatial skills. These examinations distinguish MCI, potentially preceding dementia, from more severe neurodegenerative disorders like AD (Öhman 2025).

Neuroimaging methods have proven invaluable in the diagnosis and assessment of cognitive deterioration by offering insight into structural, functional, and molecular alterations of the brain (Etekochay et al. 2024). MRI-based structural imaging is able to identify cortical atrophy, most notably in the hippocampus and medial temporal lobes, characteristic of AD. MRI is also capable of detecting vascular abnormalities, white matter lesions, and microbleeds for VCID. CT scans are less sensitive but can otherwise be utilized to eliminate other cognitive decline causes like tumors or hydrocephalus (Li et al. 2023).

Functional imaging methods, including PET and fMRI, offer information on brain function and metabolic alteration. FDG-PET (Fluorodeoxyglucose Positron Emission Tomography) quantifies glucose metabolism, with evidence of hypometabolism in areas compromised by Alzheimer's disease (Zhang et al. 2021), like the posterior cingulate cortex and temporoparietal lobes. β-amyloid plaque detection through PET imaging reflects a pathological feature of Alzheimer's disease, and it is possible to make early and precise diagnoses even before clinical manifestations. Tau PET imaging that identifies tau neurofibrillary tangles, offers complementary data regarding disease severity and progression (Pontecorvo and Mintun 2011).

Blood-based biomarkers are increasingly taking center stage as less invasive and more convenient options compared to CSF analysis. Advances in plasma biomarker research over the last few years have discovered promising candidates, such as plasma Aβ42/40 ratio, ptau181, and neurofilament light chain (NfL) (El Abiad et al. 2024). Ptau181 and NfL are associated with elevated levels of neurodegeneration and neuronal damage and hence are excellent markers for disease progression monitoring. Blood-based biomarkers can potentially allow for mass screening and early detection of cognitive impairment in clinical as well as community settings (Batzu et al. 2022).

Genetic testing has a part to play in evaluating the risk of cognitive decline, especially among people with a history of neurodegenerative disorders within their family. The presence of the APOEε4 allele can be detected by testing (Saragea 2024), which may inform one about the susceptibility to Alzheimer's disease but is not used as a diagnostic criterion. In a few instances of familial Alzheimer's disease, genetic mutations in genes like APP, PSEN1, and PSEN2 can be detected by genetic testing, which gives information about the hereditary nature of the disease (Pagnon de la Vega et al. 2022).

New technologies like digital biomarkers and AI-based solutions are transforming cognitive testing and early detection. Wearable technology, smartphone apps, and voice analysis systems can assess cognitive function continuously and identify subtle signs of early cognitive impairment. AI software can process large datasets, recognize patterns, and estimate the probability of cognitive decline, which can be used for individualized risk assessment and treatment (Sultana et al. 2023).

Combining various diagnostic modalities increases the validity of diagnosing cognitive impairment and neurodegenerative diseases. Combining clinical assessment, neuroimaging, biomarker studies, and genetic testing offers an integrated picture of a person's cognitive health and enables the formulation of customized treatment strategies (Mackey et al. 2025). With progressive research and discovery of new biomarkers and new technologies, earlier diagnosis of cognitive decline and customizing interventions on an individual's needs will increase, eventually refining outcomes for risk individuals of neurodegenerative disorders (Garg 2025).

1.7 Therapeutic Approaches and Future Directions

Therapeutic interventions for cognitive decline and neurodegenerative conditions like Alzheimer's, Parkinson's, and other dementias have hitherto centered on symptomatic management, delaying disease progression, and enhancing quality of life. Although treatments currently available yield modest benefits in halting the underlying pathology, recent advances in research and novel therapeutic approaches promise better treatments in the future. The treatment landscape of cognitive decline is changing fast, with increasing focus on disease-modifying treatments, precision medicine, and lifestyle interventions for delaying or preventing neurodegeneration (Behl et al. 2022).

Pharmacotherapies have been the pillar of therapy in Alzheimer's disease and other neurodegenerative disorders. The cholinesterase inhibitors donepezil, rivastigmine, and galantamine are the usual drugs to enhance cholinergic transmission and moderately improve cognition for a limited duration in patients with mild to moderate Alzheimer's disease (Li et al. 2019). They are effective in stopping the hydrolysis of acetylcholine, an essential neurotransmitter in learning and memory. In more progressive phases, memantine, an NMDA receptor antagonist, is utilized frequently to control the activity of glutamate and safeguard neurons against excitotoxicity. While these drugs may relieve symptoms, their impact is limited and fails to stop the inherent neurodegenerative mechanisms (Binvignat and Olloquequi 2020).

Over the past few years, the progress in disease modifying therapies has accelerated, with an emphasis on the modulation of the pathological features of Alzheimer's disease, such as β-amyloid plaques and tau neurofibrillary tangles (Ju and Tam 2022). Anti-amyloid monoclonal antibodies like aducanumab and lecanemab have been promising in lowering amyloid burden in the brain and slowing down cognitive decline in patients with early-stage Alzheimer's disease. These antibodies target amyloid plaques and enable them to be removed by the immune system. Although the therapeutic benefit of such therapies is still a matter of debate, they constitute an important advancement toward being able to change the course of Alzheimer's disease as opposed to only treating its symptoms (Yiannopoulou and Papageorgiou 2013).

Tau-directed therapies are also in the pipeline as a possible route for the treatment of Alzheimer's and other tauopathies. Tau-aggregation inhibitors, antitau antibodies, and gene-silencing strategies are designed to stop the accumulation and propagation of pathological tau within the brain. Initial studies have reported promising outcomes, but more work must be done in order to confirm their safety and effectiveness in the clinical context (Congdon and Sigurdsson 2018).

Anti-inflammatory and immunomodulatory strategies are increasingly being considered as inflammation is a key factor in the development of neurodegenerative disorders. Chronic neuroinflammation is responsible for neuronal injury and exacerbates cognitive impairment (Singhal et al. 2014). Immunomodulatory strategies targeting microglial activation as well as cytokine regulation may have the ability to minimize neuroinflammation and preserve against neuronal damage. Clinical trials are ongoing to assess the effectiveness of anti-inflammatory drugs, including NSAIDs, immunotherapies, and new compounds that act on inflammatory pathways (Hou et al. 2021).

Gene therapy and stem-cell based therapies are among the advanced techniques capable of transforming the management of neurodegenerative disorders. Gene therapy methods seek to introduce genetic material for the purpose of correcting or replacing genes that are faulty and responsible for neurodegeneration (Ingusci et al. 2019). Gene therapy strategies in Alzheimer's disease are under investigation to augment the levels of neuroprotective proteins, impede the synthesis of amyloid and tau with toxic properties, and facilitate synaptic repair. Therapy using stem cells, such as neural stem cell transplantation or the transplantation of induced pluripotent stem cells (iPSCs), promises to replace injured brain tissue and regain lost neuronal function. Though experimental in nature, such interventions present promise for reversal of cognitive deterioration and for regeneration of neurons (Rahimi Darehbagh et al. 2024).

Nonpharmacologic interventions remain instrumental in controlling cognitive decline and quality of life. Cognitive rehabilitation programs and cognitive training aimed at facilitating neural plasticity and reinforcing cognitive reserve have exhibited beneficial impacts among patients with MCI and incipient dementia (Jellinger and Attems 2013). These include specified cognitive training techniques that concentrate on memory, attention, executive ability, and the ability to resolve problems and benefit daily life, slowing down cognitive decline and advancing daily performance.

Physical exercise and lifestyle change are being increasingly identified as potent tools for maintaining cognitive function and lowering the risk of neurodegenerative diseases. Exercise in general encourages neurogenesis, improves synaptic plasticity, and enhances vascular health, all of which have positive implications for cognitive outcomes (Yau et al. 2014). The inclusion of aerobic exercise, resistance training, and mind–body interventions such as yoga and tai chi into daily activities can yield long-term brain health benefits. Furthermore, dietary interventions like the Mediterranean diet that is high in antioxidants and anti-inflammatory phytochemicals have been linked to reduced cognitive decline risk and better cognitive performance (Frisardi et al. 2010).

New technologies, such as digital therapeutics and AI-based interventions, are revolutionizing the field of cognitive health management. Digital platforms that deliver individualized cognitive training, lifestyle guidance, and cognitive function monitoring at a distance provide scalable and affordable solutions for populations at risk for cognitive decline. AI-based algorithms can process large datasets, forecast cognitive trajectories, and tailor interventions to an individual's risk profile, opening up the possibility for precision medicine strategies for cognitive care (Gutman et al. 2024).

In the future, the treatment of cognitive decline is in the potential of merging multidomain interventions that tackle the intricate dynamics of the genetic, environmental, and lifestyle risk factors that lead to neurodegeneration (Uleman 2023). Multimodal interventions featuring pharmacologic treatments, cognitive training, exercise, and lifestyle changes have the most promise to decelerate cognitive decline and promote brain resilience. Continued research into new biomarkers, therapeutic targets, and novel delivery systems will further inform the creation of next-generation treatments, leading ever closer to preventing, delaying, and eventually reversing cognitive loss.

1.8 Conclusions

Cognitive impairment and neurodegenerative disorders are enormous challenges, yet research progress and therapeutic innovation provide promise for more favorable outcomes. Although treatment has been directed toward symptom alleviation, novel disease-modifying drugs aimed at amyloid, tau, and inflammation are setting the stage for improved interventions. Nonpharmacologic interventions such as cognitive training, exercise, and lifestyle change continue to play a critical role in maintaining brain health and stalling decline. The future of cognitive care is in tailored, multidomain interventions that bring together pharmacological advances with lifestyle interventions to ultimately increase cognitive resilience and improve the quality of life for older persons.

References

Abbo Bacia J (2024) Holistic approaches to stress management: understanding, prevention, and coping strategies. Res Invent J Curr Rese Humanit Soc Sci 3:83–87

Attard R et al (2021) The impact of frequency, pattern, intensity, and type of alcohol consumption, and its combined effect with smoking on inflammation, lipid profile, and the risk of myocardial infarction. J Public Health 29:611–624

Azar J et al (2021) Precision health in Alzheimer disease: risk assessment-based strategies. Precis Med Sci 10(2):54–70

Bailey TG et al (2022) Cerebrovascular function and its association with systemic artery function and stiffness in older adults with and without mild cognitive impairment. Eur J Appl Physiol 122(8):1843–1856

Barta JR, Kim S, Park HJ (2022) Energy-dense diets lower in protein, antioxidants, and omega-3 fatty acids among US adults with a self-reported head injury with loss of consciousness: a nationwide study, NHANES 2011-2014. Nutr Res 105:147–153

Batzu L et al (2022) Plasma p-tau181, neurofilament light chain and association with cognition in Parkinson's disease. NPJ Parkinson's Dis 8(1):154

Behl T et al (2022) The road to precision medicine: eliminating the "one size fits all" approach in Alzheimer's disease. Biomed Pharmacother 153:113337

Bhaskar S (2021) Wrap your ad around media multitasking: understanding the factors that inhibit and amplify advertising effectiveness while media multitasking

Bhattacharya P, Chatterjee S, Roy D (2023) Impact of exercise on brain neurochemicals: a comprehensive review. Sport Sci Health 19(2):405–452

Bhuiyan NZ et al (2023) Prevention of Alzheimer's disease through diet: an exploratory review. Metabolism open 20:100257

Bhushan R, Goel F, Singh S (2025) Beyond dopamine: novel therapeutic pathways for Parkinson's disease through receptor signaling. CNS & Neurol Disorders-Drug Targets

Bieri G, Schroer AB, Villeda SA (2023) Blood-to-brain communication in aging and rejuvenation. Nat Neurosci 26(3):379–393

Binvignat O, Olloquequi J (2020) Excitotoxicity as a target against neurodegenerative processes. Curr Pharm Des 26(12):1251–1262

Canepa E, Fossati S (2021) Impact of tau on neurovascular pathology in Alzheimer's disease. Front Neurol 11:573324

Cardona M, Andrés P (2023) Are social isolation and loneliness associated with cognitive decline in ageing? Front Aging Neurosci 15:1075563

Cené CW et al (2022) Effects of objective and perceived social isolation on cardiovascular and brain health: a scientific statement from the American Heart Association. J Am Heart Assoc 11(16):e026493

Chauhan K, Goel F, Singh S (2024) Apigenin protects melanocytes and improve tyrosinase activity in a hydroquinone induced vitiligo mouse model targeting P38 MAP kinase signaling: histopathology and immunohistochemistry analysis. Naunyn Schmiedeberg's Arch Pharmacol 397(7):4859–4869

Chen H et al (2024) Associations of the Mediterranean-DASH intervention for neurodegenerative delay diet with brain structural markers and their changes. Alzheimers Dement 20(2):1190–1200

Choudhury P, Ramanan VK, Boeve BF (2021) Apoe ε4 allele testing and risk of Alzheimer disease. JAMA 325(5):484–485

Chromiec PA et al (2021) The proper diet and regular physical activity slow down the development of Parkinson disease. Aging Dis 12(7):1605
Clemente-Suárez VJ et al (2025) Neuro-nutrition and exercise synergy: exploring the bioengineering of cognitive enhancement and mental health optimization. Bioengineering 12(2):208
Compta Y, Revesz T (2021) Neuropathological and biomarker findings in Parkinson's disease and Alzheimer's disease: from protein aggregates to synaptic dysfunction. J Parkinsons Dis 11(1):107–121
Congdon EE, Sigurdsson EM (2018) Tau-targeting therapies for Alzheimer disease. Nat Rev Neurol 14(7):399–415
Connell E et al (2022) Microbial-derived metabolites as a risk factor of age-related cognitive decline and dementia. Mol Neurodegener 17(1):43
Culberson JW et al (2023) Urgent needs of caregiving in ageing populations with Alzheimer's disease and other chronic conditions: support our loved ones. Ageing Res Rev 90:102001
Dake M (2021) Modifiable lifestyle factors as modulators of brain parameters and cognitive decline in ageing and Alzheimer's disease. University of Sheffield
Daniele A, Lucas SJ, Rendeiro C (2022) Detrimental effects of physical inactivity on peripheral and brain vasculature in humans: insights into mechanisms, long-term health consequences and protective strategies. Front Physiol 13:998380
DeKosky ST, Asken BM (2017) Injury cascades in TBI-related neurodegeneration. Brain Inj 31(9):1177–1182
Diniz CRAF, Crestani AP (2023) The times they are a-changin': a proposal on how brain flexibility goes beyond the obvious to include the concepts of "upward" and "downward" to neuroplasticity. Mol Psychiatry 28(3):977–992
Ekundayo BE et al (2024) Oxidative stress, endoplasmic reticulum stress and apoptosis in the pathology of Alzheimer's disease. Cell Biochem Biophys 82(2):457–477
El Abiad E et al (2024) Navigating the Alzheimer's biomarker landscape: a comprehensive analysis of fluid-based diagnostics. Cells 13(22):1901
Etekochay MO et al (2024) Unveiling new strategies facilitating the implementation of artificial intelligence in neuroimaging for the early detection of Alzheimer's disease. J Alzheimer's Dis 99(1):1–20
Fathima S, Prokopiou E, Georgiou T (2023) Omega-3 polyunsaturated fatty acids and their antioxidant, anti-inflammatory and neuroprotective effects in diabetic retinopathy: a narrative review. Front Biosci Landmark 28(7):153
Frisardi V et al (2010) Nutraceutical properties of Mediterranean diet and cognitive decline: possible underlying mechanisms. J Alzheimer's Dis 22(3):715–740
Garg VK (2025) Revolutionizing early diagnosis: the role of artificial intelligence in neurodegenerative disorders. J Adv Sci Res 16(03):31–37
Gasiorowska A et al (2021) The biology and pathobiology of glutamatergic, cholinergic, and dopaminergic signaling in the aging brain. Front Aging Neurosci 13:654931
Giliberto L (2021) Neurology evaluation in dementia and neurodegenerative disease. In: Hybrid PET/MR neuroimaging: a comprehensive approach. Springer, pp 227–245
Goel F, Dobhal V, Singh S (2025) Protective effects of diacerein against quinolinic acid–induced Huntington's disease–like symptoms in adult zebrafish by targeting GSK-3β signalling. Naunyn Schmiedeberg's Arch Pharmacol:1–23
Grzywniak C (2025) The supporting role of physical exercise in the prevention end treatment of neurodegenerative and mental diseases. Baltic J Health Phys Activity 17(1):4
Guarnera J, Yuen E, Macpherson H (2023) The impact of loneliness and social isolation on cognitive aging: a narrative review. J Alzheimer's Dis Rep 7(1):699–714
Gupta UJ et al (2024) Difficulties and challenges in Alzheimer's disease: care giver support in different stages. In: Difficulties and challenges in geriatric health management, vol 303-325. IGI Global
Gutman B et al (2024) Twenty-five years of AI in neurology: the journey of predictive medicine and biological breakthroughs. JMIR Neurotechnol 3(1):e59556

Haripriyaa M, Suthindhiran K (2024) The important role of systems biology in neuroscience and neurology and its associated diseases. In: Systems biology approaches: prevention, diagnosis, and understanding mechanisms of complex diseases. Springer, pp 343–392
Hogervorst E, Temple S, O'Donnell E (2023) Sex differences in dementia. In: Sex differences in brain function and dysfunction, pp 309–331
Hou J, Karin M, Sun B (2021) Targeting cancer-promoting inflammation – have anti-inflammatory therapies come of age? Nat Rev Clin Oncol 18(5):261–279
Ingusci S et al (2019) Gene therapy tools for brain diseases. Front Pharmacol 10:724
Ivanova O, Martínez-Nicolás I, Meilán JJG (2024) Speech changes in old age: methodological considerations for speech-based discrimination of healthy ageing and Alzheimer's disease. Int J Lang Commun Disord 59(1):13–37
Jalali Z et al (2021) Smoking, alcohol and opioids effect on coronary microcirculation: an update overview. BMC Cardiovasc Disord 21:1–17
Jellinger KA, Attems J (2013) Neuropathological approaches to cerebral aging and neuroplasticity. Dialogues Clin Neurosci 15(1):29–43
Ju Y, Tam KY (2022) Pathological mechanisms and therapeutic strategies for Alzheimer's disease. Neural Regen Res 17(3):543–549
Kale MB et al (2024) AI-driven innovations in Alzheimer's disease: integrating early diagnosis, personalized treatment, and prognostic modelling. Ageing Res Rev:102497
Karunarathna I et al (2024) Navigating Alzheimer's disease: diagnosis, treatment, and care strategies
Katayama O et al (2021) Lifestyle changes and outcomes of older adults with mild cognitive impairment: a 4-year longitudinal study. Arch Gerontol Geriatr 94:104376
Kelling A (2025) Biopsychology: learning and memory
Khan SS, Khatik GL, Datusalia AK (2023) Strategies for treatment of disease-associated dementia beyond Alzheimer's disease: an update. Curr Neuropharmacol 21(2):309–339
Knight SB (2023) The role of cognitive flexibility in the relationship between cognitive reserve and cognitive functioning. Ball State University
Komleva Y et al (2021) Inflamm-aging and brain insulin resistance: new insights and role of life-style strategies on cognitive and social determinants in aging and neurodegeneration. Front Neurosci 14:618395
Koníčková D et al (2022) Biomarkers of neurodegenerative diseases: biology, taxonomy, clinical relevance, and current research status. Biomedicine 10(7):1760
Krivanek TJ et al (2021) Promoting successful cognitive aging: a ten-year update. J Alzheimer's Dis 81(3):871–920
Lee J, Kim H-J (2022) Normal aging induces changes in the brain and neurodegeneration progress: review of the structural, biochemical, metabolic, cellular, and molecular changes. Front Aging Neurosci 14:931536
Leng Q, Deng B, Ju Y (2024) Application and progress of advanced eye movement examinations in cognitive impairment. Front Aging Neurosci 16:1377406
Li D-D et al (2019) Meta-analysis of randomized controlled trials on the efficacy and safety of donepezil, galantamine, rivastigmine, and memantine for the treatment of Alzheimer's disease. Front Neurosci 13:472
Li X et al (2023) Advances in differential diagnosis of cerebrovascular diseases in magnetic resonance imaging: a narrative review. Quant Imaging Med Surg 13(4):2712
Lisko I et al (2021) How can dementia and disability be prevented in older adults: where are we today and where are we going? J Intern Med 289(6):807–830
Liss J et al (2021) Practical recommendations for timely, accurate diagnosis of symptomatic Alzheimer's disease (MCI and dementia) in primary care: a review and synthesis. J Intern Med 290(2):310–334
Lista S et al (2022) Towards personalized therapies for alzheimer's disease: challenges for translating molecular genetics and systems biology into strategies to prevent complex chronic brain disorders. Eur J Neurodegener Dis 11(2):31–41

Mackey S et al (2025) Innovations in acute and chronic pain biomarkers: enhancing diagnosis and personalized therapy. Regional Anesthes Pain Med 50(2):110–120
Mangalmurti A, Lukens JR (2022) How neurons die in Alzheimer's disease: implications for neuroinflammation. Curr Opin Neurobiol 75:102575
Meng X et al (2022) Multidomain lifestyle interventions for cognition and the risk of dementia: a systematic review and meta-analysis. Int J Nurs Stud 130:104236
Migliaccio R, Cacciamani F (2022) The temporal lobe in typical and atypical Alzheimer disease. Handb Clin Neurol 187:449–466
Mirakhori F et al (2022) Diagnosis and treatment methods in Alzheimer's patients based on modern techniques: the original article. J Pharm Negative Results:13
Misrani A, Tabassum S, Yang L (2021) Mitochondrial dysfunction and oxidative stress in Alzheimer's disease. Front Aging Neurosci 13:617588
Motillon-Toudic C et al (2022) Social isolation and suicide risk: literature review and perspectives. Eur Psychiatry 65(1):e65
Münzel T, Daiber A (2018) Environmental stressors and their impact on health and disease with focus on oxidative stress. Antioxid Redox Signal 28(9):735–740
Nelson AT (2023) Determining the role of glucose Hypometabolism as a modifier of C9orf72-ALS/FTD: potential mechanisms and downstream consequences. Thomas Jefferson University
Nepozitek J, Dusek P, Sonka K (2025) Glymphatic system, sleep, and Parkinson's disease: interconnections, research opportunities, and potential for disease modification. Sleep 48(1):zsae251
Nesin SM, Sathyamoorthy YK (2025) Psychoneuroimmunology of yoga and meditation. In: Research methodologies and practical applications in psychoneuroimmunology. IGI Global Scientific Publishing, pp 323–346
Ningthoujam DS, Singh N, Mukherjee S (2021) Possible roles of cyclic meditation in regulation of the gut-brain axis. Front Psychol 12:768031
Öhman F (2025) Advancements in the neuropsychology of Alzheimer's disease
Onyango AN (2021) Lipid peroxidation as a link between unhealthy diets and the metabolic syndrome. In: Accenting lipid peroxidation. IntechOpen
Pagnon de la Vega M et al (2022) Mutation analysis of disease causing genes in patients with early onset or familial forms of Alzheimer's disease and frontotemporal dementia. BMC Genomics 23(1):99
Paniri A, Hosseini MM, Akhavan-Niaki H (2024) Alzheimer's disease-related epigenetic changes: novel therapeutic targets. Mol Neurobiol 61(3):1282–1317
Pappalettera C et al (2024) Cognitive resilience/reserve: myth or reality? A review of definitions and measurement methods. Alzheimers Dement 20(5):3567–3586
Patel V, Edison P (2024) Cardiometabolic risk factors and neurodegeneration: a review of the mechanisms underlying diabetes, obesity and hypertension in Alzheimer's disease. J Neurol Neurosurg Psychiatry 95(6):581–589
Peng L, Bestard-Lorigados I, Song W (2022) The synapse as a treatment avenue for Alzheimer's disease. Mol Psychiatry 27(7):2940–2949
Pettit LD (2014) White matter integrity, executive dysfunction, and processing speed in Amyotrophic Lateral Sclerosis
Pinkerton JV (2020) Hormone therapy for postmenopausal women. N Engl J Med 382(5):446–455
Polidori MC (2024) Aging hallmarks, biomarkers, and clocks for personalized medicine:(re) positioning the limelight. Free Radic Biol Med
Pontecorvo MJ, Mintun MA (2011) PET amyloid imaging as a tool for early diagnosis and identifying patients at risk for progression to Alzheimer's disease. Alzheimer's Res Ther 3:1–9
Rahi V, Kaundal RK (2024) Exploring the intricacies of calcium dysregulation in ischemic stroke: insights into neuronal cell death and therapeutic strategies. Life Sci:122651
Rahimi Darehbagh R et al (2024) Stem cell therapies for neurological disorders: current progress, challenges, and future perspectives. Eur J Med Res 29(1):386
Rai SN et al (2020) Mitochondrial dysfunction: a potential therapeutic target to treat Alzheimer's disease. Mol Neurobiol 57(7):3075–3088

Ramakrishna K et al (2023) WNT-β catenin signaling as a potential therapeutic target for neurodegenerative diseases: current status and future perspective. Diseases 11(3):89
Ramakrishna K et al (2024) Indole-3 carbinol and diindolylmethane mitigated β-amyloid-induced neurotoxicity and acetylcholinesterase enzyme activity: in silico, in vitro, and network pharmacology study. Diseases 12(8):184
Reiss AB et al (2022) The physician's Alzheimer's disease management guide: early detection and diagnosis of cognitive impairment, Alzheimer's disease and related dementia. AIMS Public Health 9(4):661
Ruyle KE (2021) The neuroscience of learning agility. In: The age of agility: building learning agile leaders and organizations, pp 118–141
Saragea PD (2024) Diagnosis, genetic counseling and risk assessment in Alzheimer's disease: methodologies, implications, and ethical considerations. Int J Multidiscip Res 6(5). https://doi.org/10.36948/ijfmr.2024.v06i05.27175
Shankar SK (2010) Biology of aging brain. Indian J Pathol Microbiol 53(4):595–604
Sharan P, Vellapandian C (2024) Hypothalamic-pituitary-adrenal (HPA) Axis: unveiling the potential mechanisms involved in stress-induced Alzheimer's disease and depression. Cureus 16(8)
Singh M et al (2024) A comprehensive review and androgen deprivation therapy and its impact on Alzheimer's disease risk in older men with prostate cancer. In: Degenerative neurological and neuromuscular disease, pp 33–46
Singhal G et al (2014) Inflammasomes in neuroinflammation and changes in brain function: a focused review. Front Neurosci 8:315
Sone JY et al (2017) Helmet efficacy against concussion and traumatic brain injury: a review. J Neurosurg 126(3):768–781
Soyka M et al (2017) Guidelines for biological treatment of substance use and related disorders, part 1: alcoholism, first revision. World J Biol Psychiatry 18(2):86–119
Srivastava P et al (2019) Design and development of some phenyl benzoxazole derivatives as a potent acetylcholinesterase inhibitor with antioxidant property to enhance learning and memory. Eur J Med Chem 163:116–135
Stine-Morrow EA, Manavbasi IE (2022) Beyond "use it or lose it": the impact of engagement on cognitive aging. Ann Rev Dev Psychol 4(1):319–352
Sultana A et al (2023) AI in neurology: predictive models for early detection of cognitive decline. Rev Esp de Doc Cient 17(2):335–349
Tegiacchi RF (2023) The effects of Alzheimer's disease progression on caregivers' mental health
Thapa R et al (2024) Proteostasis disruption and senescence in Alzheimer's disease pathways to neurodegeneration. Brain Res:149202
Tripathi PN et al (2019) Biphenyl-3-oxo-1, 2, 4-triazine linked piperazine derivatives as potential cholinesterase inhibitors with anti-oxidant property to improve the learning and memory. Bioorg Chem 85:82–96
Tripathi PN et al (2024) Review of pharmacotherapeutic targets in Alzheimer's disease and its management using traditional medicinal plants. In: Degenerative neurological and neuromuscular disease, pp 47–74
Troubat R et al (2021) Neuroinflammation and depression: a review. Eur J Neurosci 53(1):151–171
Turrini S et al (2023) The multifactorial nature of healthy brain ageing: brain changes, functional decline and protective factors. Ageing Res Rev 88:101939
Uleman JF (2023) Disentangling multicausality: a system dynamics approach to Alzheimer's disease. Sl: sn
Vejandla B et al (2024) Alzheimer's disease: the past, present, and future of a globally progressive disease. Cureus 16(1)
Viola TW et al (2023) Effects of substance use disorder on oxidative and antioxidative stress markers: a systematic review and meta-analysis. Addict Biol 28(1):e13254
Won J et al (2021) Hippocampal functional connectivity and memory performance after exercise intervention in older adults with mild cognitive impairment. J Alzheimer's Dis 82(3):1015–1031
Woźniak K et al (2024) Alzheimer's disease-a comprehensive review. J Educ Health Sport 56:195–209

Xiao K et al (2021) Beneficial effects of running exercise on hippocampal microglia and neuroinflammation in chronic unpredictable stress-induced depression model rats. Transl Psychiatry 11(1):461

Xiong Y et al (2023) Social isolation and the brain: effects and mechanisms. Mol Psychiatry 28(1):191–201

Yadav AK et al (2024) Investigation of serum pro-inflammatory markers and trace elements among short stature in eastern Uttar Pradesh and Bihar populations. J Inflamm Res:6063–6073

Yau S-Y et al (2014) Physical exercise-induced adult neurogenesis: a good strategy to prevent cognitive decline in neurodegenerative diseases? Biomed Res Int 2014(1):403120

Yi T et al (2022) Glymphatic system dysfunction: a novel mediator of sleep disorders and headaches. Front Neurol 13:885020

Yiannopoulou KG, Papageorgiou SG (2013) Current and future treatments for Alzheimer's disease. Ther Adv Neurol Disord 6(1):19–33

Zhang M et al (2021) Simultaneous PET/fMRI detects distinctive alterations in functional connectivity and glucose metabolism of precuneus subregions in Alzheimer's disease. Front Aging Neurosci 13:737002

Zhao J et al (2024) The molecular mechanism of aging and the role in neurodegenerative diseases. Heliyon 10(2)

Nutraceuticals in Brain Health and Aging

2

Devyani Rajput, Mrugendra Potdar, Monika Kaushik,
Parameshwar Kondapuram, Jamal Basha,
and Sagar Pamu

Abstract

Nutraceuticals have been identified as beneficial in preserving brain function, improving cognition, and limiting the effects of aging and neurodegenerative disease. This chapter examines the complex relationship between nutraceuticals and brain function via their biochemical mechanisms, clinical action, and therapeutic uses. It describes brain aging and reduced cognitive function and identifies the role of oxidative stress, inflammation, and neurodegenerative processes in cognitive impairment. Preeminent bioactive ingredients, such as omega-3 fatty acids, polyphenols, vitamins, minerals, and plant extracts, are researched for their neuroprotective and cognition-enhancing effects. The chapter presents clinical evidence of nutraceutical therapy, describing its value in improving memory

D. Rajput · M. Kaushik
Amity Institute of Pharmacy, Amity University Madhya Pradesh,
Gwalior, Madhya Pradesh, India

M. Potdar
Shri Vile Parle Kelawani Mandal's Institute of Pharmacy, Dhule, Maharashtra, India

P. Kondapuram
School of Pharmacy, Guru Nanak Institutions Technical Campus,
Ibrahimpatnam, Telangana, India

J. Basha
Department of Pharmacognosy, Sri Padmavathi School of Pharmacy,
Tirupati, Andhra Pradesh, India

S. Pamu (✉)
Amity Institute of Pharmacy, Amity University Madhya Pradesh, Gwalior,
Madhya Pradesh, India

Department of Pharmacy Practice, Institute of Pharmacy, Nirma University, Gujarat,
Ahmedabad, India

A. Kumar Singh, S. Nand Rai (eds.), *Rejuvenating the Brain: Nutraceuticals, Autophagy, and Longevity*, Nutritional Neurosciences,
https://doi.org/10.1007/978-981-95-2790-8_2

and concentration, stabilizing moods, and supporting adaptation to stress. The chapter further elaborates on the synergistic effects of nutraceuticals with pharmacotherapy and lifestyle changes, emphasizing the need for an integrated strategy for brain health. Future research directions in nutraceutical science are highlighted, such as personalized medicine, advanced delivery systems, and the development of synergistic blends. While nutraceuticals hold vast potential for cognitive enhancement, problems of bioavailability, regulatory oversight, and the need for further large-scale clinical trials are important areas that must be pursued. In conclusion, nutraceuticals provide a safe and new approach to improving brain health and addressing aging. They are natural, evidence-based therapies for maintaining cognitive function and neuroprotection. Research and formulation technology advances will further enhance their position in prevention and therapeutic effects on brain health.

Keywords

Nutraceuticals · Rain health · Cognitive function · Neuroprotection · Aging brain · Neurodegenerative diseases · Omega-3 fatty acids · Polyphenols · Antioxidants · Inflammation

2.1 Introduction

2.1.1 Overview of Brain Health and Aging

Cognitive well-being is an integral part of general wellness that affects cognitive ability, emotional resilience, and functional performance. With advancing age, the brain undergoes physiological and structural changes affecting memory, learning, and executive function maintenance. These changes may involve neuron degeneration, decreased synaptic plasticity, and decreased levels of neurotransmitters, factors that may lead to cognitive impairment and susceptibility to neurodegenerative disorders like Alzheimer's and Parkinson's disease (PD)s (Cordeiro et al. 2024).

Aging is correlated with heightened oxidative stress, chronic inflammation, and vascular insufficiency that altogether exacerbate cognitive decline (Cordeiro et al. 2024). While some level of age-related cognitive decline is inevitable, there are certain lifestyle factors that a person partakes in, including diet, exercise, and mental stimulation, which lend themselves to cerebral health and delay the onset of neurodegenerative disease (Dominguez et al. 2021).

2.1.2 Role of Nutraceuticals in Aging and Brain Function

Nutraceuticals, or natural foods derived with potential health benefits, have garnered considerable interest due to their potential to enhance cognitive health and slow down age-related cognitive function loss. The neuroprotective effects of bioactive food nutrients, such as vitamins, minerals, polyphenols, omega-3 fatty acids, and plant extracts, vary (Mecocci et al. 2014).

Here are some ways that nutraceuticals enhance mental health:

Antioxidant activity: One of the main factors causing neuronal damage is oxidative stress, which antioxidants help to mitigate (Ionescu-Tucker and Cotman 2021).
Anti-inflammatory action: Reducing chronic neuroinflammation, a major contributor to cognitive decline, by altering inflammatory pathways (Ionescu-Tucker and Cotman 2021).
Neurotransmitter augmentation: Augmenting activity and synthesis of significant neurotransmitters such as dopamine, serotonin, and acetylcholine (Ashok et al. 2022).
Synaptic plasticity and neurogenesis: Enabling new neuron development and synaptic strength, thus enhancing learning and memory (Stuchlik 2014).

Mitochondrial protection facilitates energy metabolism in brain cells and minimizes mitochondrial dysfunction associated with aging and neurodegenerative diseases (Ionescu-Tucker and Cotman 2021)

2.1.3 Importance of Dietary Interventions in Cognitive Health

An equilibrium diet with optimum brain-enhancing nutrients is required to maintain cognitive health. Nutritional strategies based on whole foods and nutraceutical supplements can maximize brain function and prevent age-related decline (Cognitive Health and Older Adults 2024). The MIND and Mediterranean diets emphasize fruits, vegetables, whole grains, lean meats, and healthy fats, which have improved cognitive function and reduced the risk of developing neurodegenerative diseases (National Institute on Aging 2023).

Consuming a high amount of antioxidant and anti-inflammatory foods is preferred. High-nutrient foods high in flavonoids, polyphenols, and omega-3 fatty acids—like berries, dark leafy greens, nuts, and fatty fish—are the foundation for maintaining cognitive health (Kalogerakou and Antoniadou 2024). High sugar intake and ultra-processed foods can increase oxidative stress and inflammation, leading to cognitive impairment (Martínez Leo et al. 2021). Proper hydration and essential micronutrients, such as sufficient water, vitamins (e.g., B vitamins and vitamin D), and minerals (e.g., magnesium and zinc), ensure optimal brain function (Tardy et al. 2020). However, with nutraceuticals and dietary interventions in daily life, individuals can develop cognitive resilience, reduce the risk of neurodegenerative disease, and promote overall brain longevity (Lobo et al. 2022).

2.2 Understanding Nutraceuticals

2.2.1 Definition and Classification

Nutraceuticals are food-active ingredients that provide health benefits beyond mere nutrition. The term "nutraceutical" is a blend of "nutrition" and "pharmaceutical" and suggests their therapeutic potential in the prevention and treatment of diseases.

They are important in brain function, especially among the elderly, preserving thinking processes and arresting brain diseases (Vignesh et al. 2024). Nutraceuticals have been classically grouped into several categories based on their sources and bioactivity:

Dietary supplements are amino acids, vitamins, minerals, and other nutrients taken in concentrated amounts to enhance brain function. Functional foods, that is, ordinary foods fortified with bioactive compounds, such as omega-3-rich eggs or probiotic yoghurt, are developed to provide additional health benefits.

Herbal extracts such as concentrated extracts of plant-derived materials such as *Ginkgo biloba*, *Bacopa monnieri*, and *Withania somnifera* are with cognition-enhancing and neuroprotective properties. Polyphenols and antioxidants found in fruits, vegetables, and teas, have flavonoids and resveratrol. Polyphenols help fight oxidative stress and brain inflammation. Omega-3 fatty acids, primarily found in fish oil and flaxseeds, support brain function and protect against mental function loss (Puri et al. 2022). Knowing how to categorize nutraceuticals helps us understand how they might help with brain aging and wellness. Nutraceuticals function differently with the nervous system, affecting cognition, memory, and mental health.

2.2.2 Mechanisms of Action in Brain Health

Nutraceuticals benefit brain function through several mechanisms, including antioxidant activity, neuroinflammation modulation, neurotransmitter support, and neuroprotection (Chiu et al. 2020).

Antioxidant activity and free radical scavenging: The brain is most susceptible to oxidative stress due to its high metabolic rate and lipid content. Nutraceuticals like polyphenols, vitamin E, and coenzyme Q10 act to neutralize free radicals and thus prevent neuronal damage and slow cognitive decline (Lee et al. 2020).

Anti-inflammatory activities: Chronic neuroinflammation is one of the principal causes of neurodegenerative diseases. Curcumin, omega-3 fatty acids, and resveratrol are anti-inflammatory nutraceuticals that suppress excessive neuroinflammation by modulating the inflammatory pathways (Kurowska et al. 2023).

Regulation of neurotransmitters: Several nutraceuticals modulate the synthesis and activity nof neurotransmitters necessary for cognition and mood regulation. For example, B vitamins are involved in neurotransmitter synthesis; while *Bacopa monnieri* extracts augment synaptic transmission (Rajan et al. 2015).

Neurogenesis and synaptic plasticity: Certain nutraceuticals stimulate the growth of new neurons and enhance synaptic plasticity, thereby supporting cognitive functions such as learning and memory. Substances such as omega-3 fatty acids and flavonoids have been found to promote neurogenesis in the hippocampus, a significant brain area responsible for memory formation (Poulose et al. 2017).

Mitochondrial support and energy metabolism: Mitochondrial dysfunction is linked to neurodegenerative diseases. Nutraceuticals like alpha-lipoic acid and CoQ10 improve the function of mitochondria, which keeps neurons getting enough energy and keeps the brain strong (Mantle and Hargreaves 2022).

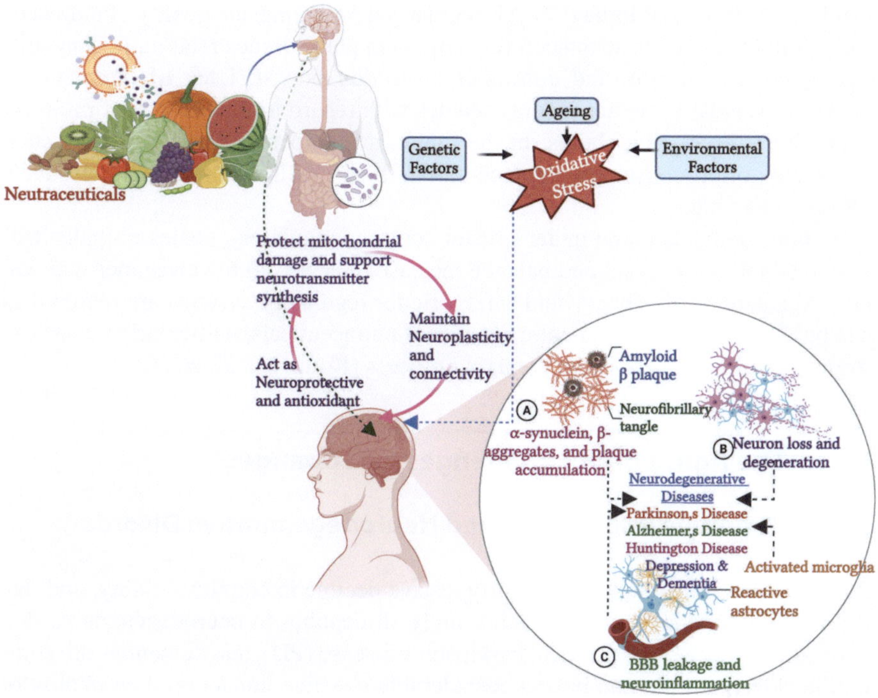

Fig. 2.1 Mechanisms of nutraceutical action in brain health

Through these avenues, nutraceuticals offer a potential way to maintain brain health and prevent age-related cognitive decline (Fig. 2.1).

2.2.3 Regulatory Issues Concerning Nutraceuticals

Regulation of nutraceuticals is rather varied by country, influencing their classification, approval, and availability in the market. Unlike pharmaceutical medications that have to undergo extensive clinical trials and strict regulatory approval, nutraceuticals are usually classified as dietary supplements or functional foods and hence come under varying ranges of regulatory control (GFRS n.d.). In America, the term nutraceutical is referred to as a dietary supplement by the 1994 Dietary Supplement Health and Education Act. The suppliers do not need FDA approval before marketing the product; but they are still in charge of its safety (Santini et al. 2018). EFSA regulates nutraceuticals, and empirical proof has to be submitted to support health claims prior to marketing. Functional foods and dietary supplements are subjected to strict labeling and safety regulations (Vettorazzi et al. 2020).

Japan and India each has its own regulatory frameworks that govern nutraceuticals. Japan's "Foods for Specific Health Uses" (FOSHU) program regulates functional foods through a licensing system. At the same time, the Food Safety and

Standards Authority of India (FSSAI) regulates the labeling and quality of nutraceuticals (Putta n.d.). Even with such regulations in place, issues relating to standardization, quality control, and consumer protection persist. Lack of international regulatory standards creates inconsistencies with regard to the potency of products and their corresponding claims. In the future, it is imperative to improve on such lapses and offer consumers science-based, high-quality nutraceutical products (OrfonlineOrg n.d.).

As nutraceuticals come under greater scrutiny, regulatory bodies globally will have to adapt to this trend and balance the ease of access to the consumer with the high expectations of efficacy and safety. Better regulatory systems are required to gain public trust in the therapeutic benefits of nutraceuticals, particularly those concerning cognitive function and the aging process (Komala et al. 2023).

2.3 The Aging Brain: Challenges and Changes

2.3.1 Cognitive Impairment and Neurodegenerative Disorders

The aging process is linked to the progressive decline in cognitive ability, and due to this, elderly people become increasingly susceptible to neurodegenerative diseases. Alzheimer's disease (AD), Parkinson's disease (PD), and dementias are common in elderly people and have a considerable negative impact on their quality of life. All these diseases are characterized by chronic neuronal damage, the deposition of pathologic proteins, and the disruption of neurotransmission, which results in cognitive and motor dysfunction (Gonzales et al. 2022).

Alzheimer's disease (AD), the most common neurodegenerative disorder, is essentially linked to the accumulation of beta-amyloid plaques and tau protein aggregates in the brain. The resulting pathological mechanisms disrupt the communication of neurons, leading to memory loss, confusion, and compromised cognitive judgement (Rajmohan and Reddy 2017). Similarly, Parkinson's disease (PD) is marked by the loss of dopamine-releasing neurons in the substantia nigra. This leads to motor impairments like tremor, stiffness, and slow movement, as well as cognitive impairment in the later stages of the disease (Zafar and Yaddanapudi 2025). Cognitive impairment due to aging transcends neurodegenerative diseases. Indeed, normal aging can be responsible for issues regarding processing speed, executive function, and working memory in the absence of disease. Yet, lifestyle factors such as diet, exercise, and mental challenge are what maintain cognitive function and preclude neurodegenerative disease from occurring (Murman 2015).

2.3.2 Structural and Functional Brain Changes with Age

The brain undergoes several structural and functional alterations as we age, which have an indirect impact on cognitive performance. Among these alterations is a decrease in brain volume, especially in areas like the prefrontal cortex and hippocampus that are implicated in memory and executive function. Loss of neuroes, a decline in synaptic connectivity, and a reduction in white matter integrity all contribute to volume loss, which in turn impacts cognitive function (Schulz et al. 2022). Cognitive impairment is significantly influenced by age-related changes in the neurotransmitter system. Dopamine levels that govern motivation and executive function, decline with age, slowing down cognitive processing and reducing cognitive flexibility. Additionally, cognitive impairment occurs in the elderly when acetylcholine, a neurotransmitter involved in learning and memory, is lost (ScienceDirect Topics n.d.).

Vascular function and cerebral blood flow are significantly altered. Narrowed blood vessel diameters and compromised blood-brain barrier function result in a decreased blood supply to the brain in older adults. As a result, these alterations could lead to a reduced flow of oxygen and vital nutrients to neurons, worsening mental impairment and increasing the risk of stroke and vascular dementia (Yang et al. 2017). Despite all of these developments, the brain is still very malleable, forming new neural connections and adapting to fill in gaps. Age-related cognitive decline can be reversed or at least stopped by increasing neuroplasticity through mental and physical exercise as well as a healthy diet (Marzola et al. 2023).

2.3.3 Consequences of Oxidative Stress and Inflammatory Responses

The two main factors that accelerate brain aging and lead to neurodegenerative diseases are oxidative stress and inflammation. An imbalance between the body's antioxidants and free radicals causes oxidative stress. The brain is particularly vulnerable to oxidative damage because of its high metabolic rate and lipid content. Overproduction of free radicals can harm proteins, DNA, and neuronal membranes, resulting in cellular dysfunction and death (Dash et al. 2025). As people age, inflammation, a normal immune response, becomes chronic and worsens cognitive impairment. In older adults, the brain's innate immune cells, known as microglia, become hyperactivated and release pro-inflammatory cytokines, which leads to neuroinflammation. Chronic inflammation is associated with the pathogenesis of Alzheimer's and Parkinson's diseases and with general cognitive impairment in the elderly (Miao et al. 2023).

Nutraceuticals rich in antioxidants and anti-inflammatory compounds, such as polyphenols, omega-3 fatty acids, and curcumin, have been reported to be beneficial against oxidative stress and neuroinflammation. By neutralizing free radicals and modulating inflammatory pathways, the agents preserve neurons from age-related injury and maintain cognitive function (Dama et al. 2024). Figuring out how neurodegenerative disease, loss of structural integrity, and oxidation happen together as we age gives us a way to approach targeted therapy. Food choice, life transition, and drugs may assist in preserving brain function and protecting cognitive resilience within aging populations (de la Fuente et al. 2023).

2.4 Nutraceuticals and Brain Function

2.4.1 Key Bioactive Compounds for Brain Health

Numerous beneficial substances found in nutraceuticals improve brain health and function. These naturally occurring substances found in food maintain the health of brain cells, enhance their ability to communicate, and shield the brain (Reza-Zaldívar and Jacobo-Velázquez 2023). The following are a few key substances that significantly impact brain health: walnuts, flaxseeds, and fatty fish all contain omega-3 fatty acids, such as docosahexaenoic acid (DHA) and eicosapentaenoic acid (EPA). They help with thinking, memory, and the prevention of brain disorders by preserving the integrity of nerve cell membranes and allowing nerve cell connections to change (Dighriri et al. n.d.).

Plant nutrients called polyphenols include curcumin, resveratrol, and flavonoids. Polyphenols have anti-inflammatory and antioxidant qualities. These characteristics control nerve impulse activity and shield the brain from free radical damage (Rudrapal et al. 2022). Minerals and vitamins such as zinc, magnesium, vitamin D, and B vitamins give our brain the nourishment it needs to produce neurotransmitters, function, and develop new brain cells. Homocysteine, a molecule linked to cognitive decline, is metabolized with the help of B vitamins (Tardy et al. 2020).

Some amino acids, such as acetyl-L-carnitine (Ferreira and McKenna 2017) and L-theanine (Dashwood and Visioli 2025) in green tea, help to balance neurotransmitters, prevent stress, and improve brain function. The neuroprotective and cognition-enhancing properties of traditional medicinal herbs such as *Ginkgo biloba*, *Bacopa monnieri*, and *Withania somnifera* (*Ashwagandha*) have been thoroughly investigated (Zieneldien et al. 2022). Additionally, they work in concert to prevent mental decline, improve brain health, and advance general well-being.

2.4.2 Antioxidant and Anti-inflammatory Activity

Two of the main causes of neurodegenerative diseases and brain aging are oxidative stress and inflammation. By scavenging free radicals and regulating the immune system, anti-inflammatory and high-antioxidant nutraceuticals eradicate these harmful effects (Uttara et al. 2009). Brain aging is caused by free radicals that are created by biochemical processes and destroy neuron cells. Polyphenols, vitamin C, and vitamin E shield brain cells from toxic substances (Ionescu-Tucker and Cotman 2021).

Prolonged inflammation in the brain harms neurons and impairs cognition. By lowering dangerous cytokines, nutraceuticals such as flavonoids, curcumin, and omega-3 fatty acids contribute to the creation of an anti-inflammatory environment (Garodia et al. 2023). Alpha-lipoic acid and coenzyme Q10 are two examples of the many antioxidants that promote mitochondrial health. This reduces oxidative damage and provides neurons with a consistent energy source (Silvestri et al. 2015).

These nutraceuticals significantly preserve cognitive function and slow neurodegeneration by lowering oxidative stress and inflammation.

2.4.3 How Nutraceuticals Protect the Brain

Nutraceuticals also improve brain function by inhibiting neurodegeneration and promoting cognitive resilience. As they regulate central proteins, induce neurogenesis, modify neurotransmission, and guard neurons from injury, natural products derived from food and medicinal plants are central to cognitive well-being (Makkar et al. 2020). One of the main methods through which nutraceuticals optimize brain function is through the improvement of neurotransmission. Nutraceuticals like amino acids, omega-3 fatty acids, and B vitamins synthesize and regulate neurotransmitters like dopamine, serotonin, and acetylcholine. These play a significant role in learning, mood, and memory. By releasing the necessary amounts of neurotransmitters, nutraceuticals optimize cognitive function and emotional stability (Raza et al. 2025).

Certain bioactive nutrients are able to initiate neurogenesis and synaptic plasticity, the basic processes that are linked to learning and memory. Omega-3 fatty acids and polyphenols increase cognitive flexibility by strengthening the synaptic link and allowing new neuron formation (Clemente-Suárez et al. 2025). This, in turn, enhances the acquisition of new information as well as the recovery of the brain from the impact of aging or damage. Nutraceuticals are important in controlling the production of a protein called brain-derived neurotrophic factor (BDNF) that is needed for neuronal cell survival and growth. Resveratrol and curcumin are found to increase BDNF levels, which are related to promoting a healthy and strong brain. Improved cognitive function and low rates of neurological disorders are connected with high levels of BDNF (Bathina and Das 2015).

Preventing protein misfolding and aggregation, the hallmark of Alzheimer's and Parkinson's diseases, is a necessary first step toward brain protection. Green tea holds a supplement referred to as epigallocatechin gallate (EGCG). EGCG inhibits proteins from misfolding and aggregating. This halts disease and nerve cell deterioration. This is a protective effect that can halt neurodegenerative diseases (Kamboj et al. 2025).

Besides, nutraceuticals inhibit excitotoxicity, a state of excess glutamate that kills and destroys neurons. Neuroprotectants such as magnesium regulate glutamate receptors to prevent overexcitation of neurons that may lead to damage. Nutraceuticals prevent cognitive impairment and neuronal loss through prevention of excitotoxicity (Mark et al. 2001). Nutraceuticals are formulated to protect against brain health, promote healthy aging, and enhance neurological function. Researchers are still exploring their possible therapeutic applications and creating new mechanisms to safeguard the brain and maximize cognitive performance (Reza-Zaldívar and Jacobo-Velázquez 2023) (Fig. 2.2).

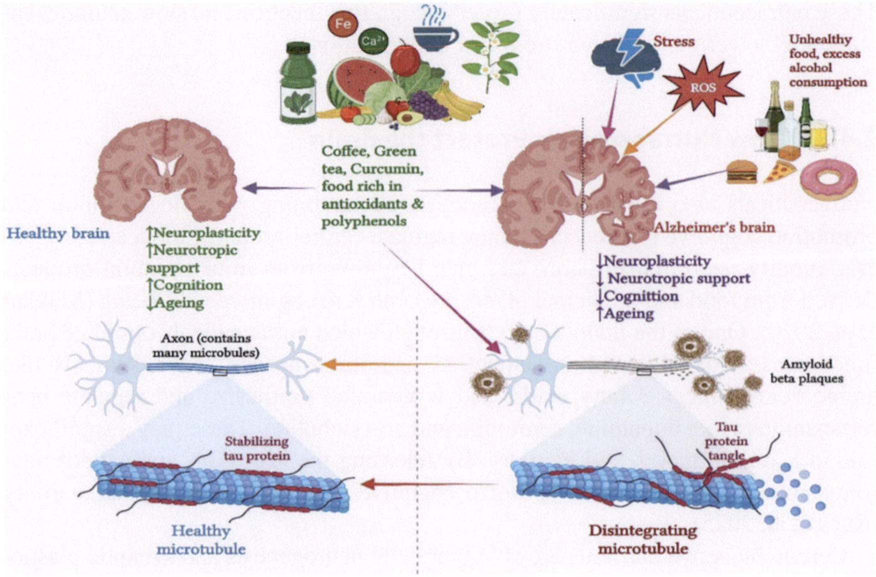

Fig. 2.2 Nutraceutical interventions across the aging brain

2.5 Overall Brain Health Supplements

2.5.1 Omega-3 Fatty Acids

Healthy brain functioning and clear cognitive processes rely on omega-3 fatty acids. Fatty fish such as sardines, mackerel, and salmon are high in the two most important ones, that is, docosahexaenoic acid (DHA) and eicosapentaenoic acid (EPA). DHA is an essential component of nerve cell membrane in the brain and is needed for communication, learning, and passing messages between these nerve cells (Dighriri et al. n.d.).

Research suggests that omega-3 fatty acids are beneficial for brain function by being anti-inflammatory, increasing a protein called brain-derived neurotrophic factor (BDNF), and preventing conditions like Alzheimer's and Parkinson's. Omega-3 s also enhance memory, learning, and mood stability through daily intake. Fish oil or DHA supplementation from algae has been conventionally recommended to low-risk patients for any complications (Thomas et al. 2015).

2.5.2 Polyphenols (Like Flavonoids and Resveratrol)

Polyphenols are a chemically diverse group of plant food origin natural antioxidants such as berries, green tea, red wine, and dark chocolate. These phytochemical flavonoids, resveratrol, and catechins are researched extensively for their neuroprotective actions (Pandey and Rizvi 2009).

Flavonoids in citrus, cocoa, and green tea also contribute to cognition by enhancing brain blood flow, minimizing oxidative damage, and sustaining synaptic plasticity. Resveratrol, a polyphenol found in red grapes and wine, also activates sirtuins, cellular longevity- and neuroprotection-related proteins (Minocha et al. 2022). Daily consumption of polyphenol-rich food has been linked to decreased cognitive impairment, improved memory retention, and protection against neuroinflammation, making it critical for brain well-being (Vauzour 2012).

2.5.3 Vitamins (e.g., B Vitamins, Vitamin D)

Vitamins are essential for brain function by facilitating neurotransmitter production, energy metabolism, and neuroprotection. B vitamins and vitamin D are among the most critical to cognitive health (Fekete et al. 2023). B vitamins (B6, B9, and B12) are crucial for brain metabolism, DNA synthesis, and the lowering of homocysteine levels, which are associated with cognitive impairment. Deficiencies in B vitamins may result in memory loss, brain fog, and a higher risk of dementia (Kennedy 2016).

Vitamin D, commonly called the "Sunshine Vitamin," plays a role in neuroprotection, mood stability, and immunity. Vitamin D deficiency has been linked to a higher risk of Alzheimer's disease, depression, and cognitive impairment (Anwar et al. 2023). Proper vitamin intake through diet or supplements is necessary to sustain brain health and avoid cognitive aging.

2.5.4 Minerals (e.g., Magnesium, Zinc)

Minerals are essential for many biochemical processes that maintain brain energy metabolism, neuronal signaling, and cognitive function (Tardy et al. 2020). Neurotransmitter release, stress adaptation, and synaptic plasticity all depend on magnesium. Cognitive decline, anxiety, and depression have all been linked to low magnesium levels. Magnesium-rich foods, such as leafy greens, seeds, and nuts, support neuroprotection (Tardy et al. 2020).

Zinc is necessary for learning, memory, and hippocampal synaptic transmission. Zinc is also essential for managing neuroinflammation and halting age-related cognitive decline. Oysters, legumes, and whole grains contain some of the dietary zinc (Tardy et al. 2020). Appropriate levels of these essential minerals must be maintained for the best possible brain function, neuronal protection, and cognitive function.

2.5.5 Plant Extracts (e.g., *Ginkgo biloba, Bacopa monnieri*)

Herbal extracts have been utilized for decades by traditional medicine to improve thinking, alleviate stress, and shield the brain (Nawrot et al. 2022). The two herbal supplements that have been studied most for their brain activity are *Ginkgo biloba* and *Bacopa monnieri*. The majority of people know that *Ginkgo biloba* can support blood flow to the brain. It also helps concentration, memory, and clear-headedness. The plant is also useful for those who might suffer from diseases that destroy the brain because it has strong antioxidants that decrease oxidative stress and inflammation (Singh et al. 2019).

The herb *Monnieri bacopa* is a long-standing ingredient in ayurvedic medicine taken to enhance memory, learning, and thought speed. It works to preserve synaptic plasticity, reducing stress hormones known as cortisol, and boost the communication between nerve cells (Walker and Pellegrini 2025). Rhodiola rosea, *Panax ginseng*, and *Withania somnifera* (*Ashwagandha*) are important herbal extracts. These adaptogens improve cognitive function and increase the brain's ability to cope with stress (Panossian and Wikman 2010). Individuals can prevent neurodegeneration, maintain a healthy brain as they age, and significantly enhance brain performance by adding these renowned supplements to their daily regimen (Loong et al. 2023). Key nutraceuticals are listed in Table 2.1, with information on their food sources, mode of action, and cognitive effect. By demonstrating neuroprotection, neurotransmitter modulation, and anti-inflammatory activity, it indicates how nutritional ingredients like omega-3 fatty acids, polyphenols, vitamins, and plant extracts help in brain function. With the help of this chapter, an awareness of the function of different nutraceuticals in enhancing cognitive function and prevention of neurodegenerative disease is provided.

2.6 Nutraceuticals and Cognitive Enhancement

2.6.1 Memory and Learning Improvements

Learning and memory are fundamental intellectual processes that decline with age and are influenced by various lifestyle factors, including diet and nutrition. Nutraceuticals are increasingly being touted as positively impacting memory storage, information processing, and cognitive function (Puri et al. 2023).

Several bioactive compounds, such as omega-3 fatty acids, polyphenols, and B vitamins, enhance memory through the facilitation of neuroplasticity and synaptic transmission (Zinkow et al. 2025). Polyphenols, such as flavonoids in berries, green tea, and dark chocolate, increase cerebral blood flow and protect against oxidative stress, thus preserving cognitive function (Yang et al. 2021). B vitamins, especially B6, B9 (folate), and B12, also regulate the level of homocysteine and prevent it from inducing neurotoxicity and cognitive function impairment (Mitchell et al. 2014). Herbal supplements like Bacopa monnieri also enhance memory by improving dendritic development and influencing neurotransmitter function. Long-term administration of *Bacopa* has been reported to improve verbal memory, processing speed, and learning ability, making it a beneficial nutraceutical for cognitive enhancement (Walker and Pellegrini 2025).

Table 2.1 Key nutraceuticals and their cognitive benefits

Nutraceutical	Source	Mechanism of action	Cognitive benefit	References
Omega-3 fatty acids	Fatty fish, flaxseeds, and walnuts	Enhances neuronal membrane integrity, reduces inflammation	Improves memory, learning, and neuroprotection	Loong et al. (2023)
Polyphenols	Berries, green tea, and dark chocolate	Antioxidant enhances blood flow to the brain	Supports cognitive function and reduces cognitive decline	Yang et al. (2021)
B vitamins	Leafy greens, eggs, and dairy	Supports neurotransmitter synthesis, reduces homocysteine	Enhances memory and prevents cognitive decline	Kennedy (2016)
Vitamin D	Sunlight, fish, and fortified dairy	Regulates neuroinflammation, supports neurogenesis	Protects against neurodegeneration	Fenercioglu (2024)
Magnesium	Nuts, seeds, and whole grains	Regulates neurotransmitter function, reduces excitotoxicity	Improves focus, learning, and mood	Wang et al. (2018)
Herbal extracts (*Ginkgo biloba*, *Bacopa monnieri*)	Traditional medicinal plants	Enhances blood circulation, supports synaptic plasticity	Boosts memory, focus, and stress resilience	Singh et al. (2019), Walker and Pellegrini (2025)

2.6.2 Focus, Attention, and Mood Regulation

Concentrating, sustaining attention, and regulating mood are required for productivity and mental well-being. Nutraceuticals can enhance concentration and attention by acting on neurotransmitters such as dopamine, acetylcholine, and serotonin, which are involved in mental alertness and emotional stability (Magomedova and Fatima 2025). L-theanine, an amino acid found in green tea, has been shown to improve attention and mental focus by stimulating alpha brain waves, which are associated with a state of relaxed alertness. Along with caffeine, L-theanine enhances attention without the jittery side effects usually associated with the intake of stimulants (Dashwood and Visioli 2025). *Panax ginseng* has been researched to enhance cognitive function and minimize mental fatigue, proving useful for sustained attention (Reay et al. 2005).

Regulation of mood is also an important aspect of cognitive function because anxiety and stress interfere with attention and decision-making. Nutraceuticals like omega-3 fatty acids and adaptogenic herbs like *Rhodiola rosea* and *Withania somnifera* (*Ashwagandha*) regulate cortisol levels to decrease stress and enhance emotional resilience. The nutrients enhance serotonin and dopamine pathways, which are implicated in mood stabilization and cognitive flexibility and are hence helpful in the management of anxiety and depression-related cognitive impairment (ScienceDirect n.d.).

2.6.3 Adaptogenic Action and Stress Reduction

Chronic stress is detrimental for brain function through elevated cortisol levels, neuroinflammation, and memory and executive function disruption. Adaptogens are a group of nutrients that help the body deal with stress and get back to a state of physiological homeostasis. They do this by changing the HPA axis and stress response pathways (Panossian and Wikman 2010). An extensively researched adaptogen, *Ashwagandha* has been demonstrated to enhance stress tolerance, lower cortisol levels, and improve cognitive function. *Ashwagandha* supplements have been shown in studies to improve information processing speed, memory, and executive function. They are, especially helpful for cognitive impairment brought on by stress (Remenapp et al. 2022).

Rhodiola rosea, a remarkable adaptogen, was investigated to see if it could reduce mental fatigue and enhance mental function under stress. By boosting serotonin and dopamine levels in the brain, it elevates mood and mental clarity. Other adaptogens with neuroprotective and stress-relieving qualities, such as *Eleutherococcus senticosus* (Siberian *ginseng*) and *Holy basil (tulsi)*, enhance mental well-being and cognitive resilience (Ivanova Stojcheva and Quintela 2022). Adding these nutraceuticals to everyday diets can enhance memory, concentration, and stress resistance, leading to better cognitive functioning and brain health.

2.7 Nutraceuticals in Neurodegenerative Diseases

2.7.1 Alzheimer's Disease

Alzheimer's disease (AD) is the most prevalent neurodegenerative illness, exhibiting a progressive decline in cognition, memory, and behavior. AD starts with the buildup of beta-amyloid plaques, tau protein tangles, oxidative stress, inflammation of the nervous system, and damage to the mitochondria. Since AD has more than one cause, treatment modalities include the application of nutraceuticals for preventing the advancement of disease and maintaining cognition (Sheppard and Coleman 2020).

Polyphenols like resveratrol and curcumin have neuroprotective effects by lowering oxidative stress, modulating neuroinflammation, and inhibiting beta-amyloid aggregation. Curcumin, the active ingredient of turmeric, has been found to cross the blood-brain barrier, exerting anti-inflammatory and antioxidant actions likely to slow down cognitive impairment in AD patients. Omega-3 fatty acids, particularly DHA, are also important in neuronal membrane stability and synaptic plasticity, which may slow cognitive impairment (Mazzanti and Di Giacomo 2016). Nutraceuticals, such as vitamins E, coenzyme Q10, and *Ginkgo biloba*, are examined for their cognitive effects in AD. They promote cerebral blood flow, enhance mitochondrial function, and protect against neurotoxicity, thereby preventing memory loss and the rate of progression (Nowak et al. 2021).

2.7.2 Parkinson's Disease

Degeneration of dopamine-secreting neurons in the substantia nigra causes Parkinson's disease (PD), a disorder of motor function that worsens over time. Tremors, stiffness, bradykinesia, and, in more severe cases, cognitive impairment are the symptoms and indicators. The neuroprotective effects of nutraceuticals against oxidative stress, mitochondrial damage, and neuroinflammation have been studied in Parkinson's disease (Kouli et al. 2018). Coenzyme Q10 (CoQ10) has received attention due to its involvement in mitochondrial energy generation and antioxidative activities. Research suggests that CoQ10 supplementation can attenuate disease progression by improving mitochondrial function and neuronal oxidative damage. Green tea polyphenols, such as epigallocatechin gallate (EGCG), have also been shown to protect neurons by stopping the death of dopaminergic neurons and changing the way inflammation works (Duberley et al. 2014).

Curcumin and omega-3 fatty acids are also important in neuroprotection in PD. Curcumin's antioxidant and anti-inflammatory properties may protect dopamine-producing neurons, whereas omega-3 s maintain neuronal function and synaptic integrity. Herbal extracts, such as *Mucuna pruriens* (a natural source of L-DOPA), have also been looked into as extra treatments to regular Parkinson's disease care (Nahar et al. 2025).

2.7.3 Other Neurodegenerative Conditions

Non-Alzheimer's and non-Parkinson's neurodegenerative diseases are illnesses like Huntington's disease (HD), amyotrophic lateral sclerosis (ALS), and multiple sclerosis (MS) that have typical pathological mechanisms in common. Neurodegenerative diseases tend to be marked by oxidative stress, mitochondrial impairment, and persistent neuroinflammation, which produces progressive neuronal deterioration. New data indicate that bioactive food-stemmed substances called nutraceuticals are likely to support neuronal health and inhibit disease progression (Ciurea et al. 2023).

Huntington's disease (HD) is a neurodegenerative condition that is passed down through families. It is caused by an abnormal expansion of the huntingtin gene, which leads to motor dysfunction, cognitive impairment, and psychiatric symptoms. Polyphenols like quercetin and resveratrol have demonstrated neuroprotective activity in HD models through the modulation of oxidative stress and inflammatory signaling pathways. The compounds listed above are very strong antioxidants that protect neurons from damage and may help keep motor and cognitive problems from getting worse in people with HD (Ajitkumar and De Jesus 2025).

Amyotrophic lateral sclerosis (ALS) is a degenerative neurologic disorder of motor neurons, which ultimately causes muscle weakness, paralysis, and failure of the respiratory muscles. The antioxidative and anti-inflammatory effects of nutraceuticals like omega-3 fatty acids, curcumin, and vitamin E have been studied in ALS (NINDS n.d.). Omega-3 fatty acids in fish oil inhibit inflammation and maintain cell membrane integrity. Curcumin from turmeric is neuroprotective by reversing oxidative stress and inhibiting neuroinflammatory pathways. Vitamin E is a potent antioxidant that has been studied to retard the advancement of the disease by inhibiting oxidative damage to neurons (Shabbir et al. 2021). Myelin sheaths of the central nervous system are the main target of multiple sclerosis (MS), an autoimmune neurodegenerative disorder characterized by interrupted transmission of nerve impulses, muscular weakness, and cognitive impairment (Tafti et al. 2025). Nutraceuticals such as vitamin D, omega-3 fatty acids, and polyphenols are presented as agents that can promote immunological balance and modulate neuroinflammation in MS patients. Because of its unique role of immune modulation, vitamin D can reduce the frequency and severity of relapses in multiple sclerosis. Polyphenols have antioxidant properties that contribute to neuronal integrity and can be associated with better outcomes in the disease, while omega-3 fatty acids have been associated with anti-inflammatory activities (Tryfonos et al. 2019). Nutraceuticals are therapeutic potential drugs that can be employed for therapeutic purposes by targeting the core pathological pathways, but cannot be utilized as monotherapies for neurodegenerative diseases. To better realize their efficacy and integrate them into comprehensive treatment regimens for these incapacitating diseases, further research is needed.

2.7.4 Efficacy and Safety of Nutraceuticals in Treatment

The establishment of the safety and clinical efficacy of nutraceuticals in neurodegenerative disease management requires strict human clinical trials. Though numerous nutraceuticals display encouraging preclinical and initial clinical outcomes, several challenges need to be addressed in order for them to be widely used as a therapeutic regimen. These challenges are standardization, dosage optimization, and increasing bioavailability. Additionally, the intricate nature of neurodegenerative diseases renders nutraceutical efficacy variable according to interindividual, interdisease, and interproduct heterogeneity (Makkar et al. 2020). The efficacy of nutraceuticals could be determined by metabolic activity, absorption, and their ability to cross the brain tissues. The therapeutic effectiveness of the majority of compounds such as resveratrol and curcumin is limited by their unreliable bioavailability. In an attempt to overcome this drawback, scientists are developing novel drug delivery systems such as liposomal and nanoparticle preparations. These new technologies are likely to extend the duration of the drug action and enhance its absorption. In order for such drugs to exert optimal neuroprotection, neurodegenerative patients need to receive the correct dose and formulation (Puri et al. 2022).

The most important consideration when using nutraceuticals for neurologic health is safety. Despite the fact that most of these supplements are safe to take in the recommended dosages, careful consideration is necessary due to potential interactions with medications. Multiple pharmacologic therapies may be administered to patients with neurodegenerative diseases, putting them at risk for negative interactions. As an example, *Ginkgo biloba* is a widely used nutraceutical that improves cognitive function and has been demonstrated to raise the risk of bleeding in patients taking anticoagulants. Medical monitoring is thus essential to permit the safe integration of nutraceuticals into treatment protocols (Makkar et al. 2020).

Regulatory issues make the clinical use of nutraceuticals more complicated. Since nutraceuticals are not held to the same strict regulatory approval procedures as pharmaceutical drugs, there are doubts regarding product consistency, quality, and scientifically based health claims. Sourcing variations, extraction methods, and processing techniques may affect the efficacy and consistency of nutraceutical products. The provision of standard formulations and running well-designed clinical trials will become essential for building acceptance in the clinical and medical environments (Santini et al. 2018).

Notwithstanding these obstacles, nutraceuticals are full of promise as adjunctive interventions for neurodegenerative illness. Continued study and development of nutraceutical formulations will further clarify their potential in neuroprotection, cognitive function preservation, and disease management. As the science behind their mechanism of action advances, nutraceuticals will be valuable assets in the overall strategy for preventing the consequences of neurodegenerative conditions and improving patients' quality of life (Dadhania et al. 2016).

2.8 Clinical Evidence and Research on Nutraceuticals

2.8.1 Studies on Brain Aging and Cognitive Function

Scientific studies have increasingly targeted the potential of nutraceuticals to counteract brain aging and maintain cognitive function. Aging is linked with neurodegeneration, oxidative stress, inflammation, and disrupted neurotransmission, all of which lead to cognitive impairment and enhanced risk of neurodegenerative diseases. Several observational and intervention studies have investigated the effects of nutraceuticals on these mechanisms (Fekete et al. 2023).

Evidence indicates that diets high in polyphenols, omega-3 fatty acids, and essential vitamins are needed to ensure cognitive health. As an illustration, the Framingham Heart Study revealed that increased omega-3 fatty acid intake correlates with lower Alzheimer's disease risk. Long-term studies of Mediterranean and MIND diet habits further show that sticking to these diet plans, high in brain-improving nutraceuticals, correlates with slowing cognitive deterioration and improved memory retention in older adults (Key and Szabo-Reed 2023).

Furthermore, curcumin, resveratrol, and flavonoid research has indicated that such substances can trigger neuroprotection, diminish oxidative damage, and enhance memory and cognitive learning. Preclinical studies on animal models show that these compounds have neurogenic and anti-inflammatory effects, which is good evidence for starting clinical trials on humans (Mamun et al. 2024).

2.8.2 Trials and Outcomes of Nutraceutical Interventions

Clinical trials have assessed nutraceuticals' effectiveness in cognitive function and neurodegenerative disorders. Although numerous studies have yielded positive results, differences in study protocols, dosages, and patient populations have contributed to variability in results. These aspects render it difficult to draw firm conclusions about the long-term efficacy of nutraceuticals in preventing or managing cognitive impairment (Puri et al. 2022).

Omega-3 fatty acids, especially docosahexaenoic acid (DHA), have been well-researched regarding their involvement in cognitive function. A randomised controlled trial (RCT) in the *Journal of Alzheimer's Disease* indicated that DHA supplementation enhanced memory and cognitive functioning in people with mild cognitive impairments (MCIs). However, in patients with advanced Alzheimer's, the effects were weaker, indicating that omega-3 fatty acids could be more effective at the onset of cognitive impairment than in more advanced neurodegeneration (Welty 2023).

Curcumin, the bioactive component of turmeric, also has potential cognitive effects. In a double-masked, placebo-controlled trial, curcumin supplementation improved working memory and attention span in healthy elderly adults. Brain imaging studies also identified decreased amyloid deposition in curcumin users, suggesting its therapeutic utility in Alzheimer's prevention. However, limitations remain in curcumin's poor bioavailability, which mandates the development of sophisticated formulations to enhance its clinical benefit (Small et al. 2018).

Ginkgo biloba, an herbal supplement rich in antioxidants and anti-inflammatory compounds, has been tested in several clinical trials for its effect on cognitive function. A meta-analysis of several RCTs showed that supplementation with *Ginkgo biloba* enhanced cognitive function and delayed the progression of dementia symptoms in Alzheimer's patients. However, individual studies have shown different efficacy levels, perhaps because of differences in dosages, treatment periods, and patient groups (Nguyen and Alzahrani 2025).

B vitamins, especially B6, B9 (folate), and B12, have been studied for their ability to lower homocysteine levels, which are correlated with cognitive impairment. The VITACOG trial showed that B-vitamin therapy significantly reduced brain atrophy in people with MCI, suggesting that it might have neuroprotective effects. Given that increased homocysteine levels correlate with a heightened risk of neurodegeneration, B-vitamin supplementation could be an effective method for maintaining cognitive function in vulnerable populations (de Jager 2014).

Resveratrol, a polyphenolic compound in red wine and grapes, has become of interest for its ability to improve cerebral blood flow and neurovascular function. Resveratrol supplementation improved memory test scores and increased hippocampal activity in an older subjects' study, supporting its potential application in maintaining brain health. However, as with the majority of other nutraceuticals, further research is required to maximize its benefits in terms of bioavailability and therapeutic dosage (Boccardi et al. 2024).

Although these clinical trials indicate the potential for nutraceuticals to support cognitive function and neuroprotection, variability in results highlights the need for more work. Standardized protocols, larger sample sizes, and long-term studies are required to verify the efficacy of these molecules and determine proper guidelines for the prevention and treatment of neurodegenerative diseases (Carrillo et al. 2025).

Table 2.2 summarizes major clinical trials investigating nutraceuticals' effects on cognitive function and memory, as well as the progression of neurodegenerative diseases. It provides the study population, intervention description, and notable findings for omega-3 s, curcumin, *Ginkgo biloba*, resveratrol, and B vitamins. The table reaffirms the evidence for nutraceuticals in brain health and identifies areas requiring additional research.

2.8.3 Challenges and Limitations in Clinical Research

Despite the growing interest in nutraceuticals for brain health, several challenges hinder their widespread acceptance in clinical practice. Differences in dosage, duration, and participant demographics across studies make it difficult to draw definitive conclusions regarding nutraceutical efficacy (Puri et al. 2022). Many nutraceuticals, such as curcumin and resveratrol, have poor bioavailability, limiting their absorption and effectiveness in human trials. Sophisticated formulations, such as liposomal and nanoparticle-based delivery systems, are now being used to optimize their therapeutic potential (Puri et al. 2022).

Table 2.2 Clinical studies on nutraceuticals and cognitive function

Nutraceutical	Study	Population	Findings	References
Omega-3 fatty acids	Framingham heart study	Older adults	Reduced risk of Alzheimer's disease and cognitive decline	Wei et al. (2023)
Curcumin	Double-masked placebo-controlled trial	Healthy older adults	Improved working memory and attention span	Francis et al. (2024)
Ginkgo biloba	Meta-analysis of RCTs	Alzheimer's patients	Improved cognitive function and delayed disease progression	Yang et al. (2016)
Resveratrol	Clinical trial on cerebral blood flow	Aging population	Enhanced memory and hippocampal activity	Hattori et al. (2024)
B vitamin supplementation	VITACOG trial	Mild cognitive impairment	Slowed brain atrophy and improved cognitive function	Wu et al. (2021)

Cognitive function tests are based on subjective self-reporting, which can be biased. Placebo effects also affect study results, making them difficult to interpret (Puri et al. 2022). In contrast to medicinal drugs, nutraceuticals do not have strict regulatory approval procedures. Therefore, there is inconsistency in the quality, purity, and activity of products among various brands and forms (Puri et al. 2022). Although short-term nutraceutical actions have been reported, limited long-term trials have investigated their drug interactions, effectiveness, and safety (Puri et al. 2022). To find out if nutraceuticals are effective as cognitive enhancers and neuroprotective agents, it will be important to overcome these problems through well-designed, large-scale, and long-term clinical trials. Future research needs to consider personalized nutritional interventions (i.e., genetic susceptibility and lifestyle) to optimize the beneficial impacts of nutraceuticals on neurological well-being (Puri et al. 2022).

2.9 Nutraceuticals in Preventive and Therapeutic Approaches

2.9.1 Role in Slowing Cognitive Decline

Aging has been linked to continuous deterioration in cognitive performance, such as memory impairment, slowed processing, and compromised executive function. Nutraceuticals have attracted significant interest due to their capacity to counteract cognitive impairments and shield cerebral function in the elderly. These bioactive compounds act through various mechanisms, including suppressing oxidative stress, alleviating neuroinflammation, and stimulating neurogenesis (Mecocci et al. 2014).

Omega-3 fatty acids, especially DHA and EPA, are among the most researched nutraceuticals for conserving cognition. The fatty acids are critical to neuronal membrane health, enhancing synaptic plasticity, and moderating inflammatory pathways, all of which help to preserve cognition. Also, polyphenols like flavonoids and resveratrol have been shown to improve blood flow to the brain, protect against oxidative damage, and encourage synaptic connectivity, all of which help the brain stay strong (Swanson et al. 2012). Several studies suggest that diets rich in brain-boosting nutraceuticals, such as the Mediterranean and MIND diets, are associated with a lower risk of cognitive impairment and neurodegenerative diseases. Regular intake of key vitamins (B, D, and E) and minerals (magnesium, zinc) further supports cognitive longevity by preventing deficiencies contributing to cognitive decline (Fekete et al. 2023).

2.9.2 Combined Approaches with Pharmacological Treatments

While nutraceuticals provide significant neuroprotective benefits, pharmacological treatments can further enhance their efficacy. Integrative approaches that incorporate conventional medications and nutraceutical interventions can optimize brain health and improve treatment outcomes for individuals with cognitive impairments (Makkar et al. 2020). In neurodegenerative diseases such as Alzheimer's and Parkinson's, pharmacological treatments primarily focus on symptom management rather than disease modification. Nutraceuticals have complementary mechanisms of action that can delay disease progression and augment drug action.

For example, because curcumin is an antioxidant and an anti-inflammatory, it has been studied as an extra treatment to cholinesterase inhibitors (like donepezil) for Alzheimer's patients to help them think and remember things better (Mishra and Palanivelu 2008). Coenzyme Q10 (CoQ10) and polyphenolic agents are candidate therapies to enhance mitochondrial function and reverse the oxidative stress that defines Parkinson's disease. They would be helpful as adjuncts to dopaminergic treatment (Sharma et al. 2025). B vitamins contribute to decreasing homocysteine, a neurodegenerative process factor, and can complement pharmacologic treatments for cognitive impairment (Orywal et al. 2025). Despite these positive synergies, more research is needed to find the best doses, interactions, and long-term safety of using nutraceuticals and conventional therapies together.

2.9.3 Lifestyle and Dietary Approaches to Brain Health

In addition to pharmacologic and nutraceutical therapies, lifestyle and dietary approaches are important for preserving cognitive function and overall brain health. A holistic approach that incorporates a nutrient-rich diet, regular physical activity, mental stimulation, and stress management can significantly enhance cognitive resilience and reduce the risk of neurodegenerative disorders (Dominguez et al. 2021).

2.9.3.1 Dietary Strategies

The Mediterranean and MIND diets emphasize nutrient-dense foods, such as leafy greens, berries, nuts, fatty fish, and whole grains, all of which provide essential vitamins, antioxidants, and anti-inflammatory compounds beneficial for brain health.

Proper hydration and a healthy gut microbiome are vital for cognitive function. Probiotic-rich foods (e.g., yoghurt or kefir) and fibre-rich foods support gut health, which are closely linked to brain function. High sugar intake and ultra-processed foods contribute to oxidative stress and inflammation, accelerating cognitive decline. Replacing refined sugars with natural sweeteners and whole foods supports long-term brain health (van Zonneveld et al. 2024).

2.9.3.2 Lifestyle Strategies

Exercise increases blood flow to the brain, enhances neurogenesis, and reduces inflammation. Aerobic exercise, strength training, and yoga have been linked to improved memory and cognitive function (Erickson et al. 2011). Engaging in mentally stimulating activities such as puzzles, reading, and learning new skills promotes neuroplasticity and cognitive aging. Social engagement and maintenance of close social relationships also lower the risk of cognitive decline (Boot et al. 2017).

Chronic stress is an etiology of neuroinflammation and cognitive dysfunction. Mindfulness, meditation, and adaptogenic herbs (e.g., *Ashwagandha*, *Rhodiola rosea*) benefit cortisol control and cognitive health (Taren et al. 2015). Integrating nutraceuticals with pharmacological interventions, dietary modification, and lifestyle modification allows individuals to adopt an integrated approach to preserving brain function and preventing cognitive decline. Several new nutraceutical preparations are under investigation to study their place in preventive and therapeutic approaches to neurodegenerative disorders (Gong et al. 2021).

2.10 Future Directions and Emerging Trends

2.10.1 Personalized Medicine and Brain Health

Personalized medicine is revolutionizing the healthcare industry by tailoring treatment to an individual's genetic predisposition, lifestyle, and environment. In brain health, personalized nutraceutical interventions are gaining popularity as scientists strive to create targeted interventions to improve cognitive capacity and reduce the risk of neurodegenerative diseases (Pandey and Gupta 2024). Advances in technology in genetic profiling and biomarker identification enable the tailoring of nutraceutical interventions to the unique needs of every individual. For instance, people carrying the APOE4 allele that they inherited, might be advised to have more polyphenols and omega-3 fatty acids to lower their risk of getting Alzheimer's disease. Second, gene mutations in the vitamin D receptor gene that affect the way a person metabolizes vitamin D might require supplements (Pérez-Beltrán et al. 2022).

Precision nutrition has grown enormously with the help of artificial intelligence (AI) and big data. Through AI, we are able to analyze a person's eating patterns, gut microbiome traits, and metabolic types and create customized nutraceutical prescriptions that cater to their individual brain and nervous system needs. As personalized medicine grows, nutraceuticals can possibly have a greater role in improving brain function with customized solutions (Kassem et al. 2025).

2.10.2 Innovations in Nutraceutical Formulations

Bioavailability, stability, and absorption problems often limit nutraceuticals' efficacy. Recent improvements in formulation technology are getting around these problems, allowing bioactive molecules to be delivered to the brain in the right amounts for therapeutic effects (Puri et al. 2022). Liposomal and nano-encapsulation technologies make chemicals like curcumin and resveratrol more bioavailable even though the body does not absorb them as well. By keeping bioactive molecules from being broken down by enzymes, new delivery systems make it easier for them to be absorbed in the gut and get into the central nervous system (Manocha et al. 2022).

Integration of probiotics and prebiotics, the gut–brain axis provides the foundation for cognitive processes, and recent research has emphasized the potential role of prebiotics and probiotics as cerebral wellness aids. Combinations of gut-supportive nutraceuticals and probiotics could increase the bioavailability of desired molecules, while supporting a healthy microbiome (Ansari et al. 2023). Extended-release formulations enable consistent delivery of nutraceuticals to the brain for a prolonged duration, enhancing their neuroprotective action. To maximize the direct action of nutraceuticals on neural tissues, new delivery methods are also being developed, such as nanoparticles that can cross the blood-brain barrier (Kamath et al. 2024).

Formulating for combinations, researchers are investigating multi-component supplements that combine synergistic nutrients such as polyphenols, omega-3 fatty acids, and adaptogens to see if they could support brain health. These blends enhance general cognitive health by utilizing the synergistic effects of several bioactive ingredients. It is anticipated that nutraceutical trends will increase the therapeutic benefit of dietary interventions, while extending their application to areas of cognitive improvement and neuroprotection (Kalogerakou and Antoniadou 2024).

2.10.3 Synergistic Interaction Potential and Combined Therapies

In order to promote neurological well-being, recent research has focused more and more on the synergistic potential of nutraceuticals when combined with complementary techniques. An integrative strategy that incorporates lifestyle changes, nutritional components, and adjuvant therapy is becoming more popular as an alternative to relying solely on one agent (Puri et al. 2022). Synergistic interaction of nutraceuticals with conventional pharmacological therapy can improve the treatment of neurodegenerative diseases. For example, coenzyme Q10 and polyphenols can improve mitochondrial function in Parkinson's disease and complement existing drug therapy (Puri et al. 2022). The incorporation of mind–body interventions, such as meditation, yoga, and mindfulness-based stress reduction (MBSR), has been demonstrated to have positive impacts on cognitive function (Farhang et al. 2019). The combination of these lifestyle interventions with nutraceutical therapies has additive or synergistic effects on reducing stress, promoting neuroplasticity, and preserving cognitive health.

Nutraceuticals are used most effectively as dietary synergies through a whole-food, high-fiber diet. Bioavailability and the best cognitive benefits are obtained with dietary supplementation and foods interacting with each other, such as omega-3 fish with polyphenol vegetables (Puri et al. 2022). Utilisation of neuroprotective adaptogens like *Ashwagandha* and *Rhodiola rosea* in foods and beverages is an effective and feasible solution for enhancing cognitive resilience and stress adaptation (Puri et al. 2022). As the science of nutraceuticals and their influence on brain aging evolves, it is anticipated that subsequent research will increasingly emphasize multi-dimensional strategies that integrate dietary changes, lifestyle modifications, and novel formulations to maximize cognitive performance and slow neurodegenerative processes. Future developments in the field of nutraceutical science point to a time when evidence-based, individualized, and holistic therapies will be the cornerstone of long-term brain health (Flanagan et al. 2020).

2.11 Conclusions

The therapeutic potential of nutraceuticals to impact aging and brain function is a rapidly developing area of research with enormous potential for the augmentation of cognitive function, prevention of neurodegenerative diseases, and their treatment. As the population ages, there is growing interest in safe, natural ways of preventing neurodegenerative diseases, and protecting cognitive function. The function of nutraceuticals in brain function has been investigated in this chapter, with specific emphasis on their quality as anti-inflammatory drugs, neuroprotective agents, and antioxidants. A number of bioactive compounds of considerable interest have been shown to be promising in protecting cognitive processes and delaying the onset of neurodegenerative disorders associated with aging like Alzheimer's and Parkinson's diseases. These include some of the useful nutrients like polyphenols, omega-3 fatty acids, vitamins, minerals, and a few plant extracts. Clinical trials have shown that such compounds are able to enhance cognitive processes like memory, learning, attention, and resistance to stress. Further extensive clinical tests are, however, needed to determine optimal treatment regimens.

The future of nutraceuticals and cognitive function will be influenced by advancements in personalized medicine, novel formulation technologies, and synergistic pairing that combines nutraceuticals with pharmacologic interventions and lifestyle changes. As science unfolds, optimization of cognitive longevity will necessitate an integrative approach that entails stress management, exercise, nutrition, and select supplements. While the vast potential offered by nutraceuticals cannot be overlooked, issues related to variations in bioavailability, regulatory oversight, and dosing standardization need to be resolved to maximize their safety and efficacy. To maximize the potential of nutraceuticals to improve cognitive health, more research, evidence-based recommendations, and consumer education are needed. In short, the significant field of nutraceuticals is seeing an increase in studies on age and brain capacity as well as the creation of safe, empirically derived techniques for enhancing brain function. The future will increasingly depend on nutraceuticals for neuroprotection and cognitive health as research breakthroughs become apparent and preventive medicine becomes more predominant.

References

Ajitkumar A, De Jesus O (2025) Huntington disease. StatPearls. StatPearls Publishing, Treasure Island (FL)

Ansari F, Neshat M, Pourjafar H, Jafari SM, Samakkhah SA, Mirzakhani E (2023) The role of probiotics and prebiotics in modulating of the gut-brain axis. Front Nutr 10:1173660. https://doi.org/10.3389/fnut.2023.1173660

Anwar MJ, Alenezi SK, Alhowail AH (2023) Molecular insights into the pathogenic impact of vitamin D deficiency in neurological disorders. Biomed Pharmacother 162:114718. https://doi.org/10.1016/j.biopha.2023.114718

Ashok A, Andrabi SS, Mansoor S, Kuang Y, Kwon BK, Labhasetwar V (2022) Antioxidant therapy in oxidative stress-induced neurodegenerative diseases: role of nanoparticle-based drug delivery systems in clinical translation. Antioxidants (Basel) 11:408. https://doi.org/10.3390/antiox11020408

Bathina S, Das UN (2015) Brain-derived neurotrophic factor and its clinical implications. Arch Med Sci 11:1164–1178. https://doi.org/10.5114/aoms.2015.56342

Boccardi V, Tagliafico L, Persia A, Page E, Ottaviani S, Cremonini AL et al (2024) The potential effects of red wine and its components on neurocognitive disorders: a narrative review. Nutrients 16:3431. https://doi.org/10.3390/nu16203431

Boot N, Baas M, van Gaal S, Cools R, De Dreu CKW (2017) Creative cognition and dopaminergic modulation of fronto-striatal networks: integrative review and research agenda. Neurosci Biobehav Rev 78:13–23. https://doi.org/10.1016/j.neubiorev.2017.04.007

Carrillo JÁ, Arcusa R, Xandri-Martínez R, Cerdá B, Zafrilla P, Marhuenda J (2025) Impact of polyphenol-rich nutraceuticals on cognitive function and neuroprotective biomarkers: a randomized, double-blind, placebo-controlled clinical trial. Nutrients 17:601. https://doi.org/10.3390/nu17040601

Chiu H-F, Venkatakrishnan K, Wang C-K (2020) The role of nutraceuticals as a complementary therapy against various neurodegenerative diseases: a mini-review. J Tradit Complement Med 10:434–439. https://doi.org/10.1016/j.jtcme.2020.03.008

Ciurea AV, Mohan AG, Covache-Busuioc R-A, Costin H-P, Glavan L-A, Corlatescu A-D et al (2023) Unraveling molecular and genetic insights into neurodegenerative diseases: advances in understanding Alzheimer's, Parkinson's, and Huntington's diseases and amyotrophic lateral sclerosis. Int J Mol Sci 24:10809. https://doi.org/10.3390/ijms241310809

Clemente-Suárez VJ, Martín-Rodríguez A, Curiel-Regueros A, Rubio-Zarapuz A, Tornero-Aguilera JF (2025) Neuro-nutrition and exercise synergy: exploring the bioengineering of cognitive enhancement and mental health optimization. Bioengineering 12:208. https://doi.org/10.3390/bioengineering12020208

Cognitive Health and Older Adults (2024) National Institute on Aging. https://www.nia.nih.gov/health/brain-health/cognitive-health-and-older-adults. Accessed 24 Mar 2025

Cordeiro A, Gomes C, Bicker J, Fortuna A (2024) Aging and cognitive resilience: molecular mechanisms as new potential therapeutic targets. Drug Discov Today 29:104093. https://doi.org/10.1016/j.drudis.2024.104093

Dadhania VP, Trivedi PP, Vikram A, Tripathi DN (2016) Nutraceuticals against neurodegeneration: a mechanistic insight. Curr Neuropharmacol 14:627–640. https://doi.org/10.2174/1570159X14666160104142223

Dama A, Shpati K, Daliu P, Dumur S, Gorica E, Santini A (2024) Targeting metabolic diseases: the role of nutraceuticals in modulating oxidative stress and inflammation. Nutrients 16:507. https://doi.org/10.3390/nu16040507

Dash UC, Bhol NK, Swain SK, Samal RR, Nayak PK, Raina V et al (2025) Oxidative stress and inflammation in the pathogenesis of neurological disorders: mechanisms and implications. Acta Pharm Sin B 15:15–34. https://doi.org/10.1016/j.apsb.2024.10.004

Dashwood R, Visioli F (2025) L-theanine: from tea leaf to trending supplement – does the science match the hype for brain health and relaxation? Nutr Res 134:39–48. https://doi.org/10.1016/j.nutres.2024.12.008

de Jager CA (2014) Critical levels of brain atrophy associated with homocysteine and cognitive decline. Neurobiol Aging 35:S35–S39. https://doi.org/10.1016/j.neurobiolaging.2014.03.040

de la Fuente AG, Pelucchi S, Mertens J, Di Luca M, Mauceri D, Marcello E (2023) Novel therapeutic approaches to target neurodegeneration. Br J Pharmacol 180:1651–1673. https://doi.org/10.1111/bph.16078

Dighriri IM, Alsubaie AM, Hakami FM, Hamithi DM, Alshekh MM, Khobrani FA et al (n.d.) Effects of Omega-3 polyunsaturated fatty acids on brain functions: a systematic review. Cureus 14:e30091. https://doi.org/10.7759/cureus.30091

Dominguez LJ, Veronese N, Vernuccio L, Catanese G, Inzerillo F, Salemi G et al (2021) Nutrition, physical activity, and other lifestyle factors in the prevention of cognitive decline and dementia. Nutrients 13:4080. https://doi.org/10.3390/nu13114080

Duberley KE, Heales SJR, Abramov AY, Chalasani A, Land JM, Rahman S et al (2014) Effect of coenzyme Q10 supplementation on mitochondrial electron transport chain activity and mitochondrial oxidative stress in coenzyme Q10 deficient human neuronal cells. Int J Biochem Cell Biol 50:60–63. https://doi.org/10.1016/j.biocel.2014.02.003

Erickson KI, Voss MW, Prakash RS, Basak C, Szabo A, Chaddock L et al (2011) Exercise training increases size of hippocampus and improves memory. Proc Natl Acad Sci 108:3017–3022. https://doi.org/10.1073/pnas.1015950108

Farhang M, Miranda-Castillo C, Rubio M, Furtado G (2019) Impact of mind-body interventions in older adults with mild cognitive impairment: a systematic review. Int Psychogeriatr 31:643–666. https://doi.org/10.1017/S1041610218002302

Fekete M, Lehoczki A, Tarantini S, Fazekas-Pongor V, Csípő T, Csizmadia Z et al (2023) Improving cognitive function with nutritional supplements in aging: a comprehensive narrative review of clinical studies investigating the effects of vitamins, minerals, antioxidants, and other dietary supplements. Nutrients 15:5116. https://doi.org/10.3390/nu15245116

Fenercioglu AK (2024) The anti-inflammatory roles of vitamin D for improving human health. Curr Issues Mol Biol 46:13514–13525. https://doi.org/10.3390/cimb46120807

Ferreira GC, McKenna MC (2017) L-carnitine and acetyl-L-carnitine roles and neuroprotection in developing brain. Neurochem Res 42:1661–1675. https://doi.org/10.1007/s11064-017-2288-7

Flanagan E, Lamport D, Brennan L, Burnet P, Calabrese V, Cunnane SC et al (2020) Nutrition and the ageing brain: moving towards clinical applications. Ageing Res Rev 62:101079. https://doi.org/10.1016/j.arr.2020.101079

Francis AJ, Sreenivasan C, Parikh A, AlQassab O, Kanthajan T, Pandey M et al (2024) Curcumin and cognitive function: a systematic review of the effects of curcumin on adults with and without neurocognitive disorders. Cureus 16:e67706. https://doi.org/10.7759/cureus.67706

Garodia P, Hegde M, Kunnumakkara AB, Aggarwal BB (2023) Curcumin, inflammation, and neurological disorders: how are they linked? Integrat Med Res 12:100968. https://doi.org/10.1016/j.imr.2023.100968

GFRS (n.d.) Regulating Health Supplements.pdf

Gong W, Zeng J, Ji J, Jia Y, Jia S, Sanders AJ et al (2021) EPLIN expression in gastric cancer and impact on prognosis and Chemoresistance. Biomolecules 11:547. https://doi.org/10.3390/biom11040547

Gonzales MM, Garbarino VR, Pollet E, Palavicini JP, Kellogg DL, Kraig E et al (2022) Biological aging processes underlying cognitive decline and neurodegenerative disease. J Clin Invest 132:e158453. https://doi.org/10.1172/JCI158453

Hattori Y, Kakino Y, Hattori Y, Iwashita M, Uchiyama H, Noda K et al (2024) Long-term resveratrol intake for cognitive and cerebral blood flow impairment in carotid artery stenosis/occlusion. J Stroke 26:64–74. https://doi.org/10.5853/jos.2023.02733

Ionescu-Tucker A, Cotman CW (2021) Emerging roles of oxidative stress in brain aging and Alzheimer's disease. Neurobiol Aging 107:86–95. https://doi.org/10.1016/j.neurobiolaging.2021.07.014

Ivanova Stojcheva E, Quintela JC (2022) The effectiveness of Rhodiola rosea L. preparations in alleviating various aspects of life-stress symptoms and stress-induced conditions – encouraging clinical evidence. Molecules 27:3902. https://doi.org/10.3390/molecules27123902

Kalogerakou T, Antoniadou M (2024) The role of dietary antioxidants, food supplements and functional foods for energy enhancement in healthcare professionals. Antioxidants (Basel) 13:1508. https://doi.org/10.3390/antiox13121508

Kamath AP, Nayak PG, John J, Mutalik S, Balaraman AK, Krishnadas N (2024) Revolutionizing neurotherapeutics: nanocarriers unveiling the potential of phytochemicals in Alzheimer's disease. Neuropharmacology 259:110096. https://doi.org/10.1016/j.neuropharm.2024.110096

Kamboj N, Sharma S, Kumar R (2025) Neuroprotective insights into epigallocatechin gallate (EGCG) for neurodegenerative disorders. Explor Neurosci 4:100673. https://doi.org/10.37349/en.2025.100673

Kassem H, Beevi AA, Basheer S, Lutfi G, Cheikh Ismail L, Papandreou D (2025) Investigation and assessment of AI'S role in nutrition – an updated narrative review of the evidence. Nutrients 17:190. https://doi.org/10.3390/nu17010190

Kennedy DO (2016) B vitamins and the brain: mechanisms, dose and efficacy – a review. Nutrients 8:68. https://doi.org/10.3390/nu8020068

Key MN, Szabo-Reed AN (2023) Impact of diet and exercise interventions on cognition and brain health in older adults: a narrative review. Nutrients 15:2495. https://doi.org/10.3390/nu15112495

Komala MG, Ong SG, Qadri MU, Elshafie LM, Pollock CA, Saad S (2023) Investigating the regulatory process, safety, efficacy and product transparency for nutraceuticals in the USA. Europe Australia Foods 12:427. https://doi.org/10.3390/foods12020427

Kouli A, Torsney KM, Kuan W-L (2018) Parkinson's disease: etiology, neuropathology, and pathogenesis. In: Stoker TB, Greenland JC (eds) Parkinson's disease: pathogenesis and clinical aspects. Codon Publications, Brisbane

Kurowska A, Ziemichód W, Herbet M, Piątkowska-Chmiel I (2023) The role of diet as a modulator of the inflammatory process in the neurological diseases. Nutrients 15:1436. https://doi.org/10.3390/nu15061436

Lee KH, Cha M, Lee BH (2020) Neuroprotective effect of antioxidants in the brain. Int J Mol Sci 21:7152. https://doi.org/10.3390/ijms21197152

Lobo F, Haase J, Brandhorst S (2022) The effects of dietary interventions on brain aging and neurological diseases. Nutrients 14:5086. https://doi.org/10.3390/nu14235086

Loong S, Barnes S, Gatto NM, Chowdhury S, Lee GJ (2023) Omega-3 fatty acids, cognition, and brain volume in older adults. Brain Sci 13:1278. https://doi.org/10.3390/brainsci13091278

Magomedova A, Fatima G (2025) Mental health and well-being in the modern era: a comprehensive review of challenges and interventions. Cureus 17:e77683. https://doi.org/10.7759/cureus.77683

Makkar R, Behl T, Bungau S, Zengin G, Mehta V, Kumar A et al (2020) Nutraceuticals in neurological disorders. Int J Mol Sci 21:4424. https://doi.org/10.3390/ijms21124424

Mamun AA, Shao C, Geng P, Wang S, Xiao J (2024) Polyphenols targeting NF-κB pathway in neurological disorders: what we know so far? Int J Biol Sci 20:1332–1355. https://doi.org/10.7150/ijbs.90982

Manocha S, Dhiman S, Grewal AS, Guarve K (2022) Nanotechnology: an approach to overcome bioavailability challenges of nutraceuticals. Jo Drug Deliv Sci Technol 72:103418. https://doi.org/10.1016/j.jddst.2022.103418

Mantle D, Hargreaves IP (2022) Mitochondrial dysfunction and neurodegenerative disorders: role of nutritional supplementation. Int J Mol Sci 23:12603. https://doi.org/10.3390/ijms232012603

Mark LP, Prost RW, Ulmer JL, Smith MM, Daniels DL, Strottmann JM et al (2001) Pictorial review of glutamate excitotoxicity: fundamental concepts for neuroimaging. AJNR Am J Neuroradiol 22:1813–1824

Martínez Leo EE, Peñafiel AM, Hernández Escalante VM, Cabrera Araujo ZM (2021) Ultra-processed diet, systemic oxidative stress, and breach of immunologic tolerance. Nutrition 91–92:111419. https://doi.org/10.1016/j.nut.2021.111419

Marzola P, Melzer T, Pavesi E, Gil-Mohapel J, Brocardo PS (2023) Exploring the role of neuroplasticity in development, aging, and neurodegeneration. Brain Sci 13:1610. https://doi.org/10.3390/brainsci13121610

Mazzanti G, Di Giacomo S (2016) Curcumin and resveratrol in the management of cognitive disorders: what is the clinical evidence? Molecules 21:1243. https://doi.org/10.3390/molecules21091243

Mecocci P, Tinarelli C, Schulz RJ, Polidori MC (2014) Nutraceuticals in cognitive impairment and Alzheimer's disease. Front Pharmacol 5:147. https://doi.org/10.3389/fphar.2014.00147

Miao J, Ma H, Yang Y, Liao Y, Lin C, Zheng J et al (2023) Microglia in Alzheimer's disease: pathogenesis, mechanisms, and therapeutic potentials. Front Aging Neurosci 15:1201982. https://doi.org/10.3389/fnagi.2023.1201982

Minocha T, Birla H, Obaid AA, Rai V, Sushma P, Shivamallu C et al (2022) Flavonoids as promising neuroprotectants and their therapeutic potential against Alzheimer's disease. Oxidative Med Cell Longev 2022:6038996. https://doi.org/10.1155/2022/6038996

Mishra S, Palanivelu K (2008) The effect of curcumin (turmeric) on Alzheimer's disease: an overview. Ann Indian Acad Neurol 11:13–19. https://doi.org/10.4103/0972-2327.40220

Mitchell ES, Conus N, Kaput J (2014) B vitamin polymorphisms and behavior: evidence of associations with neurodevelopment, depression, schizophrenia, bipolar disorder and cognitive decline. Neurosci Biobehav Rev 47:307–320. https://doi.org/10.1016/j.neubiorev.2014.08.006

Murman DL (2015) The impact of age on cognition. Semin Hear 36:111–121. https://doi.org/10.1055/s-0035-1555115

Nahar L, Charoensup R, Kalieva K, Habibi E, Guo M, Wang D et al (2025) Natural products in neurodegenerative diseases: recent advances and future outlook. Front Pharmacol 16. https://doi.org/10.3389/fphar.2025.1529194

National Institute on Aging (2023) What do we know about diet and prevention of Alzheimer's disease? https://www.nia.nih.gov/health/alzheimers-and-dementia/what-do-we-know-about-diet-and-prevention-alzheimers-disease. Accessed 25 Mar 2025

Nawrot J, Gornowicz-Porowska J, Budzianowski J, Nowak G, Schroeder G, Kurczewska J (2022) Medicinal herbs in the relief of neurological, cardiovascular, and respiratory symptoms after COVID-19 infection a literature review. Cells 11:1897. https://doi.org/10.3390/cells11121897

Nguyen T, Alzahrani T (2025) Ginkgo Biloba. StatPearls, Treasure Island (FL): StatPearls Publishing

NINDS (n.d.) Amyotrophic Lateral Sclerosis (ALS) | National Institute of Neurological Disorders and Stroke. https://www.ninds.nih.gov/health-information/disorders/amyotrophic-lateral-sclerosis-als. Accessed 31 Mar 2025

Nowak A, Kojder K, Zielonka-Brzezicka J, Wróbel J, Bosiacki M, Fabiańska M et al (2021) The use of ginkgo Biloba L. as a neuroprotective agent in the Alzheimer's disease. Front Pharmacol 12:775034. https://doi.org/10.3389/fphar.2021.775034

OrfonlineOrg (n.d.) Quality rules in India: trade, technical regulations and consumer protection. https://www.orfonline.org/research/quality-rules-in-india-trade-technical-regulations-and-consumer-protection. Accessed 25 Mar 2025

Orywal K, Socha K, Iwaniuk P, Kaczyński P, Farhan JA, Zoń W et al (2025) Vitamins in the prevention and support therapy of neurodegenerative diseases. Int J Mol Sci 26:1333. https://doi.org/10.3390/ijms26031333

Pandey A, Gupta SP (2024) Personalized medicine: a comprehensive review. Orient J Chem 40:933–944

Pandey KB, Rizvi SI (2009) Plant polyphenols as dietary antioxidants in human health and disease. Oxidative Med Cell Longev 2:270–278

Panossian A, Wikman G (2010) Effects of Adaptogens on the central nervous system and the molecular mechanisms associated with their stress – protective activity. Pharmaceuticals (Basel) 3:188–224. https://doi.org/10.3390/ph3010188

Pérez-Beltrán YE, Rivera-Iñiguez I, Gonzalez-Becerra K, Pérez-Naitoh N, Tovar J, Sáyago-Ayerdi SG et al (2022) Personalized dietary recommendations based on lipid-related genetic variants: a systematic review. Front Nutr 9:830283. https://doi.org/10.3389/fnut.2022.830283

Poulose SM, Miller MG, Scott T, Shukitt-Hale B (2017) Nutritional factors affecting adult neurogenesis and cognitive function. Adv Nutr 8:804–811. https://doi.org/10.3945/an.117.016261

Puri V, Nagpal M, Singh I, Singh M, Dhingra GA, Huanbutta K et al (2022) A comprehensive review on nutraceuticals: therapy support and formulation challenges. Nutrients 14:4637. https://doi.org/10.3390/nu14214637

Puri S, Shaheen M, Grover B (2023) Nutrition and cognitive health: a life course approach. Front Public Health 11:1023907. https://doi.org/10.3389/fpubh.2023.1023907

Putta DS (n.d.) FSSAI guidance and notification on nutraceuticals – an insight

Rajan KE, Preethi J, Singh HK (2015) Molecular and functional characterization of Bacopa monniera: a retrospective review. Evid Based Complement Alternat Med 2015:945217. https://doi.org/10.1155/2015/945217

Rajmohan R, Reddy PH (2017) Amyloid Beta and Phosphorylated tau accumulations cause abnormalities at synapses of Alzheimer's disease neurons. J Alzheimer's Dis 57:975–999. https://doi.org/10.3233/JAD-160612

Raza ML, Hassan ST, Jamil S, Fatima W, Fatima M (2025) Nutritional interventions in depression: the role of vitamin D and omega-3 fatty acids in neuropsychiatric health. Clin Nutr 45:270–280. https://doi.org/10.1016/j.clnu.2025.01.009

Reay JL, Kennedy DO, Scholey AB (2005) Single doses of Panax ginseng (G115) reduce blood glucose levels and improve cognitive performance during sustained mental activity. J Psychopharmacol 19:357–365. https://doi.org/10.1177/0269881105053286

Remenapp A, Coyle K, Orange T, Lynch T, Hooper D, Hooper S et al (2022) Efficacy of *Withania somnifera* supplementation on adult's cognition and mood. J Ayurveda Integr Med 13:100510. https://doi.org/10.1016/j.jaim.2021.08.003

Reza-Zaldívar EE, Jacobo-Velázquez DA (2023) Comprehensive review of nutraceuticals against cognitive decline associated with Alzheimer's disease. ACS Omega 8:35499–35522. https://doi.org/10.1021/acsomega.3c04855

Rudrapal M, Khairnar SJ, Khan J, Dukhyil AB, Ansari MA, Alomary MN et al (2022) Dietary polyphenols and their role in oxidative stress-induced human diseases: insights into protective effects, antioxidant potentials and mechanism(s) of action. Front Pharmacol:13. https://doi.org/10.3389/fphar.2022.806470

Santini A, Cammarata SM, Capone G, Ianaro A, Tenore GC, Pani L et al (2018) Nutraceuticals: opening the debate for a regulatory framework. Br J Clin Pharmacol 84:659–672. https://doi.org/10.1111/bcp.13496

Schulz M, Mayer C, Schlemm E, Frey BM, Malherbe C, Petersen M et al (2022) Association of age and structural brain changes with functional connectivity and executive function in a middle-aged to older population-based cohort. Front Aging Neurosci 14:782738. https://doi.org/10.3389/fnagi.2022.782738

ScienceDirect (n.d.) Adaptogenic – an overview | ScienceDirect Topics. https://www.sciencedirect.com/topics/pharmacology-toxicology-and-pharmaceutical-science/adaptogenic. Accessed 25 Mar 2025

ScienceDirect Topics (n.d.) Cognitive dysfunction – an overview. https://www.sciencedirect.com/topics/neuroscience/cognitive-dysfunction. Accessed 25 Mar 2025

Shabbir U, Rubab M, Daliri EBM, Chelliah R, Javed A, Oh DH (2021) Curcumin, quercetin, catechins and metabolic diseases: the role of gut microbiota. Nutrients 13(1):206

Sharma R, Bhate L, Agrawal Y, Aspatwar A (2025) Advanced nutraceutical approaches to Parkinson's disease: bridging nutrition and neuroprotection. Nutr Neurosci:1–17. https://doi.org/10.1080/1028415X.2025.2469170

Sheppard O, Coleman M (2020) Alzheimer's disease: etiology, neuropathology and pathogenesis. In: Huang X (ed) Alzheimer's disease: drug discovery. Exon Publications, Brisbane

Silvestri S, Orlando P, Armeni T, Padella L, Brugè F, Seddaiu G et al (2015) Coenzyme Q10 and α-lipoic acid: antioxidant and pro-oxidant effects in plasma and peripheral blood lymphocytes of supplemented subjects. J Clin Biochem Nutr 57:21–26. https://doi.org/10.3164/jcbn.14-130

Singh SK, Srivastav S, Castellani RJ, Plascencia-Villa G, Perry G (2019) Neuroprotective and antioxidant effect of Ginkgo biloba extract against AD and other neurological disorders. Neurotherapeutics 16:666–674. https://doi.org/10.1007/s13311-019-00767-8

Small GW, Siddarth P, Li Z, Miller KJ, Ercoli L, Emerson ND et al (2018) Memory and brain amyloid and tau effects of a bioavailable form of curcumin in non-demented adults: a double-blind, placebo-controlled 18-month trial. Am J Geriatr Psychiatry 26:266–277. https://doi.org/10.1016/j.jagp.2017.10.010

Stuchlik A (2014) Dynamic learning and memory, synaptic plasticity and neurogenesis: an update. Front Behav Neurosci:8. https://doi.org/10.3389/fnbeh.2014.00106

Swanson D, Block R, Mousa SA (2012) Omega-3 fatty acids EPA and DHA: health benefits throughout life. Adv Nutr 3:1–7. https://doi.org/10.3945/an.111.000893

Tafti D, Ehsan M, Xixis KL (2025) Multiple sclerosis. StatPearls, Treasure Island (FL): StatPearls Publishing

Tardy A-L, Pouteau E, Marquez D, Yilmaz C, Scholey A (2020) Vitamins and minerals for energy, fatigue and cognition: a narrative review of the biochemical and clinical evidence. Nutrients 12:228. https://doi.org/10.3390/nu12010228

Taren AA, Gianaros PJ, Greco CM, Lindsay EK, Fairgrieve A, Brown KW et al (2015) Mindfulness meditation training alters stress-related amygdala resting state functional connectivity: a randomized controlled trial. Soc Cogn Affect Neurosci 10:1758–1768. https://doi.org/10.1093/scan/nsv066

Thomas J, Thomas CJ, Radcliffe J, Itsiopoulos C (2015) Omega-3 fatty acids in early prevention of inflammatory neurodegenerative disease: a focus on Alzheimer's disease. Biomed Res Int 2015:172801. https://doi.org/10.1155/2015/172801

Tryfonos C, Mantzorou M, Fotiou D, Vrizas M, Vadikolias K, Pavlidou E et al (2019) Dietary supplements on controlling multiple sclerosis symptoms and relapses: current clinical evidence and future perspectives. Medicines (Basel) 6:95. https://doi.org/10.3390/medicines6030095

Uttara B, Singh AV, Zamboni P, Mahajan RT (2009) Oxidative stress and neurodegenerative diseases: a review of upstream and downstream antioxidant therapeutic options. Curr Neuropharmacol 7:65–74. https://doi.org/10.2174/157015909787602823

van Zonneveld SM, van den Oever EJ, Haarman BCM, Grandjean EL, Nuninga JO, van de Rest O et al (2024) An anti-inflammatory diet and its potential benefit for individuals with mental disorders and neurodegenerative diseases – a narrative review. Nutrients 16:2646. https://doi.org/10.3390/nu16162646

Vauzour D (2012) Dietary polyphenols as modulators of brain functions: biological actions and molecular mechanisms underpinning their beneficial effects. Oxidative Med Cell Longev 2012:914273. https://doi.org/10.1155/2012/914273

Vettorazzi A, López de Cerain A, Sanz-Serrano J, Gil AG, Azqueta A (2020) European regulatory framework and safety assessment of food-related bioactive compounds. Nutrients 12:613. https://doi.org/10.3390/nu12030613

Vignesh A, Amal TC, Sarvalingam A, Vasanth K (2024) A review on the influence of nutraceuticals and functional foods on health. Food Chem Adv 5:100749. https://doi.org/10.1016/j.focha.2024.100749

Walker EA, Pellegrini MV (2025) Bacopa monnieri. StatPearls, Treasure Island (FL): StatPearls Publishing

Wang J, Um P, Dickerman BA, Liu J (2018) Zinc, magnesium, selenium and depression: a review of the evidence, potential mechanisms and implications. Nutrients 10:584. https://doi.org/10.3390/nu10050584

Wei B-Z, Li L, Dong C-W, Tan C-C, Xu W (2023) The relationship of Omega-3 fatty acids with dementia and cognitive decline: evidence from prospective cohort studies of supplementation, dietary intake, and blood markers. Am J Clin Nutr 117:1096–1109. https://doi.org/10.1016/j.ajcnut.2023.04.001

Welty FK (2023) Omega-3 fatty acids and cognitive function. Curr Opin Lipidol 34:12–21. https://doi.org/10.1097/MOL.0000000000000862

Wu Y, Smith AD, Refsum H, Kwok T (2021) Effectiveness of B vitamins and their interactions with aspirin in improving cognitive functioning in older people with mild cognitive impairment: pooled post-hoc analyses of two randomized trials. J Nutr Health Aging 25:1154–1160. https://doi.org/10.1007/s12603-021-1708-1

Yang G, Wang Y, Sun J, Zhang K, Liu J (2016) Ginkgo Biloba for mild cognitive impairment and Alzheimer's disease: a systematic review and meta-analysis of randomized controlled trials. Curr Top Med Chem 16:520–528. https://doi.org/10.2174/1568026615666150813143520

Yang T, Sun Y, Lu Z, Leak RK, Zhang F (2017) The impact of cerebrovascular aging on vascular cognitive impairment and dementia. Ageing Res Rev 34:15–29. https://doi.org/10.1016/j.arr.2016.09.007

Yang W, Cui K, Li X, Zhao J, Zeng Z, Song R et al (2021) Effect of polyphenols on cognitive function: evidence from population-based studies and clinical trials. J Nutr Health Aging 25:1190–1204. https://doi.org/10.1007/s12603-021-1685-4

Zafar S, Yaddanapudi SS (2025) Parkinson disease. StatPearls, Treasure Island (FL): StatPearls Publishing

Zieneldien T, Kim J, Cao C (2022) The multifaceted role of neuroprotective plants in Alzheimer's disease treatment. Geriatrics (Basel) 7:24. https://doi.org/10.3390/geriatrics7020024

Zinkow A, Grodzicki W, Czerwińska M, Dziendzikowska K (2025) Molecular mechanisms linking Omega-3 fatty acids and the gut–brain Axis. Molecules 30:71. https://doi.org/10.3390/molecules30010071

Autophagy and Brain Longevity

3

Akhil Vijay and Ritu Varshney

Abstract

Autophagy, a critical cellular degradation process, is pivotal in maintaining neuronal homeostasis and combating age-related brain decline. Understanding autophagy's contribution to brain longevity is essential as global aging populations increase. This chapter explores the mechanisms of autophagy and its decline in aging, linking impaired autophagic flux to the accumulation of toxic proteins such as amyloid-beta and alpha-synuclein, which drive neurodegenerative diseases like Alzheimer's and Parkinson's. Recent advances in therapeutic strategies, including pharmacological agents (e.g., rapamycin, metformin), natural compounds (e.g., resveratrol, trehalose), and lifestyle interventions (caloric restriction, exercise) are reviewed for their potential to enhance autophagy and mitigate neurodegeneration. Challenges in clinical translation and future directions for targeted autophagy modulation are discussed, emphasizing the need for balanced approaches to preserve neuronal health and cognitive function in aging populations.

Keywords

Autophagy · Neurodegenerative diseases · Brain longevity · Therapeutic targets · Therapeutic strategies

A. Vijay · R. Varshney (✉)
P. D. Patel Institute of Applied Sciences, Charotar University of Science and Technology, Anand, Gujarat, India

A. Kumar Singh, S. Nand Rai (eds.), *Rejuvenating the Brain: Nutraceuticals, Autophagy, and Longevity*, Nutritional Neurosciences,
https://doi.org/10.1007/978-981-95-2790-8_3

3.1 Introduction

As global demographics shift toward an increasingly more aged population, the quest to increase not simply lifespan but also health span, and the duration of existence free from chronic disease, has become a critical subject in biomedical studies. According to the World Health Organization (WHO), by the year 2050, approximately 22% of the global population will be over 60 (WHO 2022). This demographic trend has profound implications for public health, specifically within the context of neurodegenerative illnesses, which impose huge burdens on both people and healthcare systems. Like other organs, the brain exhibits a heightened vulnerability to age-related decline, which can take place in cognitive impairments, neurodegenerative issues, and decreased standard high quality of life. The onset of diseases including Alzheimer's and Parkinson's is strongly linked to the accumulation of toxic protein aggregates, inclusive of amyloid-beta (Aβ) and alpha-synuclein (α-syn), which disrupt ordinary neuronal function and result in revolutionary cognitive and motor impairments (Goedert et al. 2010). Given the complexity of these problems, information on the cellular mechanisms that contribute to neuronal maintenance and restoration is important in growing techniques to beautify brain longevity and resilience toward disorder.

A fundamental system implicated in brain health and longevity is autophagy, an evolutionarily conserved cellular degradation pathway that plays a critical function in preserving neuronal homeostasis. Derived from the Greek words "auto" (self) and "phagy" (eating), autophagy is a biological mechanism through which cells break the damaged or dysfunctional additives, such as misfolded proteins, faulty organelles, and intracellular pathogens (Klionsky et al. 2021). This cellular recycling device runs by three fundamental mechanisms: macroautophagy, microautophagy, and chaperone-mediated autophagy (CMA), which serve to maintain cellular integrity through various yet complementary means (Stavoe and Holzbaur 2019). Increased diagnosis of autophagy dysregulation has become a major component of the pathogenesis of age-related neurodegenerative diseases owing to the stagnant flow of autophagic processes leading to protein and mitochondrial damage accumulation (Schaefer and Dikic 2021). For example, in Alzheimer's, it was noted that neuronal autophagy dysfunction had led to the buildup of autophagic vesicles within neurons, further worsening the degree of neurodegeneration. Otherwise, in Parkinson's, insufficient autophagy leads to the accumulation of α-synuclein and subsequent development of Lewy bodies, which is a representative pathology of the disease (Frake et al. 2015). Therefore, given the importance of autophagy in neuronal reorganization, its enhancement has emerged as a potential treatment method for mitigating cognitive and neurodegenerative disorders. Different treatments for boosting modulatory autophagic responses have been evaluated, including pharmacological compounds, dietary modifications, and physical activity or energy restriction known to activate pro-autophagic pathways and strengthen neuronal protection (Nieto-Torres and Hansen 2021).

Moreover, understanding genes and molecular neuroscience has turned more thoughts toward regulating autophagy and its therapeutic targeting.

This chapter discusses the overall outlines regarding the processes that modulate and contribute to the functioning of autophagy aimed at safeguarding neuron health while analyzing modern approaches directed at the utilization of autophagy for preserving an active brain. Such goals synthesized from the fields of genetics, neuroscience, and gerontology aim to demonstrate autophagy's suggestive potential to heal the damages inflicted by aging on human brain functions and neurodegenerative diseases. We want to offer a unifying view on the ways in which the mechanisms of cellular self-renewal may be used to encourage non-pathological aging of the brain in the face of increasing prevalence of neurodegenerative disorders.

3.2 Brain Aging

As people grow older, the brain experiences remarkable structural and functional changes that include a gradual volume reduction of approximately 5% each decade after the age of 40, which speed up further after 70, resulting in the enlargement of ventricles and the widening of the brain's sulci, mainly affecting the frontal and temporal lobes (Blinkouskaya and Weickenmeier 2021). Vascular modifications are important to these traits, as the stiffening of arteries leads to cerebrovascular problems and a heightened danger of dementia, such as Alzheimer's disease (AD), with higher systolic blood stress connected to gray matter loss and white matter hyperintensities because of disturbances in cerebral blood flow pulsatility (Suri et al. 2020; Schaare et al. 2019; Zimmerman et al. 2021).

On a microscopic scale, aging influences myelin integrity, where demyelination and ineffective remyelination lower neural conduction speed and hinder synaptic functionality (Suminaite et al. 2019). Biochemically, the reduction in cholinergic function, particularly the decline of nicotinic acetylcholine receptors (nAChRs), impedes neurogenesis and cognitive capabilities, worsening age-related impairments (Utkin 2019), while the reduction of neurosteroids like testosterone and estrogen further undermines blood-brain barrier integrity, mitochondrial performance, and synaptic plasticity, thereby heightening the risk of neurodegeneration (Park et al. 2025). At the cellular and molecular levels, telomere shortening and telomerase dysfunction lead to hippocampal shrinkage and cognitive deterioration, independent of cell division (Zia et al. 2021), while mitochondrial dysfunction, indicated by mutations in mitochondrial DNA (mtDNA), causes impairments in oxidative phosphorylation (OXPHOS), increased production of reactive oxygen species (ROS), and diminished ATP synthesis, all linked to glucose hypometabolism in AD (Cunnane et al. 2020). Additionally, the oxidative stress resulting from an accumulation of ROS damages crucial cellular components, including lipids, proteins, and nucleic acids, with elevated levels of markers such as 8-hydroxyguanosine associated with AD pathology (Bhatia and Sharma 2020). The aging brain also sees a reduction in autophagy effectiveness, as dysfunctions in the PINK1-parkin pathway hinder mitophagy, leading to the aggregation of toxic proteins, including α-synuclein in Parkinson's disease (PD) (Quinn et al. 2020). Moreover, changes in calcium homeostasis, especially from the activities of

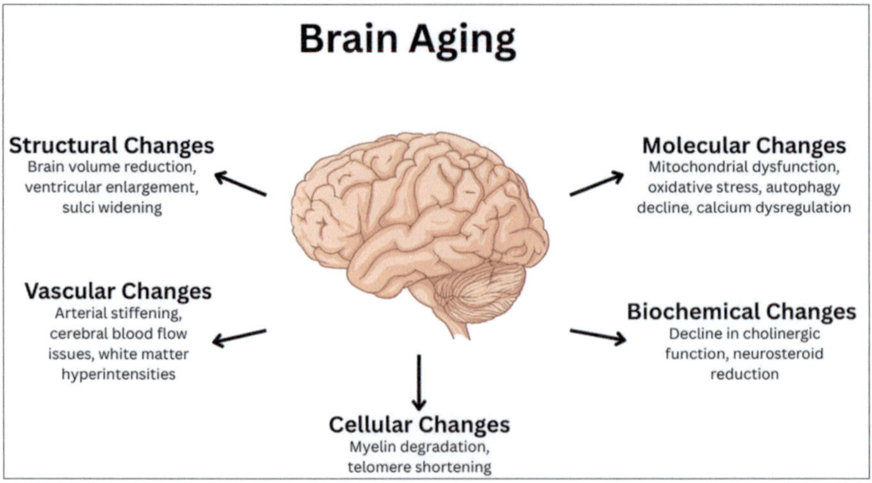

Fig. 3.1 Key changes in the aging brain. This figure highlights important age-associated changes across cellular, structural, vascular, biochemical, and molecular domains

N-methyl-D-aspartate receptors (NMDARs) and voltage-dependent calcium channels (VDCCs), affect calcium inflow and buffering systems and lead to the dysfunction of synaptic plasticity and long-term potentiation (LTP) that are important for learning and memory processes (Chandran et al. 2017). These stem from cellular, vascular, structural, biochemical, and electrophysiological components combined to bring about memory impairment associated with aging and the progression of neurodegenerative conditions (Fig. 3.1).

3.3 Autophagy and Brain Aging

Researchers found that everything regarding age-related declination issues, including brain health management and autophagy, serves exceptionally well. Recent studies underpin the importance of autophagy in cognitive development, brain healing, and, in particular, neurogenesis, as well as in neurodevelopmental disorders. Enhancing autophagy by Beclin-1 knockin mutation in aged mice prevents the loss of neural stem cells (NSC) and primes neurogenesis in the subventricular zone. Therefore, the activation of autophagy offers a desirable strategy for dealing with the consequences of aging on the brain (Wang et al. 2022).

Conversely, defective autophagy contributes to neurodevelopmental disorders; recessive *ATG7* variants impair NSC differentiation and cause structural brain abnormalities, such as corpus callosum malformations (Collier et al. 2021a, b). Pathways that are poorly regulated and TSC/mTOR-autophagy chromosomes that can be described as "disrupted" (for instance, GABARAPL1 or MAP1LC3B) are correlated with autism spectrum disease (ASD), wherein cognitive deficits stem from problems with synaptic pruning and lipid metabolism (Deng et al. 2021).

Furthermore, autophagy contributes to the maintenance of metabolic balance within the brain. For example, some metabolic lipophagy, the removal of lipid droplets via autophagy, is essential for non-stem cell (NSC) differentiation, highlighting its role in metabolic homeostasis (Wang et al. 2019). The intervention aimed at midlife neurodegenerative diseases may lie in some less risky treatments through manipulation of particular pathways of autophagy, those that remove lipid droplets or protein aggregates (Deng et al. 2021; Wang et al. 2022). The relative decrease in autophagy activity that comes with age causes the accumulation of mitochondria and proteasomes of organelles in postmitotic neurons, which hastens cell aging senescence and neurodegeneration (Chou et al. 2023). The failure of autophagy is the failure to remove aggregates of proteins, dysfunctional mitochondria, and too much wear and tear aging. A case in point is the aging model, where the accumulation of GATA4 due to failure of autophagy is mapped to senescent markers in aged human brain cells (Kang et al. 2015). Furthermore, lacking autophagy leads to a disbalance in the functioning of mitochondria, which means that the level of NAD+ is reduced compared to NADH, which is a sign of disorder in mitochondria, the brain, aging, and neurodegenerative diseases (Sun et al. 2023).

In aging brains, defective mitophagy triggers cytoplasmic mitochondrial DNA (mtDNA) release, activating the proinflammatory cGAS/STING pathway and accelerating neurodegeneration (Dwamena et al. 2024). These results highlight that altering autophagy could be a strong and promising strategy to tackle brain aging and related disorders. Autophagy is critical for things that go wrong in Parkinson's disease (PD), which mainly happens because it fails to work properly. This failure leads to the building up of unstable protein knots, worsens inflammation in the brain cells, and then makes the cells in the brain stop working, generally over time, resulting in typical PD symptoms. One of the key problems in these harmful processes in the body is when important genes that help with self-cleaning tasks inside cells get messed up. Especially, troublesome are mutations that interfere with the function of lysosomes and mitochondria. Lysosomes are like little recycling centers inside cells; mitochondria are energy powerhouses that help cells produce energy. When these parts go wrong, big-time problems can result. Something noticeable is that 50 or more mutations in the LRRK2 gene disrupt how the kinase inside the cell called LRRK2 functions. This disruption leads to troubles with a garbage disposal system inside cells called endosomal trafficking, which helps maintain lysosomes that are like recycling centers. It messes up the normal transporting of vesicles that carry important stuff into synapses where signals are regularly sent out. The subsequent increase in LRRK2 kinase activity induces lysosomal dysfunction, triggering compensatory lysophagy, a mode that ironically exacerbates neuronal injury and PD pathology (Mamais et al. 2021; Herbst et al. 2020; Rivero-Ríos et al. 2019). In addition, mutations in the GBA gene code are called glucocerebrosidase that is an enzyme that lives inside cells and happens to be important for breaking down waste products that build up inside cells called lysosomes. When GBA genes do not work properly, this crucial enzyme cannot do its job, and the cells cannot process waste product effectively. This leads to a buildup of an essential protein called α-synuclein that is key to causing Lewy Body disease, a type of neurodegenerative disorder that damages brain cells (Pradas and Martinez-Vicente 2023).

Microglial autophagy plays a vital role in neuroprotection by assisting the degradation and clearance of pathological α-Syn aggregates. Under physiological conditions, microglia employ F-actin-structured cytoskeletal networks to engulf and degrade α-Syn inclusions through autophagic pathways, thereby alleviating neuroinflammation and preventing the propagation of toxic protein aggregates (Scheiblich et al. 2021). However, in the context of PD, impaired microglial autophagy contributes to ailment development with the aid of allowing the exosome-mediated transmission of toxic α-Syn oligomers to neighboring neurons, thereby amplifying neurotoxicity and synaptic disorder (Xia et al. 2019; Guo et al. 2020). Furthermore, microglial autophagy intersects with innate immune signaling pathways, as evidenced by way of its involvement in α-Syn clearance through the TLR4-NF-κB-p62 pathway, which immediately links autophagy to synaptic phagocytosis and neuroimmune regulation (Choi et al. 2020). Conversely, immoderate activation of the NLRP3 inflammasome has been shown to inhibit autophagic mechanisms, thereby intensifying neuroinflammatory cascades and accelerating dopaminergic neuron loss (Béraud et al. 2011; Panicker et al. 2019; Wu et al. 2021a, b). Strategies targeting restoring microglial autophagy, like the p38-TFEB pathway to augment chaperone-mediated autophagy (CMA) and suppress NLRP3 inflammasome activity, have validated potential in attenuating neuroinflammation and reducing dopaminergic neuronal loss (Chen et al. 2021).

Multiple genetic and animal model studies have substantiated the direct involvement of autophagy deficiency in PD pathology. In dopamine neuron-unique autophagy-deficient mice, the buildup of p62/sequestosome-1 (SQSTM1)-containing aggregates serves as a pathological precursor to LB-like inclusions rich in α-Syn, leads to progressive neurodegeneration and motor dysfunction (Sato et al. 2018). Similarly, zebrafish models of PD have found that ATG5 deficiency impairs autophagy flux, leading to the upregulation of β-synuclein and mitophagy-related proteins, in addition to implicating autophagic dysfunction in synucleinopathy-associated neurodegeneration (Hu et al. 2023). In aging neurons, the revolutionary accumulation of neuromelanin disrupts lysosomal proteostasis, thereby contributing to LB pathology and neuronal vulnerability. Substantially, overexpression of TFEB, a master regulator of lysosomal biogenesis and autophagy, has been shown to rescue those defects and increase neuronal survival (Carballo-Carbajal et al. 2019).

Genomic and molecular investigations have further elucidated the complex relationship between autophagy-associated genetic mutations and PD pathophysiology. Mutations in genes that include SNCA (encoding α-Syn), PINK1 (PTEN-brought about kinase 1), LRRK2, VPS35 (vacuolar protein sorting-associated protein 35), PARKIN (PRKN), and PARK7 (DJ-1) had been associated in the impairment of macroautophagy, lysosomal function, and CMA underscoring the multifaceted function of autophagic dysregulation in PD progression (Nalls et al. 2019; Kluss et al. 2019). For instance, pathogenic LRRK2 mutations have been shown to inhibit CMA, thereby enhancing the accumulation of α-Syn and contributing to LB pathology (Ho et al. 2020). Likewise, mutations in PINK1 and PARKIN, each of which is central to mitophagy, the selective degradation of damaged mitochondria, were related to impaired mitochondrial turnover and oxidative pressure-induced neuronal damage, even though the phenotypic results of these mutations appear to be context-dependent (Hou et al. 2023). Additionally, posttranslational modifications such as S-nitrosylation of p62/SQSTM1 have been acknowledged as negative modulators of autophagic flux, aggravating the distribution of α-Syn aggregates and leading to neurodegeneration (Oh et al. 2022).

Considering the significant role peripheral autophagy dysfunction has in the pathogenesis of PD, there has arisen a therapeutic focus directed at altering the level of autophagic activity as a mode of intervention in the disease. Numerous pharmacological agents have shown efficacy in preclinical models of PD with the help of improving autophagic flux and restoring neuronal homeostasis. For example, corydaline, a small-molecule compound, has been proven to ameliorate PD-like signs and symptoms in murine models by way of modulating glycogen synthase *kinase*-threeβ (GSK-3β), thereby reinstating dopamine levels and enhancing motor function (Zhou and Xu 2024). In addition, the activation of TFEB has been proven to increase lysosomal proteostasis, lessen neuromelanin accumulation, and mitigate neurodegenerative pathology, highlighting its potential as a therapeutic goal (Carballo-Carbajal et al. 2019).

Autophagy is a fundamental cellular procedure accountable for preserving protein and organelle homeostasis, and its disorder has been implicated in the pathogenesis of AD. The failure of autophagy to efficaciously clean pathological protein aggregates, particularly amyloid-beta (Aβ) peptides and hyperphosphorylated tau, exacerbates neurodegeneration and contributes to cognitive impairment in AD patients (Schaefer and Dikic 2021). Aβ peptides, generated from amyloid precursor protein (APP) through proteolytic cleavage through β- and γ-secretases, acquire to form neurotoxic oligomers that disrupt synaptic integrity. In AD, impaired autophagic flux leads to defective clearance of Aβ, resulting in its intracellular accumulation inside aberrant autophagic vesicles referred to as PANTHOS (toxic Aβ tau-harboring systems), which ultimately rupture and release Aβ into the extracellular space, in addition to exacerbating plaque deposition and neuronal toxicity (Lee et al. 2022). Tau pathology, another hallmark of AD, involves the hyperphosphorylation and subsequent aggregation of tau proteins, leading to the formation of neurofibrillary tangles that impair microtubule stability and axonal transport. Experimental evidence has demonstrated that compromised autophagy significantly exacerbates tau accumulation, whereas pharmacological stimulation of autophagy can mitigate tau pathology and improve cognitive function in preclinical models (Gyparaki et al. 2021; Chinnathambi and Gorantla 2021). The role of autophagy needs special attention to maintain protein homeostasis in neurons, and failure to do so is linked to the progression of Alzheimer's disease. Studies show that the deletion of key autophagy genes, including *Beclin-1* and *ATG7*, in genetically modified mouse models of AD led to a marked acceleration of Aβ and tau tangles compared to control models. This emphasizes autophagy's putative protective effect during neurodegenerative processes (Carosi et al. 2021; Wu et al. 2021a, b).

In a proof of principle, other studies showed that neurodegenerative symptoms in mice that lacked whole body *Atg5* could be rescued by reintroducing *Atg5* in neurons. These findings demonstrate the efficacy of neuronal autophagy in sustaining normal physiological function (Deretic and Lazarou 2022). This evidence has been corroborated by examination of the brains of patients who succumbed to Alzheimer's disease. Autophagy markers such as microtubule-associated protein light chain three two (*LC3-II*) and *Beclin-1* were found in very low quantities, and there were signs of impaired lysosomal proteolysis. These deficits correspond to the degree of Aβ and tau pathology and imply that the dysfunction of autophagy is not only a consequence of Alzheimer's disease but has a major role to play in its progression (Mani et al. 2021).

Due to coping with the degradation of misfolded proteins and damaged organelles, there remains a growing concern in the therapeutic exploitation of autophagy. Pharmacological strategies aimed at boosting the activity of autophagy have been effective in preclinical models of AD (Rai et al. 2021). One of these, ganoderic acid A, was shown to improve cognition in transgenic AD mice via autophagy induction, which enhanced Aβ clearance (Qi et al. 2021). The activation of mitophagy, which is the selective degradation of damaged mitochondria, is a promising candidate for therapeutic intervention. The invention of mitophagy inducers, often referred to as Mito-AUTACs, has enabled the targeting and clearance of inefficient mitochondria, thereby diminishing the pathology associated with neurodegeneration. There is a strong likelihood that such therapies will benefit AD patients (Antico et al. 2021).

Recent studies have further elucidated the interplay among autophagy and other cellular approaches in AD pathology. For example, the formation of PANTHOS vesicles in AD represents a pathological effect of faulty autophagy, wherein collected Aβ inside swollen autophagic vesicles contributes to neuronal demise and propagates Aβ aggregation (Lee et al. 2022). Moreover, mitophagy disorder has been diagnosed as a key function of AD, with hyperactive leucine-wealthy repeat kinase 2 (LRRK2) suppressing mitochondrial clearance and leading to energy deficits and oxidative strain in neurons (Singh and Ganley 2021). Metabolic pathways additionally intersect with autophagy regulation in AD, as evidenced by research demonstrating that glycolysis-driven metabolic shifts in the course of retinal improvement rely upon programmed mitophagy, thereby underscoring the important role of mitochondrial turnover in brain maturation (Morishita 2022). Furthermore, lipid metabolism is tightly connected to autophagy characteristic, with diacylglycerol O-acyltransferase 1 (DGAT1)-dependent lipid droplet biosynthesis playing a crucial function in protecting against lipotoxicity and maintaining basal mitophagy at some point of iron depletion in Drosophila brain models (Long et al. 2022).

Beyond AD, congenital autophagy impairment has been implicated in a spectrum of neurodevelopmental issues, in addition to highlighting the significance of autophagy in brain health. Mutations in essential autophagy genes, such as ATG7 and ATG5, result in severe neurological and endocrine dysfunctions, reinforcing the essential function of autophagy in early brain development and neuronal differentiation (Collier et al. 2021a, b; Baselli et al. 2022). Together, these findings emphasize

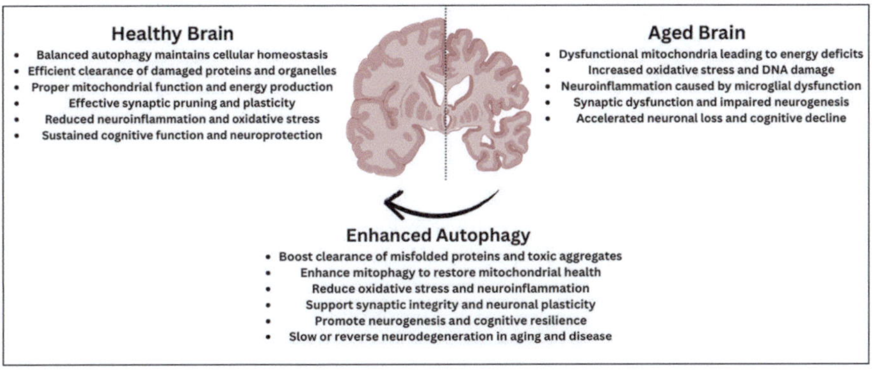

Fig. 3.2 Autophagy's impact on neuronal health and cognitive decline This schematic illustrates the key variances between a healthy brain and an aged brain, highlighting the role of autophagy in preserving brain health and potentially mitigating age-associated cognitive decline

the multifaceted role of autophagy in retaining neuronal homeostasis, preventing protein aggregation, and regulating metabolic techniques, with substantial implications for therapeutic strategies aimed toward modulating autophagy to combat neurodegenerative disease (Fig. 3.2).

3.4 Therapeutic Strategies Targeting Autophagy in Brain Aging

Despite the fact that the strong actions of autophagy disengage as one grows older, it leads to the buildup of damaged proteins and organelles which creates a pathway toward neurodegeneration and cognitive decline. Enhancing autophagy has emerged as a promising therapeutic technique to mitigate the consequences of brain aging and neurodegenerative disorders. To augment autophagy is to therapeutically engage with the brain's aging symptoms and fight progressive medical disorders of the brain. Several pharmacological agents and interventions had been explored for their ability to modulate autophagy and protect neuronal damage.

3.4.1 Pharmacological Agents Enhancing Autophagy

One of the most researched compounds that aids in increasing autophagy is rapamycin which acts as a physiologic suppressor of mTOR while helping to maintain various neuroprotective functions. Evidence exists where rapamycin was demonstrated to improve the functioning of the brain by eliminating amyloid beta clumps in conjunction with other symptoms of Alzheimer's disease in animal models. Inhibiting mTOR, rapamycin decreases protein aggregation in both the Huntington's disease (HD) model and the amyotrophic lateral sclerosis (ALS) model (Caccamo et al.

Table 3.1 Pharmacological agents enhancing autophagy

Compound	Mechanism of action	Effects on neurodegeneration	References
Rapamycin	mTOR inhibitor	Reduces Aβ, rescues cognitive deficits in AD, and reduces protein aggregation in HD and ALS	Caccamo et al. (2010)
Temsirolimus	mTOR inhibitor	Promotes autophagic clearance of Aβ, provides neuroprotection in AD models	Jiang et al. (2014)
Carbamazepine	mTOR inhibitor	Reduces Aβ, enhances autophagy, and alleviates AD pathologies	Li et al. (2013)
Latrepirdine	mTORC1-dependent autophagy inducer	Reduces Aβ accumulation	Bharadwaj et al. (2013)
Metformin	AMPK activator	Enhances cognitive function, reduces inflammation, and promotes autophagy in aging	Kodali et al. (2021)
Nilotinib	AMPK inducer	Reverses dopamine neuron loss, improves motor behavior in PD, and is implicated in Aβ clearance	Lonskaya et al. (2013)
A769662	Direct AMPK activator	Induces AMPK-mediated autophagy, clears α-synuclein in neurons	Gao et al. (2019)

2010). Research has shown that the mTOR inhibitor temsirolimus has also been demonstrated to facilitate the clearance of agglomerated beta amyloid proteins while providing some neuroprotection to AD patients (Jiang et al. 2014). One can appreciate how carbamazepine as an mTOR inhibitor showcases enhanced autophagic activity. This drug was designed for use with epilepsy patients. Better memory functioning together with Aβ pathology reduction is achieved in AD models with the help of these interventions (Li et al. 2013). Latrepirdine is a compound that induces autophagy by acting on mTORC1 to diminish Aβ deposits (Bharadwaj et al. 2013). Most professionals use metformin for their diabetes patients since it has shown potential in reducing the rate of cognitive decline. During late middle-age treatment, metformin administration, the compound modulates microglial activation, controls inflammation, and stimulates autophagy in the hippocampus (Kodali et al. 2021) (Table 3.1).

3.4.2 Natural Compounds Modulating Autophagy

A variety of natural organic compounds that may encourage autophagy are also neuroprotective. Consequently, these compounds provide a level of defense from neurodegenerative conditions. Wu et al. (2011) showed that resveratrol found on grape skin increases autophagy through AMPK/SIRT1 pathway and prevents Parkinson's disease. Additionally, resveratrol promotes autophagy by means of inhibiting the AKT/mTOR pathway, improving cognitive function in rat models of cerebral hypoperfusion (Wang et al. 2019). Zheng et al. (2021) argue that icariin

Table 3.2 Natural compounds modulating autophagy

Compound	Mechanism of action	Effects on neurodegeneration	References
Resveratrol	AMPK/SIRT1 pathway activator	Neuroprotective in PD, protects against dopamine toxicity, and enhances cognitive function	Wu et al. (2011), Wang et al. (2019)
Icariin	AMPK/mTOR/ULK1 pathway activator	Reduces amyloid pathologies, enhances autophagy	Zheng et al. (2021)
Berberine	Suggested AMPK activator	Beneficial effects in AD, PD, and HD mouse models	Yu et al. (2014)
Trehalose	Various mechanisms	Reduces Aβ accumulation, rescues learning impairment	Rodríguez-Navarro et al. (2010)
Epigallocatechin-3-Gallate (EGCG)	Nrf2 and PI3K/Akt pathways	Reduces oxidative stress, enhances autophagy	He et al. (2009), Payne et al. (2022)
Omega-3 fatty acids	Autophagy regulator	Improves cognitive function, supports brain aging resistance	Morishita (2022), Denis et al. (2013, 2015)
Cocoa-derived compounds (procyanidin B2, theobromine)	Antioxidant and autophagy inducer	Reduces oxidative stress, promotes autophagy	Sutcliffe et al. (2017), Zeli et al. (2022)
Hydroxytyrosol	Synergizes with resveratrol	Enhances autophagy, reduces inflammation	Terracina et al. (2022)

from *Epimedium* species contributes to the preservation of cognitive functions in aged rats by promoting autophagy of neurons via AMPK/mTOR/ULK1 signaling pathway. Further, berberine, an AMPK activator, has confirmed neuroprotective blessings in models of AD, PD, and Huntington's disease (HD) by promoting autophagy and reducing pathological protein accumulation (Yu et al. 2014). Trehalose induces autophagy in distinct manners and negative Aβ accumulation and enhances the learning ability in the AD mice test subjects (Rodríguez-Navarro et al. 2010). Increased levels of parkin and enhanced motor and cognitive performance in PD model organisms were observed due to the improvement of autophagic clearance of Aβ, an effect initiated by the inhibition of the tyrosine kinase nilotinib (Lonskaya et al. 2013) (Table 3.2).

3.4.3 Lifestyle Interventions and Autophagy

Caloric restriction, a standing longevity technique, induces autophagy via mTOR inhibition to reduce protein aggregation and increase the human lifespan (Chung and Chung 2019). Exercising, particularly resistance training, has also been identified as a potent autophagy activator. An eight-week resistance training regimen in the older enhanced the expression of autophagy marker genes, decreased NLRP3 inflammasome activation, and decreased apoptosis in peripheral blood mononuclear cells (Mejías-Peña et al. 2017). Prolonged sleep fragmentation disrupts autophagy in a location-precise manner, turns on microglia within the striatum, and upregulates CRFR2, suggesting that autophagy disorder may make a contribution to the cognitive and neurobiological impairments related to sleep issues (Szinyákovics et al. 2023).

Proposed new ways to stimulate autophagy seek to enhance the process of lysosomal fusion with autophagosome, which is the final step of degradation. Extension of the autophagy cycle and improvement of neuronal functions is possible using the fusion-promoting HOPS complex, SNARE proteins, and the small GTPases Rab2 and Arl8 (Szinyákovics et al. 2023). Another study showed that Kisspeptin-10 (Kp) is promising in the therapeutic landscape owing to its ability to restore mitochondrial function by inducing mitophagy via Ca2+/CaMKKβ-AMPK-ULK1 independent of mTOR signaling. Increasingly, the application of Kp has been shown to improve the microstructure of mitochondria, augment the functional reserve of complex I, and enhance ATP synthesis, thus appearing to be a candidate for treatment of aging brain and disease (Mattam et al. 2020). The enhancement of autophagy is a novel strategy in managing brain aging and neurodegenerative diseases. A set of diverse triggers like some synthetic drugs and herbal compounds, combined with caloric restriction and physical activity, can promote neuronal well-being by manipulating autophagy processes. Further research should focus on fine-tuning these strategies to formulate effective therapeutic interventions to manage age-related cognitive decline.

3.4.4 Conclusions

Autophagy is essential to maintaining neuronal health by eliminating damaged organelles, aggregated proteins, and cellular waste. The processes an individual undergoes gradually learn how to clean their body becomes weaker through aging. As aging progresses, the performance of autophagy declines, resulting in the accumulation of toxic proteins, extended oxidative pressure, mitochondrial dysfunction, and continual neuroinflammation.

Neurodegenerative diseases like Alzheimer's and Parkinson's begin to show in old age and worsen over time due to age-related issues such as Alzheimer's, Huntington's, and Parkinson's diseases. With the acquired age, the self-cleansing of cells becomes ineffective, resulting in oxidative stress, a decrease in mitochondria functions, and bolstering of proteins. This problem becomes increasingly worse due to inflammation and leads to increased neurodegeneration. Neurodegenerative issues can be partially mitigated with interventions that restore an autonomous cleansing process. There is proof of autophagy bolstering evidence through diet, medication, exercise, and intermittent fasting changes. However, there are many gaps in providing effective clinical treatment. Using the balance of the cellular system, while moderated will aid in formulating targeted therapies that can adjust autophagy levels in older neural cells. In addition to having knowledge of other processes such as mitochondria changing and the growth or breakdown of proteins, therapy strategies can be created to help implement autophagy.

With the progress of research on the effectiveness of brain autophagy in aging, devising effective strategies to maintain cognitive function and enhance the quality of life of the elderly will be increasingly possible.

References

Antico O, Ordureau A, Stevens M, Singh F, Nirujogi RS, Gierlinski M et al (2021) Global ubiquitylation analysis of mitochondria in primary neurons identifies endogenous parkin targets following activation of PINK1. Sci Adv 7(46):eabj0722. https://doi.org/10.1126/sciadv.abj0722

Baselli GA, Jamialahmadi O, Pelusi S, Ciociola E, Malvestiti F, Saracino M et al (2022) Rare ATG7 genetic variants predispose patients to severe fatty liver disease. J Hepatol 77(3):596–606. https://doi.org/10.1016/j.jhep.2022.03.031

Béraud D, Twomey M, Bloom B, Mittereder A, Ton V, Neitzke K et al (2011) α-Synuclein alters toll-like receptor expression. Front Neurosci 5:80. https://doi.org/10.3389/fnins.2011.00080

Bharadwaj P, Bates K, Porter T, Teimouri E, Perry G, Steele J et al (2013) Latrepirdine: molecular mechanisms underlying potential therapeutic roles in Alzheimer's and other neurodegenerative diseases. Transl Psychiatry 3(12):e332. https://doi.org/10.1038/tp.2013.97

Blinkouskaya Y, Weickenmeier J (2021) Brain shape changes associated with cerebral atrophy in healthyaging and Alzheimer's disease. Front Mech Eng 7:705653.https://doi.org/10.3389/fmech.2021.705653

Bhatia V, Sharma S (2020) Role of mitochondrial dysfunction, oxidative stress and autophagy in progressionof Alzheimer's disease. J Neurol Sci 421:117253. https://doi.org/10.1016/j.jns.2020.117253

Caccamo A, Majumder S, Richardson A, Strong R, Oddo S (2010) Molecular interplay between mammalian target of rapamycin (mTOR), amyloid-beta, and tau: effects on cognitive impairments. J Biol Chem 285(17):13107–13120. https://doi.org/10.1074/jbc.M110.100420

Carballo-Carbajal I, Laguna A, Romero-Giménez J, Cuadros T, Bové J, Martinez-Vicente M et al (2019) Brain tyrosinase overexpression implicates age-dependent neuromelanin production in Parkinson's disease pathogenesis. Nat Commun 10(1):1–15. https://doi.org/10.1038/s41467-019-08858-y

Carosi JM, Hein LK, Van Den Hurk M, Adams R, Milky B, Singh S et al (2021) Retromer regulates the lysosomal clearance of MAPT/tau. Autophagy 17(9):2217–2237. https://doi.org/10.1080/15548627.2020.1821545

Chandran R, Kumar M, Kesavan L, Jacob RS, Gunasekaran S, Lakshmi S, Sadasivan C, Omkumar R (2017)Cellular calcium signaling in the aging brain. J Chem Neuroanat 95:95–114.https://doi.org/10.1016/j.jchemneu.2017.11.008

Chen J, Mao K, Yu H, Wen Y, She H, Zhang H et al (2021) p38-TFEB pathways promote microglia activation through inhibiting CMA-mediated NLRP3 degradation in Parkinson's disease. J Neuroinflammation 18(1):1–18. https://doi.org/10.1186/s12974-021-02349-y

Chinnathambi S, Gorantla NV (2021) Implications of valosin-containing protein in promoting autophagy to prevent TAU aggregation. Neuroscience 476:125–134. https://doi.org/10.1016/j.neuroscience.2021.09.003

Choi I, Zhang Y, Seegobin SP, Pruvost M, Wang Q, Purtell K et al (2020) Microglia clear neuron-released α-synuclein via selective autophagy and prevent neurodegeneration. Nat Commun 11(1):1–14. https://doi.org/10.1038/s41467-020-15119-w

Chou S, Yen Y, Yuan F, Zhang S, Chong C (2023) Neuronal senescence in the aged brain. Aging Dis 14(5):1618. https://doi.org/10.14336/ad.2023.0214

Chung KW, Chung HY (2019) The effects of calorie restriction on autophagy: role on aging intervention. Nutrients 11(12):2923. https://doi.org/10.3390/nu11122923
Collier JJ, Guissart C, Oláhová M, Sasorith S, Piron-Prunier F, Suomi F et al (2021a) Developmental consequences of defective ATG7-mediated autophagy in humans. N Engl J Med 384(25):2406–2417. https://doi.org/10.1056/NEJMoa1915722
Collier JJ, Suomi F, Oláhová M, McWilliams TG, Taylor RW (2021b) Emerging roles of ATG7 in human health and disease. EMBO Mol Med 13(12):e14824. https://doi.org/10.15252/emmm.202114824
Deng Z, Zhou X, Lu JH, Yue Z (2021) Autophagy deficiency in neurodevelopmental disorders. Cell Biosci 11(1):214. https://doi.org/10.1186/s13578-021-00728-9
Cunnane SC, Trushina E, Morland C, Prigione A, Casadesus G, Andrews ZB, Beal MF, Bergersen LH, BrintonRD, Eckert A, Harvey J, Jeggo R, Jhamandas JH, Kann O, Martin WF, Mithieux G, Moreira PI, Murphy MP, NaveA, Millan MJ (2020) Brain energy rescue: An emerging therapeutic concept for neurodegenerative disorders ofageing. Nat Rev Drug Discov 19(9):609.https://doi.org/10.1038/s41573-020-0072-x
Denis I, Potier B, Vancassel S, Heberden C, Lavialle M (2013) Omega-3 fatty acids and brain resistance to ageing and stress: body of evidence and possible mechanisms. Ageing Res Rev 12(2):579–594. https://doi.org/10.1016/j.arr.2013.01.007
Denis I, Potier B, Heberden C, Vancassel S (2015) Omega-3 polyunsaturated fatty acids and brain aging. Curr Opin Clin Nutr Metab Care 18(2):139–146
Deretic V, Lazarou M (2022) A guide to membrane atg8ylation and autophagy with reflections on immunity. J Cell Biol 221(7):e202203083. https://doi.org/10.1083/jcb.202203083
Dwamena A, Asadi Y, Gilstrap E, Wang H (2024) Impaired proteasome induces mitochondrial DNA release to activate the cGAS-STING signaling pathway and cause necroptosis in mouse brain. bioRxiv. https://doi.org/10.1101/2024.11.13.623442
Frake RA, Ricketts T, Menzies FM, Rubinsztein DC (2015) Autophagy and neurodegeneration. J Clin Invest 125(1):65–74. https://doi.org/10.1172/JCI73944
Gao J, Perera G, Bhadbhade M, Halliday GM, Dzamko N (2019) Autophagy activation promotes clearance of α-synuclein inclusions in fibril-seeded human neural cells. J Biol Chem 294(39):14241–14256. https://doi.org/10.1074/jbc.RA119.008733
Goedert M, Clavaguera F, Tolnay M (2010) The propagation of prion-like protein inclusions in neurodegenerative diseases. Trends Neurosci 33(7):317–325. https://doi.org/10.1016/j.tins.2010.04.003
Guo M, Wang J, Zhao Y, Feng Y, Han S, Dong Q et al (2020) Microglial exosomes facilitate α-synuclein transmission in Parkinson's disease. Brain 143(5):1476–1497. https://doi.org/10.1093/brain/awaa090
Gyparaki MT, Arabacilar GN, Sheng M (2021) Tau and neurodegenerative diseases: a mechanistic review. Curr Opin Neurobiol 69:131–138
He M, Zhao L, Wei M-J, Yao W-F, Zhao H-S, Chen F-J (2009) Neuroprotective effects of (−)-Epigallocatechin-3-gallate on aging mice induced by D-galactose. Biol Pharm Bull 32(1):55–60
Herbst S, Campbell P, Harvey J, Bernard EM, Papayannopoulos V, Wood NW et al (2020) LRRK2 activation controls the repair of damaged endomembranes in macrophages. EMBO J 39(18):e104494. https://doi.org/10.15252/embj.2020104494
Ho PW, Leung CT, Liu H, Pang SY, Lam CS, Xian J et al (2020) Age-dependent accumulation of oligomeric SNCA/α-synuclein from impaired degradation in mutant LRRK2 knockin mouse model of Parkinson disease: role for therapeutic activation of chaperone-mediated autophagy (CMA). Autophagy 16(2):347–370. https://doi.org/10.1080/15548627.2019.1603545
Hou X, Chen TH, Koga S et al (2023) Alpha-synuclein-associated changes in PINK1-PRKN-mediated mitophagy are disease context dependent. Brain Pathol 33(5):e13175. https://doi.org/10.1111/bpa.13175
Hu J, Zhang D, Tian K, Ren C, Li H, Lin C et al (2023) Small-molecule LRRK2 inhibitors for PD therapy: current achievements and future perspectives. Eur J Med Chem 256:115475. https://doi.org/10.1016/j.ejmech.2023.115475

Jiang T, Yu J, Zhu X, Tan M, Wang H, Cao L et al (2014) Temsirolimus promotes autophagic clearance of amyloid-β and provides protective effects in cellular and animal models of Alzheimer's disease. Pharmacol Res 81:54–63. https://doi.org/10.1016/j.phrs.2014.02.008

Kang C, Xu Q, Martin TD, Li MZ, Demaria M, Aron L et al (2015) The DNA damage response induces inflammation and senescence by inhibiting autophagy of GATA4. Science 349(6255):aaa5612. https://doi.org/10.1126/science.aaa5612

Klionsky DJ, Petroni G, Amaravadi RK, Baehrecke EH, Ballabio A, Boya P et al (2021) Autophagy in major human diseases. EMBO J 40(19):e108863. https://doi.org/10.15252/embj.2021108863

Kluss JH, Mamais A, Cookson MR (2019) LRRK2 links genetic and sporadic Parkinson's disease. Biochem Soc Trans 47(2):651–661. https://doi.org/10.1042/BST20180462

Kodali M, Attaluri S, Madhu LN, Shuai B, Upadhya R, Gonzalez JJ et al (2021) Metformin treatment in late middle age improves cognitive function with alleviation of microglial activation and enhancement of autophagy in the hippocampus. Aging Cell 20(2). https://doi.org/10.1111/acel.13277

Lee J, Yang D, Goulbourne CN, Im E, Stavrides P, Pensalfini A et al (2022) Faulty autolysosome acidification in Alzheimer's disease mouse models induces autophagic build-up of aβ in neurons, yielding senile plaques. Nat Neurosci 25(6):688–701. https://doi.org/10.1038/s41593-022-01084-8

Li L, Zhang S, Zhang X, Li T, Tang Y, Liu H et al (2013) Autophagy enhancer carbamazepine alleviates memory deficits and cerebral amyloid-beta pathology in a mouse model of Alzheimer's disease. Curr Alzheimer Res 10(5):433–441. https://doi.org/10.2174/1567205011310040001

Long M, Sanchez-Martinez A, Longo M, Suomi F, Stenlund H, Johansson AI et al (2022) DGAT1 activity synchronises with mitophagy to protect cells from metabolic rewiring by iron depletion. EMBO J 41(10):e109390. https://doi.org/10.15252/embj.2021109390

Lonskaya I, Hebron ML, Desforges NM, Schachter JB, Moussa CE (2013) Nilotinib-induced autophagic changes increase endogenous parkin level and ubiquitination, leading to amyloid clearance. J Mol Med 92(4):373–386. https://doi.org/10.1007/s00109-013-1112-3

Mamais A, Kluss JH, Bonet-Ponce L, Landeck N, Langston RG, Smith N et al (2021) Mutations in LRRK2 linked to Parkinson disease sequester Rab8a to damaged lysosomes and regulate transferrin-mediated iron uptake in microglia. PLoS Biol 19(12):e3001480. https://doi.org/10.1371/journal.pbio.3001480

Mani S, Swargiary G, Singh M, Agarwal S, Dey A, Ojha S et al (2021) Mitochondrial defects: an emerging theranostic avenue towards Alzheimer's associated dysregulations. Life Sci 285:119985. https://doi.org/10.1016/j.lfs.2021.119985

Mattam U, Talari NK, Paripati AK, Krishnamoorthy T, Sepuri NBV (2020) Kisspeptin preserves mitochondrial function by inducing mitophagy and autophagy in aging rat brain hippocampus and human neuronal cell line. Biochim et Biophys Acta (BBA) Mol Cell Res 1868(1):118852. https://doi.org/10.1016/j.bbamcr.2020.118852

Mejías-Peña Y, Estébanez B, Rodriguez-Miguelez P, Fernandez-Gonzalo R, Almar M, González-Gallego J, Cuevas MJ (2017) Impact of resistance training on the autophagy-inflammation-apoptosis crosstalk in elderly subjects. Aging 9(2):408–418. https://doi.org/10.18632/aging.101167

Morishita H (2022) Role of autophagy in the eye: from physiology to disease. Curr Opin Physio 30:100592. https://doi.org/10.1016/j.cophys.2022.100592

Nalls MA, Blauwendraat C, Vallerga CL, Heilbron K, Bandres-Ciga S, Chang D et al (2019) Identification of novel risk loci, causal insights, and heritable risk for Parkinson's disease: a meta-analysis of genome-wide association studies. Lancet Neurol 18(12):1091–1102. https://doi.org/10.1016/S1474-4422(19)30320-5

Nieto-Torres JL, Hansen M (2021) Macroautophagy and aging: the impact of cellular recycling on health and longevity. Mol Asp Med:101020. https://doi.org/10.1016/J.MAM.2021.101020

Oh CK, Dolatabadi N, Cieplak P, Diaz-Meco MT, Moscat J, Nolan JP et al (2022) S-Nitrosylation of p62 inhibits autophagic flux to promote α-synuclein secretion and spread in Parkinson's disease and Lewy body dementia. J Neurosci 42(14):3011–3024. https://doi.org/10.1523/JNEUROSCI.1508-21.2022

Panicker N, Sarkar S, Harischandra DS, Neal M, Kam T, Jin H, Saminathan H, Langley M, Charli A, SamiduraiM, Rokad D, Ghaisas S, Pletnikova O, Dawson VL, Dawson TM, Anantharam V, Kanthasamy AG, Kanthasamy A(2019) Fyn kinase regulates misfolded α-synuclein uptake and NLRP3 inflammasome activation in microglia. J ExpMed 216(6):1411–1430. https://doi.org/10.1084/jem.20182191

Park GW, Kim H, Won SH, Kim NH, Choi R (2025) Neurosteroids and neurological disorders. Korean JPhysiol Pharmacol 29(2):157.https://doi.org/10.4196/kjpp.24.353

Payne A, Nahashon S, Taka E, Adinew GM, Soliman KFA (2022) Epigallocatechin-3-Gallate (EGCG): new therapeutic perspectives for neuroprotection, aging, and Neuroinflammation for the modern age. Biomolecules 12(3):371. https://doi.org/10.3390/biom12030371

Pradas E, Martinez-Vicente M (2023) The consequences of GBA deficiency in the autophagy-lysosome system in Parkinson's disease associated with GBA. Cells 12(1):191. https://doi.org/10.3390/cells12010191

Qi LF, Liu S, Liu YC, Li P, Xu X (2021) Ganoderic acid a promotes amyloid-beta clearance (in vitro) and ameliorates cognitive deficiency in Alzheimer's disease (mouse model) through autophagy induced by activating Axl. Int J Mol Sci 22(11):5559. https://doi.org/10.3390/ijms22115559

Quinn PMJ, Moreira PI, Ambrósio AF, Alves CH (2020) PINK1/PARKIN signalling in neurodegeneration andneuroinflammation. Acta Neuropathol Commun 8(1):11.https://doi.org/10.1186/s40478-020-01062-w

Rai SN, Singh P, Varshney R, Chaturvedi VK, Vamanu E, Singh MP, Singh BK (2021) Promising drug targetsand associated therapeutic interventions in Parkinson's disease. Neural Regen Res 16(9):1730–1739 https://doi.org/10.4103/1673-5374.306635

Rivero-Ríos P, Romo-Lozano M, Madero-Pérez J, Thomas AP, Biosa A, Greggio E, Hilfiker S (2019) The G2019S variant of leucine-rich repeat kinase 2 (LRRK2) alters endolysosomal trafficking by impairing the function of the GTPase RAB8A. J Biol Chem 294(13):4738–4758. https://doi.org/10.1074/jbc.RA118.005008

Rodríguez-Navarro JA, Rodríguez L, Casarejos MJ, Solano RM, Gómez A, Perucho J et al (2010) Trehalose ameliorates dopaminergic and tau pathology in parkin deleted/tau overexpressing mice through autophagy activation. Neurobiol Dis 39(3):423–438. https://doi.org/10.1016/j.nbd.2010.05.014

Sato S, Uchihara T, Fukuda T, Noda S, Kondo H, Saiki S et al (2018) Loss of autophagy in dopaminergic neurons causes Lewy pathology and motor dysfunction in aged mice. Sci Rep 8(1):1–12. https://doi.org/10.1038/s41598-018-21325-w

Schaare HL, Masouleh SK, Beyer F, Kumral D, Uhlig M, Reinelt JD, Reiter AM, Lampe L, Babayan A, ErbeyM, Roebbig J, Schroeter ML, Okon-Singer H, Müller K, Mendes N, Margulies DS, Witte AV, Gaebler M, Villringer A(2019) Association of peripheral blood pressure with gray matter volume in 19- to 40-year-old adults. Neurology92(8):e694–e704 https://doi.org/10.1212/WNL.0000000000006947

Schaefer L, Dikic I (2021) Autophagy: instructions from the extracellular matrix. Matrix Biol 100–101:1–8. https://doi.org/10.1016/j.matbio.2021.06.002

Scheiblich H, Dansokho C, Mercan D, Schmidt SV, Bousset L, Wischhof L et al (2021) Microglia jointly degrade fibrillar alpha-synuclein cargo by distribution through tunneling nanotubes. Cell 184(20):5089–5106. https://doi.org/10.1016/j.cell.2021.09.007

Singh F, Ganley IG (2021) Parkinson's disease and mitophagy: an emerging role for LRRK2. Biochem Soc Trans 49(2):551–562. https://doi.org/10.1042/BST20190236

Stavoe AKH, Holzbaur ELF (2019) Autophagy in neurons. Annu Rev Cell Dev Biol 35:477–500. https://doi.org/10.1146/annurev-cellbio-100818-125242

Suminaite D, Lyons DA, Livesey MR (2019) Myelinated axon physiology and regulation of neural circuitfunction. Glia 67(11):2050–2062. https://doi.org/10.1002/glia.23665

Sun C, Seranova E, Cohen MA, Chipara M, Roberts J, Astuti D et al (2023) NAD depletion mediates cytotoxicity in human neurons with autophagy deficiency. Cell Rep 42(5):112372. https://doi.org/10.1016/j.celrep.2023.112372

Suri S, Chiesa ST, Zsoldos E, Mackay CE, Filippini N, Griffanti L, Mahmood A, Singh-Manoux A, Shipley MJ,Brunner EJ, Kivimäki M, Deanfield JE, Ebmeier KP (2020) Associations between arterial stiffening and brainstructure, perfusion, and cognition in the Whitehall II Imaging Substudy: A retrospective cohort study. PLoS Med17(12):e1003467.https://doi.org/10.1371/journal.pmed.1003467

Sutcliffe T, Winter A, Punessen N, Linseman D (2017) Procyanidin B2 protects neurons from oxidative, nitrosative, and excitotoxic stress. Antioxidants 6(4):77. https://doi.org/10.3390/antiox6040077

Szinyákovics J, Keresztes F, Kiss EA, Falcsik G, Vellai T, Kovács T (2023) Potent new targets for autophagy enhancement to delay neuronal ageing. Cells 12(13):1753. https://doi.org/10.3390/cells12131753

Terracina S, Petrella C, Francati S, Lucarelli M, Barbato C, Minni A et al (2022) Antioxidant intervention to improve cognition in the aging brain: the example of Hydroxytyrosol and resveratrol. Int J Mol Sci 23(23):15674. https://doi.org/10.3390/ijms232315674

Utkin YN (2019) Aging affects nicotinic acetylcholine receptors in brain. CNS Agents Med Chem 19(2):119–124.https://doi.org/10.2174/1871524919666190320102834

Wang N, He J, Pan C, Wang J, Ma M, Shi X, Xu Z (2019) Resveratrol activates autophagy via the AKT/mTOR signaling pathway to improve cognitive dysfunction in rats with chronic cerebral hypoperfusion. Front Neurosci 13:859. https://doi.org/10.3389/fnins.2019.00859

Wang C, Haas M, Yeo SK, Sebti S, Fernández ÁF, Zou Z et al (2022) Enhanced autophagy in Becn1$^{F121A/F121A}$ knockin mice counteracts aging-related neural stem cell exhaustion and dysfunction. Autophagy 18(3):409–422. https://doi.org/10.1080/15548627.2021.1990671

World Health Organization (2022) Ageing and health. https://www.who.int/news-room/fact-sheets/detail/ageing-and-health

Wu Y, Li X, Zhu JX, Xie W, Le W, Fan Z et al (2011) Resveratrol-activated AMPK/SIRT1/autophagy in cellular models of Parkinson's disease. Neurosignals 19(3):163–174. https://doi.org/10.1159/000328516

Wu S, Wei Y, Li J, Bai Y, Yin P, Wang S (2021a) SIRT5 represses neurotrophic pathways and Aβ production in Alzheimer's disease by targeting autophagy. ACS Chem Neurosci 12(23):4428–4437. https://doi.org/10.1021/acschemneuro.1c00468

Wu AG, Zhou XG, Qiao G, Yu L, Tang Y, Yan L et al (2021b) Targeting microglial autophagic degradation in NLRP3 inflammasome-mediated neurodegenerative diseases. Ageing Res Rev 65:101202. https://doi.org/10.1016/j.arr.2020.101202

Xia Y, Zhang G, Han C, Ma K, Guo X, Wan F, Kou L, Yin S, Liu L, Huang J, Xiong N, Wang T (2019) Microgliaas modulators of exosomal alpha-synuclein transmission. Cell Death Dis 10(3):174.https://doi.org/10.1038/s41419-019-1404-9

Yu R, Zhang Z, Wang B, Jiang H, Cheng L, Shen L (2014) Berberine-induced apoptotic and autophagic death of HepG2 cells requires AMPK activation. Cancer Cell Int 14(1). https://doi.org/10.1186/1475-2867-14-49

Zeli C, Lombardo M, Storz MA, Ottaviani M, Rizzo G (2022) Chocolate and cocoa-derived biomolecules for brain cognition during ageing. Antioxidants 11(7):1353. https://doi.org/10.3390/antiox11071353

Zheng J, Hu S, Wang J, Zhang X, Yuan D, Zhang C et al (2021) Icariin improves brain function decline in aging rats by enhancing neuronal autophagy through the AMPK/mTOR/ULK1 pathway. Pharm Biol 59(1):181–189

Zhou K, Xu S (2024) Corydaline alleviates Parkinson's disease by regulating autophagy and GSK-3β phosphorylation. Psychopharmacology 241(5):1027–1036

Zia A, Pourbagher-Shahri AM, Farkhondeh T (2021) Molecular and cellular pathways contributing to brainaging. Behav Brain Funct 17(1):6.https://doi.org/10.1186/s12993-021-00179-9

Zimmerman B, Rypma B, Gratton G, Fabiani M (2021) Age-related changes in cerebrovascular health andtheir effects on neural function and cognition: A comprehensive review. Psychophysiology 58(7):e13796 https://doi.org/10.1111/psyp.13796

Caloric Restriction Mimetics and Their Neuroprotective Potential

4

Geetika Garg, Ishika Singh, Pranamya Prabhu, and Vishal Jain

Abstract

Neurodegenerative diseases (NDs) induced by aging are leading causes of health issues worldwide. Oxidative stress is considered the major hallmark of aging and age-associated NDs. In recent years, life expectancy has increased due to advancements in therapeutic and dietary regimes. The extension of lifespan by the modulation of various key molecular pathways is known to be achieved through caloric restriction (CR). However, it is difficult to commit to this strict regimen; thus, compounds or molecules such as caloric restriction mimetics (CRMs) that imitate pathways such as CR are emerging alternatives. CRMs regulate redox signaling and promote neurogenesis. The role of CRM in neuroprotection is an emerging area of research. This book highlights potential CRMs and their possible neuroprotective effects both preclinically and clinically. The efficacy of CRMs in humans against age-associated neurodegenerative disease and their pharmacokinetic properties have also been discussed.

Keywords

Caloric restriction mimetics · Neuroprotection · Glycolytic inhibitors · NAD^+ precursors · Polyphenols · Polyamines

G. Garg
Department of Zoology, Savitribai Phule Pune University, Pune, Maharashtra, India

I. Singh · P. Prabhu
Manipal Centre for Biotherapeutics Research, Manipal Academy of Higher Education, Manipal, Karnataka, India

V. Jain (✉)
Center of Neural Basis of Cognition, Carnegie Mellon University, Pittsburgh, PA, USA
e-mail: vishalja@andrew.cmu.edu

A. Kumar Singh, S. Nand Rai (eds.), *Rejuvenating the Brain: Nutraceuticals, Autophagy, and Longevity*, Nutritional Neurosciences,
https://doi.org/10.1007/978-981-95-2790-8_4

4.1 Introduction

In recent decades, advancements have been made in understanding the neurotrophic effects of exercise, intermittent fasting (IF), and calorie restriction (CR). Studies conducted on animal models show that CR increases the level of glutathione (GSH), which is an antioxidant, and preserves the mitochondrial function of the brain by altering neuroenergetics, thus playing a role as a neuroprotectant (Mastaloudis et al. 2020). It has been associated with attenuation of aging-associated neurodegenerative diseases and cognitive function in humans.

Neurodegenerative diseases (NDs) induced by aging potentially cause health issues worldwide. Interventions that slow the process of aging and protect against age-related disease are steadily being implemented clinically. These compounds imitate the favorable effects of CR without any rigorous dietary regime.

CR is a well-defined nongenetic intervention that has proved to increase the lifespan of yeast and other animal models and protects against inhibition of biological activities. This reduces the risk of many age-associated diseases (Taormina and Mirisola 2014). The mechanisms that include the positive effects of CR are metabolism, a reduction in reactive oxygen species, an increase in sensitivity to insulin, a reduction in inflammatory molecules, stimulation of autophagy, regulation of neuroendocrine function, and induction of hormesis (Kökten et al. 2021). Furthermore, molecular signaling pathways such as insulin-1 growth factor, G-coactivator-1α, activated AMP protein kinase known as AMPK, mammalian target of rapamycin (mTOR), and sirtuins provide the antiaging effect of CR (Rachakatla and Kalashikam 2022; Zullo et al. 2018). All these pathways form an active interaction network.

However, it becomes difficult to strictly follow a CR diet; therefore, there is a need to imitate the beneficial action of CR without alterations in diet plans, especially in old individuals. These molecules are known as CR mimetics (CRMs). Major CRM candidates, such as rapamycin, resveratrol, and 2-deoxy-D-glucose (2-DG), act on signaling pathways such as the insulin pathway, the AMPK pathway, autophagy stimulants, and other antioxidants, similar to CR (Shintani et al. 2018a). Potential CR-mimicking compounds are known to increase lifespan and protect against age-associated diseases; thus, CRMs are called "antiaging substances." Moreover, CRMs induce the process of autophagy, which is a cellular recycling mechanism that maintains homeostasis and helps degrade damaged or unwanted proteins, cells, and organelles (Madeo et al. 2019; Yin et al. 2016). Autophagy also reduces protein acetylation by activating deacetylases, inhibiting acetylases, or decreasing acetyl-CoA. CRMs also mitigate age-related neurological decline and neurodegenerative disorders by activating autophagy, reducing inflammation, and improving brain health. The majority of CRMs are considered either natural compounds, which are easily absorbed, or FDA-approved pharmacological drugs that act as therapeutic molecules against NDs (Fig. 4.1).

As a possible alternative, CRMs have been investigated through several in vivo studies for their action in mitigating the adverse effects of NDs (Kim et al. 2020). These studies help to establish clinical investigations for replicating the results in

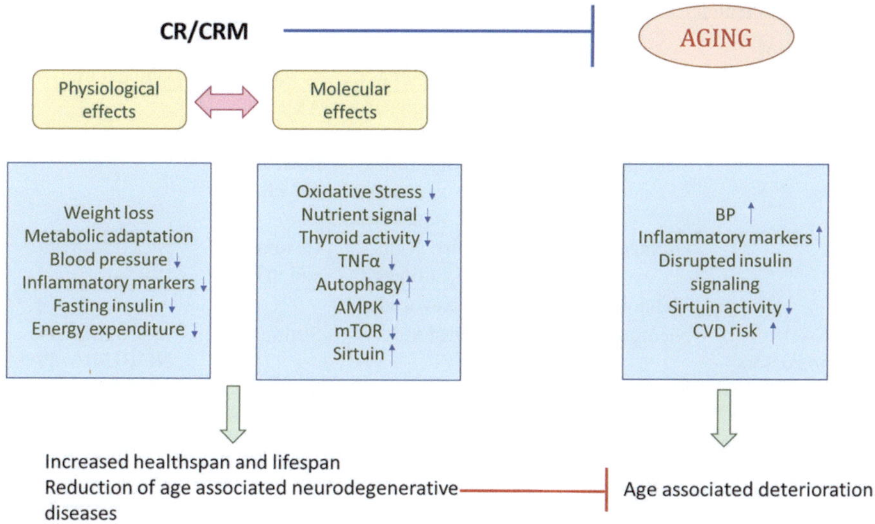

Fig. 4.1 Schematic representation of the molecular and physiological effects of CRM

human subjects. Several clinical studies have validated the role of CRMs in alleviating the effects of NDs (Trisal and Singh 2024).

Considering the limitations of CR in humans, CRMs are promising agents for pharmaceutical use. Additionally, the neuroprotective mechanisms of CRMs are not well understood. Furthermore, the effects of naturally obtained CRMs on nutrition have largely been overlooked.

4.2 Classification of CRMs and Neuroprotective Effects

Investigations to date have shown that a wide variety of molecules have been discovered as CRMs. However, exploring the action of each CRM against NDs is crucial. Characterization of the different CRMs is needed to elucidate the synergistic effects of these molecules. The ideal CRM candidate should be capable of increasing the lifespan and health span of model organisms. In this chapter, we list the compounds that either completely obey the stringent standard categorized as bona fide CRMs or partially obey it in a progressive manner known as potential CRMs. The list of CRMs categorized into various groups according to their mechanism of action is provided in Table 4.1 (Hofer et al. 2021).

4.2.1 Glycolysis Inhibitors

The use of glycolysis inhibitors such as CRMs is a compelling area of research in the context of aging, metabolism, and disease prevention. It has been proposed that

Table 4.1 Potential CRMs and their molecular targets

S. no.	Class of CRM	Compound	Dietary source	Molecular target	Clinical trial
1.	Glycolysis inhibitor	D-Allulose	Wheat, high sugar products	AMPK	
		Genistein	Legumes, fruits, nuts, soy-based items, vegetables	PI3K/Akt, NF-κB	
		D-glucosamine	Shellfish shells, fungi, cartilage	Hexokinase and mTOR	NCT02448199
		Mannoheptulose	Unripe avocados		
2.	NAD^+ precursors	Nicotinamide	Various foods	Sirtuins	NCT02213094, NCT02416739
		Nicotinamide mononucleotide	Plant and animal-based foods, peanuts, nuts	Sirtuins	NCT03151239
		Nicotinamide riboside	Various foods	Sirtuins	NCT03423342
3.	Polyamines	Spermidine	Various plant and animal-based foods, wheat germs, soybeans, and nuts	KATs (EP300), AMPK mTOR	NCT02755246, NCT03378843
4.	Polyphenols	Resveratrol	Fruits, plants, and skin of grapes	KDACs (SIRT1), AMPK, NF-kB	NCT02621554
		Curcumin	*Curcuma longa*	AMPK, mTOR, and EP300	Phase 3 and 4 studies NCT03085680, NCT01052025
		EGCG	Green tea, apples, pears, berries, cocoa, broad bean	AMPK, mTOR, HATs, and KDACs	Phase 3 and 4 studies
		Quercetin	Black tea, onions, rocket, curcuma, and fruits	SIRT1	NCT00065676
5.	Other CRMs	Rapamycin	*Streptomyces hygroscopicus*	mTOR	NCT01649960
		Metformin	*Galega officinalis*	AMPK, mTOR, HATs, and KDACs (sirtuins)	Phase 3 and 4 studies, NCT02432287

CRM achieves its beneficial effects by mimicking some of the molecular pathways activated during fasting or reduced caloric intake. Thus, glycolysis inhibitors, which disrupt the process of glycolysis, have been investigated as potential CRMs (Ingram and Roth 2021). By reducing glucose metabolism, these inhibitors can mimic the metabolic effects of CR, which is known to alter insulin/IGF-1 signaling and the mTOR and AMPK pathways, which work as nutrient sensors (Ingram and Roth 2011). Some glycolysis inhibitors include D-allulose, 2-deoxy-D-glucose (2-DG), metformin, and D-glucosamine.

D-Allulose, a monosaccharide which inhibits the metabolism of glucose and its absorption from the intestinal tract, is known as a low-calorie sweetener. D-Allulose exerts its effect by arresting the glycolytic and carbohydrate metabolic pathways in the intestines. Preclinical studies conducted on nematodes when treated with D-allulose increased their lifespan through the modulation of AMPK pathway (Shintani et al. 2017).

Another glycolysis inhibitor is known to effectively inhibit glycolysis by glucose uptake and interfering with downstream enzymatic steps, which is 2-DG. A study conducted on animal models, 2-DG, has shown to suppress tumor growth, increase lifespan, and mimic some of the actions of CR (Chen et al. 2011; Kumar et al. 2020).

Furthermore, another inhibitor of glycolysis, which acts as a precursor for protein glycosylation, called D-glucosamine, regulates glycolysis through the inhibition of hexokinase-1. This compound is considered a CRM candidate because of its ability to activate autophagy and its antiaging property in animal models (Shintani et al. 2018b; Weimer et al. 2014).

Another potential CRM, which is not a direct glycolysis inhibitor, reduces gluconeogenesis and improves insulin sensitivity. Like caloric restriction, a key regulator of energy balance, AMPK, is activated by metformin (Garg et al. 2017; Singh et al. 2017). Metformin indirectly affects glycolysis and other metabolic pathways to improve health span, thereby extending lifespan in animal models. Some other glycolysis inhibitors, such as mannoheptulose, chrysin, and genistein, act as CRM and are present naturally in plants and other foods.

4.2.1.1 Potential of Glycolytic Inhibitors for Neuroprotection

Neurodegenerative diseases (NDs) are characterized pathologically by the accumulation of toxic waste and proteins in the brain and spinal cord. NDs such as Alzheimer's disease (AD), Parkinson's disease (PD), Huntington's disease (HD), and amyotrophic lateral sclerosis (ALS) are characterized by glycolytic dysfunction, highlighting their crucial role in neurodegeneration. Thus, modulation of the glycolytic pathway has been indicated as a possible neuroprotective strategy for decreasing neuroinflammation and protecting against neurodegeneration (Vizuete et al. 2024). Glycolytic inhibitors have gained attention for their role in neuroprotection, particularly in disrupted metabolic conditions, such as neurodegenerative diseases and ischemic brain injury. During ischemic stroke, in NDs such as AD or PD, where the oxygen supply is limited, neurons depend on glycolysis for the generation of ATP (Zhang et al. 2021). Glycolytic inhibitors, such as 2-DG, might limit the detrimental effects of excessive glycolysis, reducing lactate build-up and maintaining a less acidic cellular environment. Thus, in patients with traumatic brain injury (TBI) or ischemic stroke this could prevent secondary damage.

Neuroinflammation plays an important role in disease progression in many NDs. The shift toward glycolytic metabolism in microglia may contribute to neuroinflammatory responses. Microglial activation and the associated inflammatory cascade can be reduced by inhibiting glycolysis, offering neuroprotective benefits (Cheng et al. 2021). Glycolytic inhibition might increase autophagic activity, which is involved in the clearance of damaged proteins and dysfunctional organelles, including mitochondria. By promoting the removal of dysfunctional cellular components, glycolytic inhibitors can improve cellular homeostasis and protect neurons from neurodegenerative processes (Pessoa and Duarte 2023).

They are attractive options in the field of therapeutics due to their ability to modulate energy production, reduce oxidative stress, and influence neuroinflammation and autophagy. However, careful consideration is required to avoid disrupting normal neuronal function. In ischemic and neurodegenerative conditions where metabolic disturbances contribute to neuronal damage, glycolytic inhibitors hold significant promise for neuroprotection. Further studies are needed to fully understand the long-term effects of glycolytic inhibitors in the brain and their development as potential therapeutics.

4.2.1.2 Glycolysis Inhibitors in Clinical Studies

The effectiveness of glycolysis inhibitors is difficult to track in humans. Though preclinical studies have shown the efficacy of glycolysis inhibitors. Additionally, such studies lack clinical data for most of the inhibitors. Glucosamine is a glycolysis inhibitor that has been tested in clinical trials. On average, 7.4% of the US population regularly uses glucosamine in their diet (Qato et al. 2008), which has significantly decreased mortality, as reported previously (Pocobelli et al. 2010; Satia et al. 2009). Several recent studies support this observation, both in the USA (Bell et al. 2012; King and Xiang 2020) and the UK (Yueh et al. 2022).

The effects of another glycolytic inhibitor, metformin, which can target both aging and neurodegenerative disorders, in humans, have been evaluated through several clinical trials. A phase 4 clinical study (NCT02432287) RNA sequencing of genes that were found in muscles and adipose tissue was conducted to assess alterations in the transcriptome. Another clinical trial interpreting the action of metformin on the levels of AD markers (NCT01965756) used the ADAS-Cog, known as Alzheimer's Disease Assessment Scale-Cognitive Subscale, to check its ability to alleviate cognitive decline in AD patients.

4.2.2 NAD$^+$ Precursors

NAD+ homeostasis is essential for proper cellular processing and function. The dietary precursors of NAD+ are nicotinic acid (NA), nicotinamide (NAM), nicotinamide riboside (NR), nicotinamide mononucleotide (NMN), and tryptophan, which have similar biological activities (Bogan and Brenner 2008; Chi and Sauve 2013).

These precursors exhibit CRM-like properties upon supplementation, as evident by preclinical studies, which exaggerate mitochondrial function, induce autophagy, prolong lifespan and health span, and serve as cardioprotective and neuroprotective agents.

4.2.2.1 Potential of NAD+ Precursors for Neuroprotection

Supplementing precursors of NAD+ such as nicotinamide mononucleotide (NMN) and nicotinamide riboside (NR) increases NAD^+ levels and has emerged as a promising strategy that potentially offers neuroprotective benefits. NAD^+ is crucial for ameliorating the function of mitochondria, affecting energy production and cellular metabolism while maintaining neuronal health (Amjad et al. 2021). Sirtuins, which are NAD^+-dependent deacetylases, help in regulating various cellular processes, including inflammation, stress resistance, and metabolism. The increase in NAD^+ levels results in neuroprotection by activation of sirtuins (Garg et al. 2021).

Cognitive function, synaptic plasticity, and neuronal integrity were improved with NMN supplementation in animal models of Alzheimer's disease (Ma et al. 2024). This leads to increasing mitochondrial biogenesis and sirtuin activation. Another precursor, nicotinamide riboside (NR), has demonstrated its effect in attenuating mitochondrial defects and decreasing neuronal loss in Parkinson's disease models, highlighting its neuroprotective effects (Schöndorf et al. 2018).

NMN and NR potentially mitigate age-related decline and neurodegenerative diseases which are essential to maintain neuronal integrity, showing potential for neuroprotection by elevating NAD+ levels. In AD mouse model, NR treatment showed to improve cognitive and synaptic functions by preventing neuroinflammation (Hou et al. 2021). Furthermore, in an MPTP-induced mouse model of PD, the NAM administration, another precursor, significantly attenuates neuronal loss and helps to reduce motor dysfunction, oxidative stress, and neuroinflammation (Rehman et al. 2022). Additionally, in the 3-nitropropionic acid-induced model of HD it also provides neuroprotection (Sidhu et al. 2018).

4.2.2.2 NAD^+ Precursors in Clinical Studies

Many clinical studies, in addition to these preclinical evaluations, are now being conducted on NDs to understand the effects of NAD+ precursors. NR is bioavailable orally without having drastic side effects (Martens et al. 2018; Trammell et al. 2016). A recent randomized trial indicated that supplementation with NR orally significantly increases NAD+ in neuron-enriched plasma EVs. Furthermore, NR administration in PD patients increased cerebral NAD, as evidenced by the findings of a placebo-controlled, randomized, phase I clinical trial. This effect was associated with decreased regulation of inflammatory markers (Brakedal et al. 2022). A new clinical trial (NCT03061474) is being performed for NAM among people with mild cognitive impairment to study whether phosphorylation of the tau protein can be inhibited by NAM in the cerebrospinal fluid (NCT03061474) (Ketron et al. 2025).

Some trials suggest that NAD^+ precursors may offer therapeutic benefits for age-related and neurodegenerative diseases, while human studies are limited (Reiten et al. 2021). However, more extensive clinical research is necessary to confirm these findings. Despite promising preclinical data, the efficacy and safety of NAD^+ precursor supplementation in humans require further investigation. NAD^+ precursors have potential as neuroprotective agents by enhancing cellular energy metabolism, supporting DNA repair, and activating neuroprotective sirtuins. However, more rigorous clinical studies are necessary to understand their therapeutic potential and establish evidence-based guidelines for their use in neuroprotection.

4.2.3 Polyamines

Positively charged molecules, polyamines such as putrescine, spermine, and spermidine that participate in multiple cellular processes, including cell differentiation, growth, and survival. Their ability to induce beneficial cellular responses, such as those observed during CR, including autophagy, resistance to cellular stress, and lifespan extension due to this they have gained them significant attention as potential CRMs.

In preclinical animal models, including mice, polyamine spermidine has been shown to activate autophagy (Eisenberg et al. 2009; Morselli et al. 2011).

Spermidine is produced naturally in the body. Additionally, the presence of spermidine in food products such as soybeans, wheat germs, or nuts results in adequate bioavailability (Soda et al. 2009). The concentration of spermidine decreases with age, and the supplementation with spermidine has been implicated in enhancing the health spans or lifespans of several model organisms. Spermidine is also well documented as an autophagy inducer that increases the lifespan of organisms. Spermidine, a CRM, also provides neuroprotection against natural and D-galactose-induced aging through the modulation of the autophagic pathway in male rats (Singh et al. 2021).

4.2.3.1 Potential of Polyamines for Neuroprotection

Owing to their important roles in cellular function, PAs have gained attention for their potential neuroprotective properties. Polyamines have been shown to exhibit protective effects that could help preserve neuronal function, slow disease progression, and promote brain health in neurodegenerative disorders including Alzheimer's, Parkinson's, and Huntington's diseases (Arthur et al. 2024). One of the most significant mechanisms by which polyamines, especially spermidine, provide neuroprotection is through the activation of autophagy. Polyamines can activate autophagy by inhibiting certain acetyltransferases and histone deacetylases that regulate autophagic pathways. By increasing autophagic flux, PAs can help clear these toxic protein aggregates, reducing neuronal toxicity and potentially slowing disease progression.

Chronic neuroinflammation is another hallmark of neurodegenerative diseases. Inflammation in the brain, which is mediated by glial cells such as microglia, can exacerbate neuronal damage. Polyamines reduce inflammation in the central nervous system (CNS). Spermidine has been shown to downregulate the expression of proinflammatory cytokines, thus helping maintain a more balanced and less neurotoxic inflammatory response.

Additionally, spermidine, which is known to promote cellular repair, stress resistance, and longevity, has been shown to activate the SIRT1 pathway (Ni and Liu 2021). SIRT1 activation can protect neurons by enhancing DNA repair and reducing apoptosis (programmed cell death), both of which are crucial in preventing neurodegeneration.

4.2.3.2 Polyamines in Clinical Studies

Although preclinically polyamines show promising effects, the number of clinical trials conducted on this CRM is limited. In a pilot study, elderly people were administered with decreased polyamine doses via wheat germ extract (0.6 mg/spermine, 1.2 mg spermidine, 0.2 mg putrescine per day) for a period of 3 months which improved cognition (Wirth et al. 2018). Ongoing trials are using spermidine supplementation in the diet to explore sleep quality improvement (NCT04823806), its effects on depression, hypertension (NCT04405388), and decline in cognition in older adults (NCT03094546).

4.2.4 Polyphenols

Plant polyphenols are recognized for their caloric restriction mimetic properties worldwide. It has been reported in recent publications that polyphenols may partly mimic the beneficial effects of caloric restriction, inducing similar molecular mechanisms that govern mitochondrial biogenesis and mitophagy (Davinelli et al. 2020). Polyphenols include a wide range of compounds, such as phenolic acids, flavonoids, and stilbenes. CRM-like effect showing most studied polyphenols is curcumin, quercetin, epigallocatechin gallate (EGCG), and resveratrol, among others.

A natural polyphenolic phytoalexin, resveratrol, is present in fruits and promotes health benefits, and prevents age-associated diseases, such as cancer (Pandey and Rizvi 2014; Tsai et al. 2019), atherosclerosis (Chassot et al. 2018), cardiovascular disease (Huo et al. 2019), type 2 diabetes (Cheang et al. 2019), and neurodegenerative diseases, such as Alzheimer's disease (AD) (Alkhouli et al. 2019; Cosín-Tomàs et al. 2019). Resveratrol also attenuates memory deficits, maintains blood–brain barrier integrity, ameliorates plaques, decreases tau protein levels, and decreases microglial activation (Gomes et al. 2018). In a study conducted on mouse model, it has been shown to prolong longevity and improve aging parameters. Several animal studies have also revealed the beneficial effect of supplementation with resveratrol in the prevention and treatment of AD (Ma et al. 2014). For example, resveratrol supplementation has neuroprotective effects on cells treated with Amyloid β (Rahman et al. 2020). In AD, resveratrol is associated with autophagic activation and SIRT1 and AMPK signaling, thus providing neuroprotection (Wang et al. 2019).

Resveratrol depends on the activation of SIRT1 for its mechanism of action. SIRT1 is known to deacetylate autophagy proteins such as Atg5 and Atg7, leading to the induction of autophagy (Kitada et al. 2016; Morselli et al. 2010). AMPK activates SIRT1 by indirectly increasing cellular NAD+ levels (Cantó et al. 2009), resulting in cross-talk between the sirtuin and AMPK pathways. Furthermore, resveratrol provides neuroprotection against several CNS disorders, such as AD and PD. In a study conducted on 3 × Tg-AD mice brain, treatment with resveratrol alleviated neuroinflammation and reduced the accumulation of Aβ proteins (Broderick et al. 2020). In the PD mouse model, resveratrol supplementation also ameliorated motor and cognitive impairments. It inhibits oxidative stress, astrocytosis, and microgliosis in brain tissue and also reduces the aggregation of α-synuclein (Zhang et al. 2018). Amyloid β-mediated neurotoxicity mitigated by activation of autophagy in human neural stem cells (Chiang et al. 2018), along with alleviating glucose and oxygen deprivation in human SH-SY5Y neuroblastoma cells (Lin et al. 2020).

Other polyphenols that can mimic the effects of CR and promote autophagy and mitochondrial biogenesis such as quercetin and curcumin also activate sirtuins and help regulate cellular energy homeostasis.

4.2.4.1 Potential of Polyphenols for Neuroprotection

Neurogenesis is a process in which new neurons and other brain cells are formed due to the differentiation of stem cells from the hippocampus. Numerous external and internal factors affect this process, including diet. The internal factors include traumatic brain injury, oxidative stress, aging, and neuroinflammation, as well as high sugar diet, alcohol, high fat diets, and opioids, which may adversely affect the process of neurogenesis. In adult brains, resveratrol, curcumin, blueberry, and diets enriched with PUFAs and polyphenols, along with caloric restriction, tend to induce neurogenesis. Indeed, recently natural polyphenols providing neuroprotective effects have been extensively studied and published (Shi et al. 2018; Zhao et al. 2020).

4.2.4.2 Polyphenols in Clinical Studies

While numerous preclinical data on the CR-like effects and health benefits of polyphenols are available, clinical studies are rare, because of the wide variety of compounds, their biological effects, and the complexity of their content estimation in foods, they are difficult to perform in humans. While preclinical studies in animals have shown promising neuroprotective effects of CRMs, human clinical trials are still limited. However, some early studies have suggested that compounds such as curcumin, resveratrol, and EGCG can improve cognitive function and reduce inflammation in aging individuals, indicating their potential therapeutic role in neurodegenerative diseases during aging.

Currently, clinical studies on the effect of resveratrol on NDs have been evaluated. Clinical studies reported the effect of resveratrol on individuals suffering from AD (Turner et al. 2015). Future research that includes well-designed clinical trials is needed to determine the optimal dosages and long-term effects of CRMs on brain health in humans. Additionally, the combination of different CRMs, as well as their potential synergistic effects, should be explored to maximize neuroprotection.

4.2.5 Other CRMs

Other CRMs, such as inhibitors of mTOR and rapamycin, result in extending life span by mediating SIRT1 expression and preventing age-related diseases (Singh et al. 2016a, b). Rapamycin has been used as a potential CRM that needs to be further explored to understand the targeted signaling pathways to increase its therapeutic potential. To determine the effect of rapamycin, some clinical trials have been conducted that target aging and NDs in humans (Lee et al. 2024). Clinically, the effect of rapamycin on immune function and its cardiovascular consequences has been evaluated in a study titled "Effect of mTOR Inhibition and Other Metabolism Modulating Interventions on the Elderly" (NCT02874924).

4.2.5.1 Molecular Mechanism of CRM-Based Therapeutics

The mechanism of action of CRM in its therapeutic effect includes the insulin pathway, the molecular interactions of sirtuins, and the mTOR/AMPK pathway. mTOR is sensitive to amino acid concentration and acts as a nutrient-sensing molecule. Cross-talk between the mTOR and AMPK pathways has largely been studied. Additionally, AMPK inhibits mTOR complexes by activating downstream regulators. It has already been reported that the members of the mTOR complex are directed to be inactivated and phosphorylated by AMPK. AMPK upregulates the process of autophagy via the phosphorylation of ULK1 (Garza-Lombó et al. 2018). Many age-associated neurological disorders involve upregulation of the mTOR signaling pathway which is a cellular marker of aging. Therefore, compounds inhibit the mTOR pathway, induce autophagy and prolong longevity, and act as CRMs (Ham et al. 2022). This process is crucial for neuronal health, as it helps clear damaged components that could lead to neurodegeneration. CRMs activate autophagy mediated by sirtuins, particularly SIRT1, by reducing protein acetylation.

CRMs also activate the Nrf2 pathway, resulting in reduced reactive oxygen species (ROS) generation and increased antioxidant defense (Udono and Nishida 2022). This modulation helps protect neurons from oxidative stress, which contributes to neurodegenerative diseases.

Neuronal survival via redox-sensitive signaling pathways, such as the MAPK and PI3K/Akt pathways, can be increased by CRMs (Sharma and Singh 2023).

CRMs affect insulin signaling pathways and play an important role in neuronal survival and function. CRMs can decrease the systemic levels of insulin/IGF and modulate insulin signaling (Fig. 4.2) (Lee and Min 2013). By modulating these pathways, CRMs may increase neuronal resilience and protect against age-related neurodegenerative conditions. This pathway is important that governing several cellular functions, including autophagy, apoptosis, and expression of immunogenic factors, and is also involved in cell proliferation and survival (Iams and Lovly 2015).

Sirtuins, which are histone deacetylases, furthermore, play important roles in the regulation of aging. Sirtuins aid in the maintenance of cellular homeostasis and also respond to cellular alterations such as DNA damage. NF-κB, an inflammatory molecule, which is involved in several neurodegenerative diseases, is also suppressed by sirtuins (Fernando and Wijayasinghe 2021). Sirt3, a crucial molecule for healthy aging when activated, leads to optimal oxygen consumption, thus restricting ROS production (Zhou et al. 2022).

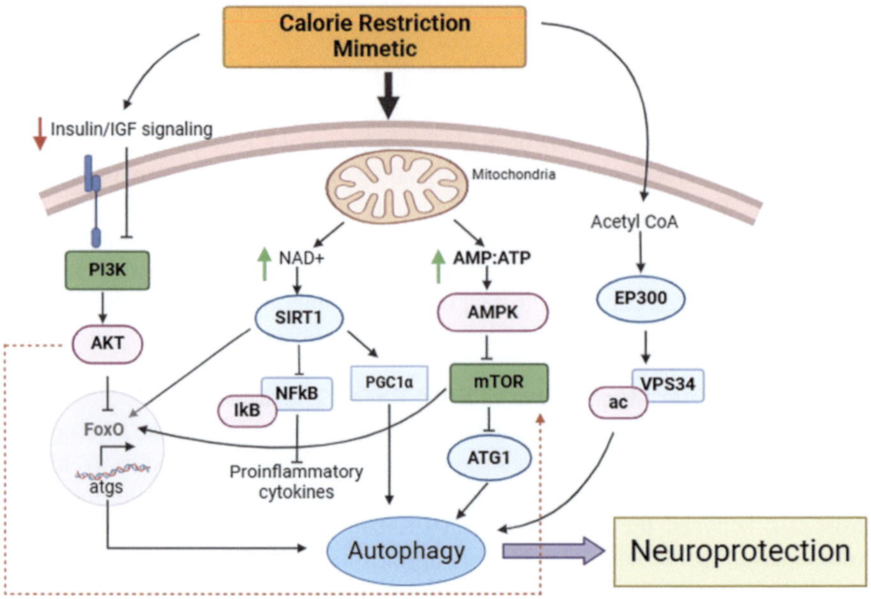

Fig. 4.2 Illustration of the mechanism of action of CRM via autophagy induction

4.3 Pharmacokinetic Properties of CRM

Both in vitro and in vivo effects of CRMs should be evaluated by transitioning these compounds from animals to humans (Yadav et al. 2021). CRMs, depending on their categories, differ in their pharmacokinetic properties, but a large number of CRMs have relatively short half-lives and are rapidly metabolized and excreted. The beneficial effects of these are often related to the modulation of key metabolic pathways. Further research will emerge to optimize the use of CRMs for health span and lifespan benefits.

Various factors, including bioavailability, efficacy, and safety, involved in the formulation of CRMs into drugs. The strategies in formulation include liposomal encapsulation, nanoparticles, and coadministration with other compounds which enhance the bioavailability and absorption of CRMs (Subramani and Ganapathyswamy 2020). CRMs encapsulated by liposomes improve stability, facilitate better absorption across cell membranes, and provide protection from degradation. Furthermore, nanoparticles are used to encapsulate CRMs, improving their solubility and absorption (Liu et al. 2024). These formulations increase the bioavailability of the drug and enhance sustained drug release.

Moreover, sustained-release technologies can be employed to increase the half-life of some CRMs (Adepu and Ramakrishna 2021). Targeted delivery systems are used to direct CRMs to specific tissues via antibodies or ligands, in turn increasing the effectiveness of drugs (Yan et al. 2024). Additionally, by chemical modification the stability of CRMs can be. These drugs can help delay aging, improve metabolic health, and prevent age-related diseases by optimizing bioavailability, targeting specific pathways, and ensuring drug stability. Ongoing research and innovation in drug delivery and formulation technologies will be key to the success of CRM-based therapeutics. Comprehensive clinical trials must be conducted to evaluate the efficacy, safety, and optimal dosing of CRM-based drugs before they can be approved for widespread use.

4.4 Conclusion

CRMs offer significant neuroprotective benefits such as reduction in oxidative stress, inflammatory proteins, mitochondrial enhancement, autophagy activation, and regulation of metabolic health through different mechanisms. These compounds act as promising candidates for the prevention and treatment of neurodegenerative diseases such as PD and AD due to their ability to mimic the beneficial effects of CR. In general, dietary CRMs, the emerging and promising field, needs to be explored scientifically. Epidemiological studies provide the majority of evidence on the effects of CRM in humans, which are generally conducted using plant-based extracts containing a number of bioactive substances. In addition, autophagy can also contribute to the beneficial effects of CRM compounds in AD patients. Furthermore, for efficient autophagy induction and prevention or treatment of AD in humans, clinical trials must be conducted to identify the optimal intervention regimens. Overall, accumulating evidence from both preclinical and clinical studies, CRMs have been shown to constitute a progressive field of research that clinical and nutritional researchers should address. While preclinical evidence is strong, to understand their efficacy and potential therapeutic applications, more human studies are needed to be performed.

Statement and Declaration of Competing Interest The authors report that there are no competing interests to declare.

Funding This research did not receive any specific grant from funding agencies.

Data Availability There is no data used in the research, as described in this paper.

References

Adepu S, Ramakrishna S (2021) Controlled drug delivery systems: current status and future directions. Molecules 26:5905. https://doi.org/10.3390/molecules26195905

Alkhouli MF, Hung J, Squire M, Anderson M, Castro M, Babu JR, Al-Nakkash L, Broderick TL, Plochocki JH (2019) Exercise and resveratrol increase fracture resistance in the 3xTg-AD mouse model of Alzheimer's disease. BMC Complement Altern Med 19:39. https://doi.org/10.1186/s12906-019-2451-6

Amjad S, Nisar S, Bhat AA, Shah AR, Frenneaux MP, Fakhro K, Haris M, Reddy R, Patay Z, Baur J, Bagga P (2021) Role of NAD+ in regulating cellular and metabolic signaling pathways. Mol Metab 49:101195. https://doi.org/10.1016/j.molmet.2021.101195

Arthur R, Jamwal S, Kumar P (2024) A review on polyamines as promising next-generation neuroprotective and anti-aging therapy. Eur J Pharmacol 978:176804. https://doi.org/10.1016/j.ejphar.2024.176804

Bell GA, Kantor ED, Lampe JW, Shen DD, White E (2012) Use of glucosamine and chondroitin in relation to mortality. Eur J Epidemiol 27:593–603. https://doi.org/10.1007/s10654-012-9714-6

Bogan KL, Brenner C (2008) Nicotinic acid, nicotinamide, and nicotinamide riboside: a molecular evaluation of NAD^+ precursor Vitamins in human nutrition. Annu Rev Nutr 28:115–130. https://doi.org/10.1146/annurev.nutr.28.061807.155443

Brakedal B, Dölle C, Riemer F, Ma Y, Nido GS, Skeie GO, Craven AR, Schwarzlmüller T, Brekke N, Diab J, Sverkeli L, Skjeie V, Varhaug K, Tysnes O-B, Peng S, Haugarvoll K, Ziegler M, Grüner R, Eidelberg D, Tzoulis C (2022) The NADPARK study: a randomized phase I trial of nicotinamide riboside supplementation in Parkinson's disease. Cell Metab 34:396–407. https://doi.org/10.1016/j.cmet.2022.02.001

Broderick TL, Rasool S, Li R, Zhang Y, Anderson M, Al-Nakkash L, Plochocki JH, Geetha T, Babu JR (2020) Neuroprotective effects of chronic resveratrol treatment and exercise training in the 3xTg-AD mouse model of Alzheimer's disease. IJMS 21:7337. https://doi.org/10.3390/ijms21197337

Cantó C, Gerhart-Hines Z, Feige JN, Lagouge M, Noriega L, Milne JC, Elliott PJ, Puigserver P, Auwerx J (2009) AMPK regulates energy expenditure by modulating NAD+ metabolism and SIRT1 activity. Nature 458:1056–1060. https://doi.org/10.1038/nature07813

Chassot LN, Scolaro B, Roschel GG, Cogliati B, Cavalcanti MF, Abdalla DSP, Castro IA (2018) Comparison between red wine and isolated trans-resveratrol on the prevention and regression of atherosclerosis in LDLr (−/−) mice. J Nutr Biochem 61:48–55. https://doi.org/10.1016/j.jnutbio.2018.07.014

Cheang WS, Wong WT, Wang L, Cheng CK, Lau CW, Ma RCW, Xu A, Wang N, Huang Y, Tian XY (2019) Resveratrol ameliorates endothelial dysfunction in diabetic and obese mice through sirtuin 1 and peroxisome proliferator-activated receptor δ. Pharmacol Res 139:384–394. https://doi.org/10.1016/j.phrs.2018.11.041

Chen K, Xu X, Kobayashi S, Timm D, Jepperson T, Liang Q (2011) Caloric restriction mimetic 2-deoxyglucose antagonizes doxorubicin-induced cardiomyocyte death by multiple mechanisms. J Biol Chem 286:21993–22006. https://doi.org/10.1074/jbc.M111.225805

Cheng J, Zhang R, Xu Z, Ke Y, Sun R, Yang H, Zhang X, Zhen X, Zheng L-T (2021) Early glycolytic reprogramming controls microglial inflammatory activation. J Neuroinflammation 18:129. https://doi.org/10.1186/s12974-021-02187-y

Chi Y, Sauve AA (2013) Nicotinamide riboside, a trace nutrient in foods, is a vitamin B3 with effects on energy metabolism and neuroprotection. Curr Opin Clin Nutr Metab Care 16:657–661. https://doi.org/10.1097/MCO.0b013e32836510c0

Chiang M-C, Nicol CJ, Cheng Y-C (2018) Resveratrol activation of AMPK-dependent pathways is neuroprotective in human neural stem cells against amyloid-beta-induced inflammation and oxidative stress. Neurochem Int 115:1–10. https://doi.org/10.1016/j.neuint.2017.10.002

Cosín-Tomàs M, Senserrich J, Arumí-Planas M, Alquézar C, Pallàs M, Martín-Requero Á, Suñol C, Kaliman P, Sanfeliu C (2019) Role of resveratrol and selenium on oxidative stress and expression of antioxidant and anti-aging genes in immortalized lymphocytes from Alzheimer's disease patients. Nutrients 11:1764. https://doi.org/10.3390/nu11081764

Davinelli S, De Stefani D, De Vivo I, Scapagnini G (2020) Polyphenols as caloric restriction Mimetics regulating mitochondrial biogenesis and Mitophagy. Trends Endocrinol Metabol 31:536–550. https://doi.org/10.1016/j.tem.2020.02.011

Eisenberg T, Knauer H, Schauer A, Büttner S, Ruckenstuhl C, Carmona-Gutierrez D, Ring J, Schroeder S, Magnes C, Antonacci L, Fussi H, Deszcz L, Hartl R, Schraml E, Criollo A, Megalou E, Weiskopf D, Laun P, Heeren G, Breitenbach M, Grubeck-Loebenstein B, Herker E, Fahrenkrog B, Fröhlich K-U, Sinner F, Tavernarakis N, Minois N, Kroemer G, Madeo F

(2009) Induction of autophagy by spermidine promotes longevity. Nat Cell Biol 11:1305–1314. https://doi.org/10.1038/ncb1975

Fernando KKM, Wijayasinghe YS (2021) Sirtuins as potential therapeutic targets for mitigating Neuroinflammation associated with Alzheimer's disease. Front Cell Neurosci 15:746631. https://doi.org/10.3389/fncel.2021.746631

Garg G, Singh S, Singh AK, Rizvi SI (2017) Antiaging effect of metformin on brain in naturally aged and accelerated senescence model of rat. Rejuvenation Res. https://doi.org/10.1089/rej.2016.1883

Garg G, Singh AK, Singh S, Rizvi SI (2021) Promising drug discovery strategies for sirtuin modulators: what lessons have we learnt? Expert Opin Drug Discov 16:915–927. https://doi.org/10.1080/17460441.2021.1915980

Garza-Lombó C, Schroder A, Reyes-Reyes EM, Franco R (2018) mTOR/AMPK signaling in the brain: cell metabolism, proteostasis and survival. Curr Opin Toxicol 8:102–110. https://doi.org/10.1016/j.cotox.2018.05.002

Gomes BAQ, Silva JPB, Romeiro CFR, Dos Santos SM, Rodrigues CA, Gonçalves PR, Sakai JT, Mendes PFS, Varela ELP, Monteiro MC (2018) Neuroprotective mechanisms of resveratrol in Alzheimer's disease: role of SIRT1. Oxidative Med Cell Longev 2018:8152373. https://doi.org/10.1155/2018/8152373

Ham DJ, Börsch A, Chojnowska K, Lin S, Leuchtmann AB, Ham AS, Thürkauf M, Delezie J, Furrer R, Burri D, Sinnreich M, Handschin C, Tintignac LA, Zavolan M, Mittal N, Rüegg MA (2022) Distinct and additive effects of calorie restriction and rapamycin in aging skeletal muscle. Nat Commun 13:2025. https://doi.org/10.1038/s41467-022-29714-6

Hofer SJ, Davinelli S, Bergmann M, Scapagnini G, Madeo F (2021) Caloric restriction Mimetics in nutrition and clinical trials. Front Nutr 8:717343. https://doi.org/10.3389/fnut.2021.717343

Hou Y, Wei Y, Lautrup S, Yang B, Wang Y, Cordonnier S, Mattson MP, Croteau DL, Bohr VA (2021) NAD^+ supplementation reduces neuroinflammation and cell senescence in a transgenic mouse model of Alzheimer's disease via cGAS–STING. Proc Natl Acad Sci USA 118:e2011226118. https://doi.org/10.1073/pnas.2011226118

Huo X, Zhang T, Meng Q, Li C, You B (2019) Resveratrol effects on a diabetic rat model with coronary heart disease. Med Sci Monit 25:540–546. https://doi.org/10.12659/MSM.910996

Iams WT, Lovly CM (2015) Molecular pathways: clinical applications and future direction of insulin-like growth Factor-1 receptor pathway blockade. Clin Cancer Res 21:4270–4277. https://doi.org/10.1158/1078-0432.CCR-14-2518

Ingram DK, Roth GS (2011) Glycolytic inhibition as a strategy for developing calorie restriction mimetics. Exp Gerontol 46:148–154. https://doi.org/10.1016/j.exger.2010.12.001

Ingram DK, Roth GS (2021) Glycolytic inhibition: an effective strategy for developing calorie restriction mimetics. Geroscience 43:1159–1169. https://doi.org/10.1007/s11357-020-00298-7

Ketron GL, Grun F, Grill JD, Feldman HH, Rissman RA, Brewer GJ (2025) Pharmacokinetic and pharmacodynamic assessment of oral nicotinamide in the NEAT clinical trial for early Alzheimer's disease. Alzheimer's Res Ther 17:59. https://doi.org/10.1186/s13195-025-01693-y

Kim DH, Bang E, Jung HJ, Noh SG, Yu BP, Choi YJ, Chung HY (2020) Anti-aging effects of calorie restriction (CR) and CR Mimetics based on the Senoinflammation concept. Nutrients 12:422. https://doi.org/10.3390/nu12020422

King DE, Xiang J (2020) Glucosamine/chondroitin and mortality in a US NHANES cohort. J Am Board Fam Med 33:842–847. https://doi.org/10.3122/jabfm.2020.06.200110

Kitada M, Ogura Y, Koya D (2016) Role of Sirt1 as a regulator of autophagy. In: Autophagy: cancer, other pathologies, inflammation, immunity, infection, and aging. Elsevier, pp 89–100. https://doi.org/10.1016/B978-0-12-802937-4.00003-X

Kökten T, Hansmannel F, Ndiaye NC, Heba A-C, Quilliot D, Dreumont N, Arnone D, Peyrin-Biroulet L (2021) Calorie restriction as a new treatment of inflammatory diseases. Adv Nutr 12:1558–1570. https://doi.org/10.1093/advances/nmaa179

Kumar R, Saraswat K, Rizvi SI (2020) 2 –Deoxy – d-glucose at chronic low dose acts as a caloric restriction mimetic through a mitohormetic induction of ROS in the brain of accelerated senescence model of rat. Arch Gerontol Geriatr 90:104133. https://doi.org/10.1016/j.archger.2020.104133

Lee S-H, Min K-J (2013) Caloric restriction and its mimetics. BMB Rep 46:181–187. https://doi.org/10.5483/bmbrep.2013.46.4.033

Lee DJW, Hodzic Kuerec A, Maier AB (2024) Targeting ageing with rapamycin and its derivatives in humans: a systematic review. Lancet Healthy Longevity 5:e152–e162. https://doi.org/10.1016/S2666-7568(23)00258-1

Lin C-H, Nicol CJB, Cheng Y-C, Yen C, Wang Y-S, Chiang M-C (2020) Neuroprotective effects of resveratrol against oxygen glucose deprivation induced mitochondrial dysfunction by activation of AMPK in SH-SY5Y cells with 3D gelatin scaffold. Brain Res 1726:146492. https://doi.org/10.1016/j.brainres.2019.146492

Liu Y, Liang Y, Yuhong J, Xin P, Han JL, Du Y, Yu X, Zhu R, Zhang M, Chen W, Ma Y (2024) Advances in nanotechnology for enhancing the solubility and bioavailability of poorly soluble drugs. Drug Des Devel Ther 18:1469–1495. https://doi.org/10.2147/DDDT.S447496

Ma T, Tan M-S, Yu J-T, Tan L (2014) Resveratrol as a therapeutic agent for Alzheimer's disease. Biomed Res Int 2014:350516. https://doi.org/10.1155/2014/350516

Ma R-Y, Li L, Yang H, Zou B, Ma R-X, Zhang Y, Wu M-M, Chen P, Yao Y, Li J (2024) Therapeutic effect of nicotinamide mononucleotide on Alzheimer's disease through activating autophagy and anti-oxidative stress. Biomed Pharmacother 178:117199. https://doi.org/10.1016/j.biopha.2024.117199

Madeo F, Carmona-Gutierrez D, Hofer SJ, Kroemer G (2019) Caloric restriction Mimetics against age-associated disease: targets, mechanisms, and therapeutic potential. Cell Metab 29:592–610. https://doi.org/10.1016/j.cmet.2019.01.018

Martens CR, Denman BA, Mazzo MR, Armstrong ML, Reisdorph N, McQueen MB, Chonchol M, Seals DR (2018) Chronic nicotinamide riboside supplementation is well-tolerated and elevates NAD+ in healthy middle-aged and older adults. Nat Commun 9:1286. https://doi.org/10.1038/s41467-018-03421-7

Mastaloudis A, Sheth C, Hester SN, Wood SM, Prescot A, McGlade E, Renshaw PF, Yurgelun-Todd DA (2020) Supplementation with a putative calorie restriction mimetic micronutrient blend increases glutathione concentrations and improves neuroenergetics in brain of healthy middle-aged men and women. Free Radic Biol Med 153:112–121. https://doi.org/10.1016/j.freeradbiomed.2020.04.017

Morselli E, Maiuri MC, Markaki M, Megalou E, Pasparaki A, Palikaras K, Criollo A, Galluzzi L, Malik SA, Vitale I, Michaud M, Madeo F, Tavernarakis N, Kroemer G (2010) Caloric restriction and resveratrol promote longevity through the Sirtuin-1-dependent induction of autophagy. Cell Death Dis 1:e10–e10. https://doi.org/10.1038/cddis.2009.8

Morselli E, Mariño G, Bennetzen MV, Eisenberg T, Megalou E, Schroeder S, Cabrera S, Bénit P, Rustin P, Criollo A, Kepp O, Galluzzi L, Shen S, Malik SA, Maiuri MC, Horio Y, López-Otín C, Andersen JS, Tavernarakis N, Madeo F, Kroemer G (2011) Spermidine and resveratrol induce autophagy by distinct pathways converging on the acetylproteome. J Cell Biol 192:615–629. https://doi.org/10.1083/jcb.201008167

Ni Y-Q, Liu Y-S (2021) New insights into the roles and mechanisms of spermidine in aging and age-related diseases. Aging Dis 12:1948–1963. https://doi.org/10.14336/AD.2021.0603

Pandey KB, Rizvi SI (2014) Role of resveratrol in regulation of membrane transporters and integrity of human erythrocytes. Biochem Biophys Res Commun 453:521–526. https://doi.org/10.1016/j.bbrc.2014.09.117

Pessoa J, Duarte AI (2023) Overcoming mitochondrial dysfunction in neurodegenerative diseases. Neural Regen Res 18:1486–1488. https://doi.org/10.4103/1673-5374.360279

Pocobelli G, Kristal AR, Patterson RE, Potter JD, Lampe JW, Kolar A, Evans I, White E (2010) Total mortality risk in relation to use of less-common dietary supplements. Am J Clin Nutr 91:1791–1800. https://doi.org/10.3945/ajcn.2009.28639

Qato DM, Alexander GC, Conti RM, Johnson M, Schumm P, Lindau ST (2008) Use of prescription and over-the-counter medications and dietary supplements among older adults in the United States. JAMA 300:2867–2878. https://doi.org/10.1001/jama.2008.892

Rachakatla A, Kalashikam RR (2022) Calorie restriction-regulated molecular pathways and its impact on various age groups: an overview. DNA Cell Biol 41:459–468. https://doi.org/10.1089/dna.2021.0922

Rahman MH, Akter R, Bhattacharya T, Abdel-Daim MM, Alkahtani S, Arafah MW, Al-Johani NS, Alhoshani NM, Alkeraishan N, Alhenaky A, Abd-Elkader OH, El-Seedi HR, Kaushik D, Mittal V (2020) Resveratrol and neuroprotection: impact and its therapeutic potential in Alzheimer's disease. Front Pharmacol 11:619024. https://doi.org/10.3389/fphar.2020.619024

Rehman IU, Khan A, Ahmad R, Choe K, Park HY, Lee HJ, Atiq A, Park J, Hahm JR, Kim MO (2022) Neuroprotective effects of nicotinamide against MPTP-induced Parkinson's disease in mice: impact on oxidative stress, Neuroinflammation, Nrf2/HO-1 and TLR4 Signaling pathways. Biomedicine 10:2929. https://doi.org/10.3390/biomedicines10112929

Reiten OK, Wilvang MA, Mitchell SJ, Hu Z, Fang EF (2021) Preclinical and clinical evidence of NAD+ precursors in health, disease, and ageing. Mech Ageing Dev 199:111567. https://doi.org/10.1016/j.mad.2021.111567

Satia JA, Littman A, Slatore CG, Galanko JA, White E (2009) Associations of herbal and specialty supplements with lung and colorectal cancer risk in the VITamins and lifestyle study. Cancer Epidemiol Biomarkers Prev 18:1419–1428. https://doi.org/10.1158/1055-9965.EPI-09-0038

Schöndorf DC, Ivanyuk D, Baden P, Sanchez-Martinez A, De Cicco S, Yu C, Giunta I, Schwarz LK, Di Napoli G, Panagiotakopoulou V, Nestel S, Keatinge M, Pruszak J, Bandmann O, Heimrich B, Gasser T, Whitworth AJ, Deleidi M (2018) The NAD+ precursor nicotinamide riboside rescues mitochondrial defects and neuronal loss in iPSC and Fly models of Parkinson's disease. Cell Rep 23:2976–2988. https://doi.org/10.1016/j.celrep.2018.05.009

Sharma A, Singh AK (2023) Molecular mechanism of caloric restriction mimetics-mediated neuroprotection of age-related neurodegenerative diseases: an emerging therapeutic approach. Biogerontology 24:679–708. https://doi.org/10.1007/s10522-023-10045-y

Shi D-D, Dong CM, Ho LC, Lam CTW, Zhou X-D, Wu EX, Zhou Z-J, Wang X-M, Zhang Z-J (2018) Resveratrol, a natural polyphenol, prevents chemotherapy-induced cognitive impairment: involvement of cytokine modulation and neuroprotection. Neurobiol Dis 114:164–173. https://doi.org/10.1016/j.nbd.2018.03.006

Shintani T, Sakoguchi H, Yoshihara A, Izumori K, Sato M (2017) D -Allulose, a stereoisomer of d -fructose, extends *Caenorhabditis elegans* lifespan through a dietary restriction mechanism: a new candidate dietary restriction mimetic. Biochem Biophys Res Commun 493:1528–1533. https://doi.org/10.1016/j.bbrc.2017.09.147

Shintani H, Shintani T, Ashida H, Sato M (2018a) Calorie restriction Mimetics: upstream-type compounds for modulating glucose metabolism. Nutrients 10:1821. https://doi.org/10.3390/nu10121821

Shintani T, Kosuge Y, Ashida H (2018b) Glucosamine extends the lifespan of *Caenorhabditis elegans* via autophagy induction. J Appl Glycosci 65:37–43. https://doi.org/10.5458/jag.jag.JAG-2018_002

Sidhu A, Diwan V, Kaur H, Bhateja D, Singh CK, Sharma S, Padi SSV (2018) Nicotinamide reverses behavioral impairments and provides neuroprotection in 3-nitropropionic acid induced animal model ofHuntington's disease: implication of oxidative stress- poly(ADP- ribose) polymerase pathway. Metab Brain Dis 33:1911–1921. https://doi.org/10.1007/s11011-018-0297-0

Singh AK, Kashyap MP, Tripathi VK, Singh S, Garg G, Rizvi SI (2016a) Neuroprotection through rapamycin-induced activation of autophagy and PI3K/Akt1/mTOR/CREB Signaling against amyloid-β-induced oxidative stress, synaptic/neurotransmission dysfunction, and neurodegeneration in adult rats. Mol Neurobiol. https://doi.org/10.1007/s12035-016-0129-3

Singh AK, Singh S, Garg G, Rizvi SI (2016b) Rapamycin alleviates oxidative stress-induced damage in rat erythrocytes. Biochem Cell Biol 94:471–479. https://doi.org/10.1139/bcb-2016-0048

Singh AK, Garg G, Singh S, Rizvi SI (2017) Synergistic effect of rapamycin and metformin against age-dependent oxidative stress in rat erythrocytes. Rejuvenation Res 20:420–429. https://doi.org/10.1089/rej.2017.1916

Singh S, Kumar R, Garg G, Singh AK, Verma AK, Bissoyi A, Rizvi SI (2021) Spermidine, a caloric restriction mimetic, provides neuroprotection against normal and D-galactose-induced

oxidative stress and apoptosis through activation of autophagy in male rats during aging. Biogerontology 22:35–47. https://doi.org/10.1007/s10522-020-09900-z
Soda K, Kano Y, Sakuragi M, Takao K, Lefor A, Konishi F (2009) Long-term oral polyamine intake increases blood polyamine concentrations. J Nutr Sci Vitaminol (Tokyo) 55:361–366. https://doi.org/10.3177/jnsv.55.361
Subramani T, Ganapathyswamy H (2020) An overview of liposomal nanoencapsulation techniques and its applications in food and nutraceutical. J Food Sci Technol 57:3545–3555. https://doi.org/10.1007/s13197-020-04360-2
Taormina G, Mirisola MG (2014) Calorie restriction in mammals and simple model organisms. Biomed Res Int 2014:308690. https://doi.org/10.1155/2014/308690
Trammell SAJ, Schmidt MS, Weidemann BJ, Redpath P, Jaksch F, Dellinger RW, Li Z, Abel ED, Migaud ME, Brenner C (2016) Nicotinamide riboside is uniquely and orally bioavailable in mice and humans. Nat Commun 7:12948. https://doi.org/10.1038/ncomms12948
Trisal A, Singh AK (2024) Clinical insights on caloric restriction Mimetics for mitigating brain aging and related neurodegeneration. Cell Mol Neurobiol 44:67. https://doi.org/10.1007/s10571-024-01493-2
Tsai Y-F, Chen C-Y, Chang W-Y, Syu Y-T, Hwang T-L (2019) Resveratrol suppresses neutrophil activation via inhibition of Src family kinases to attenuate lung injury. Free Radic Biol Med 145:67–77. https://doi.org/10.1016/j.freeradbiomed.2019.09.021
Turner RS, Thomas RG, Craft S, van Dyck CH, Mintzer J, Reynolds BA, Brewer JB, Rissman RA, Raman R, Aisen PS (2015) A randomized, double-blind, placebo-controlled trial of resveratrol for Alzheimer disease. Neurology 85:1383–1391. https://doi.org/10.1212/WNL.0000000000002035
Udono H, Nishida M (2022) Metformin-ROS-Nrf2 connection in the host defense mechanism against oxidative stress, apoptosis, cancers, and ageing. Biochim Biophys Acta Gen Subj 1866:130171. https://doi.org/10.1016/j.bbagen.2022.130171
Vizuete AFK, Fróes F, Seady M, Caurio AC, Ramires Junior OV, Leite AKO, Farias CP, Wyse ATS, Gonçalves C-A (2024) Targeting glycolysis for neuroprotection in early LPS-induced neuroinflammation. Brain Behav Immun Health 42:100901. https://doi.org/10.1016/j.bbih.2024.100901
Wang N, He J, Pan C, Wang J, Ma M, Shi X, Xu Z (2019) Resveratrol activates autophagy via the AKT/mTOR Signaling pathway to improve cognitive dysfunction in rats with chronic cerebral Hypoperfusion. Front Neurosci 13:859. https://doi.org/10.3389/fnins.2019.00859
Weimer S, Priebs J, Kuhlow D, Groth M, Priebe S, Mansfeld J, Merry TL, Dubuis S, Laube B, Pfeiffer AF, Schulz TJ, Guthke R, Platzer M, Zamboni N, Zarse K, Ristow M (2014) D-glucosamine supplementation extends life span of nematodes and of ageing mice. Nat Commun 5:3563. https://doi.org/10.1038/ncomms4563
Wirth M, Benson G, Schwarz C, Köbe T, Grittner U, Schmitz D, Sigrist SJ, Bohlken J, Stekovic S, Madeo F, Flöel A (2018) The effect of spermidine on memory performance in older adults at risk for dementia: a randomized controlled trial. Cortex 109:181–188. https://doi.org/10.1016/j.cortex.2018.09.014
Yadav J, El Hassani M, Sodhi J, Lauschke VM, Hartman JH, Russell LE (2021) Recent developments in *in vitro* and *in vivo* models for improved translation of preclinical pharmacokinetics and pharmacodynamics data. Drug Metab Rev 53:207–233. https://doi.org/10.1080/03602532.2021.1922435
Yan S, Na J, Liu X, Wu P (2024) Different targeting ligands-mediated drug delivery Systems for Tumor Therapy. Pharmaceutics 16:248. https://doi.org/10.3390/pharmaceutics16020248
Yin F, Sancheti H, Patil I, Cadenas E (2016) Energy metabolism and inflammation in brain aging and Alzheimer's disease. Free Radic Biol Med. https://doi.org/10.1016/j.freeradbiomed.2016.04.200
Yueh H-Z, Yeh C-J, Wei JC-C (2022) Correspondence to "associations of regular glucosamine use with all-cause and cause-specific mortality: a large prospective cohort study" by Li et al. Ann Rheum Dis 81:e182. https://doi.org/10.1136/annrheumdis-2020-218486

Zhang L, Yu X, Ji M, Liu S, Wu X, Wang Y, Liu R (2018) Resveratrol alleviates motor and cognitive deficits and neuropathology in the A53T α-synuclein mouse model of Parkinson's disease. Food Funct 9:6414–6426. https://doi.org/10.1039/C8FO00964C
Zhang X, Alshakhshir N, Zhao L (2021) Glycolytic metabolism, brain resilience, and Alzheimer's disease. Front Neurosci 15:662242. https://doi.org/10.3389/fnins.2021.662242
Zhao D, Simon JE, Wu Q (2020) A critical review on grape polyphenols for neuroprotection: strategies to enhance bioefficacy. Crit Rev Food Sci Nutr 60:597–625. https://doi.org/10.1080/10408398.2018.1546668
Zhou L, Pinho R, Gu Y, Radak Z (2022) The role of SIRT3 in exercise and aging. Cells 11:2596. https://doi.org/10.3390/cells11162596
Zullo A, Simone E, Grimaldi M, Musto V, Mancini FP (2018) Sirtuins as mediator of the anti-ageing effects of calorie restriction in skeletal and cardiac muscle. Int J Mol Sci 19:928. https://doi.org/10.3390/ijms19040928

Mitochondrial Dynamics and Brain Aging

5

Falguni Goel, Aditi Giri, Vishi Khattri, Payal Singh, and Syed Mohammad Nasar Ata

Abstract

Mitochondria are critical in ensuring cellular energy homeostasis and controlling vital processes like apoptosis, calcium signaling, and management of oxidative stress. Mitochondrial dynamics through ongoing fission, fusion, transport, and quality control are crucial for the maintenance of mitochondrial function and adjusting to cellular needs. Mitochondrial dynamics dysregulation has come to be a key characteristic of brain aging and neurodegenerative diseases. Aging is correlated with disturbed mitochondrial fission and fusion equilibrium, resulting in abnormal mitophagy, the presence of dysfunctional mitochondria, and augmented oxidative stress, further promoting neuronal impairment. The brain is most susceptible to mitochondrial dysfunction owing to considerable energy requirements and susceptibility to reactive oxygen species (ROS). Mitochondrial fragmentation, disrupted biogenesis, and abnormal axonal transport add to synaptic dysfunction and neuronal survival. New evidence highlights the importance of mitochondrial quality control mechanisms, such as mitophagy and biogenesis, in maintaining resistance to age-related damage and neuronal integrity. Genetic and epigenetic alterations also modulate the crucial proteins governing mitochondrial dynamics and steer the neurodegeneration course. Deciphering the association of brain aging and mitochondrial dynamics reveals insight into treat-

F. Goel · A. Giri · V. Khattri
Department of Pharmaceutical Technology, Meerut Institute of Engineering & Technology (MIET), Meerut, Uttar Pradesh, India

P. Singh
Department of Zoology, Banaras Hindu University, Varanasi, Uttar Pradesh, India

S. M. N. Ata (✉)
Department of Neurology, Henry Ford Health System, Detroit, MI, USA
e-mail: sata1@hfhs.org

A. Kumar Singh, S. Nand Rai (eds.), *Rejuvenating the Brain: Nutraceuticals, Autophagy, and Longevity*, Nutritional Neurosciences,
https://doi.org/10.1007/978-981-95-2790-8_5

ment modalities to attain mitochondrial homeostasis and avoid neuronal demise. The following chapter deals with the basic mechanisms governing mitochondrial dynamics, their importance in brain aging, and treatment modalities for targeting mitochondrial malfunction to avert cognitive impairment during aging.

Keywords

Mitochondrial dysfunction · Mitochondrial DNA mutations · Mitophagy · Neurodegeneration · Age-related diseases · Mitochondrial-targeted therapy

5.1 Introduction

Mitochondria, also commonly known as the "powerhouses of the cell," are essential in cellular homeostasis, preservation of energy, apoptosis regulation, and reactive oxygen species (ROS) management (Prajapati et al. 2025). In the brain, an organ of immense energy dependence to keep neurons functioning and synaptic transmission in place, a dynamic balance of mitochondrial function is important to maintain cognition and motor function throughout life. Mitochondrial dynamics like fission, fusion, mitophagy, and biogenesis control healthy numbers of mitochondria by removing defective organelles and forming new ones. But in brain aging, this complex homeostasis is increasingly lost, and defective mitochondria accumulate, oxidative stress follows, and neuronal injury occurs (Goel et al. 2025).

Mitochondrial fusion, mediated by MFN1 and MFN2 and OPA1, promotes intermixing of mitochondrial components, spreading damaged material, and preserving bioenergetic capacity. On the other hand, mitochondrial fission, controlled by DRP1 and FIS1, promotes detachment of damaged mitochondria for degradation by mitophagy. In older age, mitochondrial over-fragmentation through disruption of the fission-fusion balance in favor of defective bioenergetics, enhanced release of ROS, and apoptosis pathway occurs. Additionally, PINK1/Parkin-mediated dysregulated mitophagy permits damaged mitochondria to aggregate, enhancing cell stress and inducing neurodegeneration (Yadav et al. 2024).

Mitochondrial biogenesis, which is controlled by PGC-1α, controlling the transcription of genes for mitochondrial DNA replication and oxidative phosphorylation, is also influenced by aging. Dysfunction of PGC-1α due to aging leads to decreased turnover in the mitochondria and a decrease in the number of mitochondria as a whole, further impairing neuronal function. As a result, senescent neurons are also subjected to more and more metabolic stress and inflammation, which are features of age-related neurodegenerative conditions like AD, PD, and HD (Tripathi et al. 2024).

The complex interplay between mitochondrial dynamics and brain aging points to the significance of upholding mitochondrial health for retaining cognitive and motor function in older age. Interventional strategies designed to reestablish mitochondrial balance (Xie et al. 2024), increase mitophagy, and stimulate mitochondrial biogenesis could mitigate the impact of aging on the brain (Song et al. 2021).

Deciphering molecular mechanisms that regulate mitochondrial dynamics throughout aging is key to devising targeted therapeutics that will prevent or postpone neurodegenerative conditions. As scientists continue to untangle the intricacies of mitochondrial biology, new therapeutic horizons open up to enhance brain function and quality of life in patients with advancing age (Monaghan 2025).

5.2 Fundamentals of Mitochondrial Dynamics

Mitochondrial dynamics are the tightly regulated processes of mitochondrial fusion, fission, mitophagy, and biogenesis that are critical for the integrity, function, and adaptability of the mitochondria. Mitochondrial dynamics processes guarantee the quality control of mitochondria balanced against energy demand which is of critical importance for cellular homeostasis, particularly for cells that require high energy demands like neurons. Mitochondrial dynamic disturbances have been associated with neurodegenerative diseases and aging, which implies the necessity to explain the mechanisms (Chauhan et al. 2024).

Mitochondrial fusion is the mechanism of fusing adjacent mitochondria to form large networks, allowing the exchange of mtDNA, proteins, and lipids, thereby enhancing mitochondrial function and bioenergetics. MFN1 and MFN2 on the outer mitochondrial membrane and OPA1 on the inner mitochondrial membrane control this process primarily. Fusion allows healthy mitochondria to compensate for defective organelles, preserving mitochondrial integrity and preventing oxidative stress-induced damage (Bhushan et al. 2025).

Mitochondrial fission, however, is the opposite process, where mitochondria split to create smaller organelles. DRP1 is responsible for mediating this through movement from the cytosol to the outer mitochondrial membrane and interaction with FIS1, causing mitochondrial fragmentation (Ghani 2025). Fission is necessary to divide damaged mitochondria so that they can be removed through mitophagy and to ensure an optimal population of healthy mitochondria. It is also responsible for delivering mitochondria to daughter cells during cell division and the transport of mitochondria to areas where energy is needed in bulk, i.e., synapses of neurons (Ghani 2025).

Mitophagy is a type of selective autophagy that clears out dysfunctional or damaged mitochondria to avoid the accumulation of harmful byproducts and ensure cellular health. The PINK1/Parkin pathway is a key process for identifying and marking dysfunctional mitochondria to be degraded. Defective mitophagy results in the buildup of faulty mitochondria, leading to oxidative stress, inflammation, and cellular dysfunction (Antunes et al. 2021).

Mitochondrial biogenesis refers to the process of producing new mitochondria to replace dysfunctional organelles and respond to heightened cellular energy requirements. PGC-1α regulates this process, activating NRF1 and NRF2 and TFAM for the induction of mtDNA replication and expression of mitochondrial genes (Casanova et al. 2023). The fine balance between these processes ensures mitochondrial homeostasis, cellular adaptation to metabolic stress, and demand. Mitochondrial dynamics dysregulation causes cellular dysfunction, oxidative damage, and apoptosis, which contribute to the development of age-related diseases and neurodegenerative disorders (Rani et al. 2022).

5.3 Mitochondrial Dysfunction in Aging

Mitochondrial dysfunction is a characteristic feature of aging, playing a crucial role in impairing cellular energy production, promoting oxidative stress, and triggering pathways of cell death. Mitochondria, which are the central sites of generation of ATP via OXPHOS, are vital for ensuring neuronal health. Nevertheless, upon aging, mitochondria become increasingly less efficient and, as a result, create bioenergetic deficits, form ROS, and become more vulnerable to cellular damage (Casanova et al. 2023).

One of the main causes of mitochondrial dysfunction with aging is a buildup of mtDNA mutations. Unlike nuclear DNA, mtDNA does not have efficient repair capabilities and is in constant contact with high concentrations of ROS generated as byproducts during mitochondrial respiration (Stevic et al. 2022). These gradually cause defects in ETC, reducing ATP yields and promoting greater leakage of electrons that enhance the production of ROS. This forms a vicious circle where oxidative damage weakens mitochondrial integrity further, causing cellular senescence and apoptosis (Fig. 5.1).

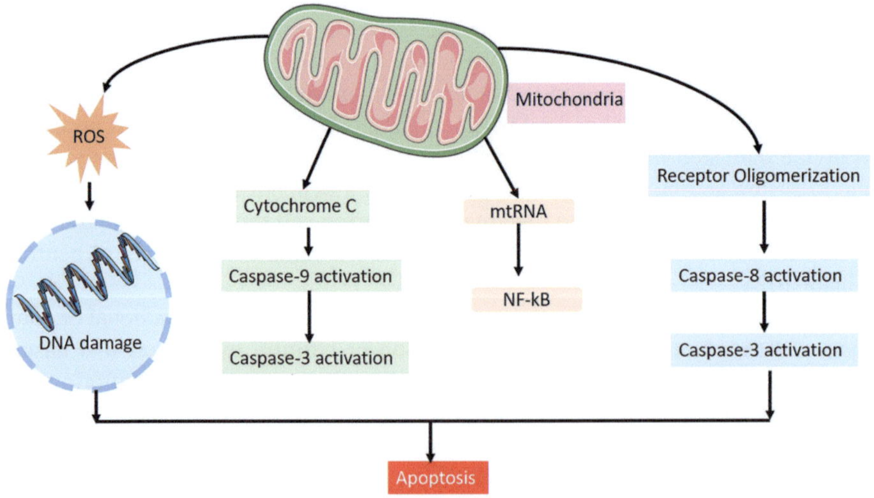

Fig. 5.1 Schematic illustration of mitochondrial dysfunction during aging. Mitochondrial dynamic changes with aging, such as defective fusion, hyperfission, and defective mitophagy, result in bioenergetic failure, elevated production of reactive oxygen species (ROS), and the formation of dysfunctional mitochondria. Such dysfunctions are responsible for oxidative stress, neuroinflammation, and apoptosis of cells, which are primary drivers of cognitive aging and neurodegenerative diseases. Identification of these pathways may help develop targeted therapeutic approaches to ensure healthy aging and neuroprotection

A second important feature of mitochondrial dysfunction during aging is mitochondrial dynamic imbalance, namely the disruption of fission, fusion, and mitophagy. During aging, there is a loss of balance between mitochondrial fusion, which facilitates the transfer of mitochondrial material to preserve function, and fission, which isolates dysfunctional mitochondria for destruction (Green et al. 2022). Enhanced mitochondrial fragmentation as a consequence of overactive fission and defective fusion leads to the buildup of dysfunctional mitochondria that cannot fulfill the bioenergetic needs of senescent cells. In addition, diminished mitophagy in aging does not allow for effective removal of dysfunctional mitochondria, leading to cell stress and neurodegeneration (Picca et al. 2023).

Mitochondrial biogenesis, or the process of refilling the mitochondrial pool through new organelle generation, is also affected by aging. Function of PGC-1α, the master coordinator of mitochondrial biogenesis, diminishes with age, and it reduces mitochondrial turnover along with adaptability toward metabolic stress (Krantz 2021). The aged cells hence show decreased amounts of viable mitochondria, contributing to increased sensitivity to oxidative damage and inflammation. The combined effect of mitochondrial dysfunction in aging is especially harmful to neurons, which consume large amounts of energy and are very sensitive to mitochondrial injury (Mayorga-Weber et al. 2022). Dysfunctional mitochondria cause synaptic dysfunction, neuroinflammation, and consequent neuronal loss, leading to the progression of age-related neurodegenerative diseases like AD, PD, and HD. Targeting mitochondrial impairments by interventions that promote mitochondrial biogenesis, reconstitute mitochondrial dynamics, and alleviate oxidative stress is promising in minimizing the impact of aging and maintaining neuronal well-being.

5.4 Mitochondria and Neurodegenerative Diseases

Mitochondria have a central function in supporting the health of neurons through control of energy metabolism, calcium homeostasis, and apoptosis. Neurons, because of their high energy requirements and restricted ability to regenerate, are especially susceptible to malfunctioning mitochondria (Slater et al. 2022). Dysfunctional mitochondria are a central pathological feature in various neurodegenerative disorders, such as AD, PD, HD, and ALS (Table 5.1). Dysregulation of mitochondrial dynamics, enhanced oxidative stress, mitophagy impairment, and bioenergetic failure are responsible for the pathogenesis of these disorders to result in neuronal dysfunction and death (Singh et al. 2024).

In AD, mitochondrial impairment is an early phenomenon preceding the onset of clinical signs and is in close association with Aβ and hyperphosphorylated tau protein deposition. Aβ fibrils impair mitochondrial function by disrupting the ETC, increasing ROS production, and inducing mitochondrial fragmentation (Dey et al. 2024). Tau pathology also helps to cause additional mitochondrial dysfunction through disruption of axonal transport and disruption of mitochondrial dynamics, culminating in synaptic dysfunction and loss of neurons (Ramakrishna et al. 2023). PD is a neurodegenerative condition that is characterized by the selective

Table 5.1 Mitochondria and their role in neurodegenerative diseases

Neurodegenerative disease	Mitochondrial dysfunction	Pathological consequences	Key molecular pathways affected	References
Alzheimer's disease (AD)	Impaired oxidative phosphorylation, increased ROS production, reduced ATP levels, and mitochondrial DNA mutations	Synaptic dysfunction, neuronal loss, and amyloid-beta (Aβ) accumulation	Dysregulation of PGC-1α, mitophagy, and impaired mitochondrial biogenesis	Bandaru et al. (2022), Dhapola et al. (2022)
Parkinson's disease (PD)	Defective mitophagy, impaired complex I activity, and accumulation of dysfunctional mitochondria	Dopaminergic neuron degeneration, oxidative stress, and α-synuclein aggregation	PINK1/parkin pathway dysregulation, decreased mitochondrial clearance, and altered fission/fusion balance	Xiao et al. (2022), Wang et al. (2023a)
Huntington's disease (HD)	Mitochondrial fragmentation, impaired calcium buffering, and defective mitochondrial transport	Neuronal death, striatal degeneration, and altered cellular energy metabolism	DRP1-mediated fission, impaired mitochondrial trafficking, and altered bioenergetics	Mohamed Yusoff and Mohd Khair (2025), Bhatti et al. (2022)
Amyotrophic lateral sclerosis (ALS)	Increased mitochondrial ROS, altered calcium homeostasis, and disruption of mitochondrial dynamics	Motor neuron degeneration, protein aggregation, and synaptic failure	SOD1 and TDP-43 mutations affecting mitochondrial integrity and quality control	Jhanji et al. (2021), Zhong et al. (2024)
Multiple sclerosis (MS)	Mitochondrial respiratory chain dysfunction, oxidative stress, and inflammation	Axonal degeneration, demyelination, and neuronal death	Impaired mitochondrial biogenesis, increased ROS, and mitochondrial DNA damage	Shadab et al. (2024), Wang et al. (2024)

(continued)

Table 5.1 (continued)

Neurodegenerative disease	Mitochondrial dysfunction	Pathological consequences	Key molecular pathways affected	References
Friedreich's ataxia (FRDA)	Defective iron-sulfur cluster biogenesis, mitochondrial iron accumulation, and oxidative stress	Neurodegeneration, ataxia, and cardiomyopathy	Decreased frataxin expression, impaired mitochondrial function, and oxidative damage	Ercanbrack et al. (2025)
Leigh syndrome	Impaired oxidative phosphorylation and defective mitochondrial enzyme activity	Neurodevelopmental delays, motor dysfunction, and lactic acidosis	Mutations in mitochondrial DNA and nuclear-encoded genes affecting energy metabolism	Magro et al. (2025)

dopaminergic loss of neurons from the substantia nigra, and the malfunction of mitochondria is at its core. Genomic mutations of genes involved in mitochondrial quality control and mitophagy, such as PINK1, Parkin, DJ-1, and LRRK2, result in an accumulation of impaired mitochondria in PD. Impaired mitophagy results in an accumulation of impaired mitochondria, leading to increased oxidative stress, impaired production of ATP, and neuronal cell loss. A deficiency of complex I in the ETC is also observed in PD, which contributes to bioenergetic shortfall and oxidative insult (Tripathi et al. 2019).

HD is due to a mutation of the HTT gene, which results in the creation of mutant huntingtin protein that damages mitochondrial function. Mitochondrial damage in HD involves disrupted calcium homeostasis, decreased ATP production, and enhanced ROS production (Sharma et al. 2021). Mutant huntingtin also interferes with microtubule-mitochondrial interaction, resulting in mitochondrial transport defects and synaptic dysfunction. Abnormal mitodynamics and defective mitophagy are also involved in the neuronal degeneration of HD. In ALS, mitochondrial injury results in cumulative damage to motor neurons. Genetic mutations in SOD1, TDP-43, and FUS genes lead to ALS and are characterized by enhanced oxidative stress, mitofragmentation, and impaired calcium buffering. Mitochondria of ALS are structurally abnormal, have disrupted transport, and decreased ATP generation, which causes synaptic dysfunction and neuronal degeneration (Srivastava et al. 2019).

5.5 Mitochondrial Dynamics and Cellular Stress Responses

Mitochondrial dynamics are essential for cellular homeostasis, especially in the context of diverse forms of cellular stress. Mitochondrial fusion, fission, mitophagy, and biogenesis interact to control mitochondrial quality, repairing or eliminating damaged organelles and meeting cellular energy needs. In physiological conditions, these processes achieve a homeostatic balance between energy generation, oxidative stress, and apoptosis control. But under conditions of cellular stress due to oxidative damage, nutrient starvation, or inflammation, mitochondrial dynamics are regulated to safeguard the cell and restore homeostasis (Rai et al. 2020).

Mitochondrial fusion is a protective process in stress where mitochondria fuse, enabling the mixing of contents like mtDNA, proteins, and lipids. The process, mediated by MFN1 and MFN2 and OPA1, serves to dilute defective mitochondrial contents and ensure bioenergetic efficacy (Xu et al. 2025). The process of promoting elongation of mitochondria inhibits the storage of dysfunctional and fragmented mitochondria, promoting cellular stress resilience. (Ghani 2025). Fission creates smaller mitochondria that can be targeted selectively for degradation by mitophagy to ensure the removal of damaged organelles and curb the spread of oxidative stress. Nonetheless, over- or extended fission may cause mitochondrial fragmentation, dysfunctional bioenergetics, and heightened sensitivity to apoptosis, ultimately leading to cell death (Fig. 5.2).

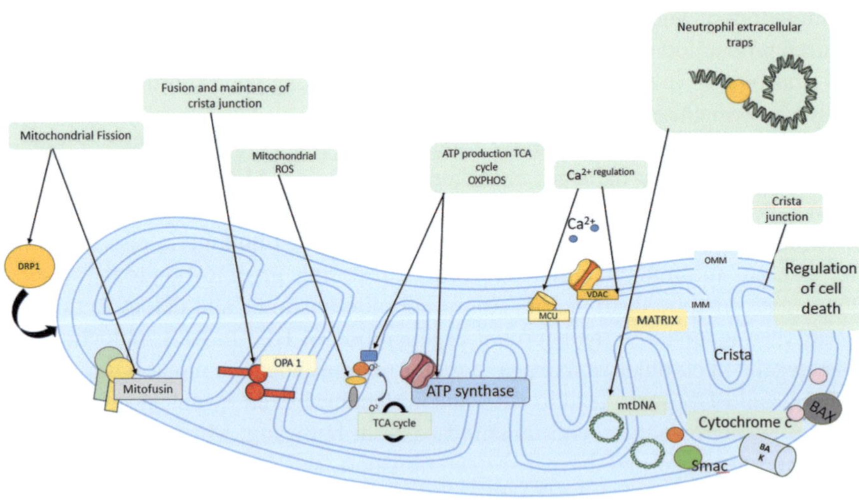

Fig. 5.2 Schematic illustration of mitochondrial dynamics and cellular stress responses. Mitochondria respond dynamically to cellular stress conditions through fission and fusion processes to ensure mitochondrial quality and function. Mitochondrial fission, which is Drp1-mediated, allows damaged mitochondria to be eliminated by mitophagy, whereas mitochondrial fusion, OPA1- and MFN1/2-regulated, restores mitochondrial integrity and ensures energy homeostasis. Chronic stress causes mitochondrial damage, heightened ROS generation, and cellular stress pathway activation, which ultimately result in apoptosis and neurodegeneration

Mitophagy is the selective type of autophagy responsible for the elimination of dysfunctional mitochondria during cell stress. One of the prime regulators of mitophagy is the PINK1/Parkin pathway, identifying dysfunctional mitochondria and marking them for destruction. In stressed environments, dysfunctional mitochondria accumulate on their outer membranes, PINK1, which causes recruitment of the E3 ubiquitin ligase Parkin. Parkin adds ubiquitin chains to mitochondrial proteins, marking them for autophagic degradation within the lysosome. Effective mitophagy avoids the accumulation of defective mitochondria and restricts oxidative damage, thus saving the cell from apoptosis (Masaldan et al. 2022).

Mitochondrial biogenesis is induced upon cellular stress to restore mitochondrial numbers and bioenergetic potential. PGC-1α plays an important role in the regulation of mitochondrial biogenesis and stimulates the expression of genes responsible for mitochondrial DNA synthesis and oxidative phosphorylation. Enhanced mitochondrial biogenesis maintains mitochondrial homeostasis and increases cellular plasticity to stress (Lou et al. 2024). When mitochondrial dynamics become dysregulated following chronic stress, the consequent accumulation of defective mitochondria plays a role in enhanced oxidative stress, inflammation, and apoptosis. This disruption is involved in a number of pathological conditions, such as neurodegenerative disorders, metabolic syndromes, and aging. Thus, cellular adaptation to stress and cell survival depend on preserving the fine balance of mitochondrial dynamics.

5.6 Mitochondria, Inflammation, and Oxidative Stress

Mitochondria are at the center of modulating inflammation and oxidative stress, two interrelated processes that significantly contribute to cell damage and the pathogenesis of many diseases, including neurodegenerative disorders. As the major source of ATP via OXPHOS, mitochondria also generate ROS as a natural byproduct of respiration (Yadav et al. 2022). Though ROS at low levels are signaling molecules that play a role in cellular homeostasis, an excessive generation of ROS interferes with this balance to cause oxidative stress, mitochondrial dysfunction, and inflammation pathway activation.

Oxidative stress arises when the production of ROS is not balanced with the cellular antioxidant defense system. In physiological conditions, ROS produced in the mitochondrial ETC, mainly at complexes I and III, are quenched by endogenous antioxidant enzymes like SOD, CAT, and GPx. Yet, under states of mitochondrial dysfunction, the levels of ROS are elevated, overwhelming the antioxidant defenses and resulting in oxidative damage to mtDNA, proteins, and lipids (Jaffri 2023). The mtDNA is especially susceptible to oxidative damage because of its location close to the ETC and lack of protective histones, resulting in mutations and disrupted mitochondrial function.

Inflammation has a close association with mitochondrial failure and oxidative stress. Dysfunctional mitochondria also release DAMPs, including mitochondrial DNA, cytochrome c, and cardiolipin, that stimulate the innate immune response by activating PRRs like TLRs and NLRs (Chen et al. 2023a). The activation promotes the assembly of the inflammasome, which is a multiprotein complex that also activates caspase-1 to mature and secrete pro-inflammatory cytokines IL-1β and IL-18. The interaction among mitochondria, inflammation, and oxidative stress forms a cycle of causation, where oxidative stress causes mitochondrial damage, resulting in the liberation of DAMPs and stimulation of inflammatory pathways. This condition of chronic inflammation also increases mitochondrial injury, leading to tissue damage and the development of age-related and neurodegenerative diseases like AD, PD, and ALS (Gąssowska-Dobrowolska et al. 2024).

Recovery of mitochondrial function, improvement of antioxidant defenses, and modulation of inflammatory pathways are of great therapeutic interest for the disruption of this cycle and prevention of the detrimental effects of mitochondrial impairment. Interventions to diminish oxidative stress, enhance mitophagy, and regulate inflammasome activation are promising avenues to reduce inflammation and prevent neurodegeneration (Sinha et al. 2024).

5.7 Mitochondrial Dynamics and Synaptic Function

Mitochondrial dynamics, including fusion, fission, mitophagy, and biogenesis, are important for synaptic function and neuronal homeostasis. As a result of high energy requirements and intricate morphology, neurons heavily depend on healthy mitochondria to provide ATP, calcium homeostasis regulation, and synaptic plasticity (Gupta et al. 2025). Mitochondria are moving along axons and dendrites actively to provide synapses with their metabolic demands, where they mediate neurotransmitter release, synaptic transmission, and plasticity. Mitochondrial dynamics disruption weakens synaptic function and causes synaptic degeneration, leading to the development of neurodegenerative diseases.

Mitochondrial fusion plays a critical role in ensuring the health of mitochondria by facilitating the transfer of mitochondrial contents, such as mtDNA, proteins, and lipids. It is mediated by MFN1 and MFN2 on the OPA1 on the inner membrane and helps to improve mitochondrial function and defend against oxidative stress (Cetin and Akbay 2024). Within synapses, mitochondrial fusion maintains broken mitochondria so that they are repaired and the bioenergetic requirements are provided, maintaining the continued synaptic plasticity and activity. Mutations in mitochondrial fusion cause fragmented and defective mitochondria that are not able to address the metabolic requirement of synaptic transmission, which ends up resulting in impaired release of neurotransmitters and synaptic malfunction (Baytas et al. 2024).

Mitochondrial fission, which is controlled by DRP1 and FIS1, is just as important for synaptic function. Fission allows for the separation of dysfunctional mitochondria and the development of smaller organelles that are more transportable to synaptic terminals. During increased synaptic activity, mitochondrial fission is involved in supplying sufficient mitochondria to synapses to meet the greater energy requirements. However, pathological or excessive fission causes mitochondrial fragmentation, bioenergetic dysfunction, and impaired calcium buffering, resulting in synaptic dysfunction and neuronal death (Verma et al. 2023).

Mitophagy, the autophagic degradation of abnormal mitochondria, is vital for mitochondrial quality and synaptic function. The PINK1/Parkin pathway is essential to mark and destroy abnormal mitochondria (Wang et al. 2023b). In synaptic regions, efficient mitophagy prevents the accumulation of abnormal mitochondria that would otherwise interfere with neurotransmission and synaptic plasticity. Abnormal mitophagy results in the accumulation of abnormal mitochondria, resulting in synaptic dysfunction and neurodegeneration. Mitochondrial biogenesis, controlled by PGC-1α, replenishes the mitochondrial pool to preserve synaptic function. Increased mitochondrial biogenesis ensures that synaptic terminals are provided with adequate numbers of healthy mitochondria, especially when synaptic activity is heightened. Decreased mitochondrial biogenesis, as often observed in aging and neurodegenerative disorders, causes synaptic failure and neuronal degeneration (Trigo et al. 2022).

The delicate balance between mitochondrial dynamics is crucial for synaptic transmission, synaptic plasticity, and cognitive learning and memory functions. Disruption of normal mitochondrial fusion, fission, mitophagy, or biogenesis results in synaptic failure and cognitive deficiency as well as neurodegenerative disorders like AD, PD, and HD. Modulation of mitochondrial dynamics to augment synaptic function and ensure neuronal health is a novel therapeutic strategy to reduce neurodegeneration and maintain cognitive functions (Li et al. 2022).

5.8 Genetic and Epigenetic Regulation of Mitochondrial Dynamics

Mitochondrial dynamics, including fusion, fission, mitophagy, and biogenesis, are under exact genetic and epigenetic control. These mechanisms are critical to maintaining cellular mitochondrial quality, bioenergetic efficacy, and homeostasis, particularly in highly energetic tissues, for instance, the brain. Dysregulation of these controlling events leads to mitochondrial pathology, oxidative stress, and neuronal injury that are hallmarks of neurodegenerative disease, metabolic disease, and diseases of aging (Ren et al. 2020).

Mitochondrial dynamics are regulated by nuclear-coded genes that oversee key proteins responsible for maintaining mitochondrial form and function. Fusion, in which the merger of mitochondria is used to transfer mtDNA, proteins, and lipids, is regulated by MFN1 and MFN2 on the outer mitochondrial membrane and OPA1 on the inner mitochondrial membrane (Tasakis and Tsolaki 2015). MFN2 mutations are associated with CMT2A, which is an axonal degeneration and peripheral neuropathy condition, and OPA1 defects result in dominant optic atrophy, demonstrating the vital importance of fusion proteins in the health of neurons. In contrast, fission is controlled by DRP1, which is coded by the DNM1L gene, and the proteins that are associated with it like FIS1, MFF, and MiD49/51 (Lambiri and Levin 2024).

DNM1L mutations compromise mitochondrial and peroxisomal fission, leading to critical developmental and neurological defects (Table 5.2).

Mitophagy, the selective degradation of dysfunctional mitochondria by autophagic breakdown, is yet another central aspect of mitochondrial quality control. Orchestration of this process involves the PINK1/Parkin pathway (Gustafsson and Dorn 2019), whereby PINK1 stabilizes on depolarized mitochondria and recruits Parkin, an E3 ubiquitin ligase, to mark the dysfunctional mitochondria for degradation. PINK1 and PARK2 mutations are linked to early-onset Parkinson's disease, highlighting the significance of mitophagy in ensuring mitochondrial and neuronal integrity. Mitochondrial biogenesis, the production of new mitochondria, is also controlled by PGC-1α, which binds to NRF1 and NRF2 and TFAM to enhance the transcription of mitochondrial genes. Dysregulation of PGC-1α disrupts mitochondrial biogenesis and leads to mitochondrial dysfunction in numerous neurodegenerative disorders (Khot et al. 2022).

Epigenetic regulation, such as DNA methylation, histone modifications, and non-coding RNA, also modulates gene expression of genes that contribute to mitochondrial dynamics. DNA methylation, specifically hypermethylation of promoters for genes such as PGC-1α, NRF1, and TFAM, represses gene expression and inhibits mitochondrial biogenesis (Li et al. 2024). Abnormal patterns of DNA methylation have been detected in a number of neurodegenerative diseases and have been implicated in mitochondrial pathology and neuronal injury. Histone modifications, including acetylation, methylation, and phosphorylation, regulate chromatin accessibility and influence mitochondrial gene transcription. SIRT1 and SIRT3, NAD + -dependent deacetylases called sirtuins, are essential regulators of mitochondrial gene expression. SIRT1, by activating PGC-1α, enhances mitochondrial biogenesis, while SIRT3 regulates mitochondrial protein acetylation for metabolic homeostasis (Abdel Khalek et al. 2014).

The non-coding RNAs, notably miRNAs and lncRNAs, add another aspect of regulation of mitochondrial dynamics. miRNAs miR-34a lower the levels of PGC-1α mRNA, which impact mitochondrial biogenesis, while miR-195 and miR-499 regulate DRP1 and MFN2 and suppress them, and therefore modulate fission-fusion equilibrium. Abnormalization of these non-coding RNAs has roles in mitochondrial pathophysiology, oxidative stress, and neurodegeneration (Kumar et al. 2024).

The cross-talk between genetic and epigenetic regulation of mitochondrial dynamics is essential for the integrity of mitochondria and cellular homeostasis. Dysregulation of these control processes underlies the pathogenesis of neurodegenerative diseases such as Alzheimer's, Parkinson's, and Huntington's disease, where dysfunctional mitochondrial quality control leads to synaptic impairment and neuronal death (Chen et al. 2023b). In metabolic diseases such as diabetes, pathological mitochondrial dynamics exacerbate insulin resistance and oxidative stress. Targeting mitochondrial dynamic modulators, both genetic and epigenetic, offers a thrilling new therapeutic approach toward the repair of mitochondrial function, prevention of cell damage, and inhibition of disease progression. Both modulation of significant mitochondrial gene expression or epigenetic alteration reversals are means by which

Table 5.2 Genetic and epigenetic regulation of mitochondrial dynamics

Regulatory mechanism	Key players/ genes	Function/impact on mitochondrial dynamics	Relevance to neurodegeneration	References
Mitochondrial fusion	MFN1, MFN2, OPA1	Promote fusion, maintain mitochondrial integrity	Mutations linked to Charcot-Marie-tooth disease and neurodegeneration	Zhang et al. (2024)
Mitochondrial fission	DRP1, FIS1	Facilitate mitochondrial fission and quality control	Excessive fission linked to Parkinson's and Alzheimer's diseases	Jenkins et al. (2024)
Mitophagy regulation	PINK1, parkin, BNIP3	Mediate clearance of damaged mitochondria	Impaired mitophagy contributes to Parkinson's disease and aging	Mishra and Thakur (2023), Markaki et al. (2021)
Biogenesis and transcription	PGC-1α, NRF1, NRF2, TFAM	Regulate mitochondrial biogenesis and gene expression	Downregulation linked to age-related mitochondrial dysfunction	Shaito et al. (2023), Hyttinen et al. (2021)
mtDNA maintenance and repair	POLG, TWNK, MGME1	Ensure mtDNA replication and repair	Mutations lead to mtDNA depletion and neurodegeneration	Pizzamiglio et al. (2024), Tanushi (2024)
Epigenetic DNA methylation	DNMT1, DNMT3A/B	Regulate methylation of nuclear-encoded mitochondrial genes	Altered methylation patterns linked to aging and cognitive decline	Huang et al. (2022)
Histone modifications	SIRT1, SIRT3, HDACs	Modulate chromatin structure and mitochondrial gene expression	Dysregulation affects mitochondrial biogenesis and oxidative stress	Xu et al. (2023), Shelke et al. (2023)
Non-coding RNAs	miRNAs, lncRNAs	Regulate post-transcriptional control of mitochondrial dynamics	Aberrant expression contributes to impaired mitochondrial homeostasis	Abed et al. (2024)
Oxidative stress response	Nrf2, Keap1	Regulate antioxidant responses and mitochondrial protection	Impaired Nrf2 signaling increases oxidative damage in aging	Cai et al. (2021)
Inflammatory modulation	NF-κB, STAT3	Control inflammatory responses impacting mitochondrial dynamics	Chronic inflammation exacerbates mitochondrial dysfunction	de Oliveira et al. (2021)

mitochondria may be enhanced for good health and survival in multiple aging and neurodegenerative disease states (Donato et al. 2024).

5.9 Therapeutic Strategies Targeting Mitochondrial Dynamics

Therapeutic interventions targeting mitochondrial dynamics attempt to restore mitochondrial homeostasis by regulating the dynamic balance between fission and fusion, activating mitophagy, optimizing biogenesis, and removing oxidative stress (Lin et al. 2024). Dysregulation of these processes is involved in mitochondrial failure, which contributes to neurodegenerative disease, metabolic disease, and senescence. Rebalancing these deficits has therapeutic potential to stabilize neurons, preserve cellular energy balance, and prevent disease progression.

One of the encouraging strategies is supplementing mitochondrial fusion to counteract over-fragmentation seen in neurodegenerative diseases. Pharmacological agents such as leflunomide and M1 peptide derivatives facilitate mitochondrial fusion through MFN1 and MFN2 and OPA1 induction, thereby preserving mitochondrial integrity and promoting oxidative phosphorylation. Such interventions have been shown to be effective in reducing neuronal injury and augmenting cognitive gains in preclinical models of PD and AD (Verma et al. 2023).

On the other hand, blocking overactive mitochondrial fission is able to avert fragmentation and induce elongation of mitochondria, and thus shield against apoptosis and energy loss. Small molecules like Mdivi-1, which is a specific DRP1 inhibitor, are capable of diminishing mitochondrial fragmentation and oxidative stress in neurodegenerative models (Lin et al., 2024). Through the modulation of DRP1 activity, Mdivi-1 alleviates mitochondrial impairment and induces neuronal survival and hence is a hopeful drug candidate for neurodegenerative conditions like Huntington's disease and ALS (Lin et al., 2024).

Fostering mitophagy, the selective breakdown of impaired mitochondria, is yet another essential strategy for ensuring mitochondrial quality and avoiding the buildup of defective organelles. Augmenting the PINK1/Parkin pathway may enhance the removal of defective mitochondria and lower oxidative damage (Wu et al., 2022, 449). Pharmacological stimulation of Parkin or drugs like urolithin A and nicotinamide riboside have shown the potential to induce mitophagy, thus enhancing mitochondrial turnover as well as neuronal function (Pradeepkiran 2022, 450). These drugs have the potential to cure diseases that have defective mitophagy, for example, PD.

Mitochondrial biogenesis, the production of new mitochondria to replace old or damaged ones, is crucial for energy homeostasis and cellular well-being. Stimulation of peroxisome PGC-1α, a master gene of mitochondrial biogenesis, has been shown to improve mitochondrial function and decrease oxidative stress. Resveratrol and AICAR (5-aminoimidazole-4-carboxamide ribonucleotide) have been demonstrated to stimulate PGC-1α, resulting in enhanced mitochondrial biogenesis and neuroprotection in preclinical models (Sabzevari Rad 2024, 451).

Another promising approach is modulating sirtuins, specifically SIRT1 and SIRT3, which control mitochondrial function and cellular metabolism by deacetylating important transcription factors and enzymes. Activation of SIRT1 increases PGC-1α activity, stimulating mitochondrial biogenesis, whereas SIRT3 controls mitochondrial protein acetylation, enhancing mitochondrial respiration and decreasing ROS production (Li et al. 2022, 452). Pharmacologic agents such as NMN and SIRT1 activators have been shown to provide neuroprotection through enhanced mitochondrial function and disease deceleration. Gene therapy approaches focusing on the dynamics of mitochondria have also been recognized as prospective options for the rectification of genetic mutations responsible for mitochondrial disorders (Soldatov 2022, 453). Viral vector-mediated transfer of MFN2-, OPA1-, or PINK1-encoding genes has the potential to reverse mitochondrial dynamics and suppress neuronal degeneration in preclinical models. These therapies have the potential to offer therapeutic benefit in patients with inherited mitochondrial disorders or neurodegenerative disease resulting from mitochondrial dysfunction (Rey 2022, 454).

Finally, antioxidant therapies to counteract oxidative stress and preserve mitochondrial integrity are also in the limelight. Mitochondria-targeted antioxidants such as MitoQ, SkQ1, and SS31 are found to scavenge ROS, preserve mitochondrial membranes, and inhibit mitochondrial permeability transition pore (mPTP) opening, reducing neuronal injury and promoting cell survival (Sacks et al., 2021). By inhibiting mitochondrial dynamics with these multifaceted approaches, one can enhance mitochondrial quality control, restore bioenergetic balance, and diminish cellular injury. These approaches have great therapeutic potential in the treatment of neurodegenerative diseases, metabolic disorders, and age-related diseases, ultimately resulting in improved health and prolonged cellular longevity Grel (2023).

5.10 Future Directions in Mitochondrial Research and Brain Aging

Directions of future research in the field of mitochondrial biology and aging brain focus on deciphering the multifaceted relationship between mitochondrial dynamics, energy metabolism, cellular homeostasis, and age, as well as neurodegeneration. Because the brain is significantly dependent on mitochondrial metabolism to fuel its colossal energy expenditure, understanding the ways in which mitochondrial failure drives age-related cognitive dysfunction and neurodegenerative disease will inform the development of focused therapeutic intervention (Li 2021, 457).

One of the greatest potential avenues of future investigation is the discovery of molecular cross-talk among mitochondrial dynamics and other cellular processes such as autophagy, inflammation, and synaptic transmission within the aging brain (Xie 2024, 458). Investigation into the mechanisms of age-dependent changes in mitochondrial fusion, fission, mitophagy, and biogenesis to produce disruption in neuronal homeostasis can advance understanding of Alzheimer's, Parkinson's, and other neurodegenerative disease pathophysiology. Sophisticated imaging techniques, such as super-resolution microscopy and in vivo tracking of mitochondrial dynamics, will allow real-time monitoring of mitochondrial activity in neurons and disclose points of key dysfunction during aging (Fang et al., 2024, 459).

Another promising avenue is to elucidate the contribution of mtDNA mutations and heteroplasmy to brain aging. Time-dependent accumulation of mtDNA mutations disrupts oxidative phosphorylation, enhances production of ROS, and promotes neuronal degeneration (Liao 2022, 460). Future research characterizing mtDNA mutation patterns in aged neurons and how they affect mitochondrial function can set the stage for new therapeutic avenues, including mitochondrial gene editing through CRISPR-based tools or mitochondrial replacement therapy to remove aberrant mtDNA (Clemente-Suárez et al., 2023, 461).

Epigenetic control of mitochondrial gene expression is another area that needs to be investigated. Determining how age-dependent alterations in DNA methylation, histone modification, and non-coding RNAs affect the expression of genes that regulate mitochondrial dynamics can lead to novel targets for the restoration of mitochondrial homeostasis (Kumar 2024, 462). Recent studies on the role of miRNAs and lncRNAs in the modulation of mitochondrial function indicate that pharmacological manipulation of these molecules holds the potential for therapeutic reversal of mitochondrial dysfunction and maintenance of neuronal health upon aging (Chen 2022, 463).

A greater understanding of how mitochondria operate in neuroinflammation and immune responses during aging brain would be another very important direction. Mitochondrial malfunction in astrocytes and microglia causes chronic neuroinflammation that worsens the damage to neurons during aging and neurodegeneration. There is potential that future research identifying modulating the function of mitochondria in glial cells will give new prospects to decrease the level of neuroinflammation and prevent loss of cognition (Singh 2022, 464).

Technological advances in precision and personalized medicine hold promising potential for customizing mitochondrial-based interventions to persons with neurodegenerative diseases and age-related cognitive impairment (Marques-Aleixo 2021, 465). Mitochondrial biomarkers such as plasma mitochondrial DNA, markers of oxidative stress, and mitochondrial-derived peptides could be used as presymptomatic predictors of brain aging and useful markers of neural mitochondrial dysfunction progression. Combining these biomarkers with high-end omics technologies, including transcriptomics, proteomics, and metabolomics, can facilitate the identification of patient-specific mitochondrial signatures and inform the design of targeted interventions (Singh 2025, 466).

The future of mitochondria-targeted therapies, such as mitochondrial antioxidants, mitophagy inducers, and small molecules that modulate mitochondrial dynamics, is an enticing horizon. Their translation from model systems into the clinic, however, requires a better appreciation of their long-term safety, efficacy, and dosing optimization. The development of new drug delivery systems that possess the capability of penetrating the blood-brain barrier and selectively targeting mitochondria in neurons will be critical in ensuring the maximal therapeutic utility (Rawal et al., 2022, 467).

Lastly, studies on the interaction between lifestyle variables, mitochondrial function, and brain aging provide avenues for prevention interventions. Understanding how dietary intake, exercise, caloric restriction, and intermittent fasting influence mitochondrial function can provide the basis upon which lifestyle interventions will maintain cognitive function and avert brain aging (Wang 2024, 468). With the progress in the science of mitochondrial research, the use of multidisciplinary methods and deployment of novel technologies will be critical to deciphering the intricate mechanisms behind the interlink between mitochondrial dysfunction and brain aging (Liao et al., 2025, 469). Such advances promise to revolutionize age-related neurodegenerative disease treatment and pave the way to healthy brain aging.

5.11 Conclusion

Mitochondrial dynamics are vital in maintaining cellular homeostasis, especially in the energy-requiring microclimate of the brain. Dysregulation of the dynamic equilibrium between mitochondrial fission, fusion, mitophagy, and biogenesis initiates the onset of age-related cognitive dysfunction and neurodegenerative disease states such as Alzheimer's, Parkinson's, and Huntington's diseases. Acceleration of oxidative stress, initiation of neuroinflammation, and disruption of synaptic function by mitochondrial dysfunction eventually lead to neuronal degeneration as the brain gets older.

New data unveil the therapeutics of manipulating mitochondrial dynamics to restore cellular energy homeostasis, minimize oxidative damage, and promote neuronal plasticity. Pharmacologic modulators of mitochondrial fusion and fission, mitophagy activators, and mitochondrially incorporated modulators of biogenesis are potential approaches to rescue the pathogenic effects of mitochondrial dysfunction in neurodegenerative disease and aging. Gene therapy, mitochondrial transplant, and precision medicine strategies remain promising to remediate mitochondrial defects and preserve brain health.

In spite of notable advances, challenges persist in translating these therapeutic modalities from the bench to the bedside. Increased knowledge of the complex molecular pathways governing mitochondrial dynamics and their interplay with cellular mechanisms is necessary. Moreover, the creation of mitochondria-targeted therapies capable of effectively penetrating the blood-brain barrier and selectively modifying mitochondrial function in neurons will be critical to therapeutic efficacy.

Future studies would then need to direct their efforts at the identification of mitochondrial biomarkers to facilitate early diagnosis, assessment of disease progress, and individualized treatment interventions. Investigation on the impact of lifestyle interventions like caloric restriction and exercise on mitochondrial function represents another promising front for maintaining cognition and retarding brain aging. As a whole, emphasis on mitochondrial dynamics is a potential future in the war against neurodegenerative disease and brain aging. Ongoing progress in mitochondrial biology, along with innovative therapeutic approaches, is going to revolutionize the field of age-related neurological disorders and enhance the health condition of the elderly population.

References

Abdel Khalek W et al (2014) SIRT3, a mitochondrial NAD+-dependent deacetylase, is involved in the regulation of myoblast differentiation. PLoS One 9(12):e114388

Abed S et al (2024) The role of non-coding RNAs in mitochondrial dysfunction of Alzheimer's disease. J Mol Neurosci 74(4):100

Antunes MA, Lopes-Pacheco M, Rocco PR (2021) Oxidative stress-derived mitochondrial dysfunction in chronic obstructive pulmonary disease: a concise review. Oxidative Med Cell Longev 2021(1):6644002

Bandaru LJM et al (2022) Mechanisms associated with the dysregulation of mitochondrial function due to lead exposure and possible implications on the development of Alzheimer's disease. Biometals 35(1):1–25

Baytas O et al (2024) Loss of mitochondrial enzyme GPT2 leads to reprogramming of synaptic glutamate metabolism. Mol Brain 17(1):87

Bhatti GK et al (2022) Targeting mitochondrial bioenergetics as a promising therapeutic strategy in metabolic and neurodegenerative diseases. Biom J 45(5):733–748

Bhushan R, Goel F, Singh S (2025) Beyond dopamine: novel therapeutic pathways for Parkinson's disease through receptor signaling. CNS Neurol Disord Drug Targets

Cai Z-Y et al (2021) Therapeutic effect of Keap1-Nrf2-ARE pathway-related drugs on age-related eye diseases through anti-oxidative stress. Int J Ophthalmol 14(8):1260

Casanova A et al (2023) Mitochondria: it is all about energy. Front Physiol 14:1114231

Cetin EA, Akbay D (2024) Mitochondrial dynamics and mitochondrial biogenesis. In: Mitochondrial transplantation and transfer. Elsevier, pp 35–58

Chauhan K, Goel F, Singh S (2024) Apigenin protects melanocytes and improve tyrosinase activity in a hydroquinone induced vitiligo mouse model targeting P38 MAP kinase signaling: histopathology and immunohistochemistry analysis. Naunyn Schmiedeberg's Arch Pharmacol 397(7):4859–4869

Chen Z, Li Y, Zhang H, Huang P, Luthra R (2022) Hypoxia-regulated microRNA-210 modulatesmitochondrial function and decreases ISCU and COX10 expression. J of Cellular Physiology, 237(5):1982–1994. https://doi.org/10.1002/jcp.30034

Chen S, Liao Z, Xu P (2023a) Mitochondrial control of innate immune responses. Front Immunol 14:1166214

Chen W, Zhao H, Li Y (2023b) Mitochondrial dynamics in health and disease: mechanisms and potential targets. Signal Transduct Target Ther 8(1):333

Clemente-Suárez VJ, et al (2023) Mitochondria and brain disease: A comprehensive review of pathologicalmechanisms and therapeutic opportunities. Biomedicines, 11(9):2488. https://doi.org/10.3390/biomedicines11092488

de Oliveira LG et al (2021) Unraveling the link between mitochondrial dynamics and neuroinflammation. Front Immunol 12:624919

Dey A et al (2024) Alzheimer's disease pathology and assistive nanotheranostic approaches for its therapeutic interventions. Int J Mol Sci 25(17):9690

Dhapola R et al (2022) Recent advances in molecular pathways and therapeutic implications targeting mitochondrial dysfunction for Alzheimer's disease. Mol Neurobiol 59(1):535–555

Donato L et al (2024) From powerhouse to regulator: the role of mitoepigenetics in mitochondrion-related cellular functions and human diseases. Free Radic Biol Med

Ercanbrack WS et al (2025) Frataxin deficiency and the pathology of Friedreich's ataxia across tissues. Tissue Barriers:2462357

Fang H, Zhang Y, Li J (2024) Molecular probes for super-resolution imaging of drug dynamics in living cells.J Microscopy, 276(2):85–92. https://doi.org/10.1111/jmi.13010
Gąssowska-Dobrowolska M et al (2024) Novel insights into parkin–mediated mitochondrial dysfunction and "Mito-inflammation" in α-Synuclein toxicity. The role of the cGAS–STING signalling pathway. J Inflamm Res:4549–4574
Ghani M (2025) The role of DRP1-a mitochondrial fission protein-in mitochondrial morphology rearrangement during neuronal differentiation
Goel F, Dobhal V, Singh S (2025) Protective effects of diacerein against quinolinic acid–induced Huntington's disease–like symptoms in adult zebrafish by targeting GSK-3β signalling. Naunyn Schmiedeberg's Arch Pharmacol:1–23
Grel H (2023) Mitochondrial dynamics in neurodegenerative diseases. Frontiers in Neuroscience, 17, 10487704. https://doi.org/10.3389/fnins.2023.10487704
Green A, Hossain T, Eckmann DM (2022) Mitochondrial dynamics involves molecular and mechanical events in motility, fusion and fission. Front Cell Dev Biol 10:1010232
Gupta S et al (2025) Mitochondria and its epigenetic dynamics: insight into synaptic regulation and synaptopathies. Funct Integr Genomics 25(1):26
Gustafsson ÅB, Dorn GW (2019) Evolving and expanding the roles of mitophagy as a homeostatic and pathogenic process. Physiol Rev 99(1):853–892
Huang W et al (2022) LncRNA-mediated DNA methylation: an emerging mechanism in cancer and beyond. J Exp Clin Cancer Res 41(1):100
Hyttinen J et al (2021) Therapeutic potential of PGC-1α in age-related macular degeneration (AMD)–the involvement of mitochondrial quality control, autophagy, and antioxidant response. Expert Opin Ther Targets 25(9):773–785
Jaffri JM (2023) Reactive oxygen species and antioxidant system in selected skin disorders. MJMS 30(1):7
Jenkins BC et al (2024) Mitochondria in disease: changes in shapes and dynamics. Trends Biochem Sci
Jhanji R et al (2021) Mitochondrial dysfunction and traffic jams in amyotrophic lateral sclerosis. Mitochondrion 58:102–110
Khot M et al (2022) NLRP3 inflammasomes: a potential target to improve mitochondrial biogenesis in Parkinson's disease. Eur J Pharmacol 934:175300
Krantz S (2021) The role of mitophagy in the differentiation of human iPSCs. University of Illinois at Chicago
Kumar A, Choudhary A, Munshi A (2024) Epigenetic reprogramming of mtDNA and its etiology in mitochondrial diseases. J Physiol Biochem:1–15
Lambiri DW, Levin LA (2024) Maculopapillary bundle degeneration in optic neuropathies. Curr Neurol Neurosci Rep 24(7):203–218
Li R-L et al (2022) Regulation of mitochondrial dysfunction induced cell apoptosis is a potential therapeutic strategy for herbal medicine to treat neurodegenerative diseases. Front Pharmacol 13:937289
Li Q, Zhang Y, Wang H (2024) Hypomethylation in promoters of PGC-1α involved in mitochondrialbiogenesis in aged skeletal muscle. Aging Cell, 23(1):e13788. https://doi.org/10.1111/acel.13788
Liao S (2022) The fate of damaged mitochondrial DNA in the cell. Cellular and Molecular Life Sciences, 79(1), 1–15. https://doi.org/10.1007/s00018-022-04299-2
Liao J, Zhang C, Wang J (2025) Mitochondria in brain diseases: Bridging structural insights intomitochondrial dysfunction with advanced therapeutic technologies. Cell Biomaterials, 1(1), 1–12. https://doi.org/10.1016/j.cbm.2025.01.001
Lin X et al (2024) The role of polyphenols in modulating mitophagy: implications for therapeutic interventions. Pharmacol Res:107324
Lin Y, et al (2024) Targeting DRP1 with Mdivi-1 to correct mitochondrial abnormalities in ADOA+ models. JCIInsight, 9(4), e180582. https://doi.org/10.1172/jci.insight.180582

Lou H et al (2024) Potential effect of acupuncture on mitochondrial biogenesis, energy metabolism and oxidation stress in MCAO rat via PGC-1α/NRF1/TFAM pathway. J Stroke Cerebrovasc Dis 33(11):107636
Magro G, Laterza V, Tosto F (2025) Leigh syndrome: a comprehensive review of the disease and present and future treatments. Biomedicine 13(3):733
Markaki M, Tsagkari D, Tavernarakis N (2021) Mitophagy mechanisms in neuronal physiology and pathology during ageing. Biophys Rev 13(6):955–965
Marques-Aleixo I, et al (2021) Preventive and therapeutic potential of physical exercise in neurodegenerativediseases: A focus on mitochondrial dynamics. Neurobiology of Disease, 148:105-118. https://doi.org/10.1016/j.nbd.2020.105118
Masaldan S, Callegari S, Dewson G (2022) Therapeutic targeting of mitophagy in Parkinson's disease. Biochem Soc Trans 50(2):783–797
Mayorga-Weber G, Rivera FJ, Castro MA (2022) Neuron-glia (mis) interactions in brain energy metabolism during aging. J Neurosci Res 100(3):835–854
Mishra E, Thakur MK (2023) Mitophagy: a promising therapeutic target for neuroprotection during ageing and age-related diseases. Br J Pharmacol 180(12):1542–1561
Mohamed Yusoff AA, Mohd Khair SZN (2025) Unraveling mitochondrial dysfunction: comprehensive perspectives on its impact on neurodegenerative diseases. Rev Neurosci 36(1):53–90
Monaghan RM (2025) The fundamental role of mitochondria–endoplasmic reticulum contacts in ageing and declining healthspan. Open Biol 15(2):240287
Picca A et al (2023) Mitophagy in human health, ageing and disease. Nat Metab 5(12):2047–2061
Pradeepkiran JA (2022) Are mitophagy enhancers therapeutic targets for neurodegenerative diseases?Frontiers in Aging Neuroscience, 14, 914841. https://doi.org/10.3389/fnagi.2022.914841
Pizzamiglio C, Hanna MG, Pitceathly RD (2024) Primary mitochondrial diseases. Handb Clin Neurol 204:53–76
Prajapati C et al (2025) An update of fungal endophyte diversity and strategies for augmenting therapeutic potential of their potent metabolites: recent advancement. Appl Biochem Biotechnol:1–68
Rai SN et al (2020) Mitochondrial dysfunction: a potential therapeutic target to treat Alzheimer's disease. Mol Neurobiol 57(7):3075–3088
Ramakrishna K et al (2023) WNT-β catenin signaling as a potential therapeutic target for neurodegenerative diseases: current status and future perspective. Diseases 11(3):89
Rani L, Sahu MR, Mondal AC (2022) Age-related mitochondrial dysfunction in Parkinson's disease: new insights into the disease pathology. Neuroscience 499:152–169
Rawal P, Sharma S, Verma R (2022) Mitochondria-targeted drug delivery across the blood-brain barrier:Emerging strategies for neurodegenerative disease therapy. J Controlled Release, 350, 500–518. https://doi.org/10.1016/j.jconrel.2022.06.015
Ren L et al (2020) Mitochondrial dynamics: fission and fusion in fate determination of mesenchymal stem cells. Front Cell Dev Biol 8:580070
Rey F (2022) Mitochondrial dysfunction in neurodegenerative diseases: Potential therapeutic application ofmitochondrial transfer to central nervous system-residing cells. Neural Regeneration Research, 17(4):757–758.https://doi.org/10.4103/1673-5374.331409
Sabzevari Rad, R. (2024). Effects of exercise mimetics as putative therapeutics on brain health, aging, andneurodegenerative diseases. Sport Sciences for Health, 20(4), 1207–1218. https://doi.org/10.1007/s11332-024-01187-8
Sacks B, Onal H, Martorana R, Sehgal A, Harvey A, Wastella C, Ahmad H, Ross E, Pjetergjoka A,Prasad S, Barsotti RJ, Young LH, Chen Q (2021) Mitochondrial targeted antioxidants, mitoquinone andSKQ1, not vitamin C, mitigate doxorubicin-induced damage in H9c2 myoblast: pretreatment vs. co-treatment.PCOM Scholarly Works, 2121. https://digitalcommons.pcom.edu/scholarly_papers/2121
Shadab A et al (2024) The interplay between mitochondrial dysfunction and Nlrp3 inflammasome in multiple sclerosis: therapeutic implications and animal model studies. Biomed Pharmacother 175:116673

Shaito A et al (2023) Resveratrol-mediated regulation of mitochondria biogenesis-associated pathways in neurodegenerative diseases: molecular insights and potential therapeutic applications. Curr Neuropharmacol 21(5):1184–1201

Sharma A et al (2021) Mitochondrial dysfunction in Huntington's disease: pathogenesis and therapeutic opportunities. Curr Drug Targets 22(14):1637–1667

Shelke V et al (2023) Epigenetic regulation of mitochondrial-endoplasmic reticulum dynamics in kidney diseases. J Cell Physiol 238(8):1716–1731

Singh D (2022) Astrocytic and microglial cells as the modulators of neuroinflammation in Alzheimer's disease.Journal of Neuroinflammation, 19(1):206. https://doi.org/10.1186/s12974-022-02565-0

Singh M et al (2024) A comprehensive review and androgen deprivation therapy and its impact on Alzheimer's disease risk in older men with prostate cancer. In: Degenerative neurological and neuromuscular disease, pp 33–46

Singh S (2025) Unveiling the future of metabolic medicine: Omics approaches to personalized therapeuticstrategies. J Clinical Medicine, 14(5):1234–1245. https://doi.org/10.1016/j.jcm.2025.03.012

Sinha JK et al (2024) The potential of mitochondrial therapeutics in the treatment of oxidative stress and inflammation in aging. Mol Neurobiol:1–16

Slater PG et al (2022) Mitochondrial function in spinal cord injury and regeneration. Cell Mol Life Sci 79(5):239

Soldatov VO, Kubekina MV, Skorkina MY, Belykh AE, Egorova TV, Korokin MV, Pokrovskiy MV,Deykin AV, Angelova PR (2022) Current advances in gene therapy of mitochondrial diseases. JTranslational Medicine, 20(1):562. https://doi.org/10.1186/s12967-022-03685-0

Song M, Zhao X, Song F (2021) Aging-dependent mitophagy dysfunction in Alzheimer's disease. Mol Neurobiol 58(5):2362–2378

Srivastava P et al (2019) Design and development of some phenyl benzoxazole derivatives as a potent acetylcholinesterase inhibitor with antioxidant property to enhance learning and memory. Eur J Med Chem 163:116–135

Stevic N et al (2022) Cooling uncouples differentially ROS production from respiration and Ca2+ homeostasis dynamic in brain and heart mitochondria. Cells 11(6):989

Tanushi X (2024) Impairing the repair of the oxidative DNA damage 8-oxoG in nuclear and mitochondrial genomes, which consequences for the cell? Université Paris-Saclay

Tasakis RNS, Tsolaki M (2015) Mitochondria pathogenesis and dysfunction in Alzheimer's disease. Hell J Nucl Med 18:10

Trigo D et al (2022) Mitochondria, energy, and metabolism in neuronal health and disease. FEBS Lett 596(9):1095–1110

Tripathi PN et al (2019) Biphenyl-3-oxo-1, 2, 4-triazine linked piperazine derivatives as potential cholinesterase inhibitors with anti-oxidant property to improve the learning and memory. Bioorg Chem 85:82–96

Tripathi PN et al (2024) Review of pharmacotherapeutic targets in Alzheimer's disease and its management using traditional medicinal plants. Degener Neurol Neuromuscul Dis:47–74

Verma H et al (2023) Understanding the neuronal synapse and challenges associated with the mitochondrial dysfunction in mild cognitive impairment and Alzheimer's disease. Mitochondrion 73:19–29

Wang Z et al (2023a) Outlook of PINK1/parkin signaling in molecular etiology of Parkinson's disease, with insights into Pink1 knockout models. Zool Res 44(3):559

Wang S et al (2023b) The mitophagy pathway and its implications in human diseases. Signal Transduct Target Ther 8(1):304

Wang P-F et al (2024) Mitochondrial and metabolic dysfunction of peripheral immune cells in multiple sclerosis. J Neuroinflammation 21(1):28

Wang J (2024) Effects of caloric restriction and intermittent fasting on cognitive function and mitochondrial health.Frontiers in Aging Neuroscience, 16, 123–135. https://doi.org/10.3389/fnagi.2024.123456

Wu Y, et al (2022) PINK1/Parkin-mediated mitophagy in cardiovascular disease. International Journal ofCardiology, 348, 63–71. https://doi.org/10.1016/j.ijcard.2022.02.018
Xiao B, Kuruvilla J, Tan E-K (2022) Mitophagy and reactive oxygen species interplay in Parkinson's disease. NPJ Parkinson's Dis 8(1):135
Xie W et al (2024) Unraveling the nexus of age, epilepsy, and mitochondria: exploring the dynamics of cellular energy and excitability. Front Pharmacol 15:1469053
Xie XC, Liu H, Li C (2024) The route of random process to ultraslow aging phenomena. arXiv. https://doi.org/10.48550/arXiv.2409.14422
Xu H et al (2023) Sirtuins at the crossroads between mitochondrial quality control and neurodegenerative diseases: structure, regulation, modifications, and modulators. Aging Dis 14(3):794
Xu W et al (2025) The interconnective role of the UPS and autophagy in the quality control of cancer mitochondria. Cell Mol Life Sci 82(1):42
Yadav E et al (2022) Resveratrol: a potential therapeutic natural polyphenol for neurodegenerative diseases associated with mitochondrial dysfunction. Front Pharmacol 13:922232
Yadav AK et al (2024) Investigation of serum pro-inflammatory markers and trace elements among short stature in eastern Uttar Pradesh and Bihar populations. J Inflamm Res:6063–6073
Zhang Y et al (2024) Mfn2R364W, Mfn2G176S, and Mfn2H165R mutations drive Charcot-Marie-tooth type 2A disease by inducing apoptosis and mitochondrial oxidative phosphorylation damage. Int J Biol Macromol 278:134673
Zhong R, Rua MT, Wei-LaPierre L (2024) Targeting mitochondrial Ca2+ uptake for the treatment of amyotrophic lateral sclerosis. J Physiol 602(8):1519–1549

6 Inflammation and Brain Ageing: Anti-inflammatory Nutraceuticals

Monika Kaushik, Parameshwar Kondapuram, Devyani Rajput, Sagar Pamu, and Mrugendra Potdar

Abstract

According to the research titled 'World Population Prospects: the 2019 Revision, United Nation', the sum of individuals aged 65 and older is expected to climb by 16% by the year 2050, which is equivalent to one in every six individuals. According to this data, the world population that is becoming older has been expanding at an exponential rate. The ageing of the people has a number of important repercussions, one of which is an augmented risk of cognitive damage and age-linked neurodegenerative disorders. These diseases include Alzheimer's disease, mild cognitive impairment, Parkinson's disease, vascular dementia and stroke.

These conditions are associated with advancing age, and it is anticipated that by the year 2040, there will be roughly 81.1 million people who are afflicted with these diseases or conditions. The fact that there is a direct connection between lifestyle choices, dietary patterns and neurodegeneration brings to light the possibility of nutraceuticals serving as preventative therapy for those who suffer from neurological diseases. It is also important to take into consideration nutraceuticals that contain bioactive components that have antioxidant and anti-inflammatory characteristics because these complexes can help reduce the effects of cognitive impairment.

M. Kaushik · D. Rajput · S. Pamu
Amity Institute of Pharmacy, Amity University Madhya Pradesh, Gwalior, Madhya Pradesh, India

P. Kondapuram
School of Pharmacy, Guru Nanak Institutions Technical Campus, Ibrahimpatnam, Telangana, India

M. Potdar (✉)
Shri Vile Parle Kelawani Mandal's Institute of Pharmacy, Dhule, Maharashtra, India

A. Kumar Singh, S. Nand Rai (eds.), *Rejuvenating the Brain: Nutraceuticals, Autophagy, and Longevity*, Nutritional Neurosciences,
https://doi.org/10.1007/978-981-95-2790-8_6

Nutraceuticals are chemicals or compounds that have a beneficial outcome on the body's physiological processes and offer protection against chronic diseases. Nutraceuticals, when combined with contemporary medications that are currently available, have the potential to improve our immune system and overall health by reducing cognitive deficiencies in older persons and providing protection against age-related cognitive diseases. The authors have provided a synopsis of a few of the nutraceuticals that have been the subject of extensive investigation. These nutraceuticals include curcumin, berberine, resveratrol and others that are established to play a role in age-linked illnesses and brain ageing.

Keywords

Nutraceutical · Plant-derived compound · Neuroinflammation · Neurodegenerative diseases · Supplementary diets

6.1 Introduction

6.1.1 Overview of Brain Ageing and the Role of Inflammation

Ageing is inevitable, causing physical decline and increased risk of illness and death. Ageing populations increase the financial burden of age-related health issues; thus, reducing or preventing them is crucial (Blinkouskaya et al. 2021). Elderly people often have neurodegenerative illnesses and cognitive deficits, affecting their longevity and quality of life. A complicated brain disease, neurodegeneration, is still poorly understood. Elderly traits, including mitophagy, cellular senescence and protein aggregation, have been related to neurodegenerative diseases like Alzheimer's disease (AD) and Parkinson's disease (PD) (Currais 2015).

Brain swelling, characterised by microglial and astrocytic initiation, increases with age and is a hallmark of neurodegenerative ailments like AD. Astrocytes, the main central nervous system (CNS) glial cells, cause neuroinflammation (Takasugi 2018). Free saturated fatty acids, infections, LPS and oxidative stress excite astrocytes. Astrocytes produce inflammatory cytokines and cyclooxygenase-2, leading to increased Aβ production (Jęśko et al. 2017).

The brain's response to trauma, illness and infection is mediated by microglia, the CNS's inherent immune cells. Age-linked alterations in the brain's response to and recovery from damage may be caused by increased neuroinflammation and responsiveness of activated microglia (Hagberg et al. 2015). In addition to resting microglia, a subgroup has shorter processes and cell surface markers like activated microglia, but no cytokine release. The term 'primed' refers to microglia that activate faster and release more cytokines than non-primed microglia (Calder et al. 2017). The brains of older people have more inflammatory cytokines and microglia that are active and ready to attack. Interleukin (IL)-6 levels are high in the mind after the peripheral immune system is activated or when the brain is injured. They may also be high during regular ageing and play a role in neuroinflammation that comes with getting older (Peirce and Alviña 2019).

6.1.2 Importance of Anti-inflammatory Nutraceuticals in Brain Health

Bioactive nutraceuticals and neuropathology food supplements are unique ways to improve the elderly value of life. The primary causes of neurodegenerative diseases are oxidative stress, mitotoxicity, calcium overload and sweling. Nutraceuticals regulate multiple biological pathways, making them appropriate for long-term usage to mitigate or prevent these procedures (Elbandy 2022).

Neuroinflammation and neurological diseases like AD, PD, autism, brain tumors and multiple sclerosis can be treated with diet and metabolism. Chronic neuroinflammation and cerebral metabolic stress (Pandareesh et al. 2018). Nutraceuticals regulate energy metabolism and neurotransmission and neuroinflammation signaling pathways. Nutritional supplements include creatine, carnitine, vitamins, unsaturated fatty acids, polyphenols, stilbenes, coenzyme Q10, and phytoestrogens reduce neuroinflammation and neurodegeneration (Pekdemir et al. 2024). Many enzyme and receptor functions use nutraceuticals as substrates and cofactors. Pantothenic acid (vitamin B5) is necessary for carbohydrate and lipid oxidation. It produces hormones and neurotransmitters (Onaolapo et al. 2022).

Sulphur-containing complexes, unsaturated fatty acids, polyphenols, vitamins C, D and E, coenzyme Q10, stilbenes, creatine and phytoestrogens can affect Parkinson's disease symptoms, including α-synuclein aggregation, ubiquitin-proteasome function, mTOR signaling, lysosomal-mediated autophagy, energy deficits and impaired catecholamine and dopamine transport (Kocsis et al. 2025). Research indicates that polyphenols, flavonoids and diets abundant in flavonoids may impede neurodegenerative illnesses and protect neurones. Polyphenols elevate glutathione concentrations and the efficacy of antioxidant enzymes, including glutathione peroxidase, superoxide dismutase and catalase, which neutralize free radicals and protect cellular integrity. Cells use polyphenols and other nutraceuticals, including vitamins C and E, to impede the oxidation of proteins, lipids and nucleic acids (McCarty and Lerner 2021).

6.2 The Ageing Brain and Inflammation

6.2.1 Age-Related Neuroinflammation and Its Effects on Brain Function

The brain has many immunological properties, but it is not an immunologically separate organ. The brain lacks a lymphatic system to catch prospective antigens and expresses less major histocompatibility complexes I and II than the periphery, indicating no specific antigen presentation mechanism (Delage et al. 2021; Rea et al. 2018). The blood-brain barrier (BBB) normally blocks cells and larger molecules from entering. Neuroinflammation is now related to various neurodegenerative ailments, including AD. Some writers, although not all, believe chronic neuroinflammation may trigger the disease's neuronal dysfunction. Even healthy brains develop neuroinflammatory changes with ageing. Given the growing older population and age-related neurodegenerative illnesses, understanding brain inflammatory changes is crucial (Brahadeeswaran et al. 2022).

6.2.2 The Role of the Blood-Brain Barrier in Neuroinflammation

The BBB is an extremely selective structural and functional barrier that isolates blood from the CNS, establishing a microenvironment essential for brain function and homeostasis. Brain endothelial cells engage with astrocytes, nerve terminals, microglia, pericytes and perivascular macrophages within the BBB. To highlight the complex connections between BBB cells, the neurovascular unit concept was proposed (Candelario-Jalil et al. 2022).

The BBB protects the CNS; therefore, its breakdown can harm an organism. Initial cytokine study showed quick disruption, but additional investigation suggested that insufficient solvent removal from early cytokine preparations may have contributed to this (Yang and Chen 2022). Lipopolysaccharide (LPS) may penetrate the BBB. Gram-negative bacteria produce LPS, which causes an increase in the levels of cytokines in the blood and the brain. Recent study suggests that breaking the BBB with cytokines and LPS is difficult. Systems are expected to have evolved to prevent BBB rupture, making this teleologically consistent. Neuroimmune events disrupt the BBB through unknown processes. There is growing evidence that the BBB's limiting qualities may change physiologically. Certain cytokines may affect brain endothelial cells, altering the BBB. These impacts include actin cytoskeleton and tight junction changes (Brandl and Reindl 2023; Angiulli et al. 2021).

6.3 Impact of Inflammation on Brain Health

6.3.1 Effects on Cognitive Decline and Neurodegenerative Diseases

Neuroinflammation and cognitive impairment are important topics in modern neuroscience. Microglia and astrocyte activation cause CNS inflammation, known as neuroinflammation. This inflammatory reaction may cause several neurodegenerative and cognitive problems (Stephenson et al. 2018). Brain inflammation is associated with chronic neurodegenerative disorders, and elevated baseline inflammatory profiles may heighten the risk of dementia over time. In addition, amyloid plaque deposits and neurofibrillary tangles (NFT), the neuropathological correlates of AD, which cause progressive and persistent cognitive decline, notably memory, contain specific inflammatory markers (Chen et al. 2016).

Peripheral inflammation facilitates peripheral immune cell infiltration into the CNS and activates central innate immune cells, causing neurocognitive impairment (Rauf et al. 2022). Neuroinflammation is related to numerous neurodegenerative diseases and contributes to AD. Inflammation in the nervous system may be associated with various causes, including infection, oxidative agents, trauma, redox-active iron, tau oligomers and amyloid-beta (Aβ) levels (Farì et al. 2021). Proinflammatory cytokines released during neuroinflammation activate signaling pathways, hyperphosphorylating tau protein at residues not altered physiologically (Spagnuolo et al. 2018).

Activated microglia release proinflammatory cytokines such as IL-6, IL-1β, IFN-γ, IL-12 and TNF-α. Therefore, they can create ROS, NO and other neurodegenerative disease-related chemicals. Astrocytes, the most abundant glial cells, support neuronal cell activity and health (Wang et al. 2018).

6.3.2 Inflammatory Biomarkers and Their Relationship with Brain Ageing

Reduced levels of SAP and some inflammatory biomarkers, including PTX3, IL-2 and ET-1, were linked to cognitive decline, indicating a possible link between endothelial dysfunction, systemic inflammation and vascular inflammation and cognitive decline (Jin et al. 2022). There are ten biomarkers that indicate different causes of inflammation: vascular inflammation (PTX3 and SAP), oxidative stress (RAGE), endothelial function (ET-1), metabolic function (adiponectin, resistin and PAI-1) and general inflammation (IL-6, IL-2 and IL-10) (Currais 2015) (Fig. 6.1).

The relationship between inflammation and memory, psychomotor speed and brain function is directly related. Inflammation is associated with decreased neurogenesis and disruption of long-term potentiation (LTP) in the hippocampus, an important region for memory and learning (Mou et al. 2022). In addition to reducing the survival and growth of new cells, activated (proinflammatory) microglia may

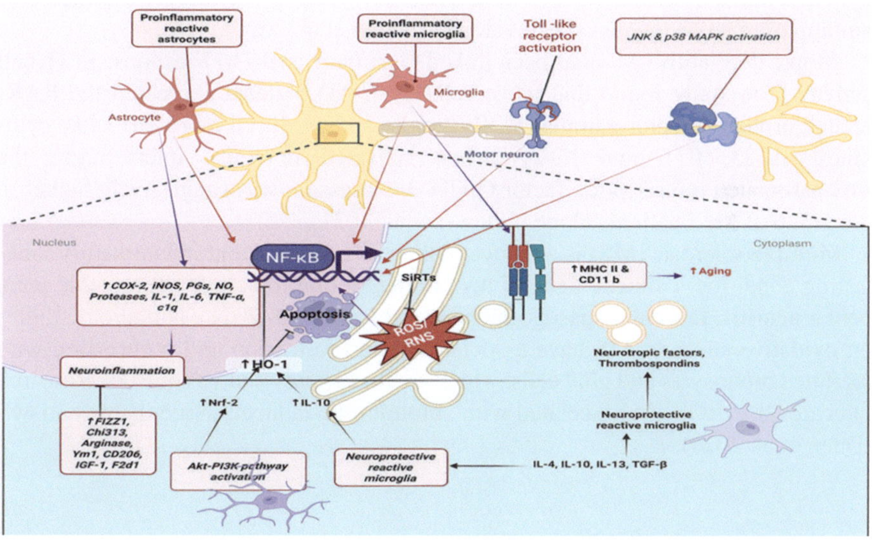

Fig. 6.1 Inflammatory biomarkers of neuroinflammation

also affect how well-established brain networks integrate them. By stimulating brain vascular endothelial cells or vagal afferent nerves, peripheral cytokines can interact through the CNS and produce proinflammatory cytokines, which can cause symptoms like poor memory consolidation (Dunleavy et al. 2022).

The relevance of a vascular component in cognitive decay was supported by the correlation between PTX3 and SAP, indicators of vascular inflammation and changes in cognition over time (Mac Giollabhui et al. 2021). Reduced psychomotor speed and linguistic ability were associated with elevated peripheral ET-1 levels, indicating that endothelial dysfunction may be a major factor in cognitive impairment and its development (Wan et al. 2023). A decline in language and memory skills was linked to IL-2, a measure of complete inflammation. IL-2 receptors are found in rodents' hippocampal and dentate gyrus, signifying that IL-2 may have an effect on memory and learning (Park et al. 2020).

6.3.3 The Connection Between Inflammation, Oxidative Stress and Neuronal Damage

Aβ buildup is associated with oxidative stress, which is acknowledged as an important element in the aetiology of AD. Aβ plaques have been linked to a number of cellular effects, such as (1) the p38 mitogen-activated protein kinase (MAPK) signaling pathway being activated, which causes tau hyperphosphorylation and the subsequent development of intracellular NFTs (Fischer and Maier 2015); (2) apoptotic pathways being mediated by the pro-apoptotic factor Bcl-2, which causes the release of cytochrome C from the mitochondria. T lymphocyte infiltrates into the cerebral parenchyma (Lassmann and van Horssen 2016; Cassano et al. 2016). On the other hand, systemic or CNS inflammation favorably encourages the excessive buildup of reactive oxygen species (ROS) (Ismail et al. 2020).

Atopic dermatitis (AD) has been linked to an increased Th17 response in Th cell activity. The study found that naïve cells from AD patients expressed the RAR-related orphan receptor gamma (RORγt) more and produced more Th17-like cytokines (IL-23, IL-6 and IL-21) when activated in vitro. Interestingly, the Th2-associated transcription factor GATA-3 expression was suggestively higher in the group of MCI patients alone (Ghowsi et al. 2021).

Multiple sclerosis (MS) is a demyelinating, progressive and inflammatory long-lasting condition. Clinical signs of myelin deterioration include neuropathic pain, ocular neuritis, muscular spasms and paralysis (Yao et al. 2021). Numerous studies on oxidative stress in MS have associated lesion formation and progression with recruited monocytes and glial cells, which are the primary sources of ROS. Research indicates that ROS are associated with autoimmune-induced tissue damage in MS (Duan et al. 2016).

6.4 Anti-inflammatory Mechanisms in Nutraceuticals

6.4.1 How Nutraceuticals Modulate Inflammatory Pathways in the Brain

According to molecular and cellular neuroscience, oxidative stress, excitotoxicity and their aftereffects are some of the mechanisms that lead to neurodegenerative diseases and brain damage. Neuronal injury is concomitantly linked to inflammation (activation of micro- and macroglial signaling), reduced cerebral blood flow and changes in the BBB (McCarty and Lerner 2021). Therefore, it is reasonable to propose that a certain occurring and bioavailable compound among a combination of possibly bioactive molecules may engage with several marks to elicit a variety of beneficial effects on the CNS (Ooi and Pak 2021). Herbal dietary supplements have attracted considerable attention as neuronutraceuticals because of their balanced array of important components beneficial to the brain, such as carotenoids, probiotics, saponins, polyunsaturated fatty acids (PUFA), flavonoids, marine natural products and vitamins. Numerous nutraceuticals may modulate cellular processes, and inflammation plays a critical role in ageing and chronic degenerative diseases (Pandareesh et al. 2018).

The age-associated rise in oxidative stress and chronic inflammation warrants the utilization of dietary supplements abundant in polyphenols as a regulator of age-linked cognitive decline. Phytochemicals' natural anti-inflammatory and antioxidant qualities often make them regarded as healthy (Kurowska et al. 2023). At modest concentrations, phytochemicals can have stimulatory or helpful effects on animal cells, but in large quantities, they can be harmful. Research indicates that hormonal phytochemicals can enhance cellular resilience against injury and disease. Resveratrol, hypericin, catechins, sulphoraphane, allicin and curcumin are some of the components (Dama et al. 2024; Catorce and Gevorkian 2020).

6.4.2 Anti-inflammatory Signaling Pathways

6.4.2.1 NF-κB Signaling Pathway

The nuclear factor kappa-light-chain-enhancer of activated B cells (NF-κB) pathway serves as the principal regulator of inflammation in both central and peripheral immune systems. Activation of the NF-κB pathway in astrocytes, microglia and neurones results in the production of chemokines, pro-inflammatory cytokines and adhesion molecules inside the CNS (Saha et al. 2020). These chemicals increase the inflammatory response and recruit immune cells. The NF-κB pathway may be triggered by several signals, including Toll-like receptors (TLRs), TNF-α and IL-1β. The dysfunction of this system results in persistent inflammation in neurodegenerative diseases (Yu et al. 2016). Extreme NF-κB initiation has been associated with AD because it encourages the development of tau pathology and Aβ plaques. Targeting NF-κB signaling is a useful tactic for lowering neuroinflammation and delaying the course of disease (Laurindo et al. 2023).

6.4.2.2 JAK-STAT Pathway

Janus kinase-signal transducer and activator of transcription (JAK-STAT) is a key biological process that underlies neuroinflammation. Cytokines such as IL-6 and IL-10 activate the JAK-STAT pathway, resulting in transcriptional programs that influence immune cell survival, activation and production (Jain et al. 2021). The activation of STAT3 in astrocytes is associated with the development of the neurotoxic A1 astrocyte phenotype in the central nervous system, exacerbating neuronal damage. The potential of roxolitinib and other JAK-STAT pathway inhibitors to lessen neuroinflammation in conditions like MS and amyotrophic lateral sclerosis is now being studied therapeutically (Rusek et al. 2023).

6.4.2.3 NLRP3 Inflammasome

An essential component of the innate immune response is the multiprotein compound known as the NLRP3 inflammasome. When the NLRP3 inflammasome receives signals from ATP, ROS or misfolded proteins, it activates caspase-1 (Brahadeeswaran et al. 2022). This mechanism leads to the synthesis and release of the pro-inflammatory cytokines IL-18 and IL-1β. The NLRP3 inflammasome is highly active in microglia during neuroinflammation. Various neurodegenerative disorders, including Parkinson's disease and Alzheimer's disease, are associated with it (Zhang et al. 2024). The NLRP3 inflammasome is activated by Aβ plaques in AD, causing persistent inflammation and neuronal damage. Pharmacological inhibition of the NLRP3 inflammasome is now being investigated as a potential treatment strategy to reduce neuroinflammation in certain illnesses (Yuan et al. 2023).

6.4.2.4 MAPK Signaling Pathway

The MAPK signaling pathway is a robust mechanism that regulates cellular responses to many stimuli, including inflammation, oxidative stress and infection (Li et al. 2023). The MAPK pathway is essential for regulating neuroinflammation in the CNS because it disturbs the synthesis of chemokines, pro-inflammatory cytokines and other immune mediators. Its activation is especially significant during microglial and astrocytic responses, which are critical in the onset of neurodegenerative illnesses like AD, PD and MS (Peixoto et al. 2015). The three main branches of the MAPK pathway—p38 MAPK, JNK (c-Jun N-terminal kinase) and ERK (extracellular signal-regulated kinase)—are each activated by different stimuli and are in charge of different cellular reactions. In order to translate these discoveries into efficient therapies for neurodegenerative illnesses, numerous MAPK inhibitors are presently being investigated in clinical settings (Jazvinšćak Jembrek et al. 2021; Zhang and Yin 2023).

6.4.3 The Role of Antioxidants and Bioactive Compounds in Inflammation Reduction

Through the respiratory chain, ROS are continuously produced by all cells through energy consumption. When kept at low to moderate levels, ROS can carry out significant physiological functions and act as vital second messengers in cellular communication (Grodzicki and Dziendzikowska 2020). Excessive reactive oxygen species (ROS) pose a threat to living organisms, since long-term exposure may damage vital cellular macromolecules such as proteins, lipids and DNA, leading to necrotic and apoptotic cell death (Kessas et al. 2022). Oxidative stress is produced by a disparity between the creation of ROS and the antioxidant capability of cells. Increased ROS generation or a weakened endogenous antioxidant system could be the source of this imbalance. Biochemical changes in macromolecular components may lead to various human clinical states and disorders. The CNS is particularly vulnerable to oxidative stress due to its elevated oxygen consumption, inadequate antioxidant defences and the final differentiation of neurones. Therefore, it is held that oxidative stress is the primary cause of many neurodegenerative disorders (Elbandy 2022; Cichon et al. 2021).

Several nutraceutical compounds, such as carotenoids, polyphenols and omega-3 fatty acids, have been revealed in numerous trials to decrease the risk of chronic neurological problems when taken daily (Dhalaria et al. 2020). A certain food category may contain a range of neuroprotective substances. Mangosteen pericarp water (MPW) extract protected neurones from harm caused by free radicals and stopped cell death in a mice model of dementia. After getting MPW, mice that had been given scopolamine did better at remembering where things they had been shown (Mladenov et al. 2023). It was found in the leaves of the grass species Acanthus mollis L., which is in the Acanthaceae family. Verbascoside showed promise as a medicine because it stops MAO-A from working, which slows down the breakdown of monoaminergic neurotransmitters. This action is boosted by its capability to remove free radicals, which lowers oxidative stress in brain cells. Thymol and carvacrol are the main phenolic monoterpenes found in Thymus vulgaris essential oil, which is characterised by a high concentration of these compounds (Islam et al. 2022). In the zebrafish model of cognitive impairment, this natural oil shows promise for refining the cholinergic nervous system and offering oxidative protection (Wang et al. 2018). A study assessing a combination of extracts from Hibiscus sabdariffa L. calyces and Olea europaea L. leaves was published by Chiaino and associates. In rat brain slices and the SH-SY5Y cellular model, the researchers found that these extracts demonstrated neuroprotective qualities. Because their different components interacted, their combination had a synergistic effect (Miller et al. 2019). Regular consumption of phytochemicals may strengthen the antioxidant system, which would increase cognitive function and support the survival of brain cells. Natural antioxidants have a useful outcome on lowering cognitive dysfunction and memory loss (Henriques et al. 2020). The extract of Juglans regia L. made the anti-inflammatory response stronger at the BBB. This lowered oxidative stress in brain tissue and made it easier to think and remember things (Ramli et al. 2020).

6.5 Key Anti-inflammatory Nutraceuticals for Brain Ageing

6.5.1 Omega-3 Fatty Acids and Their Anti-inflammatory Properties

Omega-3 dietary additions may aid your brain work better and make you less likely to experience cognitive loss if you take them regularly. In addition to being important for the immune system, omega-3 PUFA might also help with memory and lower the risk of long-term degenerative diseases like AD and trouble paying attention (Cutuli 2017). Omega-3 fatty acids may benefit people with different types of dementia in remembering things and keeping their brains working well. They may also lower the risk of getting these diseases. Scientists have found that omega-3 fatty acids keep brains healthy and make strong brains work better (Mora et al. 2022). People who took 1 gram of fish oil every day, either by itself or with vitamin E and carotenoids, had better feelings and memories, did better on memory tasks and were able to pay more attention and think more clearly (Ahmmed et al. 2020).

Brain swelling from microglia and astrocytes becoming more polarised can cause brain cells to die and brain function problems. PUFAs in omega-3 can lower inflammation and keep neurones healthy (Chen et al. 2024). However, different n-3 PUFA treatments have had different results. The difference in microglial M1/M2 polarization was fixed in the hippocampus by both n-3 PUFAs, with DHA having a stronger effect on CD206 (Troesch et al. 2020). This difference was shown by the increase in IBA1 and the downregulation of CD206 and arginase-1 in these cells. Brain pro-inflammatory cytokines may be higher when microglial M1 and astrocytic A1 polarizations are increased, but DHA is better than EPA at reducing age-related neuroimmunological changes, which makes memory loss worse (Gareri 2022) (Fig. 6.2).

6.5.2 Polyphenols

There are a lot of different kinds of molecules found in nature that are phenolic chemicals. Polyphenols and phenolic compounds are significant secondary metabolites that can be found in many plants, such as fruits, tea, chocolate, herbs, whole grains, spices, edible mushrooms, coffee, fungal fruiting bodies, veggies and red wine (Devi and Chamoli 2020). They show these traits in a number of ways, such as by restoring antioxidant activity, blocking pro-inflammatory enzymes and controling molecules and transcription factors that are complex in inflammatory processes (Hong et al. 2022).

Phenolic chemicals reduce inflammation by lowering the production, activity or gene expression of molecules that cause inflammation. They do this in a number of different ways. They go after COX-2, eicosanoids and iNOS, which stop immune cells from activating and change transcription factors like NF-κB and Nrf-2 (Sarubbo et al. 2017). This has anti-inflammatory and antioxidant effects. In addition to

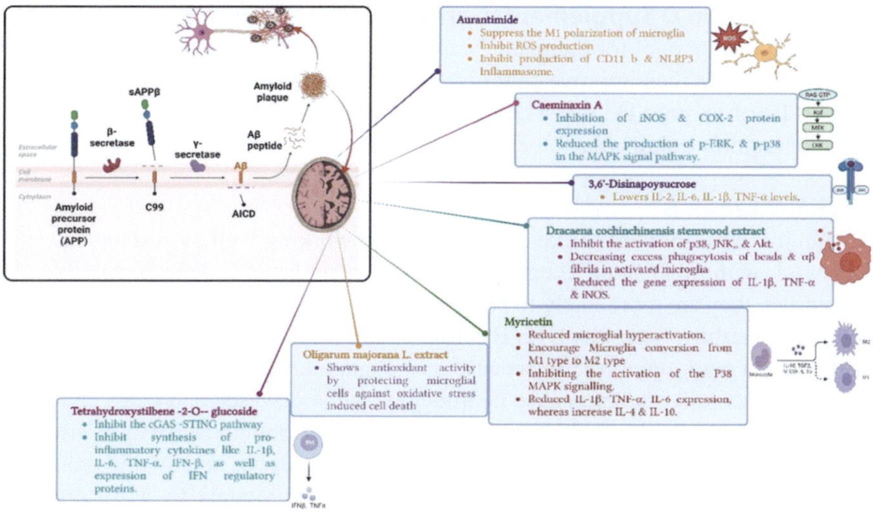

Fig. 6.2 Effects of aurantiamide, caeminaxin A and other neutraceuticals in neuroinflammation and brain ageing

lowering pro-inflammatory markers like IL-1β, IL-6 and TNF-α, phenolic compounds also lower LDL oxidation. This lowers vascular inflammation, the chance of platelets adhering and the effects of oxidative stress and nitric oxide (NO) (Sarubbo et al. 2018).

Some chemicals called flavonoids can stop PLA2, COX and LOX enzymes from doing their jobs. They can also stop COX and LOX proteins from being made. The primary flavonoid discovered to prevent PLA2 in human polymorphonuclear leukocytes was quercetin (Ziegler et al. 2023). It is believed that flavonoids can disrupt signaling networks, such as the NF-KB, protein kinase C and tyrosine kinase pathways, which can prevent the COX gene from being produced (Figueira et al. 2017). Both in the lab and in living organisms, flavones such as apigenin and luteolin, isoflavones, flavonols such as genistein and flavanols such as catechin have been demonstrated to prevent the action of COX and/or LOX enzymes (Rivas et al. 2022).

6.5.3 Vitamins and Minerals

6.5.3.1 Vitamin B Supplementation and Cognitive Function

Folic acid, B6 and B12 are some of the B vitamins that are very important for keeping your brain healthy. Taking vitamin B supplements alone or along with vitamin D, magnesium, vitamin C and docosahexaenoic acid (DHA) led to big improvements in memory and brain power in healthy adults ($p < 0.001$), especially when antioxidant vitamins and magnesium were added (Ford and Almeida 2019).

6.5.3.2 Vitamin D Supplementation and Cognitive Function

Vitamin D is important for brain and heart health. Vitamin D supplements may help prevent dementia then improve cognitive function. However, a study that gave people 600 IU of vitamin D3 every day found that it had no significant effect on their mood or cognitive skills (Aspell et al. 2018).

6.5.3.3 Vitamin B

B vitamins are important for keeping your brain working well as you age because they work with other brain chemicals to make many brain processes possible. As people age, they are more likely to not get enough B vitamins, which can make their brains less healthy. Vitamins B12, B9 (folate) and B6 are very significant for keeping the brain working normally because they change the breakdown of homocysteine, an amino acid that can cause neurodegenerative diseases and heart and brain problems when it is present in large amounts (Ford and Almeida 2019). B vitamins are very important for making neurotransmitters and keeping myelin, the protective covering around nerve cells that is needed for brain signals to work properly (Mikkelsen et al. 2016). B vitamins protect against cognitive loss, which is particularly helpful for older people who are more likely to develop dementia. This suggests that getting enough of these nutrients may be very important for keeping older people's brains healthy (Zhang et al. 2017).

6.5.3.4 Vitamin C, Vitamin E and Other Antioxidants

Taking extra vitamin E, C and other antioxidants may help keep your brain healthy as you age by lowering oxidative stress. Oxidative stress is an important part of cellular ageing and the start of heart and brain vessel problems, neurodegeneration and cognitive decline that comes with getting older (Monacelli et al. 2017).

Vitamin C is a strong antioxidant that protects the brain from reactive stress, which can make it harder to think and remember things. Additionally, vitamin C is necessary for the brain chemicals dopamine and noradrenaline to be made and work (Martin et al. 2002).

Vitamin E is a strong antioxidant that dissolves in fat. Glutathione, selenium, vitamin C, carotenes and carotenoids are some of the other antioxidants that it works with. Vitamin E keeps neural cells from getting damaged by oxidative stress and does a great job of keeping neurones healthy and working well. People know that it can stop lipid peroxidation in cell walls, which is important for keeping brain cells working well (Zaidi and Banu 2004).

Polyphenols, which include resveratrol and curcumin, are a group of chemicals that have powerful anti-inflammatory and antioxidant properties. They play a big role in keeping your brain healthy as you age. Resveratrol, which is created in grapes and red wine, has gotten a lot of attention because it might help protect neurones (Zaidi and Banu 2004). It is thought to stimulate pathways that protect cells from harm and increase blood flow to the brain, which could make thinking better and lower the risk of neurodegenerative diseases. Clinical study recommends that taking resveratrol supplements may help older people remember things and think more clearly (Martin et al. 2000).

Curcumin, the main chemical in turmeric, is a polyphenol that has been studied a lot and is known for having strong anti-inflammatory and antioxidant properties. It has been revealed to be able to cross the BBB and have neuroprotective benefits, which could help stop the cognitive decline that comes with getting older. According to research, curcumin may help people whose memories and mood are getting worse due to normal ageing. Flavonoids, which are found in berries, tea and chocolate, are another type of protein that is important for brain health (Naziroglu et al. 2011).

6.5.3.5 Vitamin D

Vitamin D is very significant for your health and does many things, like keeping your bones healthy and helping your immune system work. It makes it easier for the body to absorb calcium and phosphorus, which are needed for strong bones. It is also important for muscle activity and glucose metabolism (Khairy and Attia 2021). During early childhood, it helps the brain and nervous system grow, and later in life, it is necessary for cognitive processes like remembering, reasoning and attention. Nerve chemicals like dopamine and serotonin are made possible by vitamin D. Mood and feeling are controlled by these chemicals (Terock et al. 2022).

6.5.3.6 Vitamin K

Vitamin K is a significant fat-soluble vitamin that the body needs to work properly. There are two types of it: K1 (phylloquinone) and K2 (menaquinone). Both are necessary for many bodily processes. The main thing it does is control haemostasis by making clotting factors. Vitamin K is important for many processes in cell growth, proliferation, origin and apoptosis, in addition to its main roles (Popa et al. 2021). Keeping your vitamin K levels at a healthy level may also help older people keep their memories. Researchers have found a link between lower amounts of vitamin K in the blood and cognitive decline in persons aged 65 and up. Preclinical studies show that vitamin K2 may protect brain cells from damage caused by Aβ (Ferland 2012; Tsaioun 1999).

6.5.3.7 Mineral Supplementation

Mineral supplements are very significant for improving brain and cognitive function, particularly in older people. Minerals are inorganic substances that are important for the body's functions. The body can't make them itself; it has to get them from food or supplements. It is becoming more and more clear that they have an effect on brain abilities (Fekete et al. 2023). Taking magnesium supplements may help your brain work better and lower your risk of getting dementia. Researchers have found that people whose magnesium levels are high are much less likely to develop dementia. As a protective agent for neural cells and a positive regulator of blood pressure, magnesium is especially good for brain health. It may maintain cognitive function and lower the risk of dementia (Naureen et al. 2022).

Iron is important for many bodily functions, such as supporting brain function and cell processes. Mild iron deficiency can make it hard to focus, lower your immune system and is linked to slower brain development and poor school success.

Researchers have found that taking extra iron can raise IQ scores. On the other hand, iron-deficiency anaemia in older people is related to a progressive risk of dementia and AD (Schättin et al. 2016). Glutathione peroxidase and other antioxidant enzymes can't work without selenium. Oxidative stress is linked to getting older and neurological diseases like PD and AD. These enzymes shield the brain from it. Oxidative stress hurts neuronal cells and lowers cognitive function (Lemon et al. 2016). Because of this, selenium's ability to lower oxidative stress is very important for protecting brain health. Selenium boosts the body's antioxidant protection and lowers the risk of getting sick by making the immune system stronger. Higher levels of selenium in the blood are linked to a lower chance of dementia and some types of cancer. Selenium is important for the health of your immune system and for the thyroid gland to work properly (Gualano et al. 2016).

Zinc is important for growth and immune system function. It is also good for the nervous system and may benefit people who are depressed because it calms them down. Not getting enough zinc can make it harder to think and remember things. Zinc is needed for the protective enzyme superoxide dismutase to work. This enzyme helps get rid of free radicals (White et al. 2016). Copper is needed to make energy, build connective tissue and keep the heart and nervous system healthy. It needs to be taken in balance with zinc. Copper can be found in many foods and can also be bought as a dietary supplement. Copper is needed by the body for many important functions, but there is a complicated link amongst its intake and the growth of diseases like AD. Studies show that both too much and too little copper can be harmful (Dror et al. 2014; Nuttall and Oteiza 2014).

6.5.4 Herbal Extracts

6.5.4.1 Ginkgo biloba

Ginkgo biloba (GB) leaf extracts have been used for thousands of years to treat health problems and are one of the most popular plant products in the world. In some parts of Europe, they are often recommended as a nootropic for older people and people with dementia (Barbalho et al. 2022). GB has several terpenes that are biologically active and unique to this species. These include bilobalide, ginkgolides A, B, C and J and several flavonoid glycosides. These parts of the plant have insect-killing, antifeedant and antibacterial qualities, and environmental stresses make them work harder (Zuo et al. 2017).

There are several possible ways that GB works in the CNS. These include blocking platelet-activating factor, making more nitric oxide bioavailable, and improving blood movement in the brain and other parts of the body (Diamond et al. 2000). It also changes an amount of neurotransmitter systems, such as monoamine oxidase A and the uptake of dopamine, serotonin and norepinephrine by synaptosomes. GB can also get rid of and stop free radicals. It has neuroprotective effects in both lab and living things, like stopping Aβ neurotoxicity, protecting against hypoxic challenges and lowering the effects of higher oxidative stress (Sastre et al. 1998).

6.5.4.2 Melissa officinalis (Lemon Balm)

For over two thousand years, Melissa officinalis (MO) has been utilised as a medication for the purpose of enhancing memory and alleviating anxiety. The majority of the potentially active components consist of monoterpenoids and sesquiterpenes, which include neral, 6-methyl-5-hepten-2-one, geranial, citronellal, β-caryophyllene, β-caryophyllene oxide, geranyl acetate and 1,8-cineole (Kennedy et al. 2003). MO extracts have been shown to have a number of effects on the CNS in vitro. These effects include antioxidant characteristics, direct binding to nicotinic and muscarinic cholinergic receptors in human brain tissue and acetylcholinesterase (AChE)-inhibiting effects due to the combined effects of the components. Extracts of MO bind to GABAA receptors, which consequently prevents the enzyme GABA transaminase from performing its function. Because of this, GABA activity is increased (Mathews et al. 2024; Yoo et al. 2011).

6.5.4.3 Panax ginseng

Root extracts from Panax ginseng, the most common species in the Araliaceae family, have been used for healing for over 5000 years. Ginsenosides, which are 40 or more species-specific triterpene saponins, are supposed to be the main active ingredients. In addition to killing insects, molluscs, fungi, viruses and bacteria, ginsenosides also stop other plants from feeding on them and stop them from spreading (Choi et al. 2021).

Ginseng extracts (GE) defend neurones and the heart by affecting the hypothalamic-pituitary-adrenal axis and neurotransmission through many cellular pathways, such as controlling the production of nitric oxide (de Oliveira Zanuso et al. 2022). Animal behaviour studies show that ginsenosides can help with worry, depression and anxiety, as well as moderate fatigue. They can also improve memory in mice that have had their memories damaged, and they can make learning easier by encouraging neurogenesis and changing LTP in the hippocampus (Lin et al. 2024).

6.5.4.4 Salvia officinalis and Salvia lavandulifolia (Sage)

Sage has been used for a long time to improve memory and slow down brain decline. It was first used in ancient Greece. Salvia officinalis and Salvia lavandulaefolia (SL), the two most common types of sage, have similar chemical makeups. The only difference is that SL has much less of the possibly dangerous monoterpenoid thujone, which blocks GABAA receptors (Lopresti 2017). Some of the most important active ingredients are polyphenolic compounds and a group of monoterpenes, including camphor, borneol, α-pinene, limonene and geraniol. Both essential oils and hydro-alcoholic solutions of sage have proved the ability to prevent human acetylcholinesterase and butyrylcholinesterase in laboratory settings (Edwards et al. 2021). This inhibition was also observed in rodent brains following oral treatment. 1,8-Cineole has demonstrated the most efficacy as a singular component in inhibiting cholinesterase. Nonetheless, the overall efficacy is greater when the components function in unison (or in opposition) (Kennedy and Scholey 2005).

6.5.4.5 Valeriana officinalis L. (Valerian)

Valerian extracts (VE) have been used for a long time to help people relax and calm down. These effects are caused by a group of monoterpenes and sesquiterpenes, especially valepotriates and valerenic acid, which are unique to the Valerian genus. There are also large amounts of GABA in root products (Malva et al. 2004). A lot of neurotransmitter receptors are affected by constituents. One of these is the 5-HT5A receptor, which is linked to anxiety and diurnal rhythms. As allosteric modulators of subunit-specific GABAA channels and adenosine A1 receptors, they do their job. Valerenic acid and its derivatives have strong antianxiety effects in rodent models. These effects are backed by the fact that they bind to a precise subunit (β3) of the GABAA receptor while they are in living things (Nam et al. 2013; Sudati et al. 2009).

6.5.4.6 Phenolics

A diverse array of phenolic pharmaceuticals interacts directly with neurotransmitter systems. In animal models, various flavonoids present in traditional medicinal extracts exhibit anxiolytic and calming effects through binding to GABAA receptors, enhance cognitive function by inhibiting GABAA receptors and promoting cholinergic upregulation and demonstrate antidepressant properties by inhibiting monoamine oxidase, resulting in raised levels of serotonin, dopamine and noradrenaline in precise brain regions (Sudati et al. 2009). Some of the phenolic phytoestrogens that plants make are isoflavonones, flavones, stilbenes and lignans. These chemicals protect plants from herbivores by messing with insects' endocrine systems and changing their life cycles (Mungali et al. 2021).

6.5.4.7 Curcumin

Curcumin affects a huge number of molecular targets. This encompasses transcription factors such as NF2, which regulates the antioxidant response; protein kinase enzymes, crucial for numerous cellular pathways, particularly those associated with signal transduction; enzymes like heme oxygenase 1, a stress response protein whose expression increases following curcumin consumption and is linked with neuroprotection; invasion and angiogenesis biomarkers such as matrix metalloproteinase 9, involved in various functions including tissue repair; as well as inflammatory mediators like NF-κB and cytokines such as TNFα, IL-1 and IL-6. Curcumin has demonstrated the ability to prevent the degradation of amyloid precursor protein and, contingent upon dosage, impede the production and proliferation of Aβ fibrils while also destabilising existing fibrils (Benameur et al. 2021; Shen et al. 2013).

6.5.4.8 EGCG

Several catechin antioxidants found in tea (Camellia sinensis L.) are thought to have belongings that make them biologically active. There are four main flavanols in tea: (−)-epigallocatechin, EGCG, (−)-epicatechin and (−)-epicatechin-3-gallate. EGCG is usually thought to be the most active ingredient in green tea (Davinelli et al. 2013).

In vitro activities of EGCG that might protect neurones comprise metal chelation, anti-inflammatory properties and lowering Aβ levels. It may also mitigate the harm that Aβ inflicts on neurones (Luo et al. 2022). Certain advantageous effects occur via activating cell survival genes and altering the signaling mechanisms of protein kinase C. EGCG has demonstrated the ability to enhance cholinergic signaling, promote neurite outgrowth and maintain stable cerebral blood flow in healthy individuals (Hong et al. 2022).

6.5.4.9 Hypericum perforatum (St. John's Wort)

Hypericum perforatum (HP) extracts have been utilised for medicinal purposes since the era of the ancient Greeks. HP comprises many compounds that may exert biological effects. This encompasses phenolic acids (such as chlorogenic acid) and flavonoids (including rutin, quercetin, quercitrin, hyperoside, isoquercitrin and epigenanin), along with organizationally analogous phloroglucinol derivatives like hyperflorin and naphthodianthrones such as hypericin (Zirak et al. 2019). At first, it was thought that naphthodianthrones and, more recently, hyperforin and other flavonoid components gave HP its sedative and anti-inflammatory effects (Greeson et al. 2001).

HP has influential anti-inflammatory and antibacterial properties. It also affects the way mammals' brains work by stopping neurones from reabsorbing serotonin (5-HT), gamma-aminobutyric acid (GABA), dopamine (DA), norepinephrine and L-glutamate. It does this while also making neurotransmitters more sensitive and increasing receptor binding. Neuroprotection, less cognitive impairment and better cognitive performance in rodents are some of the functional benefits (Barnes et al. 2001; Valvassori et al. 2018).

6.5.4.10 Resveratrol

Resveratrol (3, 4′, 5-trihydroxystilbene), a phytoalexin, is made by many edible plants when they are exposed to environmental pressures and tissue damage, such as fungal and viral attacks (Sarubbo et al. 2018).

Some possible ways that these substances protect neurones are by increasing blood flow and perfusion and making antioxidant defences stronger. In living things, these benefits are probably caused by resveratrol making the body's natural antioxidant defences stronger by increasing the activity of numerous antioxidant enzymes (Sarubbo et al. 2017). This might be partly because the Nrf2 transcription factor is being activated. This factor is important for the regulatory cellular redox state and changing protein kinases that protect neurones from Aβ-induced damage in both the lab and the brain, especially in the hippocampus. In vivo oral treatment has been shown to stop the formation of Aβ plaques in a precise area in transgenic mice (Zhou et al. 2021).

6.5.4.11 Soy Isoflavones

Soy extracts comprise several soy isoflavones (SI), with genistein, daidzein and glycitein, which demonstrate structural resemblances to oestrogen and possess oestrogenic characteristics, such as direct binding to oestrogen receptors, aromatase inhibition and disruption of oestrogen signaling (Lee et al. 2005).

Animal models have shown improvements in memory ability in ovariectomised mice and middle-aged or older rats after supplement with both genistein and SI, including boosts in cholinergic and prefrontal dopaminergic determination (Hsieh et al. 2009).

6.6 Mechanisms of Action of Anti-inflammatory Nutraceuticals

6.6.1 Neuroinflammation and Microglia Dysregulated Genes in AD and the Effects of Phytochemicals

The role of glial initiation and immune cell dysfunction is well-documented in AD. Single-cell RNA sequencing exposed transcriptomic anomalies in the brain related to AD, specifically among certain microglial subtypes. Single-cell RNA sequencing of peripheral blood revealed augmented immune cell signs and diminished B cell-associated molecular markers in AD (Sharifi-Rad et al. 2022; Abdolmaleky and Zhou 2023). Furthermore, to elucidate the processes behind immune cell dysfunction, an extensive whole-genome DNA methylation study of post-mortem brain and blood samples revealed that alterations in DNA methylation linked to AD result in immune system dysregulation (Prasanth et al. 2024). Supplementary findings indicate that heightened DNA methylation of TREM2 in the larger temporal gyrus is related to the aetiology of Alzheimer's disease, particularly with significant microglial innate immune receptor genes. A study indicates elevated TREM2 expression in the blood cells of AD patients, accompanied by DNA hypomethylation of numerous CpGs in its intron 1 (Davinelli et al. 2016).

Multiple studies have shown that phytochemicals may reduce neuroinflammation and progress cognitive function associated with TREM2 in mouse models of AD. In APP/PSEN1 transgenic mice models of AD, Bilberry anthocyanins were revealed to mitigate cognitive damage and hippocampal neuroinflammation, while promoting microglial phagocytosis of Aβ plaques via the upregulation of the TREM2/TYROBP/CD33 signaling pathway (Uddin et al. 2021). A current study revealed that Cyanidin-3-O-glucoside (C3G), a phytochemical generated in fruits and vegetables, may decrease neuroinflammation and ROS, while enhancing microglial Aβ42 phagocytosis through TREM2 activation in a mice model of AD (Limanaqi et al. 2020).

6.6.2 Brain Neuronal Stem Cells (NSCs) in Ageing and the Effects of Phytochemicals

Ageing is related to a decline in the overall number of neural stem cells in the hippocampus, diminishing from over 20,000 per square millimetre at 3 months to approximately 7000 at 12 months in mice. Moreover, quiescent neural stem cells (qNSCs), which shift from a dormant condition to a primed quiescent state upon activation and eventually generate neurones and glia, attain a deeper quiescent state due to ageing. As a result, the restorative ability of the aged brain is reduced (Abdolmaleky and Zhou 2023; Hamedi et al. 2015).

Research indicates that homolocarpum seed oil, abundant in α-linolenic acid and β-sitosterol, enhances NSC proliferation, suggesting its potential as a technique to mitigate brain ageing. The seed extract of Alyssum homolocarpum has been observed to enhance neural stem cell proliferation in murine brains. Kuwanon V, extracted from the mulberry tree (Morus bombycis), has demonstrated the ability to enhance neurogenesis in rat neural stem cells and promote cell survival (Zong et al. 2021). This process is enabled by decreased phosphorylation of extracellular signal-regulated kinase 1/2, increased expression of p21, downregulation of Notch/Hairy and upregulation of the microRNAs miR-181a, miR-29a and miR-9. Silibinin, a polyphenolic flavonoid derived from Silybum marianum, enhances neural stem cell proliferation in mice via BDNF/TrkB signaling transduction (Li et al. 2024).

6.6.3 Telomere Attrition and Ageing and the Effects of Phytochemicals

The diminution of telomere length, frequently noted during cellular replication, is an additional factor affecting ageing. Psychological stress, chronic infections, mitochondrial dysfunction and ROS are recognised to reduce telomere length (Ismaila et al. 2025). Chronic inflammation induced by HIV, herpes, CMV and hepatitis B and C diseases diminishes telomere length. Elongated telomeres have a crucial role in the development of long non-coding RNAs that modulate the expression of NF-kB, VEGF, MYC and DNMTs associated with inflammation. Telomere length disturbs the expression of multiple other genes through looping and different mechanisms (D'Angelo 2023; Praengam et al. 2024).

Telomerase enzymes safeguard against telomere attrition, and cells with elevated telomerase levels display elongated telomeres; nonetheless, the temporary consumption of fried brassica green vegetables may enhance telomerase activity in human CD8+ lymphocytes (Alibakhshi et al. 2016). Furthermore, data indicate that the chemical constituents of Astragalus membranaceus stimulate telomerase, impede senescence and confer neuroprotective benefits. The bark extract of Eucalyptus camaldulensis enhances telomerase expression, demonstrating anti-ageing, anti-apoptotic and anti-senescence properties (Sharma and Martins 2020).

6.6.4 Gut Microbiome, the Dysfunction of Brain Microglia and Astrocytes in Brain Ageing and Phytochemical Effects

The enteric peripheral nervous system, akin to the brain, contains glial cells that respond to the makeup of gut bacteria, as well as to infection and inflammation. Glial cell dysfunction has been observed in neurodegenerative illnesses, and changes in gut microbiota have often been described in AD and PD (Abdolmaleky and Zhou 2023). Moreover, research indicates that faecal transplantation from aged mice to juvenile mice affects their performance in spatial learning and memory assessments. This pertains to the modified expression of hippocampal proteins associated with neurotransmission and synaptic plasticity, and the ageing-like characteristics of the receiver mice's microglia in the hippocampal fimbria (Kim et al. 2024).

Clostridium butyricum may inhibit microglial activation and Aβ buildup, correlating with decreased inflammatory cytokines and enhanced cognitive performance in a rat model of AD (Wang et al. 2022).

This evidence together suggests that dysbiotic gut microbiota may provoke astroglial inflammation and inflammation-mediated oxidative stress, potentially modifying the epigenetic profiles of microglia or astrocytes, hence initiating neurological disorders (Sarubbo et al. 2023). Conversely, accumulating data indicate that probiotics, dietary choices and phytochemicals influence gut microbial arrangement and may assist in the inhibition or treatment of AD and age-related neurodegenerative disorders (Wang et al. 2018).

6.6.5 Metabolic Disease, Caloric Restriction, Physical Exercise and Ageing

Overeating causes insulin resistance and mTOR activation, which contribute to AD. Caloric restriction mitigates ageing by decreasing glucose, lipids and amino acids while augmenting NAD+ and AMP, which moderate SIRT1, IGF1, mTOR and AMPK (Normandin et al. 2017). It enhances mitochondrial physiology and haemostasis by regulating FOXO and PGC1α. Metformin, a widely used antidiabetic medication sourced from the Galega officinalis plant, modulates many pathways associated with ageing and age-linked conditions by activating AMPK, enhancing SIRT1 expression and preventing mTOR, ROS and NF-kB signaling (Fontana and Klein 2007). Epicatechin, resveratrol and the NAD+ precursor nicotinamide riboside (a derivative of vitamin B3) enhance mitochondrial function, similar to metformin. Similar to metformin, physical activity elevates AMPK levels, which therefore enhance IL-15 expression in muscle tissue. Nonetheless, the activities of both AMPK and IL-15 diminish with advancing age. Exercise mitigates age-linked chronic and sterile inflammation as well as immunosenescence, which diminish mitochondrial functions and exacerbate mitochondrial damage (Broskey et al. 2019; Mercken et al. 2012).

6.6.6 X Inactivation and Neurodegeneration

XIST, a long non-coding RNA that deactivates one X chromosome in female cells, is a major cause of female AD. In female AD patients, the brain and entorhinal cortex express more XIST (Chanda and Mukhopadhyay 2020). Single-nuclei RNA sequencing showed that female mice's hypothalamus neurones produce more XIST with age, a strong predictor of neuronal ageing. Research shows that XIST causes Aβ accumulation and neuroinflammation in AD animal models (Voskuhl and Itoh 2022). This advises that calorie restriction through fibre-rich foods like the Mediterranean diet and nutritional control of diabetes (which increases AD risk) may suppress XIST synthesis. Further research is needed to determine how phytochemicals affect female XIST expression (Bajic et al. 2015).

6.6.7 Vascular System and Neurodegeneration

Neuronal degeneration is linked to vascular dysfunction and endothelial cell deterioration. In old or irradiated mice, endothelial cells produce more TGFB1, which kills brain stem/progenitor cells. Targeted TGFB signaling inhibitors can prevent this. Dyslipidaemia may cause endothelial cell senescence, which, together with atherosclerosis, disturbs their integrity and permeability, affecting the BBB and cerebral functions (Ahmad et al. 2020). Complement C3, raised in aged hippocampal astrocytes, affects endothelial cell surface C3aR1, producing inflammation and vascular permeability. In contrast to withaferin A and active portions of golden-flowered tea, black chokeberry (Aronia melanocarpa L.) active ingredients inhibit brain cell C3 receptor expression and TGFB1 expression (Akinyemi et al. 2013).

6.7 Clinical Evidence and Research on Anti-inflammatory Nutraceuticals

Numerous studies has examined the Mediterranean Diet's neurodegenerative disease prevention and treatment. These studies found that the Mediterranean Diet decreases neurodegenerative disorders, improves cognitive performance and slows cerebral ageing (Table 6.1).

6.8 Challenges and Future Directions

Previous studies have shown that inflammatory reactions are the main cause and strongly impact AD. Early diagnosis and medication using illness biomarkers helps avoid inflammatory reactions and produce selective inhibitors targeting inflammatory cascade mediators (Lustig et al. 2009). Astrocytes and microglia are crucial to brain development and physiology. Astrocytes stabilise, insulate and cushion extracellular compartments to protect the CNS. These supportive cells fight inflammation and proliferate to block damaged areas. Astroglial cells protect neurones from toxicity and provide metabolic support for optimal performance via influencing synaptic plasticity and transmission. Also, astrocytes help start and advance AD (Bonfanti and Charvet 2021).

Anti-inflammatory drugs have been used in AD models, but they haven't provided any protection. Additional research is desired to determine anti-inflammatory responses and find a model that enhances therapeutic agent benefits (Gutchess 2014). Novel AD preventive medicines using phytochemicals have received attention recently. Polyphenolic substances, nutraceuticals and antioxidants have been shown to reduce neuroinflammation in AD and other neurodegenerative illnesses. Scientists believe phytochemicals and polyphenols may treat AD and prevent it by modulating inflammation-related pathways (Bijsterbosch et al. 2020). Many polyphenols have been shown to be effective AD preventatives. However, phytochemicals need further investigation to become new AD preventives. Anti-inflammatory

Table 6.1 Status and rationale of selected nutraceuticals targeting various neurodegenerative diseases clinical trials

Nutraceutical	Neurodegenerative disease targeted	Current clinical trial status	Rationale for use	References
Curcumin	AD	Ongoing phase II	Recognized for its anti-inflammatory and antioxidant properties, it reduces neuroinflammation	Benameur et al. (2021)
Omega-3 fatty acids	AD, PD	Completed phase III	Supports neuronal integrity and reduces neuroinflammation	Ahmmed et al. (2020)
GB	AD, dementia	Ongoing phase II	Antioxidant properties that shield brain cells from oxidative stress	Barbalho et al. (2022)
Resveratrol	AD, PD	Ongoing phase I	Modulates inflammatory pathways and protects against oxidative stress	Gomes et al. (2018)
Vitamin D	AD	Completed phase II	Modulates immune response and neuroinflammation	Ford and Almeida (2019
Panax ginseng	AD, PD	Ongoing phase I	Recovers cognitive function and reduces inflammation in neurodegenerative diseases	de Oliveira Zanuso et al. (2022)
Bacopa monnieri	AD, cognitive decline	Ongoing phase III	Enhances memory, reduces oxidative stress and protects against neurodegeneration	Abdul Manap et al. (2019)
Green tea extract (EGCG)	AD, PD	Ongoing phase II	Antioxidant, reduces amyloid plaques and oxidative damage	Fernandes et al. (2021)
Ashwagandha	AD, PD	Ongoing phase II	Reduces inflammation and promotes neurogenesis and stress resistance	Mikulska et al. (2023)
Magnesium	AD, cognitive decline	Completed phase II	Improves cognitive function and protects against neuronal damage	Fekete et al. (2023
Alpha-lipoic acid	AD, PD	Completed phase III	Antioxidant properties that shield the brain from oxidative stress	Gomes et al. (2025)
Selenium	AD, PD	Ongoing phase II	Reduces oxidative stress and protects neurones	Pyka et al. (2024)
Nigella sativa	AD, PD	Ongoing phase II	Exhibits anti-inflammatory and neuroprotective effects	Khazdair (2015)
Coenzyme Q10	PD, AD	Ongoing phase III	Enhances mitochondrial function and energy production	Mantle et al. (2022)

drugs have been used in AD models, but they haven't provided any protection (Mohammadi and Karwowski 2024). Additional research is needed to determine anti-inflammatory responses and find a model that enhances therapeutic agent benefits. Novel AD preventive medicines using phytochemicals have received attention recently. Polyphenolic substances, nutraceuticals and antioxidants have been shown to reduce neuroinflammation in AD and other neurodegenerative illnesses. Scientists believe phytochemicals and polyphenols may treat AD and prevent it by modulating inflammation-related pathways. Many polyphenols are effective AD preventatives. However, phytochemicals need further investigation to become new AD preventives (Dent et al. 2019; Nilsson and Lövdén 2018).

6.9 Conclusion

Several natural compounds have beneficial effects on neurodegenerative diseases, including AD and PD, according to the clinical, preclinical and experimental results offered here. The chemicals that were tested all show strong effects on the immune system, neuroprotection and biology. Consequently, the results show that these nutraceuticals are quite effective in treating neurodegenerative diseases, and it is highly recommended to use them as a supplement to traditional pharmaceutical treatments. However, further studies are needed to determine how effective these nutraceuticals are as a treatment. This is especially true in clinical trials where small sample numbers or toxicological consequences are not taken into account. In order to manage neurodegenerative disorders and their related treatments, it is important to assess the safety, bioavailability and tolerability of these natural compounds across a large cohort. This will help determine which nutraceuticals are most effective, while also encouraging a change in lifestyle toward a more nature-oriented model.

References

Abdolmaleky HM, Zhou J-R (2023) Underlying mechanisms of brain aging and neurodegenerative diseases as potential targets for preventive or therapeutic strategies using phytochemicals. Nutrients 15(15):3456

Abdul Manap AS et al (2019) Bacopa monnieri, a neuroprotective Lead in Alzheimer disease: a review on its properties, mechanisms of action, and preclinical and clinical studies. Drug Target Insights 13:1177392819866412

Ahmad A et al (2020) The role of neurovascular system in neurodegenerative diseases. Mol Neurobiol 57:4373–4393

Ahmmed MK et al (2020) Marine omega-3 (n-3) phospholipids: a comprehensive review of their properties, sources, bioavailability, and relation to brain health. Compr Rev Food Sci Food Saf 19(1):64–123

Akinyemi O et al (2013) Vascular risk factors and neurodegeneration in ageing related dementias: Alzheimer's disease and vascular dementia. Curr Alzheimer Res 10(6):642–653

Alibakhshi A et al (2016) An update on phytochemicals in molecular target therapy of cancer: potential inhibitory effect on telomerase activity. Curr Med Chem 23(22):2380–2393

Angiulli F et al (2021) Blood-based biomarkers of neuroinflammation in Alzheimer's disease: a central role for periphery? Diagnostics 11(9):1525
Aspell N, Lawlor B, O'Sullivan M (2018) Is there a role for vitamin D in supporting cognitive function as we age? Proc Nutr Soc 77(2):124–134
Bajic V et al (2015) Skewed X-chromosome inactivation in women affected by Alzheimer's disease. J Alzheimer's Dis 43(4):1251–1259
Barbalho SM et al (2022) Ginkgo biloba in the aging process: a narrative review. Antioxidants 11(3):525
Barnes J, Anderson LA, Phillipson JD (2001) St John's wort (Hypericum perforatum L.): a review of its chemistry, pharmacology and clinical properties. J Pharm Pharmacol 53(5):583–600
Benameur T et al (2021) Curcumin as prospective anti-aging natural compound: focus on brain. Molecules 26(16):4794
Bijsterbosch J et al (2020) Challenges and future directions for representations of functional brain organization. Nat Neurosci 23(12):1484–1495
Blinkouskaya Y et al (2021) Brain aging mechanisms with mechanical manifestations. Mech Ageing Dev 200:111575
Bonfanti L, Charvet CJ (2021) Brain plasticity in humans and model systems: advances, challenges, and future directions. Int J Mol Sci 22(17):9358
Brahadeeswaran S, Sivagurunathan N, Calivarathan L (2022) Inflammasome signaling in the aging brain and age-related neurodegenerative diseases. Mol Neurobiol 59(4):2288–2304
Brandl S, Reindl M (2023) Blood–brain barrier breakdown in Neuroinflammation: current in vitro models. Int J Mol Sci 24(16):12699
Broskey NT et al (2019) The panacea of human aging: calorie restriction versus exercise. Exerc Sport Sci Rev 47(3):169–175
Calder PC et al (2017) Health relevance of the modification of low grade inflammation in ageing (inflammageing) and the role of nutrition. Ageing Res Rev 40:95–119
Candelario-Jalil E, Dijkhuizen RM, Magnus T (2022) Neuroinflammation, stroke, blood-brain barrier dysfunction, and imaging modalities. Stroke 53(5):1473–1486
Cassano P et al (2016) Review of transcranial photobiomodulation for major depressive disorder: targeting brain metabolism, inflammation, oxidative stress, and neurogenesis. Neurophotonics 3(3):031404
Catorce MN, Gevorkian G (2020) Evaluation of anti-inflammatory nutraceuticals in lps-induced mouse neuroinflammation model: an update. Curr Neuropharmacol 18(7):636–654
Chanda K, Mukhopadhyay D (2020) LncRNA Xist, X-chromosome instability and Alzheimer's disease. Curr Alzheimer Res 17(6):499–507
Chen WW, Zhang X, Huang WJ (2016) Role of neuroinflammation in neurodegenerative diseases. Mol Med Rep 13(4):3391–3396
Chen T-B et al (2024) Neuroimmunological effects of omega-3 fatty acids on migraine: a review. Front Neurol 15:1366372
Choi S-H et al (2021) Ginseng gintonin, aging societies, and geriatric brain diseases. Integrative Medicine Research 10(1):100450
Cichon N et al (2021) Unusual bioactive compounds with antioxidant properties in adjuvant therapy supporting cognition impairment in age-related neurodegenerative disorders. Int J Mol Sci 22(19):10707
Currais A (2015) Ageing and inflammation–a central role for mitochondria in brain health and disease. Ageing Res Rev 21:30–42
Cutuli D (2017) Functional and structural benefits induced by omega-3 polyunsaturated fatty acids during aging. Curr Neuropharmacol 15(4):534–542
D'Angelo S (2023) Diet and aging: the role of polyphenol-rich diets in slow down the shortening of telomeres: a review. Antioxidants 12(12):2086
Dama A et al (2024) Targeting metabolic diseases: the role of nutraceuticals in modulating oxidative stress and inflammation. Nutrients 16(4):507
Davinelli S et al (2013) Synergistic effect of L-carnosine and EGCG in the prevention of physiological brain aging. Curr Pharm Des 19(15):2722–2727

Davinelli S et al (2016) Dietary phytochemicals and neuro-inflammaging: from mechanistic insights to translational challenges. Immun Ageing 13:1–17

de Oliveira Zanuso B et al (2022) Panax ginseng and aging related disorders: a systematic review. Exp Gerontol 161:111731

Delage C et al (2021) Traumatic brain injury: an age-dependent view of post-traumatic neuroinflammation and its treatment. Pharmaceutics 13(10):1624

Dent E et al (2019) Management of frailty: opportunities, challenges, and future directions. Lancet 394(10206):1376–1386

Devi SA, Chamoli A (2020) Polyphenols as an effective therapeutic intervention against cognitive decline during normal and pathological brain aging. In: Reviews on new drug targets in age-related disorders, pp 159–174

Dhalaria R et al (2020) Bioactive compounds of edible fruits with their anti-aging properties: a comprehensive review to prolong human life. Antioxidants 9(11):1123

Diamond BJ et al (2000) Ginkgo biloba extract: mechanisms and clinical indications. Arch Phys Med Rehabil 81(5):668–678

Dror Y, Stern F, Gomori MJ (2014) Vitamins in the prevention or delay of cognitive disability of aging. Current Aging Science 7(3):187–213

Duan X et al (2016) Intracerebral hemorrhage, oxidative stress, and antioxidant therapy. Oxidative Med Cell Longev 2016(1):1203285

Dunleavy C et al (2022) Inflammation in first-episode psychosis: the contribution of inflammatory biomarkers to the emergence of negative symptoms, a systematic review and meta-analysis. Acta Psychiatr Scand 146(1):6–20

Edwards KD et al (2021) Assessment of the effects of a sage (Salvia officinalis) extract on cognitive performance in adolescents and young adults. Medrxiv, p. 2021.05. 28.21257776

Elbandy M (2022) Anti-inflammatory effects of marine bioactive compounds and their potential as functional food ingredients in the prevention and treatment of neuroinflammatory disorders. Molecules 28(1):2

Farì G et al (2021) The effect of physical exercise on cognitive impairment in neurodegenerative disease: from pathophysiology to clinical and rehabilitative aspects. Int J Mol Sci 22(21):11632

Fekete M et al (2023) Improving cognitive function with nutritional supplements in aging: a comprehensive narrative review of clinical studies investigating the effects of vitamins, minerals, antioxidants, and other dietary supplements. Nutrients 15(24):5116

Ferland G (2012) Vitamin K, an emerging nutrient in brain function. Biofactors 38(2):151–157

Fernandes L et al (2021) Green tea polyphenol epigallocatechin-Gallate in amyloid aggregation and neurodegenerative diseases. Front Neurosci 15:718188

Figueira I et al (2017) Polyphenols beyond barriers: a glimpse into the brain. Curr Neuropharmacol 15(4):562–594

Fischer R, Maier O (2015) Interrelation of oxidative stress and inflammation in neurodegenerative disease: role of TNF. Oxidative Med Cell Longev 2015(1):610813

Fontana L, Klein S (2007) Aging, adiposity, and calorie restriction. JAMA 297(9):986–994

Ford AH, Almeida OP (2019) Effect of vitamin B supplementation on cognitive function in the elderly: a systematic review and meta-analysis. Drugs Aging 36:419–434

Gareri P (2022) Omega-3 long-chain polyunsaturated fatty acids in the elderly: a review. OBM Geriatrics 6(2):1–28

Ghowsi M et al (2021) Inflammation, oxidative stress, insulin resistance, and hypertension as mediators for adverse effects of obesity on the brain: a review. Biomedicine 11(4):13

Gomes BAQ et al (2018) Neuroprotective mechanisms of resveratrol in Alzheimer's disease: role of SIRT1. Oxidative Med Cell Longev 2018:8152373

Gomes BAQ et al (2025) Alpha-Lipoic acid reduces Neuroinflammation and oxidative stress induced by Dapsone in an animal model. Nutrients 17(5):791

Greeson JM, Sanford B, Monti DA, St. (2001) John's wort (Hypericum perforatum): a review of the current pharmacological, toxicological, and clinical literature. Psychopharmacology 153:402–414

Grodzicki W, Dziendzikowska K (2020) The role of selected bioactive compounds in the prevention of Alzheimer's disease. Antioxidants 9(3):229

Gualano B et al (2016) Creatine supplementation in the aging population: effects on skeletal muscle, bone and brain. Amino Acids 48:1793–1805

Gutchess A (2014) Plasticity of the aging brain: new directions in cognitive neuroscience. Science 346(6209):579–582

Hagberg H et al (2015) The role of inflammation in perinatal brain injury. Nat Rev Neurol 11(4):192–208

Hamedi A et al (2015) Alyssum homolocarpum seeds: phytochemical analysis and effects of the seed oil on neural stem cell proliferation and differentiation. J Nat Med 69:387–396

Henriques JF et al (2020) The anti-neuroinflammatory role of anthocyanins and their metabolites for the prevention and treatment of brain disorders. Int J Mol Sci 21(22):8653

Hong M et al (2022) Tea polyphenols as prospective natural attenuators of brain aging. Nutrients 14(15):3012

Hsieh H-M, Wu W-M, Hu M-L (2009) Soy isoflavones attenuate oxidative stress and improve parameters related to aging and Alzheimer's disease in C57BL/6J mice treated with D-galactose. Food Chem Toxicol 47(3):625–632

Islam F et al (2022) Bioactive compounds and their derivatives: an insight into prospective phytotherapeutic approach against alzheimer's disease. Oxidative Med Cell Longev 2022(1):5100904

Ismail H et al (2020) Traumatic brain injury: oxidative stress and novel anti-oxidants such as mitoquinone and edaravone. Antioxidants 9(10):943

Ismaila M, Adnan M, Sasidharan S (2025) Exploring medicinal plants and phytochemicals for telomere maintenance and anti-aging: insights into telomerase activation and longevity. Rev Bras 35:1–16

Jain M et al (2021) Role of JAK/STAT in the neuroinflammation and its association with neurological disorders. Ann Neurosci 28(3–4):191–200

Jazvinšćak Jembrek M et al (2021) Anti-oxidative, anti-inflammatory and anti-apoptotic effects of flavonols: targeting Nrf2, NF-κB and p53 pathways in neurodegeneration. Antioxidants 10(10):1628

Jęśko H et al (2017) S irtuins and their roles in brain aging and neurodegenerative disorders. Neurochem Res 42:876–890

Jin R et al (2022) Relationships between inflammation and age-related neurocognitive changes. Int J Mol Sci 23(20):12573

Kennedy DO, Scholey AB (2005) Sage and brain function

Kennedy D et al (2003) Modulation of mood and cognitive performance following acute administration of single doses of Melissa officinalis (Lemon balm) with human CNS nicotinic and muscarinic receptor-binding properties. Neuropsychopharmacology 28(10):1871–1881

Kessas K et al (2022) Role of bioactive compounds in the regulation of mitochondrial dysfunctions in brain and age-related neurodegenerative diseases. Cells 11(2):257

Khairy EY, Attia MM (2021) Protective effects of vitamin D on neurophysiologic alterations in brain aging: role of brain-derived neurotrophic factor (BDNF). Nutr Neurosci 24(8):650–659

Khazdair MR (2015) The protective effects of Nigella sativa and its constituents on induced neurotoxicity. J Toxicol 2015:841823

Kim Y, Lim J, Oh J (2024) Taming neuroinflammation in Alzheimer's disease: the protective role of phytochemicals through the gut– brain axis. Biomed Pharmacother 178:117277

Kocsis AE et al (2025) Much more than nutrients: the protective effects of nutraceuticals on the blood–brain barrier in diseases. Nutrients 17(5):766

Kurowska A et al (2023) The role of diet as a modulator of the inflammatory process in the neurological diseases. Nutrients 15(6):1436

Lassmann H, van Horssen J (2016) Oxidative stress and its impact on neurons and glia in multiple sclerosis lesions. Biochim et Biophys Acta (BBA)-Mol Basis Disease 1862(3):506–510

Laurindo LF et al (2023) Phytochemicals and regulation of NF-kB in inflammatory bowel diseases: an overview of in vitro and in vivo effects. Meta 13(1):96

Lee Y-B, Lee HJ, Sohn HS (2005) Soy isoflavones and cognitive function. J Nutr Biochem 16(11):641–649
Lemon J et al (2016) A multi-ingredient dietary supplement abolishes large-scale brain cell loss, improves sensory function, and prevents neuronal atrophy in aging mice. Environ Mol Mutagen 57(5):382–404
Li X et al (2023) Inflammation and aging: signaling pathways and intervention therapies. Signal Transduct Target Ther 8(1):239
Li L, Wang L, Zhang L (2024) Therapeutic potential of natural compounds from herbs and nutraceuticals in alleviating neurological disorders: targeting the Wnt signaling pathway. J Agric Food Chem 72(5):2411–2433
Limanaqi F et al (2020) Merging the multi-target effects of phytochemicals in neurodegeneration: from oxidative stress to protein aggregation and inflammation. Antioxidants 9(10):1022
Lin L et al (2024) Research on the anti-aging mechanisms of Panax ginseng extract in mice: a gut microbiome and metabolomics approach. Front Pharmacol 15:1415844
Lopresti AL (2017) Salvia (sage): a review of its potential cognitive-enhancing and protective effects. Drugs R&D 17(1):53–64
Luo Y-P et al (2022) Epigallocatechin-3-gallate alleviates galactose-induced aging impairment via gut–brain communication. Food Funct 13(21):11200–11209
Lustig C et al (2009) Aging, training, and the brain: a review and future directions. Neuropsychol Rev 19:504–522
Mac Giollabhui N et al (2021) The longitudinal associations of inflammatory biomarkers and depression revisited: systematic review, meta-analysis, and meta-regression. Mol Psychiatry 26(7):3302–3314
Malva JO, Santos S, Macedo T (2004) Neuroprotective properties of Valeriana officinalis extracts. Neurotox Res 6:131–140
Mantle D, Heaton RA, Hargreaves IP (2022) Coenzyme Q10, ageing and the nervous system: an overview. Antioxidants 11(1):2
Martin A et al (2000) Effect of fruits, vegetables, or vitamin E–rich diet on vitamins E and C distribution in peripheral and brain tissues: implications for brain function. J Gerontol Ser A Biol Med Sci 55(3):B144–B151
Martin A et al (2002) Roles of vitamins E and C on neurodegenerative diseases and cognitive performance. Nutr Rev 60(10):308–326
Mathews IM et al (2024) Clinical efficacy and tolerability of lemon balm (Melissa officinalis L.) in psychological well-being: a review. Nutrients 16(20):3545
McCarty MF, Lerner A (2021) The second phase of brain trauma can be controlled by nutraceuticals that suppress DAMP-mediated microglial activation. Expert Rev Neurother 21(5):559–570
Mercken EM et al (2012) Of mice and men: the benefits of caloric restriction, exercise, and mimetics. Ageing Res Rev 11(3):390–398
Mikkelsen K et al (2016) Cognitive decline: a vitamin B perspective. Maturitas 93:108–113
Mikulska P et al (2023) Ashwagandha (Withania somnifera)-current research on the health-promoting activities: a narrative review. Pharmaceutics 15(4)
Miller ED et al (2019) A review of various antioxidant compounds and their potential utility as complementary therapy in multiple sclerosis. Nutrients 11(7):1528
Mladenov M et al (2023) Oxidative stress, reductive stress and antioxidants in vascular pathogenesis and aging. Antioxidants 12(5):1126
Mohammadi H, Karwowski W (2024) Graph neural networks in brain connectivity studies: methods, challenges, and future directions. Brain Sci 15(1):17
Monacelli F et al (2017) Vitamin C, aging and Alzheimer's disease. Nutrients 9(7):670
Mora I et al (2022) Structured long-chain omega-3 fatty acids for improvement of cognitive function during aging. Int J Mol Sci 23(7):3472
Mou Y et al (2022) Gut microbiota interact with the brain through systemic chronic inflammation: implications on neuroinflammation, neurodegeneration, and aging. Front Immunol 13:796288
Mungali M, Tripathi A, Singhal S (2021) Valeriana officinalis (valerian). In: Naturally occurring chemicals against Alzheimer's disease. Elsevier, pp 283–291

Nam SM et al (2013) Valeriana officinalis extract and its main component, valerenic acid, ameliorate D-galactose-induced reductions in memory, cell proliferation, and neuroblast differentiation by reducing corticosterone levels and lipid peroxidation. Exp Gerontol 48(11):1369–1377
Naureen Z et al (2022) Dietary supplements in neurological diseases and brain aging. J Prev Med Hyg 63(2 Suppl 3):E174
Naziroglu M, Butterworth PJ, Sonmez TT (2011) Dietary vitamin C and E modulates antioxidant levels in blood, brain, liver, muscle, and testes in diabetic aged rats. Int J Vitam Nutr Res 81(6):347
Nilsson J, Lövdén M (2018) Naming is not explaining: future directions for the "cognitive reserve" and "brain maintenance" theories. Alzheimer's Res Ther 10:1–7
Normandin E et al (2017) Effect of resistance training and caloric restriction on the metabolic syndrome. Med Sci Sports Exerc 49(3):413
Nuttall JR, Oteiza PI (2014) Zinc and the aging brain. Genes Nutr 9:1–11
Onaolapo OJ et al (2022) Neuroinflammation and oxidative stress in Alzheimer's disease; can nutraceuticals and functional foods come to the rescue? Antiinflamm Antiallergy Agents Med Chem 21(2):75–89
Ooi S-L, Pak S-C (2021) Nutraceuticals in immune function. Molecules 26(17):5310
Pandareesh M et al (2018) Nutrition and nutraceuticals in neuroinflammatory and brain metabolic stress: implications for neurodegenerative disorders. CNS Neurol Disorders Drug Targets 17(9):680–688
Park J-C, Han S-H, Mook-Jung I (2020) Peripheral inflammatory biomarkers in Alzheimer's disease: a brief review. BMB Rep 53(1):10
Peirce JM, Alviña K (2019) The role of inflammation and the gut microbiome in depression and anxiety. J Neurosci Res 97(10):1223–1241
Peixoto CA, Nunes AKS, Garcia-Osta A (2015) Phosphodiesterase-5 inhibitors: action on the signaling pathways of neuroinflammation, neurodegeneration, and cognition. Mediat Inflamm 2015(1):940207
Pekdemir B et al (2024) Mechanisms and potential benefits of neuroprotective agents in neurological health. Nutrients 16(24):4368
Popa D-S, Bigman G, Rusu ME (2021) The role of vitamin K in humans: implication in aging and age-associated diseases. Antioxidants 10(4):566
Praengam K et al (2024) Efficacy of a dietary supplement derived from five edible plants on telomere length in Thai adults: a randomized, double-blind, placebo-controlled trial. Food Sci Nutr 12(3):1592–1604
Prasanth MI et al (2024) Role of epigenetic modulation in neurodegenerative diseases: implications of phytochemical interventions. Antioxidants 13(5):606
Pyka P et al (2024) Selenium-containing compounds: a new hope for innovative treatments in Alzheimer's disease and Parkinson's disease. Drug Discov Today 29(8):104062
Ramli NZ et al (2020) A mechanistic evaluation of antioxidant nutraceuticals on their potential against age-associated neurodegenerative diseases. Antioxidants 9(10):1019
Rauf A et al (2022) Neuroinflammatory markers: key indicators in the pathology of neurodegenerative diseases. Molecules 27(10):3194
Rea IM et al (2018) Age and age-related diseases: role of inflammation triggers and cytokines. Front Immunol 9:586
Rivas F et al (2022) Effects of polyphenols in aging and neurodegeneration associated with oxidative stress. Curr Med Chem 29(6):1045–1060
Rusek M et al (2023) The role of the JAK/STAT signaling pathway in the pathogenesis of Alzheimer's disease: new potential treatment target. Int J Mol Sci 24(1):864
Saha S et al (2020) An overview of Nrf2 signaling pathway and its role in inflammation. Molecules 25(22):5474
Sarubbo F et al (2017) Effects of resveratrol and other polyphenols on the most common brain age-related diseases. Curr Med Chem 24(38):4245–4266
Sarubbo F et al (2018) Effects of resveratrol and other polyphenols on Sirt1: relevance to brain function during aging. Curr Neuropharmacol 16(2):126–136

Sarubbo F et al (2023) Impact of gut microbiota in brain ageing: polyphenols as beneficial modulators. Antioxidants 12(4):812
Sastre J et al (1998) A Ginkgo biloba extract (EGb 761) prevents mitochondrial aging by protecting against oxidative stress. Free Radic Biol Med 24(2):298–304
Schättin A et al (2016) Effects of physical exercise combined with nutritional supplements on aging brain related structures and functions: a systematic review. Front Aging Neurosci 8:161
Sharifi-Rad J et al (2022) Multi-target mechanisms of phytochemicals in Alzheimer's disease: effects on oxidative stress, neuroinflammation and protein aggregation. J Personal Med 12(9):1515
Sharma R, Martins N (2020) Telomeres, DNA damage and ageing: potential leads from ayurvedic rasayana (anti-ageing) drugs. J Clin Med 9(8):2544
Shen LR et al (2013) Curcumin and aging. Biofactors 39(1):133–140
Spagnuolo C, Moccia S, Russo GL (2018) Anti-inflammatory effects of flavonoids in neurodegenerative disorders. Eur J Med Chem 153:105–115
Stephenson J et al (2018) Inflammation in CNS neurodegenerative diseases. Immunology 154(2):204–219
Sudati JH et al (2009) In vitro antioxidant activity of Valeriana officinalis against different neurotoxic agents. Neurochem Res 34:1372–1379
Takasugi M (2018) Emerging roles of extracellular vesicles in cellular senescence and aging. Aging Cell 17(2):e12734
Terock J et al (2022) Vitamin D deficit is associated with accelerated brain aging in the general population. Psychiatry Res Neuroimaging 327:111558
Troesch B et al (2020) Expert opinion on benefits of long-chain omega-3 fatty acids (DHA and EPA) in aging and clinical nutrition. Nutrients 12(9):2555
Tsaioun KI (1999) Vitamin K–dependent proteins in the developing and aging nervous system. Nutr Rev 57(8):231–240
Uddin MS et al (2021) Exploring potential of alkaloidal phytochemicals targeting neuroinflammatory signaling of Alzheimer's disease. Curr Pharm Des 27(3):357–366
Valvassori SS et al (2018) Hypericum perforatum chronic treatment affects cognitive parameters and brain neurotrophic factor levels. Rev Bras Psiquiatr 40(4):367–375
Voskuhl R, Itoh Y (2022) The X factor in neurodegeneration. J Exp Med 219(12):e20211488
Wan S et al (2023) Plasma inflammatory biomarkers in cerebral small vessel disease: a review. CNS Neurosci Ther 29(2):498–515
Wang J et al (2018) Connection between systemic inflammation and neuroinflammation underlies neuroprotective mechanism of several phytochemicals in neurodegenerative diseases. Oxidative Med Cell Longev 2018(1):1972714
Wang Y et al (2022) Dietary phytochemicals that influence gut microbiota: roles and actions as anti-Alzheimer agents. Crit Rev Food Sci Nutr 62(19):5140–5166
White DJ et al (2016) Functional brain activity changes after 4 weeks supplementation with a multi-vitamin/mineral combination: a randomized, double-blind, placebo-controlled trial exploring functional magnetic resonance imaging and steady-state visual evoked potentials during working memory. Front Aging Neurosci 8:288
Yang X, Chen X (2022) The crosstalk between the blood–brain barrier dysfunction and neuroinflammation after general anaesthesia. Curr Issues Mol Biol 44(11):5700–5717
Yao Z, Bai Q, Wang G (2021) Mechanisms of oxidative stress and therapeutic targets following intracerebral hemorrhage. Oxidative Med Cell Longev 2021(1):8815441
Yoo DY et al (2011) Effects of Melissa officinalis L.(lemon balm) extract on neurogenesis associated with serum corticosterone and GABA in the mouse dentate gyrus. Neurochem Res 36:250–257
Yu T, Lao X, Zheng H (2016) Influencing COX-2 activity by COX related pathways in inflammation and cancer. Mini Rev Med Chem 16(15):1230–1243
Yuan Z et al (2023) Research progress of NLRP3 inflammasome and its inhibitors with aging diseases. Eur J Pharmacol 957:175931

Zaidi SKR, Banu N (2004) Antioxidant potential of vitamins A, E and C in modulating oxidative stress in rat brain. Clin Chim Acta 340(1–2):229–233

Zhang M, Yin Y (2023) Dual roles of anesthetics in postoperative cognitive dysfunction: regulation of microglial activation through inflammatory signaling pathways. Front Immunol 14:1102312

Zhang D-M et al (2017) Efficacy of vitamin B supplementation on cognition in elderly patients with cognitive-related diseases: a systematic review and meta-analysis. J Geriatr Psychiatry Neurol 30(1):50–59

Zhang J et al (2024) From inflammatory signaling to neuronal damage: exploring NLR inflammasomes in ageing neurological disorders. Heliyon 10(12)

Zhou D-D et al (2021) Effects and mechanisms of resveratrol on aging and age-related diseases. Oxidative Med Cell Longev 2021(1):9932218

Ziegler T et al (2023) Polyphenols' impact on selected biomarkers of brain aging in healthy middle-aged and elderly subjects: a review of clinical trials. Nutrients 15(17):3770

Zirak N et al (2019) Hypericum perforatum in the treatment of psychiatric and neurodegenerative disorders: current evidence and potential mechanisms of action. J Cell Physiol 234(6):8496–8508

Zong W et al (2021) The antioxidant phytochemical schisandrin a promotes neural cell proliferation and differentiation after ischemic brain injury. Molecules 26(24):7466

Zuo W et al (2017) Advances in the studies of Ginkgo biloba leaves extract on aging-related diseases. Aging Dis 8(6):812

7 Oxidative Stress and Antioxidants in the Aging Brain

Sagar Pamu, Mrugendra Potdar, Parameshwar Kondapuram, Devyani Rajput, and Monika Kaushik

Abstract

The scope of this chapter is to present an overview of oxidative stress, antioxidants, and the beneficial role of free radicals in the brain. Studies have revealed an alarming increase in oxidative stress due to aging; thus, it initiates damage to cells. A unique defense mechanism known as antioxidants has been developed to counteract the adverse effects of excessive free radical amounts within the body. These naturally occurring substances are readily available within the body or are complementary additions, such as the flavonoids and vitamin E found in food. These substances function to ensure that homeostasis is maintained as well as to protect the body from a range of complications. Cognitive deficiencies, vision loss, responses to stress, immune responses, infertility, and neurological dysfunctions are all complications resulting from the imbalance of free radicals within the body. The brain is deemed at greater risk than other organs, as 20% of the body's oxygen requirement is allocated to the brain, resulting in an unprecedented amount of free radicals. The brain, as the most metabolically active organ, is adversely affected by an imbalance in free radicals, which are generated due to oxidative stress. Research has revealed the brain's complex susceptibility to oxidative stress-induced cellular damage that initiates a range of neurological disorders, such as multifactorial aging and neurodegenerative diseases, stroke,

S. Pamu · D. Rajput · M. Kaushik (✉)
Amity Institute of Pharmacy, Amity University Madhya Pradesh, Gwalior, Madhya Pradesh, India

M. Potdar
Shri Vile Parle Kelawani Mandal's Institute of Pharmacy, Dhule, Maharashtra, India

P. Kondapuram
School of Pharmacy, Guru Nanak Institutions Technical Campus, Ibrahimpatnam, Telangana, India

A. Kumar Singh, S. Nand Rai (eds.), *Rejuvenating the Brain: Nutraceuticals, Autophagy, and Longevity*, Nutritional Neurosciences,
https://doi.org/10.1007/978-981-95-2790-8_7

migraine headaches, concussions, and neurodevelopmental anomalies. Such cognitive deficiencies are directly linked to the affected brain regions. Consequently, cognitive deficiencies such as memory loss are observed when the neocortex or hippocampus is affected. Antioxidants, derived from food or endogenously present in the body, are necessary for steroidogenesis and neurological pathology. These antioxidants promote cognitive function but could also potentially contribute to treating various neurological disorders and the recovery or aging process.

Keywords

Oxidative stress · Antioxidants · Neurodegeneration · Free radicals · Cognitive function

Abbreviations

ROS	Reactive Oxygen Species
RNS	Reactive Nitrogen Species
OS	Oxidative Stress
CNS	Central Nervous System
SOD	Superoxide Dismutase
GSH	Reduced Glutathione
NO	Nitric Oxide
H_2O_2	Hydrogen Peroxide
O2	Superoxide Anion.
OH	Hydroxyl Radical
AD	Alzheimer's Disease
PD	Parkinson's Disease
MS	Multiple Sclerosis
BBB	Blood-Brain Barrier
Aβ	Amyloid Beta

7.1 Introduction

The aging of the brain is not identical to the aging of the body; it is the biological process of accumulation over time of many sources of structural and physiological damage to the cells and systems (Sies 2015). Among those, the changes in neuronal and glial functions related to glucose metabolism demonstrate a clear association. Intriguingly, processes involving the production of endogenous cellular toxins resulting from the oxidative metabolism of glucose are implicated in the basic etiology of age-related degenerative changes in the brain and other organs (Muchová et al. 2014). However, repair systems are dependent on the presence of acid-repairable proteins and oligopeptides, and the systems capable of making repairs, including DNA and oligopeptides synthesis, do not correlate with molecular turnover, DNA replication, or the actual aging of the neuron (Salminen and Paul 2014). The aging of the brain, or the loss of its faculties that allow a person's continued participation in daily life activities, is a lifelong process with a typical duration from age 20 to death at 60 years. This lengthy process is subjected to a multitude of aging influences, from the direct effects of the genetic nature of the individual, possibly related to the developmental age of the individual during growth or postgrowth adult life, to the influences of the surroundings of the individual, including social and cultural factors. Many age-associated impairments, such as the loss of cognitive function and the development of neurodegenerative diseases, have been theorized to be caused by the accumulating effects of excessive production of reactive oxygen species in the brain (Valko et al. 2007). However, different tissues have different thresholds of vulnerability to reactive oxygen species. These effects depend on the tissue or organ involved, with the brain seemingly the most vulnerable to oxidative stress. Four main reasons are usually invoked to qualify the brain as a target for oxidative stress. First, the central nervous system consumes a significant volume of oxygen despite the relatively small weight of the organ. Second, the brain's cell membranes are rich in polyunsaturated fatty acids and dominated by docosahexaenoic and arachidonic acids. Third, the brain has a limited capacity for antioxidant defense. Finally, the brain lacks the high levels of DNA repair mechanisms found in other organs.

7.1.1 Overview of Brain Aging and the Role of Oxidative Stress

Primarily due to improvements in medical conditions, life expectancy continues to increase. One promising interpretation of age-related cognitive incompetence is a detrimental collision of the intensified instability within the native environments of the neurons, which signal transduction pathways (Daniele et al. 2018). Reactive oxygen species may carry the neurotoxic means of these destructive assaults. The brain endures a particularly severe hazard from reactive oxygen species due to their outstanding yields during normal functions, like synaptic, predominantly associated with high conservation of the prooxidant metals (Hashim and Makpol 2022). Typically short-lived in the brain, requisite radicals are long-lived due to a weakening of the cellular antioxidant equilibrium recognized in aging and age-linked cognitive dysfunction. Analysis of age-related oxidative elevation corroborates the oxidative menace concept, whereby the neuro psycho-neuropathology is only discernible when both oxidative damages to a pathway and cognitive failures linked with terminal ischemia are present (Dwivedi et al. 2020).

One of the postulated explanations of the fundamental cognitive deterioration in age-linked neurodegenerative syndromes is characteristic of gray matter oxidative lesions, which mainly involve substances important for the processes of cognitive memory (Alcendor 2020). The free radical-mediated molecular damage has been shown to play a role in several aspects of the structures important for memory and sensory discrimination, such as acetylcholine synthesis and release, acetylcholine

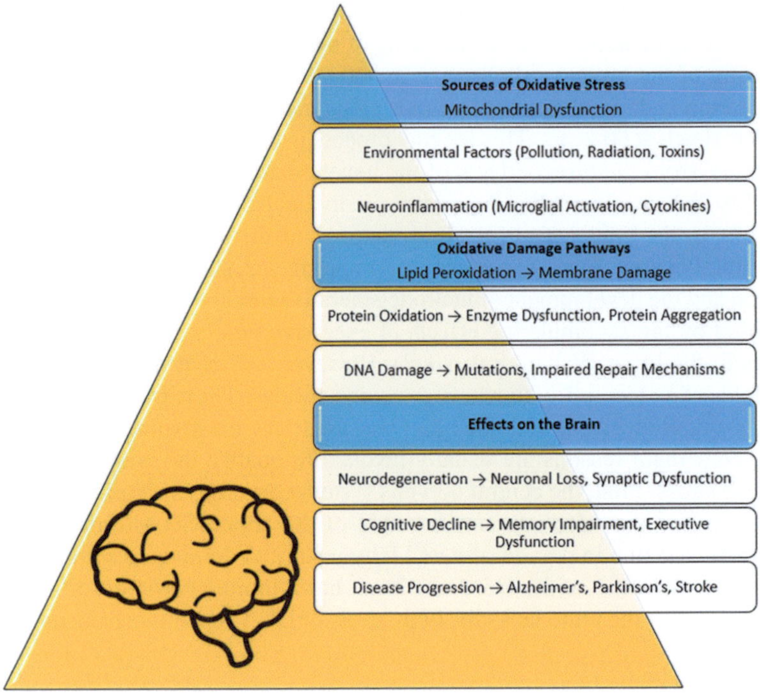

Fig. 7.1 Oxidative stress and brain aging pathways

and dopamine receptor density and functioning, astrocyte and dendrite interactions, synaptic expansion, and long-term structural changes represented in Fig. 7.1. The involved molecular chains are highly complicated and intricate interactions (Li et al. 2022). Free radicals might be involved directly in inducing permanent gene mutations or by gradual, consecutive adaptations of the genetic systems. Free radicals might act indirectly by presenting as epitope-less nerve cells that are unable to survive the trophic environment through alterations, causing them to activate the apoptosis apparatus (Shacka et al. 2008). Similarly, these models of thought extinction would be replicated in the general nonsurvivability outside of the target zone, following depletion reactionally to specific kinds of cellular damage. Antioxidants can reduce peroxide-derived activation of the surrounding lethal independent pathways and suppress nerve cell death.

7.1.2 The Connection Between Oxidative Stress and Neurodegenerative Diseases

Age is well-established as the primary risk factor for neurodegenerative diseases. Thus, age-related changes in the CNS may play an important role in the initiation or progression of these diseases (Zia et al. 2021). A rather generic and significant characteristic of aging is an increased susceptibility to various forms of stress, including genotoxic, metabolic, and environmental stress. All of them lead to damage of macromolecules and disruption of several metabolic and signaling pathways; therefore, they can compromise the functional integrity of a cell (Saretzki 2023). In the brain, such stress can have multiple deleterious effects, affect cellular functions and synaptic physiology, promote cognitive loss, and predispose individuals to neurodegenerative disease.

Oxidative stress has been suggested as being initially added as a consequence of aging since old organisms exhibit increased levels of oxidative damage to macromolecules in several organs, including the brain (Cervellati and Bergamini 2016). Furthermore, redox raging has been used to describe oxidative stress, abundantly present in the aging brain, and its negative impact on cell signaling, cell function, neurovascular unit integrity, neurotransmission, and memory. The levels of antioxidant defenses in the brain, such as superoxide dismutase, glutathione-dependent enzymes, and the reduced amount of α-tocopherol, increase, remain the same, or decrease slightly during aging, while those of catalase and glutathione reductase decrease (Scheiblich et al. 2024). These reports raise the possibility that oxidatively mediated damage could contribute to the decline in brain function observed in the elderly, the progression of brain aging, and overall higher susceptibility to neurodegenerative diseases (Aguilar et al. 2024).

7.1.3 Importance of Antioxidants in Protecting Brain Health During Aging

Given the importance of the brain in normal human activities and interactions, the mechanisms involved in maintaining its health during aging require study (Grabska-Kobyłecka et al. 2023). Advances in medical knowledge and practice have resulted in a rapidly increasing elderly population, leading to a demographic shift from a pyramid with a young base and relatively few older adults to a more rectangular form (Hyun et al. 2006). Consequently, the need to ensure healthy aging has increased interest in the influence of the senescence process on the nervous system, particularly regarding the maintenance of cognitive functions. During aging, several alterations, particularly lower antioxidant defense and increased production of reactive oxygen species, are often found in the brain after middle age. These changes are frequently accompanied by a progressive decrease in volume and weight, increasing the risk of diseases, including neurodegenerative diseases (Lövdén et al. 2020). Although aging is not a disease, the mechanisms and causes for these senescence-associated alterations in neuron and astrocyte functions have yet to be clarified.

To slow down the aging process, there is a growing interest in modifying lifestyles or searching for compounds that reduce oxidative stress or reactive oxygen species production or facilitate the removal of damage produced by reactive oxygen species (Szczuko et al. 2021). In this context, several drugs with antioxidant properties have been administered to old animals to maintain normal brain functions. Additionally, in naturally occurring aged animals presenting locomotion disorders, natural lifespans related to illnesses involving oxidative stress and cognitive problems, chronic treatment with physiological doses of antioxidants appears to produce beneficial effects (Aaseth et al. 2021).

7.2 Understanding Oxidative Stress in the Aging Brain

Oxidative stress is an imbalance in the generation and detoxification of reactive oxygen species, damaging cellular components, such as DNA and lipids (Regenhardt et al. 2018). It is an overarching hypothesis that explains cognitive and memory decline in normal aging and provides a model for the role of oxidative stress in aging brain diseases (Andrade-Guerrero et al. 2023). This chapter reviews the mechanisms and balance of oxidative stress in aging, the current evidence for the role of oxidative stress in cognitive decline, and the relationship with Alzheimer's disease. We also review the evidence for the augmentation by aging of the reduction in brain levels of antioxidants through diet and activity (Andrade-Guerrero et al. 2023).

Metabolism of oxygen involves generating and detoxifying several reactive oxygen species, the majority being by-products of physiological processes (Angelova and Abramov 2016). An imbalance in this generation and destruction results in oxidative stress, wherein reactive oxygen species damage DNA, proteins, and lipids. The aging brain is one of the tissues particularly vulnerable to oxidative stress, as it consumes large amounts of oxygen and generates a large number of reactive oxygen species, which are more than its antioxidant capacity (NCBI 2025a). The resulting damage is a central feature of the aging brain, and claims about antioxidants as treatments for aging have escalated (Vanmassenhove et al. 2016). Furthermore, the pivotal role that oxidative stress plays in neurodegenerative diseases has been shown in Table 7.1.

7.2.1 Definition and Mechanisms of Oxidative Stress

Oxidative stress was first termed in the 1980s to describe the imbalance between oxidants and antioxidants in biological systems that would lead to damage. Excessive stress arises when free radicals overwhelm the cells' antioxidant defenses (Terao 2023). Although an essential part of life, free radicals mediate hormone receptor signaling, and inter- and intracellular communications provide defense from pathogens. Accumulating these small and unstable molecules can cause damage at the cellular level (Terao 2023). Once they come in contact with cell membranes' polyunsaturated fatty acids, they can cause local structural alterations. Free radicals can also interact with proteins, lipids, and DNA (Evans and Lawrenson 2023). When they damage key structures, this can become a big problem for the cells, mainly if they are postmitotic, such as those that make up the brain.

The production of free radicals is unavoidable: they are a by-product of mitochondrial activity, and oxygen is, in a way, a waste molecule produced by complex

Table 7.1 Major sources and functions of endogenous and dietary antioxidants in the brain

Antioxidant type	Example	Source	Function in the brain	References
Endogenous antioxidants	Superoxide dismutase (SOD)	Mitochondria, cytoplasm	Converts superoxide radicals into hydrogen peroxide, reducing oxidative stress and neuronal damage.	Mirończuk-Chodakowska et al. (2018)
Endogenous antioxidants	Catalase	Peroxisomes	Breaks down hydrogen peroxide into water and oxygen, preventing oxidative injury.	Aldini et al. (2018)
Endogenous antioxidants	Glutathione (GSH)	Liver, neurons	Detoxifies free radicals, maintains redox balance, and protects against neurodegeneration.	Aldini et al. (2018)
Dietary antioxidants	Vitamin E (tocopherol)	Nuts, seeds, vegetable oils	Protects neuronal membranes from lipid peroxidation and oxidative stress.	Xiong et al. (2023)
Dietary antioxidants	Vitamin C (ascorbic acid)	Citrus fruits, berries, peppers	Scavenges free radicals, regenerates vitamin E, and supports neurotransmitter synthesis.	Shen et al. (2022)
Dietary antioxidants	Polyphenols (flavonoids)	Green tea, berries, dark chocolate	Modulate oxidative stress, reduce neuroinflammation, and improve cognitive function.	Shen et al. (2022)
Dietary antioxidants	Carotenoids (beta-carotene, lutein)	Carrots, spinach, tomatoes	Protect neurons from oxidative stress, enhance visual and cognitive function.	NCBI (2025b)

IV. Other enzymes produce free radicals, such as NADPH oxidase, xanthine oxidase, and the biosynthetic enzymes porphyrinogen oxidase, P450, or prostaglandin H synthase. When subjected to stress (be it oxidative stress or other factors such as pathogen recognition, hypoxia, or physical damage), other peroxisomes, endoplasmic reticulum, and cell membrane enzymes can start producing free radicals (Evans and Lawrenson 2023). Certain metals such as iron, copper, or zinc can promote free radical production. Finally, exogenous factors like radiation, some drugs, toxins, or smoking can cause the overproduction of free radicals.

7.2.2 Sources of Oxidative Stress in Brain Aging

In all metabolic reactions used to generate ATP in cells, an inevitable leakage of electrons reduces oxygen to create O_2. Because molecular oxygen is the ultimate acceptor for electrons, the mitochondrial energy-generating machinery operates without the direct participation of scavengers, which are absent in a suitable concentration to detoxify reactive oxygen species in the mitochondrial matrix (Singh et al. 2019). Hence, during mitochondrial oxidative phosphorylation, reactive oxygen species are an inevitable by-product and are the most important source of reactive oxygen species in physiological conditions (Wang et al. 2021). When complexes I and III of the electron transport chain predominantly work in the forward mode, NADH dehydrogenase continuously reduces oxygen to O_2 and H_2O_2, generating about 90% of reactive oxygen species in vivo. With the aging process, there is a decrease in the number and function of mitochondrial electron transport chain complexes.

Under conditions where the regular O_2 consumption rate is decreased, complexes I and III can work in reverse mode, thus decreasing the production of free radicals (Shore and Bianchi 2009). This indicates that a threshold imbalance between electron flow in the mitochondrial respiratory chain and the O_2-reducing capacity of the chain is responsible for an increase in reactive oxygen species production. Indeed, for both lifespan and neurological function in C. elegans, it was found that the condition for healthy homeostasis in adults requires ensuring a degree of suppression of mitochondrial oxidative stress (Martín-Fernández and Gredilla 2018). Low levels of reactive oxygen species-based signals are not harmful to health. By modulating gene expression, they can act as true messengers between mitochondrial operation and cellular defense systems, thus increasing mitochondrial longevity (Yamada et al. 2020). At micromolar concentrations, H_2O_2 restrains insulin/insulin-like growth factor activity, further triggering cold shock protein expression, inhibiting eukaryotic translation initiation factor 4E, and reducing global translation, decreasing reactive oxygen species production (Pettersson et al. 2014). As such, it was demonstrated that a low-calorie diet increases the lifespan of rodents through decreased production of free radicals from the tissue, and they live shorter lives when the mitochondrial production of reactive oxygen species is increased from the skin.

7.2.3 The Impact of Oxidative Stress on Cellular Function and Neurodegeneration

The impact of oxidative stress on cell and tissue function differs. It resonates fairly high magnitude in tissues with higher oxygen utilization. Tissue function is greatly affected by the impaired ability of cells to generate bioenergetic molecules, i.e., altered mitochondrial function (Goicoechea et al. 2023). In addition to bioenergetics, mitochondria notably regulate apoptosis and participate in the regulation of cytosolic calcium by generating reactive oxygen species (ROS)-mediated signaling. Thus, any disruption of these functions or increase in oxidative damage to a cell's mitochondrial or nuclear DNA is detrimental, particularly to cells such as neurons with a postmitotic and limited capacity for self-renewal (NCBI 2025c). Oxidative stress contributes not only to aging but also to the development and progression of neurodegenerative diseases (Volkert and Crowley 2020). The central nervous system, which contains neurons and glial cells, is an energy-demanding and highly oxygen-dependent organ; it will benefit from a greater understanding of the risk factors and the processes by which cells counteract oxidative stress.

The necessity for cellular response mechanisms is vital. Small animal models hint at the importance of oxidative stress as a driving force for aging and age-associated functional decline. Nonetheless, exposure of various animal models to antioxidants and various gene manipulations led to largely inconclusive results. Though there is uncertainty about the effect of antioxidants in aging, the preservation of function may result from changes to specific genes. Even mild, single-gene interventions might delay specific aspects of the aging process (Hazell et al. 2013). For instance, manipulating the list of gene targets changes hepatic lipid metabolism in mice, ultimately suppressing the expression of inflammatory genes. Reducing levels of oxidative stress in murine nervous systems through central nervous system-specific manganese SOD overexpression may have other CNS benefits, such as a reduction in ataxia and improved performance on a rotarod test. Such relatively small effect sizes with various uncertainties should not detract from the fact that oxidative stress is a major driving force for the aging processes and may be essential for the onset of age-associated diseases (Mouat et al. 2022). Prominent in CNS aging, oxidative stress is the dysregulation caused by an imbalance between the production of ROS and/or reactive nitrogen species and the capabilities of the antioxidant scavenging enzymes.

7.2.4 The Role of Oxidative Stress in Brain Aging

Although the exact mechanisms of brain aging (or aging in general) are far from being elucidated, it has been proposed that the damaging effects of free radicals on cells are important factors in this process (Trist et al. 2019). It is generally assumed that oxygen and oxygen-derived species, including free radicals, may cause cell dysfunction and aging by inducing oxidative damage to cellular macromolecules (Liu 2022). The oxidatively modified molecules have the potential to accumulate in the brain during aging and may eventually compromise the survival of the cells (Picca et al. 2023). This statement may well apply to the brain. It is sensitive to oxidative stress because of its high rate of oxygen consumption, high-energy metabolic rate, and prooxidant transition metal concentrations. In addition, the human brain has high levels of polyunsaturated fatty acids, which are more susceptible to peroxidation than other fatty acids (Bocheva et al. 2024). The depletion of antioxidant defenses and the increase in oxidized molecules found in the brains of old animals might be considered as seminal evidence that supports the oxidative stress hypothesis of brain aging.

There is strong evidence that oxidative stress is linked to the pathological changes in the aged brain. Oxidative damage has been demonstrated to affect all classes of biomolecules in the aged brain, including lipids, proteins, and DNA. Many reviews have provided an extensive understanding of the underlying biology of oxidative stress in aging, and in particular, of the processes of oxidative brain aging (Poulose et al. 2017). The complexity of aging itself and, above all, the multidimensional nature of oxidative stress, which encompasses everything from free radical biology to general brain signaling, is described.

7.2.5 Effects on Neuronal Health, Synaptic Plasticity, and Cognitive Function

Oxidative stress is commonly associated with reduced antioxidant defense mechanisms, which include enzymatic compounds like superoxide dismutase, catalase, reduced glutathione, and antioxidant nutrients such as vitamins C and E. Aging is related to augmented oxidative stress, which tends to compromise neuronal signaling, reduce neurogenesis in the dentate gyrus, and exacerbate the neuroinflammatory response (Cassilhas et al. 2016). Neurons are vulnerable to oxidative stress because of their relatively low regenerative capacity; thus, most of the resistance against oxidative injury is lowered by redox catalysts and antioxidants (Chow et al. 2021). In order to cope with oxidative stress, neurons need to regenerate glutathione, keep a strong network, maintain synaptic plasticity, and transmit signals at optimal levels; they are vulnerable to low antioxidant capacity. Oxidative stress deregulates postsynaptic receptor mobility, thereby affecting synaptic plasticity. To combat these changes, cognitive functions may efficiently activate protein kinase signaling pathways to preserve structure (Colcombe and Kramer 2003).

These modifications are the result of intracellular signaling cascades linked to a variety of proteins, including calcium calmodulin-dependent protein kinase, protein kinase A, B, and C. This is due to the fact that oxidative stress activates mitochondrial pathways and declines the Bcl-w expression (Olivares-Bañuelos et al. 2019). However, some changes increase the cell's vulnerability to oxidative stress. Dendritic atrophy and synaptic lesions are seen in animal models, emphasized by studies of cognition (Jafar-Nejad et al. 2010). Synaptic P38 MAP kinase may stimulate dendritic spines. There is evidence that traumatic brain injury can lead to long-term changes in hippocampal function in older adults. Increased cognitive vibration in old transgenic mice overexpressing the oxidative enzyme NADPH oxidase, generating peroxide molecules, is being studied. This manifested as a cognitive motor deficit in animals that lacked normal arm function (Boldrini et al. 2018). Transgenic animals also showed elevated oxidative injury to the lipid and protein linked to oxidative stress directly by the brain homogenate (Foret et al. 2024).

7.2.6 The Relationship Between Oxidative Stress and Neurodegenerative Diseases (e.g., Alzheimer's, Parkinson's)

Generally, in old age, while there is an increase in the number of free radicals produced, there is a decrease in the activity of the antioxidant system defenses. Of note, since the human brain is very rich in lipids, these oxidative environmental influences are of considerable importance to the aging brain and are a cause of functional and cognitive decline in old age, as well as the establishment of neurodegenerative diseases. The onset of aging-related neurologic defects and several neurodegenerative conditions is also related to the loss of homeostasis in the intracellular redox environment (Jomova et al. 2025).

The oxidative stress status endured by individual cells is related to the concentration of prooxidant species and the capability of antioxidant defense mechanisms in various cellular regions and can also symbolize the preceding state of the brain's procurements shown in Table 7.2. If one of the redox-controlled signal transduction paths is vulnerable to oxidative modifications, consequently the oxidative state is likely to affect many aspects of cell physiology, and the integrity of the antioxidant system inside neurons and/or support cells in the brain may also lead to these pathophysiologic alterations (Cardinali 2021). Several antioxidant-based therapies and gene-manipulated animal studies now highlight the potential of antioxidant

Table 7.2 Oxidative stress-related neurological disorders and their associated biomarkers

Neurological disorder	Oxidative stress biomarkers	Pathological implications	References
Alzheimer's disease (AD)	Malondialdehyde (MDA), 4-hydroxynonenal (4-HNE), protein carbonyls, reduced glutathione (GSH)	Increased lipid peroxidation, oxidative protein damage, amyloid-beta aggregation, neuronal loss	Park et al. (2021)
Parkinson's disease (PD)	8-hydroxy-2′-deoxyguanosine (8-OHdG), nitrotyrosine, reduced glutathione (GSH), α-synuclein aggregation	Mitochondrial dysfunction, dopaminergic neuron degeneration, impaired redox homeostasis	Mehra et al. (2019)
Stroke (ischemic and hemorrhagic)	Superoxide dismutase (SOD), catalase, glutathione peroxidase (GPx), F2-isoprostanes	Ischemia-induced ROS production, blood-brain barrier disruption, and neuronal apoptosis	Li et al. (2021)
Amyotrophic lateral sclerosis (ALS)	3-nitrotyrosine, oxidized DNA bases, SOD1 mutations, protein carbonyls	Motor neuron degeneration, mitochondrial dysfunction, and neuroinflammation	Grad et al. (2017)
Multiple sclerosis (MS)	Oxidized low-density lipoprotein (oxLDL), 8-OHdG, myeloperoxidase (MPO), nitric oxide metabolites	Demyelination, immune-mediated oxidative damage, and neuroinflammation	Doshi and Chataway (2016)
Migraine	Nitric oxide (NO), peroxynitrite, superoxide anion, mitochondrial DNA damage	Vasodilation, neurovascular dysfunction, mitochondrial stress	Silberstein (2004)
Depression and anxiety	Lipid peroxidation markers, nitric oxide (NO), protein carbonylation, antioxidant enzyme activity	Altered neurotransmitter function, increased oxidative damage, and neuronal inflammation	Kalueff and Nutt (2007)

resources in the therapy of aging brain decline and neurodegenerative diseases (Temple 2023).

7.2.7 The Blood-Brain Barrier and Oxidative Stress

The BBB isolates most of the adult CNS from blood-borne compounds, while allowing some molecules, such as oxygen and glucose, to pass into the brain (Chen et al. 2022). Electrolyte and osmotic balance are maintained by the BBB in the CNS to protect the neurons from environmental fluctuations, as well as to provide the required nutrients for proper neuronal function. Capillaries in the brain are quite distinct from those in other organs, and, in addition to expressing a very low level of pinocytotic vesicles compared to those of the endothelium in other tissues, the brain endothelium contains specific features such as intercellular tight junctions involving zonula occludens proteins and claudins, and heat shock proteins (Ortiz et al. 2014). Therefore, the brain endothelium presents a large barrier, which is much more restrictive to the passage of polar molecules than the bulk of the peripheral vasculature. This low permeability of the BBB to the circulation, the low abundance or absence of active efflux systems, makes it an ideal barrier to the brain from drugs and other xenobiotics, including redox-active compounds (Yang et al. 2019).

Under normal conditions, CAT and SOD protect the BBB from oxidative stress-induced injury, but other cellular defense systems also act. In cases of ischemia/reperfusion injury, some preclinical studies indicate that defective activity of GSH, coumarin hydroxylase, cytochrome P-450 isoenzyme, and quinone reductases may cause the BBB to become vulnerable. Moreover, since the brain uses a large amount of oxygen, massive amounts of cellular ROS are generated during metabolic reactions, and the BBB represents the best. Accompanied by an abundance of iron, the increased ROS and the decline in antioxidants at the BBB suggest that oxidative stress facilitates the aging and pathogenesis of the BBB. Once the BBB is broken, the normally homeostatic microenvironment in the brain becomes vulnerable to toxins, just as if the systemic blood circulation was accessible to the brain. High levels of ROS, oxidative stress, or agents that induce oxidative stress result in BBB destruction (Ronaldson and Davis 2020). As aging increases ROS, oxidative stress may lead to increases in BBB permeability or apoptosis. In this section, we discuss a few examples of the findings that aging can damage the BBB and that treatment with antioxidants can alleviate aging-induced increases in BBB permeability. Oxidative stress not only represents a part of the damage to the BBB, but it is also an inducer that activates leukocytes. The activation of leukocytes increases oxidative damage to endothelial cells. In addition, ROS impairs transport at the BBB, including the function of P-glycoproteins, nutrient carriers, and efflux proteins. Some antioxidants may exert protective effects by acting directly on the BBB to preserve the barrier function. To summarize, aging more or less increases the vulnerability of the BBB, and many antioxidants act to guard this major gate to the brain against oxidative stress, which has a significant impact on maintaining brain function.

7.2.8 Antioxidants: Mechanisms and Functions

Oxidative stress may result from a disturbance in the balance of endogenous defense mechanisms of oxidant metabolism across the blood-brain barrier and may be a substantial pathogenic factor in the aging brain as well as the etiology of various age-related cognitive dysfunctions. ROS formation is unwanted and needs to be eliminated. The generation of ROS protects against viral, bacterial, and fungal infections or conveys a death signal in response to many types of nuclear DNA damage and mitochondrial damage. Free radicals are very reactive intermediates that damage biological targets represented in Fig. 7.2, and living beings must minimize their damaging action. An internal antioxidant network effectively uses molecular species such as thioredoxin, reduced glutathione, and ubiquinol to eliminate ROS, and it acts both in the cytoplasm and in the mitochondria of cells (Lochhead et al. 2024). The modification of protein targets by oxidative changes could result in additional information that could assist in the development of possible novel disease prevention strategies and modes of action from the growing pool of natural phytochemicals. Current evidence indicates that various food antioxidants are molecular preventers of oxidative stress. These antioxidants are putative therapeutic agents to

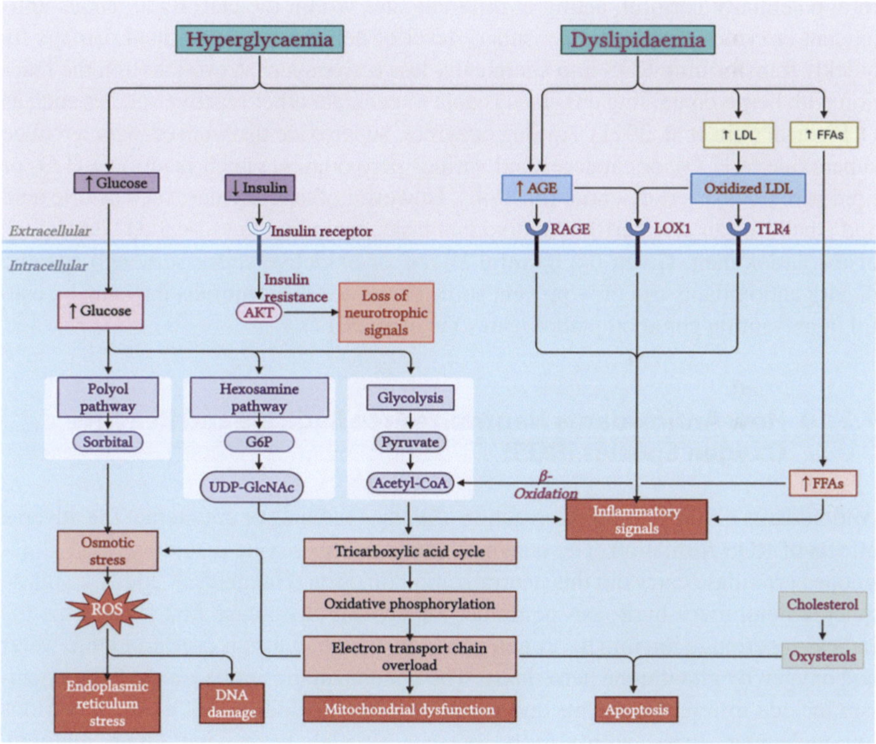

Fig. 7.2 Mechanisms of oxidative stress and free radical generation in the aging brain

prevent free radical-mediated aging and memory damage. Forestalling any stage of the ROS-initiated cascade of molecular events will help mitigate oxidative dysfunction.

7.2.9 Definition and Types of Antioxidants

A given antioxidant can interfere with free radical-mediated phenomena through several nonmutually exclusive mechanisms: by preventing the hydrocarbon oxidation chain from breaking, through chelating unbound transition metal ions responsible for radical generation, or by soaking the excited state of the released oxygen or peroxides (Sies et al. 2022). According to the place where the antioxidant activity is carried out, we can differentiate two primary types of antioxidants, whose only structural and functional paradigm is provided by tocopherol: hydrophilic antioxidants, located in the hydrophilic environment, acting in the phagosome, or associable with NH_2 groups of proteins and lipids; and lipophilic antioxidants, residing in the hydrophobic environment, effective at the membrane level (Zhang et al. 2021). The distinction of the primary antioxidant type is, however, often difficult to define, as antioxidants generally reside in the two environments, interpolating between the two activities, being used by the cell to neutralize radicals of different natures and, therefore, potentially harmful, acting at different sites within the cell. Endogenous antioxidant enzymes represent a secondary level of defense against radical damage by quickly transforming ROS into chemically less active species, even though the reaction with heme-containing enzymes is able to generate other reactive species such as OH radical (Yin et al. 2011). Among enzymes, superoxide dismutases, which reduce superoxide to H_2O_2, or catalases and various peroxidases, which neutralize H_2O_2 or lipid peroxides, exert essential functions. However, other enzymes, such as uric acid and albumin, can also perform antioxidant tasks, including the so-called "dark side" of the antioxidant. Given the harmful effects of oxidative stress, although specific dietary antioxidants can only prevent some ROS-mediated injuries, they can be useful in preventing common pathologies (Yin et al. 2011).

7.2.10 How Antioxidants Neutralize Free Radicals and Reactive Oxygen Species (ROS)

Antioxidants either prevent the reaction with free radicals or counteract the adverse effects of ROS formation. The enzymes catalase, superoxide dismutase, and glutathione peroxidase carry out this neutralization function (Hajam et al. 2022). Catalase quickly neutralizes hydrogen peroxide. Superoxide dismutase first transforms the radical superoxide ion into hydrogen peroxide, which in turn is converted into water and oxygen by glutathione peroxidase. The nonenzymatic antioxidants of the organism include different elements and metallothionein. Additionally, there are various low-molecular mass antioxidants present in the brain, including albumin, ceruloplasmin, transferrin, ferritin, melatonin, uric acid, ascorbic acid, vitamin E, and many other compounds.

The importance of these compounds is based on the fact that antioxidants, by means of their reducing properties, serve as a basis for neutralizing free radicals and modulating oxidative stress. Substances with antioxidant properties either prevent the reaction with free radicals or act in an antagonistic fashion with ROS formation. Three enzymatic systems have been essentially scrutinized, such as catalases, superoxide dismutases, and glutathione peroxidases, as their activities have been observed to decrease during aging (He et al. 2017). Furthermore, the brain possesses a high antioxidative potential that is less strictly controlled. In addition to having various nonenzymatic antioxidants, the brain contains different elements and metallothionein, as well as various liposoluble antioxidants, including vitamin E, carotenoids, and liposoluble polyphenols. Dietary supplementation with substances that have antioxidant properties improves the appearance in regard to learning ability and memory in aging animals.

7.2.11 Cellular Defense Mechanisms: Enzymatic Versus Nonenzymatic Antioxidants

A wide variety of low-molecular-weight endogenous scavengers have evolved in order to prevent oxidative damage. Antioxidants are classified into different types, depending on their involvement in either enzymatic or nonenzymatic cellular defense mechanisms. Enzymatic antioxidants include both single enzymes, such as superoxide dismutases, catalase, and glutathione peroxidases, and enzymes requiring the aid of specific cofactors and metal ions for radical scavenging activity (Kőszegi et al. 2017). Catalases, glutathione peroxidases, and thioredoxin peroxidases, which possess different cofactor preferences, rapidly reduce hydrogen peroxide within tissue compartments. Additionally, nonselenoenzyme peroxiredoxins are also involved in the reduction of hydroperoxides, depending on disulfide bond formation. Superoxide dismutases are pivotal enzymes that utilize metals such as copper and zinc for the cytosolic and mitochondrial forms, while manganese is used in the mitochondrial matrix. Iron-oxidizing extracellular enzymes are utilized by extracellular organisms.

Nonenzymatic antioxidants are normally present at high concentrations. These include vitamins such as C, E, and K, as well as reduced forms of glutathione, ubiquinol, NADPH, and thioredoxin. Sulfur-containing amino acids are crucial cofactors for the oxidation defense against reactive oxygen species and reactive nitrogen species. Ascorbic acid is a powerful intracellular ROS scavenger and has the capacity to reduce oxidized forms of low-molecular-weight compounds, including alpha-tocopherol (Alp et al. 2019). Vitamin E is the major antioxidant of the biological membrane that contributes to the first line of antioxidant defense. Along with ascorbic acid, vitamin E selectively accumulates in mitochondria, where ROS are primarily produced. In addition to these vitamins, polyphenolics and carotenoids also show excellent antioxidant activity. Flavonoids and melatonin protect nervous tissue from oxidative damage caused by ROS. Plant carotenoids, astaxanthin, and lycopene are strong antioxidants, while beta-carotene is essential for its conversion to retinol, a form of vitamin A that protects against radiation and oxidants. Dietary polyphenolics have multiple functions, such as antiradicals, antioxidants, or prooxidants, and they may also chelate enzymes to exhibit in vitro or subcellular effects. Minerals and antioxidants regulate the activities of many of the intracellular redox enzymes (Ornatowski et al. 2020). Excess intake of antioxidants with supplements should be avoided, as this could produce potentially harmful prooxidants.

7.2.12 Key Antioxidants for Brain Health

Brain homeostasis is regulated by the continuous renewal and support of a consorted group of cells, including neurons, astrocytes, oligodendrocytes, and microglia. However, biological processes such as oxygen supply, facilitation of neuromuscular activity, and regulation of synaptic plasticity also contribute to increased superoxide production and cognitive dysfunction if they are not regulated (Ionescu-Tucker and Cotman 2021). To protect the brain from these damaging processes, several mechanisms have been described in detail, including the upregulation or suppression of pro- or antioxidant enzymes, respectively. Particularly, the activity of enzymes such as superoxide dismutase, catalase, and glutathione peroxidases, along with a pool of small molecules (scavengers) such as vitamin C, E, and glutathione, has been shown to neutralize the effects of reactive oxygen species and, therefore, maintain redox balance. Unfortunately, with aging, these systems are disrupted, resulting in an imbalance of these reactive species, leading to damage to membrane lipids, proteins, and DNA in the brain (Baranowski et al. 2020).

The brain is sensitive to the effects of oxidative stress due to its high level of oxygen use and redox-relevant fatty acid content. To regulate oxygen utilization and limit reactive oxygen species formation, the brain utilizes several key antioxidants, including ascorbate, alpha-tocopherol, and glutathione. Additionally, both lipophilic and hydrophilic antioxidants, along with brain antioxidant enzymes such as superoxide dismutase, catalase, and glutathione peroxidases, help to limit reactive oxygen species formation and reduce oxidative stress. Moreover, because reactive oxygen species can be generated in various subcellular compartments, it is necessary to have a robust and diverse antioxidant network to assist with reactive oxygen species neutralization (Kubben et al. 2016). However, with increasing age, the levels of these antioxidants decrease and, simultaneously, oxidative stress increases, often resulting in the impairment of cognitive and sensorimotor function.

7.2.13 Endogenous Antioxidants (e.g., Glutathione, Superoxide Dismutase, Catalase)

Endogenous antioxidants are functional components of the body that exert antioxidant effects against diseases by being or becoming present in the tissue or blood without additional intake, including glutathione, superoxide dismutase, and catalase, which exist in the body even in the absence of substantial amounts of dietary antioxidants. Glutathione is a vital intracellular water-soluble antioxidant that functionally protects cells from oxidative damage against various reactive oxygen species and radical species by scavenging hydrogen peroxide, oxygen, and hydroxyl radicals (Valko et al. 2006). Glutathione is an essential low-molecular-weight antioxidant and is often considered the most important antioxidant within the oxidative system. Superoxide dismutase is an enzyme present in all animals and plants, which can catalyze the dismutation of superoxide into oxygen and hydrogen peroxide. Superoxide dismutase is a key antioxidant enzyme and can destroy active oxygen free radical damage, while also inhibiting the conversion of active oxygen-free radicals to hydrogen peroxide (Radomska-Leśniewska et al. 2017).

Catalase is a common enzyme found in nearly all living organisms exposed to oxygen. Catalase is a catalyst for the decomposition of hydrogen peroxide into oxygen and water, thereby mitigating the effects of oxidative stress. Glutathione, superoxide dismutase, and catalase are more focused intracellular scavenging enzymes. The glutathione system is the primary intracellular antioxidant system for scavenging reactive oxygen species and resistance to oxidative stress, that is involved in many aging and neurodegenerative diseases. Similarly, the upregulation of enzymes that are part of the glutathione peroxidase system is an important defense mechanism against reactive oxygen species. Additionally, the age-related decrease in superoxide dismutase expression is more evident in brain mitochondria, leading to an increase in tissue oxidative damage and neuronal apoptosis and contributing to the susceptibility of the senescent brain to increased oxidative stress (Wołonciej et al. 2016). In addition, the age-associated overexpression of glutathione peroxidase and superoxide dismutase levels can be reversed by exercise, which can further help improve learning and memory in the elderly, at least partially signaling through the Sirt1/PGC1a/NR4A axis pathway.

7.2.14 Exogenous Antioxidants (e.g., Vitamins C and E, Polyphenols, Flavonoids)

Although no evidence suggests that exogenous antioxidants alone are useful in reversing or counteracting oxidative stress in normal and degenerating neural cells, some preliminary findings highlight their potential beneficial effects on specific subsets of reactive oxygen species that are generated by neurons in the course of their activities and interactions with other brain cells. It should be stressed that if ROS generation in the aging brain were restricted to such specific subdomains, exogenous antioxidants might indeed help neurons to repair and avoid direct severe toxicity (Neha et al. 2019). Nonetheless, current study findings firmly indicate a partial benefit only after irreversible oxidative damage has taken place, such as when mutant neurons may have reached their false "point of no return." Indiscriminate in vivo use of exogenous antioxidants might even exacerbate neuronal ROS-related death processes. In addition, exogenously administered antioxidants, either alone or in combination with some endogenous antioxidant mechanisms, might expand their effects as organizational components for adaptive and long-lasting antioxidant protection (Pisoschi and Pop 2015).

7.2.15 Minerals with Antioxidant Properties (e.g., Selenium, Zinc, Magnesium)

Selenium, as an integral part of the enzyme glutathione peroxidase, also plays a role in the metabolism of ethanol and, therefore, may protect the brain against chronic alcohol-induced oxidative stress (Kiełczykowska et al. 2018). Using complementary bisenzyme B of Clostridium, it was shown that the blood glucose-lowering effect of zinc bis (maltolato) complex could be related to its insulin-enhancing properties, decreasing insulin resistance, or its mimetic actions. In pharmacological doses, selenium has the ability to counteract lead, mercury, and arsenic poisoning, and such properties might be associated with its modulatory actions on the induction of proinflammatory cytokines and nitric oxide production, as well as proapoptotic-related protein levels. Since selenium reverses the ability of lead to disrupt the phasic remodeling of bone, it may also reverse the effect of lead exposure on the brain (Hariharan and Dharmaraj 2020).

Magnesium is involved in many biochemical processes, including the propagation of the calcium-induced immune response through neuromodulation and communication, and this can be mediated, in part, through inhibition of mast cell degranulation. Chelation is able to perform a protective role at high doses against oxygen toxicity, indicating that zinc, like selenium, can modulate the expression of inflammatory signaling proteins and confer elevated resistance to CNS inflammatory insults. Pregnant rats exposed to a zinc-deficient diet give birth to offspring with fewer taste buds. Since zinc was found to act as a suicide inhibitor/covalent modifier during reactions in the presence of reduced FMN, it may delay flavoprotein activity, particularly for flavin nucleotide-containing enzymes such as nitroreductase, nitric oxide synthase, and neutral sphingomyelinase (Rapa et al. 2019). Thus, the zinc-free nitroreductase may have a reduced ability to generate products of reductive activation processes, and using such a strategy, intracellular zinc can potentially modulate the biological activity of a flavoprotein.

7.2.16 The Role of Antioxidants in Brain Aging

The brain is susceptible to oxidative stress and to damage by free radicals because it consumes a great amount of oxygen, contains large quantities of easily peroxidizable lipids, and has a high content of redox-active transition metal ions. The aging of the brain is accompanied by an increasing amount of free radical-induced oxidative stress and cellular damage, and the occurrence of neurobehavioral declines (Melzer et al. 2021). Since the inverse relationship between free radical-induced damage and antioxidant capacity in the aged brain is well established, antioxidant therapy presents an attractive target for the treatment of age-related progressive neurodegenerative disorders. Research over the last few years has revealed that brain aging can be significantly ameliorated, and the lifespan extended, by dietary supplementation of antioxidants.

Numerous data suggest that antioxidants such as vitamin E, plant-derived antioxidants, or polyunsaturated fatty acids slow down or reduce the free radical-mediated age-related neurochemical and cellular changes in the brains of various experimental animals, and modulate age-associated impairments in learning, memory, and motor performance. This chapter summarizes the current level of knowledge of brain aging, discusses the role of oxidative stress in brain aging, and outlines the knowledge about antioxidative nutritional habits (Martínez-Guardado et al. 2022). Administration of antioxidants may improve cognitive function and delay the progression of progressive neurodegenerative disorders. Since the impact of antioxidants may depend on a combination of several antioxidants, well-designed and validated multicenter studies on the optimal composition of antioxidants are warranted.

7.2.17 Protection Against Oxidative Damage to Neurons and Glial Cells

The prevalence of Alzheimer's disease, Parkinson's disease, and the ongoing loss of cognitive elements in the normal aging process emphasize the requirement for a better understanding of the aging brain and the processes that exist within their age-relevant decline. In understanding oxidative stress, certain age-associated adjustments in the relative range of CNS antioxidant mechanisms that may contribute to CNS health and help understand these disorders are observed in a collection of species, like rodents, nonhuman primates, and humans (Goodman et al. 2024). These data demonstrate that as age increases, the levels of several key enzymatic and nonenzymatic antioxidant systems reduce, while an increase in oxidative damage becomes apparent. It is presently unclear whether antioxidant maintenance in response to oxidative damage during senescence differs from that observed in Alzheimer's disease, Parkinson's disease, and other age-associated neurodegenerative diseases.

Antioxidants provide protection in additional processes against harmful waste released from the CNS, neuromodulation, neurotransmission, and pathological mechanisms due to heavy metals and other factors during periods of injury and cell death. With a rapid rise in the employment of these cells, revascularization and recapitulation with new tissue should occur alongside a rapid increase in the removal and disposal of such toxic accumulations. Studies must be conducted to understand how antioxidants can accomplish neuroprotection or be properly employed in the design of experiments for long-term treatment of Alzheimer's disease and Parkinson's disease (Cabezas et al. 2019).

7.2.18 Reducing the Risk of Cognitive Decline and Neurodegeneration

A recurrent idea in the preceding sections is that prevention, by living a healthy life inside and outside the body, is important in reducing the risk of cognitive decline and neurodegeneration. Cognitive decline may not be brain damage, but the neuron, the basic unit of the brain, does not seem to have a backup that replaces the neuron if it dies. Cognitive reserve might help reduce the perception of cognitive decline, but the declining brain needs the help of the human or animal that houses it, especially if the household does not have inhabitants to protect the brain in any necessary health problem (Jessen et al. 2020). Dementia is an umbrella that includes a vast number of different etiologies that are probably related by a clinical spectrum. Surgery, trauma, infections, poisoning, metabolic problems, cancer, and neurodegenerative diseases are some of the problems that can lead to the wrong equation: age with cognitive decline or neoplastic comorbidity with cognitive decline. The etiology of decline is still unclear; nevertheless, the ability to confront the problem and evaluate it in the correct way is important for an efficient approach (Bektas et al. 2018).

Despite age, the brain is a privileged organ: disconnected from lasting nervous stimuli, the brain continues to expend energy and requires the same blood flow as it would if it were working. The blood-brain barrier helps prevent toxic products from entering the brain, but when toxic products increase in the blood, they cross the blood-brain barrier and reduce the ability of the shield to protect the brain. The balance between the stripper and the toxic product could be different in an old brain, or the clearance of the toxin could be ineffective, or even the source of oxidative stress with consequent neuronal death. Preventing the formation of these products and helping the barrier to defend the brain is desirable in dementia. An oxidized brain looks like a brain affected by different forms of cognitive decline; however, better oxygen trafficking and optimal nutrition are essential to differentiate between normal brain activity and the brain that is at the center of the problems. Oxygen in the wrong place, or too much food, is counterproductive to all cells, not only brain cells.

7.2.19 Antioxidant Effects on Mitochondrial Function and Energy Metabolism

Mitochondria are the major source of reactive oxygen species and are considered to play a crucial role in brain aging, neurodegenerative disorders, and traumatic injury. Because mitochondria are also the main target of free radical attack, it is believed that oxidative damage to mitochondrial components leads to a progressive loss of function and further increases the generation of reactive oxygen species (Unsal et al. 2021). Oxidative damage to mitochondrial DNA may lead to mutations in the polypeptide-encoding gene, which, in turn, may decrease the efficiency of its oxidative phosphorylation system because many of the subunits of the oxidative phosphorylation complexes are encoded by nuclear DNA and synthesized in the cytoplasm. In this context, a positive correlation exists between the lifespan of a species and the stability of mitochondrial DNA. Moreover, the metabolic needs of the brain make that organ particularly sensitive to oxidative stress. During neurodegenerative processes, an impairment of energy and redox metabolism can lead to neuronal dysfunction and degeneration. The fact that both age-related and injury-related neurodegeneration share similar signaling events only increases the importance of such a pathophysiological event.

Administration of antioxidants to experimental animals, either early in life to delay neurodegeneration or after the onset of age-related changes, has often been noted to protect neurons and preserve their function. The beneficial effects of antioxidants can modify several molecular reactions leading from energetic failure to degeneration and inhibition of complex IV of the electron transport chain, neuronal death, and impairment of memory. The need to protect mitochondria reinforces the importance of encouraging results not only of antioxidant treatment but also of the use of dietary antioxidants. They may be particularly useful if the whole antioxidative system is strengthened. Indeed, the use of antioxidants that act at different levels of radical chain reaction or of combined antioxidants that have not only no prooxidant effect but also display increased efficacy would limit the oxidative damage and could be very beneficial. Because soluble preparations of natural antioxidants have poor permeability across the blood-brain barrier, and direct administration to the brain is unlikely to be tolerated therapeutically, a nontoxic protection or regeneration in response to oxidative changes allows the mitochondria to fulfill their pivotal role in developing neurodegenerative disorders (Unsal et al. 2021). Data suggest that biophysical generic modifications of the cell membrane might increase the resistance of neuronal plasma membrane sensitivity to peroxidizable molecules, and that some of the therapeutic efficacy of dietary antioxidants may derive from this property. The protective effect of antioxidants seems not always to be related to the protection of neurons; indeed, a presynaptic effect has been hypothesized. In fact, an antioxidant specifically targeted to mitochondria has been localized at synapses, inhibiting the apoptosis of synaptic mitochondria. This selective localization within neurons could establish compounds that protect the presynaptic terminal, which consumes massive energy, to be particularly useful in neurodegenerative disorders. Taken together, this information suggests that nontoxic bioactive compounds might be a powerful tool to modulate lipid components of membranes, potentially interfering with their integrity and modulating susceptibility to peroxidative damage. A new nonwater-soluble form of the potent antioxidant compound merits attention for its potential role in brain aging and its correlation with age-related neurodegenerative disorders (Gupta et al. 2024).

7.2.20 Antioxidant-Rich Nutraceuticals for Brain Health

There has been an increasing interest in research focusing on the aging brain and the prevention or even reversal of aging-related cognitive decline and dementia. These are likely to have a considerable personal, economic, and social impact as they will lead to a growing demand for health or social care services. At the same time, there is an increasing interest in using diet and food components as well as nutraceuticals to preserve brain structure and function over the lifespan. In particular, dietary phytochemicals such as those derived from the consumption of fruits, berries, vegetables, herbs, spices, teas, algae, seeds, and other plant foods have been shown to play a potential role as antioxidants and modulators of the signaling pathways linked to synaptic plasticity and neurotransmission, two types of molecular mechanisms responsible for the improvement of neuronal signaling during aging (Rehman et al. 2023).

7.2.21 Omega-3 Fatty Acids and Their Antioxidant Effects

The profile of the fatty acids from the omega-3 family in triglyceride-enriched neural membranes has been shown to confer high membrane resistance to different free radical derivatives. The greater resistance of DHA to free radicals and singlet oxygen damage also arises at the level of lipid peroxidation (Stavrinou et al. 2020). This fatty acid, in comparison with other polyunsaturated fatty acids such as arachidonic and linoleic acid, has more efficient peroxidic elongation mechanisms, which take place in the initial phase of peroxidation. Thus, during the propagation phase, an event of repair occurs in the phospholipid chain, and this can prevent the physicochemical changes in the surface of the plasma membrane, restrain the alteration of enzyme activity and consequent neurotransmission interruption, guarantee cell survival, and also promote the stability of the membrane lipophilic structures.

Omega-3, particularly DHA, promotes peroxidic protection at the cellular and mitochondrial levels after having led to the same at the plasma membrane level (Santos et al. 2020). Perhaps the metabolic effects associated with EPA and DHA, regarding amelioration of insulin resistance, anti-inflammatory activity, improvement of endothelial function, enhancement of vascular homeostasis, and lipid droplet storage in adipocytes, are linked to the increased membrane fluidity resulting from the presence of larger concentrations of respective fatty acids.

7.2.22 Polyphenols (e.g., Curcumin, Resveratrol, Green Tea Catechins)

Polyphenols, especially flavonoids such as curcumin and catechins, have gained a reputation for effectively combating aging and Alzheimer's disease. Curcumin, present in the spice turmeric, is consumed daily in India for its anti-inflammatory effects. Curcumin downregulates amyloid precursor protein levels and secretase activity that otherwise increases amyloid peptide generation. In addition, curcumin increases levels of the newly described A1–42-degrading enzyme MMP-9 and upregulates the acetyltransferase enzyme TIP60, which forms a complex with the

A1–42-degrading proteasome determined by lower DUB activity. The resveratrol found in red wine helps mice maintain spatial memory and decreases A1–42 levels. Resveratrol requires the activation of AMP-activated protein kinase and silent information regulator T1 to elicit its positive effects (Luca et al. 2020).

Green tea catechins like (−)-epigallocatechin-3-gallate upregulate secretase inhibitor protein and Tollip in microglia at noncytotoxic levels, as well as ameliorating A pathology when injected peripherally into younger A precursor/PS1 mice or administered directly to the Alzheimer's disease animal model. In the 3XTg-AD aged mouse model, chronic dietary administration of a green tea polyphenol blend significantly reduces oxidative stress and Alzheimer's disease pathology while improving synaptic function. In all three Alzheimer's disease mouse models, treatment with green tea reverses preexisting memory decline. Although disputes continue regarding the clinical use of these and other polyphenolic agents despite the plethora of their reported bioactivities, strategies continue to improve their bioavailability (NCBI 2025d).

7.2.23 Herbal Extracts with Antioxidant Properties (e.g., Ginseng, Turmeric, Rosemary)

Ginseng, the root of Panax ginseng, has been used as a traditional medicine in Asia for thousands of years. Extracts of ginseng have been shown to have numerous effects, including nootropic activities, which are defined as cognitive function enhancers. The study of ginseng using an animal model suggested that the increase in BDNF mRNA correlates with improved hippocampus-dependent learning and memory tasks. In addition, a diluted ginseng berry juice diet improved cognitive function in both mice with age-related and high-fat diet-induced cognitive disorders. This could be a valuable finding and worth further testing for its clinical significance (Sharifi-Rad et al. 2018). Turmeric has been used in traditional medicine for promoting healthy brain function. Curcuminoids obtained from the rhizomes of Curcuma longa have been reported to have antioxidant properties and reduce cognitive deficits in animal models.

Rosemary, an herb, has been widely used in traditional medicine for the alleviation of neuropsychological symptoms. An extract of rosemary ameliorated learning and memory impairment induced by focal cerebral ischemia in rats. It has also been pointed out that rosemary has antioxidant capacity and the ability to hamper β-amyloid fibril formation. The antioxidant mechanism of rosemary includes the scavenging of free radicals, interruption of cellular chain reactions, and chelation activities of transition metals. Furthermore, rosmarinic acid, another potent antioxidant isolated from rosemary, inhibits direct and α-tocopherol-induced LDL oxidation (Tian et al. 2019). Therefore, rosemary has the potential to become a valuable agent for the prevention of age-related neurodegenerative diseases.

7.2.24 Antioxidants in Neurodegenerative Diseases

The amount of damage that occurs in all cellular components as a result of outer oxidants, ROS, and oxidative stress reactions increases due to limited removal actions. To prevent excessive production of these molecules, an equilibrium must exist between oxidant production and the defense system. Maintenance of this balance implies having an efficient system to scavenge ROS and to repair the damage induced by oxidative modifications of proteins, nucleic acids, and endogenous lipids. This can be carried out by the action of antioxidant enzymes and nonenzymatic substances (Tian et al. 2019). The balance for protecting the cell and eliminating or repairing previous oxidative damage is important to maintain the proper functioning of the cell represented in Fig. 7.3. Moreover, there are many research studies about the relationship that exists among antioxidants, poloquinonols, and the brain in the elderly. Many of these studies establish that poloquinonols are capable of reducing oxidative stress in the body and, in particular, in the aging brain. Also, in this regard, biochemical and histological information is available at the molecular level regarding alterations of biomarkers related to aging due to oxidative stress and their relationship with poloquinonols. Of the studies conducted so far, there have been few clinical and drug applications of these properties of poloquinonols. More research and development of drugs based on scientific discoveries will be necessary for these compounds to be applied.

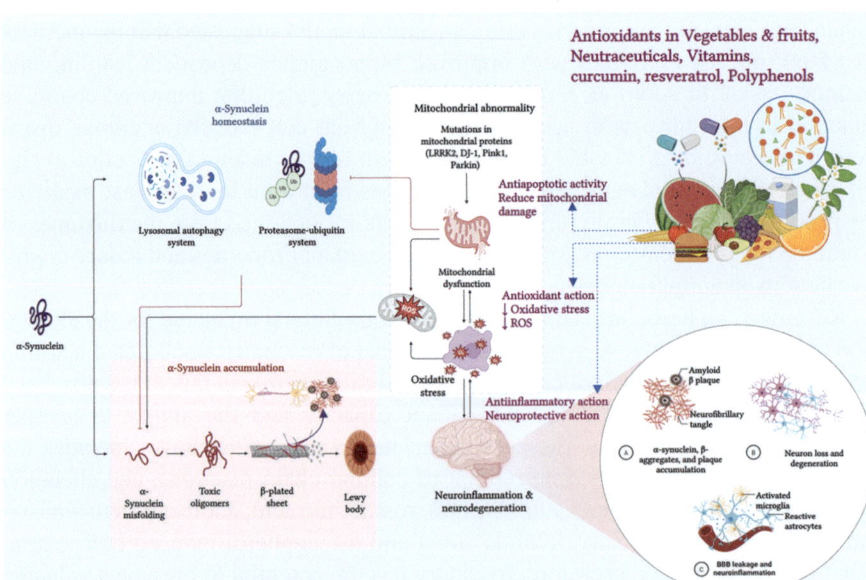

Fig. 7.3 Protective role of antioxidants in preventing neurodegenerative damage

7.2.25 Role of Antioxidants in Alzheimer's Disease and Parkinson's Disease

The human brain is approximately 2% of the entire human body, but it uses about 20% of the total oxygen in the blood. To efficiently utilize this oxygen, the brain exhibits many unique characteristics with regard to energy-consuming biomolecules and metabolic processes (Jomova et al. 2023). This high consumption of oxygen can lead to the production of excessive free radicals, leading to oxidative stress in the body. Oxidative stress in the aging brain can bring about aging and nervous system diseases such as Alzheimer's disease and Parkinson's disease. In this review, we will discuss the relationship between oxidative stress and the aging brain. We will also summarize the substances that have been proven effective in reducing neurodegeneration in Alzheimer's and Parkinson's diseases and discuss their interaction with oxidative stress.

Oxidative stress plays a significant role in the aging process and several neurodegenerative diseases, such as Parkinson's and Alzheimer's diseases. For this reason, researchers have paid considerable attention to the effects of antioxidants present in various foods on neurodegenerative diseases. Recent studies with plant polyphenol extracts show the potential for delaying or even preventing Alzheimer's and Parkinson's diseases (Kumar et al. 2022). Animal studies indicate these polyphenolic substances act through many pathways to delay or prevent Alzheimer's and Parkinson's diseases. Nevertheless, these conclusions are largely based on additional studies. The parameters and evaluation criteria vary in many of the studies. Additional research is necessary to design a polyphenol compound for production and therapy in terms of efficiency, safety, absorption, pharmacokinetics, and selective tissue distribution, as well as metabolism.

7.2.26 Neuroprotective Effects of Antioxidants in Other Neurodegenerative Conditions

In general, the literature on neuroprotection for alpha-tocopherol or beta-carotene alone, as well as for naturally occurring mixtures of antioxidants such as those found in diets rich in fruits and vegetables, is reviewed for Parkinson's disease, Huntington's disease, amyotrophic lateral sclerosis, and glaucoma, with particular emphasis on environmental and genetic influences thought to be relevant to the etiopathology of these neurodegenerative diseases (Johnson et al. 2022). Specifically, for each of these neurodegenerative diseases, the possibility that oxidative injury is an important contributing factor is reviewed, as is evidence for the involvement of the mitochondrial respiratory chain and other biological, and in many cases, genetic risk factors such as mutations or polymorphisms in key mitochondrial and antioxidant enzymes important in the neutralization of toxic oxidants. In this context, evidence for neuroprotective effects exerted by a diet rich in fruits and vegetables for Parkinson's disease, Parkinson's disease risk and progression, or Huntington's disease is selectively described, as well as potential lifestyle factors in the etiopathology of these diseases, such as high levels of aerobic exercise and moderate alcohol consumption (Johnson et al. 2022). It is hoped that this review will raise awareness of other neurodegenerative diseases in which oxidative stress and antioxidants may be important, and thereby stimulate further interest and research activity to better understand their relevance to pathogenesis, with the eventual aim of designing more rational preventive and therapeutic measures (Hong et al. 2022).

7.2.27 Current Research and Clinical Studies on Antioxidant Supplementation

A primary area of aging research today is devoted to determining the effects of dietary antioxidants on aging and age-associated neurodegeneration. Epidemiological and clinical findings regarding antioxidative vitamins with a particular focus on the central nervous system are presented (Power et al. 2022). The idea is that animal and human studies may be evaluated for clinical relevance through the use of surrogate markers and controlled experiments. It is necessary to think of nutrients in the same fashion as drugs when evaluating their therapeutic potential. Proper assessment of the therapeutic potential of natural products should be relevant and practical (NCBI 2025e).

7.3 Combining Antioxidants with Other Nutraceuticals for Synergistic Effects

Although intervention with antioxidants is a practical strategy for the prevention of age-related brain disorders, it is still unclear which particular antioxidants are the most efficacious for such intervention. This is because nutrients in our diet do not act individually within our body, but jointly as a network (Fernandez and Marette 2017). Several studies have shown that the combination of an antioxidant with another micronutrient often shows synergistic efficacy. For example, certain combinations of vitamin B6, vitamin A, and selenium reduced inflammation in monkeys more effectively than individual nutrients, and omega-3 fatty acids reduced pulmonary diseases in an experiment in which animals were exposed to both ozone and fire smoke. Several other studies also showed that a combination of antioxidants with other nutrients provided a significant benefit.

The results of our study using gamma-tocotrienol as the model suggest that the combined use of antioxidants and other nutrients may also prevent degenerative brain disorders. We have found that the combination of gamma-tocotrienol with two other nutrients, carnosine and green tea, produced significant synergistic neuroprotection in our aged rat model. These agents as the first step toward identifying a mixture of natural antioxidants for the prevention of neurovascular aging because all three of them have shown favorable effects on age-related brain endothelial dysfunction in aged rats, and all three have compounds that are neuroprotective and can easily cross the blood-brain barrier. Since the present study is an in vitro approach, future in vivo studies should be designed to confirm that our combination is more efficacious and has fewer side effects than the compounds when used alone (Mandelker 2004).

7.3.1 The Potential for Combining Antioxidants with Anti-inflammatory and Neuroprotective Compounds

The knowledge gained over the last few years regarding the antioxidant and anti-inflammatory efficiency of dietary compounds provides a basis for further investigating their ability to prevent, delay, and possibly treat various age-related neurological and neurodegenerative disorders. Dietary supplementation with natural compounds, such as resveratrol, curcumin, polyphenols, and other antioxidants from various fruits and vegetables, is now being tested in clinical trials. A recent study described the beneficial effect of epigallocatechin-3-gallate, the major catechin in green tea, on learning and memory function in a mouse model for Alzheimer's disease. The beneficial effect of green tea polyphenols, tetrahydrocurcumin, and resveratrol on cognitive dysfunction was reported. Additional studies described the positive effects of vitamins and omega-3 fatty acids on the cognitive function of individuals aged over 65 years. There are many ongoing debates over whether the increased intake of dietary antioxidants may be the cause of the reduced incidence of Alzheimer's disease, Parkinson's disease, or other neurodegenerative disorders (Pan et al. 2019).

The induction of a slight transcriptionally mediated stress response under conditions of a mild challenge is considered to be beneficial for diverse tissues and species, causing enhanced tolerance to a subsequent more challenging stress. Such a type of hormesis (neurohormesis) was reported for the nervous system in invertebrates and can be achieved by exposure to weak environmental, oxidative, metabolic, and other challenges associated with aging and age-related diseases. Indeed, the accumulation of tissue damage caused by chronic oxidative stress decreases with age and is dependent on the antioxidant capacity of an organism. As a single type of antioxidant cannot cover all the different aspects of oxidative stress, the increased total antioxidant status achieved by using a combination of dietary compounds provides significant neuroprotection for the old brain, enhancing its coping capacity for age-related oxidative, proinflammatory, and immune challenges. Awakening natural defense pathways through moderate and safe challenges results in the reduction of biomolecular and oxidative stress, improved cellular repair, a more efficient interference with age-related oxidative, immune, and proinflammatory challenges, and, preferably, the preservation of optimal cognitive function with aging (Al-Shami et al. 2023).

7.3.2 Synergistic Effects in Slowing Brain Aging and Preventing Neurodegeneration

Several interventions designed to test the hypothesis that oxidative stress is a significant cause of aging consist of a "cocktail" of antioxidants with other potential pharmacological strategies, including an antioxidant drug, exercise, dietary modification, or hormone replacement therapy. In all situations, combinations of treatment display synergistic effects in slowing brain aging, preventing major risk factors for brain aging or neurodegeneration, and occasionally enhancing the outcome (NCBI 2025f). Among the most effective and synergistic therapies against oxidative stress is the concomitant administration of several antioxidants, simultaneously modifying several factors leading to oxidative stress in the aged brain and, by doing so, enhancing the effectiveness of individual trophic and protective signals, thus having vast potential to defend the brain against physiological brain aging or chronic neurodegeneration. Thus, from this point of view, such therapies would fail to fulfill the reproductive criterion and break the aging process only in an indirect manner. Similarly, therapies involving exercise, dietary choices, or lifestyle modifications, initially designed to prevent cardiovascular diseases, have all shown results against age-related brain dysfunction, acting only indirectly and incompletely against factors promoting oxidative stress. However, if used in addition to therapeutic strategies, these modifications can be therapeutic and provide complementary benefits through their action on risk factors and inducing neurotrophic effects (Higgins-Chen et al. 2021).

7.3.3 Nutraceutical Formulations and Combination Therapies

There is evidence showing that substances that are poor antioxidants, such as estrogen and resveratrol, still show a lot of good actions in conditions of neuronal damage, like Down syndrome or Alzheimer's disease. There is a need to test not only parts of oxidative stress but the complete state of redox homeostasis in the models showing improvement when substances are administered, as well as a complete drug characterization. The idea is to test all the well-characterized antioxidant substances in combination in order to prevent oxidative damage (Hill et al. 2016). Not only the evaluation of the brain's redox conditions but also early biomarkers of damage in flies and worms are still needed to maximize the benefits of tested antioxidants in humans in a combined way. People consume not only food supplements but also food, and we are exposed throughout our lives to numerous substances in our diet with antioxidant properties. Therefore, studying the redox state of the brain and the antioxidant defense system in old animals treated with nutraceuticals is necessary to better understand the possible benefits and side effects of these substances in the prevention of dementia.

The therapeutic potential of melatonin, coenzyme Q10, nicotinamide adenine dinucleotide, and resveratrol in combination with a potent antioxidant cocktail has been successfully tested in transgenic flies that have the human protein beta-amyloid 42 expressed constitutively in their eyes, in combination with infrared light and nanoparticles for several weeks. The Drosophila are indeed genetically engineered and therefore contain defects, and the development of such an accurate fly model system based on old Drosophila could be used for neuroprotection prediction. Data obtained from in vivo experiments with Drosophila show the power of using this biological platform for aging and neuronal-related issues. Drosophila is a high-complexity model with a short lifespan and an easy breeding protocol, which will allow faster testing on many different transgenic fly lines that may show Aβ42 or tau pathology in holes and specific neuronal systems, such as Alzheimer's. The results are remarkable because the combination of natural antioxidants shows potential neuroprotective properties, and further studies are recommended in order to enhance the beneficial effect of these nutrients administered alternatively to formulations of antioxidants (Di Liegro et al. 2019).

7.4 Clinical Evidence and Research on Antioxidants for Brain Aging

Potential drug targets related to the aging-related changes observed in microglia are the enzymes through which they regulate their extracellular environment. Presented here is an overview of the evidence, often mixed, derived from clinical work on the use of antioxidants to alleviate age-related neurological and vascular diseases. The available randomized clinical trial data on several major antioxidants have been conflicting. It is possible that the divergent results could be due to the confounding factors on study demographics, drug properties, the age of the patients, the timing of treatment relative to the disease stage, and other genetic factors. However, these uncertainties might also suggest that the deleterious effects of microglial oxidative stress are worse in the preclinical than in the clinical phase or that normalizing oxidative stress is not a bona fide antiaging strategy (Costa et al. 2022).

We need better diagnostics of phenotypic and functional mitochondrial density and redox status, with correlative imaging markers, for evaluations of drug effects on microglial mitochondrial function. The possible benefits and risks should be compared in exercise, low-calorie diet, pharmaceutical interventions, adult tissue stem cell transplants, and the regulation of microglial population dynamics. System preconditioning with mild oxidative stress-inducing agents to alleviate an acute or chronic oxidative injury or disease represents an emerging field whose therapeutic windows would benefit from a better understanding of mitochondrial redox signaling.

7.4.1 Review of Clinical Trials and Studies on Antioxidants for Brain Health

Alzheimer's disease (AD) is a major neurodegenerative disorder that has a series of changes, including progressive memory loss and the pathological accumulation of beta-amyloid (Aβ). Antioxidant therapy has been one of the most promising topics in treating Aβ-induced memory impairment. However, study conclusions have been inconsistent. This paper aimed to review the effects of antioxidants on amyloid-induced memory deficits and further explore potential factors leading to the inconsistency.

Clinical studies have been carried out to explore the potential role of antioxidants in reducing AD risks or treating AD patients. Although most of the studies did not support antioxidants to treat AD, a recent comprehensive meta-analysis suggested that antioxidants can effectively treat AD. Furthermore, in various animal models of AD, antioxidants were reported to be effective in attenuating the amyloid-induced memory deficits. However, some studies still showed that antioxidants had no effectiveness (Chauhan and Chauhan 2020). Thus, because the conclusions of preclinical research and clinical studies were inconsistent, more studies are required to demonstrate the effects and potential mechanisms of antioxidants on memory, especially on amyloid-protein-impaired memory.

7.4.2 Efficacy of Antioxidant Interventions in Cognitive Function and Neuroprotection

The consistent evidence across the different experiments in rodents suggested that a combined approach that includes dietary interventions and pharmacological supplementation with standardized extracts might be the best means of providing a meaningful potential treatment for age-related cognitive decline, although further studies are needed. It is essential to clarify whether long-term intake of antioxidants, including those in foods, is necessary in order to prevent cognitive decline. These are points to be further elucidated. Several clinical trials in Alzheimer's disease and mild cognitive impairment subjects using antioxidants and other food supplements are now in progress, and assuming that oxidative stress is one of the mechanisms involved in the pathogenesis of Alzheimer's disease and mild cognitive impairment, we expect promising results from them (NCBI 2025g).

Given that a large number of antioxidants are accessible, we should now evaluate the efficacy of long-term use of appropriate antioxidants, including supplements and food, for improving the quality of brain function and for the prevention of age-related cognitive decline. Responsiveness of antioxidant enzyme activities and thiobarbituric acid reactive substances formation in the brain to oxidative stress induced by did not change in aged rats, in spite of the increase in brain thiobarbituric acid reactive substances.

7.4.3 Safety, Dosage, and Potential Side Effects of Antioxidant Supplementation

Dose accuracy is crucial in pharmacology. Despite the absence of upper safety levels for assessing vitamins, a dosage above 25 IU vitamin E/kg is not recommended without medical supervision. A recent meta-analysis reported significantly increased mortality at vitamin E dosages above 400 IU/day. The safety of antioxidants requires more attention, as minimal effects of antioxidants on normal physiology can become deleterious when animals are subjected to situations of concurrent stress. Similarly, for other individual or combination antioxidants, lower dosages promise to be more effective, but double-blinded placebo-controlled clinical trials in which the most appropriate combinations, dosages, treatment duration, and criteria for therapeutic response are investigated are essential (NCBI 2025g).

Most clinical trials examining the preventive or therapeutic effects of dietary supplementation with large dosages of vitamin C, vitamin E, or beta-carotene have based their dosages on previously suggested intake levels without considering their antioxidative effectiveness. Unfortunately, the antioxidative effectiveness of a high dosage of vitamin E is diminished as a consequence of increased free radical production. Results from randomized controlled trials consistently indicate that prevention with a combination of lower dosages of antioxidants is particularly beneficial in younger individuals. Ideally, we should endeavor to give the right antioxidant to the right patient at the right time in the right dose. The greatest societal benefits of a witnessed reduction in the morbidity of diseases will be increased through a preventive rather than a therapeutic approach. However, clinical research should primarily uncover evidence-based prevention rather than unsubstantiated hype. The combination of empirical and basic research could help debunk the current unhealthy antioxidant myths and result in improved public health policy (Adıgüzel and Ülger 2024).

7.5 Challenges and Future Directions

As described in earlier sections of this review, there is evidence that aging is a chronic, low-grade inflammatory state that may indirectly predispose the human brain to neurological conditions with varying degrees of cognitive decline. Nevertheless, the potential efficacy of personalized antioxidant therapies is very much alive. A greater challenge, and one that has received much less attention, is the potential that long-term chronic oxidative stress may retard this approach. Neuroprotection is automatically a crucial aspect of any proposed treatment. This is a very long time to be chronically affecting the biochemistry of the brain, and is something that cannot be ignored when proposing novel treatments (Li et al. 2025). The field has made a lot of advances in recent years, but the challenge still remains to find more viable alternatives to regular methods. This could be through repeated low doses of pharmacological agents or through a diet rich in antioxidant compounds. Indeed, any new findings that link individual antioxidants, complexes of antioxidants with slightly differing cellular localizations, new sources of antioxidant compounds, or novel treatments are likely to prove greatly advantageous in terms of maintaining health for an aging population.

7.5.1 Limitations in Current Research and Clinical Application of Antioxidants

Unfortunately, many in the lay public refer to antioxidants as cure-alls for disease or aging, and in response, some scientists are casting doubt on the overall effectiveness of these compounds. Although some findings appear contradictory, an analysis of these studies reveals important limitations that cast some doubt on the conclusions. Some of these limitations are related to the doses of antioxidants used, the types of antioxidants, as well as the methods of delivery to the human brain, which are not optimized (Jia et al. 2023). The absence of daily activities that require strong and powerful supplementation due to increased energy requirements as antioxidant support can even produce harmful episodes. On the contrary, ingestion of antioxidants that do not specify and the use of nonspecific drugs in age-related pathogenesis in the brain contribute to these limitations. Additionally, the lack of standardization of protocols, such as the use of compounds that can act on a multiplicity of targets and signaling pathways, in conjunction with some natural antioxidants, may not enhance synergism.

Relevant information on these limitations, along with a large number of combinations and interactions with other drugs from different families with therapeutic usefulness used in neurodegenerative diseases, raises doubts about the conclusions that antioxidants cannot yet be used to treat patients. Many questions about these supplements need further empirical studies, particularly in humans, with clearly defined doses of antioxidants and well-characterized pharmacokinetics in the rated segments, i.e., the temporal and regional patterns of uptake, which are often poorly defined in terms of central effects. Since several antioxidants of the central nervous system are involved in several enzymes, consuming levels of specific antioxidants or in signaling pathways, understanding these responses and dose adjustments in these roles will allow for more successful supplementation in certain pathological conditions.

7.5.2 The Need for Personalized Antioxidant Therapies

Human aging is a highly personalized process. There is a profound variation in lifespan among individuals due to the combined effects of genetic, epigenetic, lifestyle, and ecological factors. The time to the onset and the rates of all morbidities, including age-related neurodegenerative diseases, again vary among individuals. Even identical twins may age at different rates. The need for personalized antiaging or prohealthy lifespan management strategies is well recognized in the aging literature. Oxidative stress is a part of aging (Sehrawat et al. 2020). Oxidative stress is a phenomenon of the imbalance in redox reactions in cells and tissues and could thus be regulated by antioxidants. The availability of specific molecular tools opens the possibility to design selective antioxidant therapies. Such therapies will have to consider the type of mitochondrial oxidants, age, sex, and genetic background of the recipients and thus be highly personalized.

7.5.3 Emerging Trends in Antioxidant Research for Brain Aging

Recent years have seen innovations in antioxidant research for brain aging. While traditional antioxidants focused on eliminating toxic free radicals from the brain, an emerging trend is to target mitochondrial reactive oxygen species. Mitochondria are direct free radical sources, and laboratory studies indicated that the accumulation of mitochondrial ROS in aging brain cells ultimately impaired mitochondrial function, the so-called "mitochondrial cascade hypothesis." Damage to neuron mitochondria precedes many neuropathological changes and can be an early marker of brain senescence (Huhn et al. 2018). Due to the essential role of brain mitochondrial defects in the aging process and since neurons, glial cells, endothelial cells, and the blood-brain barrier involve different defense strategies, it is plausible to modify mitochondrial ROS directly by targeted therapeutic agents. Fortunately, many drugs have been introduced to absorb excess mitochondrial ROS generation to improve mitochondrial function.

As the role of hormones in aging and age-related diseases has been widely recognized, growing attention has been paid to the antioxidant activities of endogenous hormonal mediators. It has been suggested that these antioxidative properties contribute to hormone-mediated beneficial neuroprotection against age-related cognitive decline or neurodegenerative dysfunction. Recently, several hormone derivatives have been prepared and have shown obvious antioxidant effects in the brain; their beneficial role in the brain senescence process has been further confirmed by experimental data. Hormone-related drugs, with well-known safety and strong antioxidant effects, are expected to be potential candidates for treating the aging brain in the near future. Different signaling pathways are activated during the aging process and ultimately result in neuronal apoptosis or damage. However, all the signaling pathways or proteins respond to ROS at the same early stage. Furthermore, some of them cross talk with each other and stimulate positive or negative feedback, thus influencing the expression of other aging-related factors. These observations suggest pathways or factors that may be potential targets for drugs to alleviate ROS damage during the brain aging process (Jallow et al. 2024).

7.6 Conclusion

They strongly suggest that antioxidants will reduce the overproduction of oxidatively damaged DNA with age. The clinical trial showed that modest supplementation with antioxidant vitamins and alpha-lipoic acid can reduce overproduction. Experimental trial and previous clinical trials of short-term increases in antioxidant status all suggest strongly that the principal route of accumulation of DNA damage from a dietary deficiency of antioxidants will, only occasionally, be mutations. The most likely routes for this nonviral gene expression problem are premature differentiation of stem cells or stem cell apoptosis below some threshold stem cell population size for optimal brain function. There could also be some mutations from the inflammatory route of damage accumulation. In either case, a long-term deficiency of antioxidants will slowly degrade brain function. The history of an aging population in many affluent countries over the last hundred years shows that this breakdown in brain function has outpaced other improvements in public health.

References

Aaseth J, Alexander J, Alehagen U (2021) Coenzyme Q10 supplementation – in ageing and disease. Mech Ageing Dev 197:111521

Adıgüzel E, Ülger TG (2024) A marine-derived antioxidant astaxanthin as a potential neuroprotective and neurotherapeutic agent: a review of its efficacy on neurodegenerative conditions. Eur J Pharmacol 977:176706

Aguilar K, Jakubek P, Zorzano A, Wieckowski MR (2024) Primary mitochondrial diseases: the intertwined pathophysiology of bioenergetic dysregulation, oxidative stress and neuroinflammation. Eur J Clin Investig 54(7):e14217

Alcendor DJ (2020) Interactions between amyloid-B proteins and human brain Pericytes: implications for the pathobiology of Alzheimer's disease. J Clin Med 9(5):1490

Aldini G, Altomare A, Baron G, Vistoli G, Carini M, Borsani L et al (2018) N-Acetylcysteine as an antioxidant and disulphide breaking agent: the reasons why. Free Radic Res 52(7):751–762

Alp NB, Akdağ G, Karduz G, Vardar K, Aksu U (2019) Biochemical markers decrease and increase disproportionally in A1 pulley tissue of type 2 diabetic trigger finger patients. Eklem Hastalik Cerrahisi 30(2):117–123

Al-Shami AS, Essawy AE, Elkader HTAEA (2023) Molecular mechanisms underlying the potential neuroprotective effects of Trifolium pratense and its phytoestrogen-isoflavones in neurodegenerative disorders. Phytother Res 37(6):2693–2737

Andrade-Guerrero J, Santiago-Balmaseda A, Jeronimo-Aguilar P, Vargas-Rodríguez I, Cadena-Suárez AR, Sánchez-Garibay C et al (2023) Alzheimer's disease: an updated overview of its genetics. Int J Mol Sci 24(4):3754

Angelova PR, Abramov AY (2016) Functional role of mitochondrial reactive oxygen species in physiology. Free Radic Biol Med 100:81–85

Baranowski BJ, Marko DM, Fenech RK, Yang AJT, MacPherson REK (2020) Healthy brain, healthy life: a review of diet and exercise interventions to promote brain health and reduce Alzheimer's disease risk. Appl Physiol Nutr Metab 45(10):1055–1065

Bektas A, Schurman SH, Sen R, Ferrucci L (2018) Aging, inflammation and the environment. Exp Gerontol 105:10–18

Bocheva G, Bakalov D, Iliev P, Tafradjiiska-Hadjiolova R (2024) The vital role of melatonin and its metabolites in the neuroprotection and retardation of brain aging. Int J Mol Sci 25(10):5122

Boldrini M, Fulmore CA, Tartt AN, Simeon LR, Pavlova I, Poposka V et al (2018) Human hippocampal neurogenesis persists throughout aging. Cell Stem Cell 22(4):589–599

Cabezas R, Baez-Jurado E, Hidalgo-Lanussa O, Echeverria V, Ashrad GM, Sahebkar A et al (2019) Growth factors and Neuroglobin in astrocyte protection against neurodegeneration and oxidative stress. Mol Neurobiol 56(4):2339–2351

Cardinali DP (2021) Melatonin and healthy aging. Vitam Horm 115:67–88

Cassilhas RC, Tufik S, de Mello MT (2016) Physical exercise, neuroplasticity, spatial learning and memory. Cell Mol Life Sci 73(5):975–983

Cervellati C, Bergamini CM (2016) Oxidative damage and the pathogenesis of menopause related disturbances and diseases. Clin Chem Lab Med 54(5):739–753

Chauhan A, Chauhan V (2020) Beneficial effects of walnuts on cognition and brain health. Nutrients 12(2):550

Chen S, Li L, Peng C, Bian C, Ocak PE, Zhang JH et al (2022) Targeting oxidative stress and inflammatory response for blood-brain barrier protection in intracerebral Hemorrhage. Antioxid Redox Signal 37(1–3):115–134

Chow ZS, Moreland AT, Macpherson H, Teo WP (2021) The central mechanisms of resistance training and its effects on cognitive function. Sports Med 51(12):2483–2506
Colcombe S, Kramer AF (2003) Fitness effects on the cognitive function of older adults: a meta-analytic study. Psychol Sci 14(2):125–130
Costa AC, Joaquim HPG, Pedrazzi JFC et al (2022) Cannabinoids in late life Parkinson's disease and dementia: biological pathways and clinical challenges. Brain Sci 12(12):1596
Daniele S, Giacomelli C, Martini C (2018) Brain ageing and neurodegenerative disease: the role of cellular waste management. Biochem Pharmacol 158:207–216
Di Liegro CM, Schiera G, Proia P, Di Liegro I (2019) Physical activity and brain health. Genes (Basel) 10(9):720
Doshi A, Chataway J (2016) Multiple sclerosis, a treatable disease. Clin Med (Lond) 16(Suppl 6):s53–s59
Dwivedi D, Megha K, Mishra R, Mandal PK (2020) Glutathione in brain: overview of its conformations, functions, biochemical characteristics, quantitation and potential therapeutic role in brain disorders. Neurochem Res 45(7):1461–1480
Evans JR, Lawrenson JG (2023) Antioxidant vitamin and mineral supplements for slowing the progression of age-related macular degeneration. Cochrane Database Syst Rev 9(9):CD000254
Fernandez MA, Marette A (2017) Potential health benefits of combining yogurt and fruits based on their probiotic and prebiotic properties. Adv Nutr 8(1):155S–164S
Foret MK, Orciani C, Welikovitch LA, Huang C, Cuello AC, Do CS (2024) Early oxidative stress and DNA damage in Aβ-burdened hippocampal neurons in an Alzheimer's-like transgenic rat model. Commun Biol 7(1):861
Goicoechea L, Conde de la Rosa L, Torres S, García-Ruiz C, Fernández-Checa JC (2023) Mitochondrial cholesterol: metabolism and impact on redox biology and disease. Redox Biol 61:102643
Goodman LD, Ralhan I, Li X, Lu S, Moulton MJ, Park YJ et al (2024) Tau is required for glial lipid droplet formation and resistance to neuronal oxidative stress. Nat Neurosci 27(10):1918–1933
Grabska-Kobyłecka I, Szpakowski P, Król A, Książek-Winiarek D, Kobyłecki A, Głąbiński A et al (2023) Polyphenols and their impact on the prevention of neurodegenerative diseases and development. Nutrients 15(15):3454
Grad LI, Rouleau GA, Ravits J, Cashman NR (2017) Clinical spectrum of amyotrophic lateral sclerosis (ALS). Cold Spring Harb Perspect Med 7(8):a024117
Gupta V, Chitranshi N, Gupta VB (2024) Genetic risk, inflammation, and therapeutics: an editorial overview of recent advances in aging brains and neurodegeneration. Aging Dis 15(5):1989–1993
Hajam YA, Rani R, Ganie SY, Sheikh TA, Javaid D, Qadri SS et al (2022) Oxidative stress in human pathology and aging: molecular mechanisms and perspectives. Cells 11(3):552
Hariharan S, Dharmaraj S (2020) Selenium and selenoproteins: it's role in regulation of inflammation. Inflammopharmacology 28(3):667–695
Hashim HM, Makpol S (2022) A review of the preclinical and clinical studies on the role of the gut microbiome in aging and neurodegenerative diseases and its modulation. Front Cell Neurosci 16:1007166
Hazell AS, Faim S, Wertheimer G, Silva VR, Marques CS (2013) The impact of oxidative stress in thiamine deficiency: a multifactorial targeting issue. Neurochem Int 62(5):796–802
He L, He T, Farrar S, Ji L, Liu T, Ma X (2017) Antioxidants maintain cellular redox homeostasis by elimination of reactive oxygen species. Cell Physiol Biochem 44(2):532–553
Higgins-Chen AT, Thrush KL, Levine ME (2021) Aging biomarkers and the brain. Semin Cell Dev Biol 116:180–193
Hill DA, Crider M, Hill SR (2016) Hormone therapy and other treatments for symptoms of menopause. Am Fam Physician 94(11):884–889
Hong DG, Lee S, Kim J, Yang S, Lee M, Ahn J et al (2022) Anti-inflammatory and neuroprotective effects of Morin in an MPTP-induced Parkinson's disease model. Int J Mol Sci 23(18):10578
Huhn S, Beyer F, Zhang R, Lampe L, Grothe J, Kratzsch J, Willenberg A, Breitfeld J, Kovacs P, Stumvoll M, Trampel R (2018) Effects of resveratrol on memory performance, hippocampus

connectivity and microstructure in older adults–a randomized controlled trial. NeuroImage 174:177–190
Hyun DH, Emerson SS, Jo DG, Mattson MP, De Cabo R (2006) Calorie restriction up-regulates the plasma membrane redox system in brain cells and suppresses oxidative stress during aging. Proc Natl Acad Sci 103(52):19908–19912
Ionescu-Tucker A, Cotman CW (2021) Emerging roles of oxidative stress in brain aging and Alzheimer's disease. Neurobiol Aging 107:86–95
Jafar-Nejad H, Leonardi J, Fernandez-Valdivia R (2010) Role of glycans and glycosyltransferases in the regulation of notch signaling. Glycobiology 20(8):931–949
Jallow AW, Nguyen DPQ, Sanotra MR, Hsu CH, Lin YF, Lin YF (2024) A comprehensive bibliometric analysis of global research on the role of acrolein in Alzheimer's disease pathogenesis: involvement of amyloid-beta. Front Aging Neurosci 16:1378260
Jessen F, Amariglio RE, Buckley RF, van der Flier WM, Han Y, Molinuevo JL et al (2020) The characterisation of subjective cognitive decline. Lancet Neurol 19(3):271–278
Jia J, Chen J, Wang G, Li M, Zheng Q, Li D (2023) Progress of research into the pharmacological effect and clinical application of the traditional Chinese medicine Rehmanniae radix. Biomed Pharmacother 168:115809
Johnson SA, Fang T, De Marchi F, Neel D, Van Weehaeghe D, Berry JD et al (2022) Pharmacotherapy for amyotrophic lateral sclerosis: a review of approved and upcoming agents. Drugs 82(13):1367–1388
Jomova K, Raptova R, Alomar SY, Alwasel SH, Nepovimova E, Kuca K et al (2023) Reactive oxygen species, toxicity, oxidative stress, and antioxidants: chronic diseases and aging. Arch Toxicol 97(10):2499–2574
Jomova K, Alomar SY, Valko R, Liska J, Nepovimova E, Kuca K et al (2025) Flavonoids and their role in oxidative stress, inflammation, and human diseases. Chem Biol Interact 413:111489
Kalueff AV, Nutt DJ (2007) Role of GABA in anxiety and depression. Depress Anxiety 24(7):495–517
Kiełczykowska M, Kocot J, Paździor M, Musik I (2018) Selenium – a fascinating antioxidant of protective properties. Adv Clin Exp Med 27(2):245–255
Kőszegi T, Sali N, Raknić M, Horváth-Szalai Z, Csepregi R, Končić MZ et al (2017) A novel luminol-based enhanced chemiluminescence antioxidant capacity microplate assay for use in different biological matrices. J Pharmacol Toxicol Methods 88(Pt 2):153–159
Kubben N, Zhang W, Wang L, Voss TC, Yang J, Qu J et al (2016) Repression of the antioxidant NRF2 pathway in premature aging. Cell 165(6):1361–1374
Kumar RR, Singh L, Thakur A, Singh S, Kumar B (2022) Role of vitamins in neurodegenerative diseases: a review. CNS Neurol Disord Drug Targets 21(9):766–773
Li C, Sun T, Jiang C (2021) Recent advances in nanomedicines for the treatment of ischemic stroke. Acta Pharm Sin B 11(7):1767–1788
Li R, Robinson M, Ding X, Geetha T, Al-Nakkash L, Broderick TL et al (2022) Genistein: a focus on several neurodegenerative diseases. J Food Biochem 46(7):e14155
Li C, Li B, Zhang J, Liu K, Du G, Guo C et al (2025) Bone marrow mesenchymal stem cell senescence in the development of osteoporosis: mechanisms, interventions, and future directions. Curr Mol Med
Liu RM (2022) Aging, cellular senescence, and Alzheimer's disease. Int J Mol Sci 23(4):1989
Lochhead JJ, Ronaldson PT, Davis TP (2024) The role of oxidative stress in blood-brain barrier disruption during ischemic stroke: antioxidants in clinical trials. Biochem Pharmacol 228:116186
Lövdén M, Fratiglioni L, Glymour MM, Lindenberger U, Tucker-Drob EM (2020) Education and cognitive functioning across the life span. Psychol Sci Public Interest 21(1):6–41
Luca SV, Macovei I, Bujor A, Miron A, Skalicka-Woźniak K, Aprotosoaie AC et al (2020) Bioactivity of dietary polyphenols: the role of metabolites. Crit Rev Food Sci Nutr 60(4):626–659
Mandelker L (2004) The natural activities of cells, the role of reactive oxygen species, and their relation to antioxidants, nutraceuticals, botanicals, and other biologic therapies. Vet Clin North Am Small Anim Pract 34(1):39–66

Martínez-Guardado I, Arboleya S, Grijota FJ, Kaliszewska A, Gueimonde M, Arias N (2022) The therapeutic role of exercise and probiotics in stressful brain conditions. Int J Mol Sci 23(7):3610
Martín-Fernández B, Gredilla R (2018) Mitochondrial oxidative stress and cardiac ageing. Clin Investig Arterioscler 30(2):74–83
Mehra S, Sahay S, Maji SK (2019) α-Synuclein misfolding and aggregation: implications in Parkinson's disease pathogenesis. Biochim Biophys Acta Proteins Proteom 1867(10):890–908
Melzer TM, Manosso LM, Yau SY, Gil-Mohapel J, Brocardo PS (2021) In pursuit of healthy aging: effects of nutrition on brain function. Int J Mol Sci 22(9):5026
Mirończuk-Chodakowska I, Witkowska AM, Zujko ME (2018) Endogenous non-enzymatic antioxidants in the human body. Adv Med Sci 63(1):68–78
Mouat IC, Goldberg E, Horwitz MS (2022) Age-associated B cells in autoimmune diseases. Cell Mol Life Sci 79(8):402
Muchová J, Žitňanová I, Ďuračková Z (2014) Oxidative stress and down syndrome. Do antioxidants play a role in therapy? Physiol Res 63(5):535–542
NCBI (2025a) Oxidative stress, dysfunctional glucose metabolism and Alzheimer disease – PubMed [Internet]. Available from: https://pubmed.ncbi.nlm.nih.gov/30737462/
NCBI (2025b) Carotenoids: biochemistry, pharmacology and treatment – PubMed [Internet]. Available from: https://pubmed.ncbi.nlm.nih.gov/27638711/
NCBI (2025c) Principles of fluoride toxicity and the cellular response: a review – PubMed [Internet]. Available from: https://pubmed.ncbi.nlm.nih.gov/32152649/
NCBI (2025d) Effects of Supplementation with Omega-3 and Omega-6 Polyunsaturated Fatty Acids and Antioxidant Vitamins, Combined with High-Intensity Functional Training, on Exercise Performance and Body Composition: A Randomized, Double-Blind, Placebo-Controlled Trial – PubMed [Internet]. Available from: https://pubmed.ncbi.nlm.nih.gov/39275230/
NCBI (2025e) Effect of food sources of nitrate, polyphenols, L-arginine and L-citrulline on endurance exercise performance: a systematic review and meta-analysis of randomised controlled trials – PubMed [Internet]. Available from: https://pubmed.ncbi.nlm.nih.gov/34965876/
NCBI (2025f) NAD+ in brain aging and neurodegenerative disorders – PubMed [Internet]. Available from: https://pubmed.ncbi.nlm.nih.gov/31577933/
NCBI (2025g) Neuroprotective Herbs for the Management of Alzheimer's Disease – PubMed [Internet]. Available from: https://pubmed.ncbi.nlm.nih.gov/33917843/
Neha K, Haider MR, Pathak A, Yar MS (2019) Medicinal prospects of antioxidants: a review. Eur J Med Chem 178:687–704
Olivares-Bañuelos TN, Chí-Castañeda D, Ortega A (2019) Glutamate transporters: gene expression regulation and signaling properties. Neuropharmacology 161:107550
Ornatowski W, Lu Q, Yegambaram M, Garcia AE, Zemskov EA, Maltepe E et al (2020) Complex interplay between autophagy and oxidative stress in the development of pulmonary disease. Redox Biol 36:101679
Ortiz GG, Pacheco-Moisés FP, Macías-Islas MÁ, Flores-Alvarado LJ, Mireles-Ramírez MA, González-Renovato ED et al (2014) Role of the blood-brain barrier in multiple sclerosis. Arch Med Res 45(8):687–697
Pan P, Huang YW, Oshima K, Yearsley M, Zhang J, Arnold M et al (2019) The immunomodulatory potential of natural compounds in tumor-bearing mice and humans. Crit Rev Food Sci Nutr 59(6):992–1007
Park MW, Cha HW, Kim J, Kim JH, Yang H, Yoon S et al (2021) NOX4 promotes ferroptosis of astrocytes by oxidative stress-induced lipid peroxidation via the impairment of mitochondrial metabolism in Alzheimer's diseases. Redox Biol 41:101947
Pettersson F, Del Rincon SV, Miller WH (2014) Eukaryotic translation initiation factor 4E as a novel therapeutic target in hematological malignancies and beyond. Expert Opin Ther Targets 18(9):1035–1048
Picca A, Faitg J, Auwerx J, Ferrucci L, D'Amico D (2023) Mitophagy in human health, ageing and disease. Nat Metab 5(12):2047–2061
Pisoschi AM, Pop A (2015) The role of antioxidants in the chemistry of oxidative stress: a review. Eur J Med Chem 97:55–74

Poulose SM, Miller MG, Scott T, Shukitt-Hale B (2017) Nutritional factors affecting adult neurogenesis and cognitive function. Adv Nutr 8(6):804–811
Power R, Nolan JM, Prado-Cabrero A, Roche W, Coen R, Power T et al (2022) Omega-3 fatty acid, carotenoid and vitamin E supplementation improves working memory in older adults: a randomised clinical trial. Clin Nutr 41(2):405–414
Radomska-Leśniewska DM, Bałan BJ, Skopiński P (2017) Angiogenesis modulation by exogenous antioxidants. Cent Eur J Immunol 42(4):370–376
Rapa SF, Di Iorio BR, Campiglia P, Heidland A, Marzocco S (2019) Inflammation and oxidative stress in chronic kidney disease-potential therapeutic role of minerals, vitamins and plant-derived metabolites. Int J Mol Sci 21(1):263
Regenhardt RW, Das AS, Lo EH, Caplan LR (2018) Advances in understanding the pathophysiology of lacunar stroke: a review. JAMA Neurol 75(10):1273–1281
Rehman MU, Sehar N, Dar NJ, Khan A, Arafah A, Rashid S et al (2023) Mitochondrial dysfunctions, oxidative stress and neuroinflammation as therapeutic targets for neurodegenerative diseases: an update on current advances and impediments. Neurosci Biobehav Rev 144:104961
Ronaldson PT, Davis TP (2020) Regulation of blood-brain barrier integrity by microglia in health and disease: a therapeutic opportunity. J Cereb Blood Flow Metab 40(1_suppl):S6–S24
Salminen LE, Paul RH (2014) Oxidative stress and genetic markers of suboptimal antioxidant defense in the aging brain: a theoretical review. Rev Neurosci 25(6):805–819
Santos HO, Price JC, Bueno AA (2020) Beyond fish oil supplementation: the effects of alternative plant sources of Omega-3 polyunsaturated fatty acids upon lipid indexes and Cardiometabolic biomarkers-an overview. Nutrients 12(10):3159
Saretzki G (2023) The telomerase connection of the brain and its implications for neurodegenerative diseases. Stem Cells 41(3):233–241
Scheiblich H, Eikens F, Wischhof L, Opitz S, Jüngling K, Cserép C et al (2024) Microglia rescue neurons from aggregate-induced neuronal dysfunction and death through tunneling nanotubes. Neuron 112(18):3106–3125.e8
Sehrawat TS, Liu M, Shah VH (2020) The knowns and unknowns of treatment for alcoholic hepatitis. Lancet Gastroenterol Hepatol 5(5):494–506
Shacka JJ, Roth KA, Zhang J (2008) The autophagy-lysosomal degradation pathway: role in neurodegenerative disease and therapy. Front Biosci 13:718–736
Sharifi-Rad M, Mnayer D, Morais-Braga MFB, Carneiro JNP, Bezerra CF, Coutinho HDM et al (2018) Echinacea plants as antioxidant and antibacterial agents: from traditional medicine to biotechnological applications. Phytother Res 32(9):1653–1663
Shen N, Wang T, Gan Q, Liu S, Wang L, Jin B (2022) Plant flavonoids: classification, distribution, biosynthesis, and antioxidant activity. Food Chem 383:132531
Shore D, Bianchi A (2009) Telomere length regulation: coupling DNA end processing to feedback regulation of telomerase. EMBO J 28(16):2309–2322
Sies H (2015) Oxidative stress: a concept in redox biology and medicine. Redox Biol 4:180–183
Sies H, Belousov VV, Chandel NS, Davies MJ, Jones DP, Mann GE et al (2022) Defining roles of specific reactive oxygen species (ROS) in cell biology and physiology. Nat Rev Mol Cell Biol 23(7):499–515
Silberstein SD (2004) Migraine. Lancet 363(9406):381–391
Singh A, Kukreti R, Saso L, Kukreti S (2019) Oxidative stress: a key modulator in neurodegenerative diseases. Molecules 24(8):1583
Stavrinou PS, Andreou E, Aphamis G, Pantzaris M, Ioannou M, Patrikios IS et al (2020) The effects of a 6-month high dose Omega-3 and Omega-6 polyunsaturated fatty acids and antioxidant vitamins supplementation on cognitive function and functional capacity in older adults with mild cognitive impairment. Nutrients 12(2):325
Szczuko M, Kikut J, Szczuko U, Szydłowska I, Nawrocka-Rutkowska J, Ziętek M et al (2021) Nutrition strategy and life style in polycystic ovary syndrome-narrative review. Nutrients 13(7):2452
Temple S (2023) Advancing cell therapy for neurodegenerative diseases. Cell Stem Cell 30(5):512–529

Terao J (2023) Revisiting carotenoids as dietary antioxidants for human health and disease prevention. Food Funct 14(17):7799–7824
Tian X, Liang T, Liu Y, Ding G, Zhang F, Ma Z (2019) Extraction, structural characterization, and biological functions of Lycium Barbarum polysaccharides: a review. Biomolecules 9(9):389
Trist BG, Hare DJ, Double KL (2019) Oxidative stress in the aging substantia nigra and the etiology of Parkinson's disease. Aging Cell 18(6):e13031
Unsal V, Cicek M, Sabancilar İ (2021) Toxicity of carbon tetrachloride, free radicals and role of antioxidants. Rev Environ Health 36(2):279–295
Valko M, Rhodes CJ, Moncol J, Izakovic M, Mazur M (2006) Free radicals, metals and antioxidants in oxidative stress-induced cancer. Chem Biol Interact 160(1):1–40
Valko M, Leibfritz D, Moncol J, Cronin MTD, Mazur M, Telser J (2007) Free radicals and antioxidants in normal physiological functions and human disease. Int J Biochem Cell Biol 39(1):44–84
Vanmassenhove J, Veys N, Van Biesen W (2016) Prevention and conservative management of acute kidney injury. Minerva Urol Nefrol 68(1):58–71
Volkert MR, Crowley DJ (2020) Preventing neurodegeneration by controlling oxidative stress: the role of OXR1. Front Neurosci 14:611904
Wang L, Chen P, Xiao W (2021) β-Hydroxybutyrate as an anti-aging metabolite. Nutrients 13(10):3420
Wołonciej M, Milewska E, Roszkowska-Jakimiec W (2016) Trace elements as an activator of antioxidant enzymes. Postepy Hig Med Dosw 70:1483–1498
Xiong Z, Liu L, Jian Z, Ma Y, Li H, Jin X et al (2023) Vitamin E and multiple health outcomes: an umbrella review of meta-analyses. Nutrients 15(15):3301
Yamada Y, Takano Y et al (2020) Therapeutic strategies for regulating mitochondrial oxidative stress. Biomolecules 10(1):83
Yang C, Hawkins KE, Doré S, Candelario-Jalil E (2019) Neuroinflammatory mechanisms of blood-brain barrier damage in ischemic stroke. Am J Physiol Cell Physiol 316(2):C135–C153
Yin H, Xu L, Porter NA (2011) Free radical lipid peroxidation: mechanisms and analysis. Chem Rev 111(10):5944–5972
Zhang M, Hou ZK, Huang ZB, Chen XL, Liu FB (2021) Dietary and lifestyle factors related to gastroesophageal reflux disease: a systematic review. Ther Clin Risk Manag 17:305–323
Zia A, Farkhondeh T, Pourbagher-Shahri AM, Samarghandian S (2021) The role of curcumin in aging and senescence: molecular mechanisms. Biomed Pharmacother 134:111119

8 Nutrient-Sensing Pathways: Gatekeepers of Longevity

Falguni Goel, Vipin Kumar Garg, Payal Singh, and Prabhash Nath Tripathi

Abstract

Aging is a dynamic biological process with varied influences from genetic, metabolic, and environmental factors. Nutrient-sensing pathways are the most important regulators of aging and longevity because they assess and respond to energy and nutrient availability in and around the cell. Nutrient-sensing pathways that exist in evolutionarily conserved signaling networks in cells (e.g., insulin/IGF-1 signaling (IIS), target of rapamycin (mTOR), AMP-activated protein kinase (AMPK), and sirtuins (SIRT1–7)) act as molecular gatekeepers of the availability of cellular growth, stress resistance, autophagy, mitochondrial function, and metabolic homeostasis. Dysregulation in nutrient-sensing pathways causes age-related functional decline and chronic diseases such as cancer, diabetes, cardiovascular disorders, and neurodegeneration. This chapter aims to present a broad overview of nutrient-sensing pathways and their importance in regulating longevity. We essentially sift through some of the molecular mechanisms that underlie nutrient-sensing pathways, their signaling and cross talk, and their eventual influence on cell signaling. We also examine how dietary interventions such as

F. Goel · V. K. Garg
Department of Pharmaceutical Technology, Meerut Institute of Engineering & Technology (MIET), Meerut, Uttar Pradesh, India

P. Singh
Department of Zoology, MMV, Banaras Hindu University, Varanasi, Uttar Pradesh, India

P. N. Tripathi (✉)
Department of Pharmaceutical Technology, Meerut Institute of Engineering & Technology (MIET), Meerut, Uttar Pradesh, India

Department of Pharmaceutical Sciences, College of Pharmacy, University of Arkansas for Medical Sciences, Little Rock, AR, USA
e-mail: pntripathi@uams.edu

A. Kumar Singh, S. Nand Rai (eds.), *Rejuvenating the Brain: Nutraceuticals, Autophagy, and Longevity*, Nutritional Neurosciences,
https://doi.org/10.1007/978-981-95-2790-8_8

caloric restriction, intermittent fasting, and macronutrient manipulation influence nutrient sensing in relation to fulfilling and extending lifespan and health span across species. We also explore a pharmacological approach with potential to influence these nutrient-sensing pathways with mTOR inhibitors, AMPK activators, and sirtuin enhancers while developing the note of translational potential and clinical significance of such pathways. We combine results from model organisms and from human studies to show the importance of nutrition sensing in healthy aging and the prevention of age-associated diseases. We also highlight contemporary tools such as omics technologies, artificial intelligence, and computational modeling that offer exciting opportunities to better understand longevity networks and to improve personalized interventions for healthy aging. We discuss ethical and societal implications of extending lifespan with an emphasis on reducing inequities and being thoughtful and responsible in the application of innovations that extend lifespan. By providing an integrated perspective of how nutrient-sensing pathways regulate aging and longevity, we suggest the potential of these pathways as therapeutic targets/ foundational components in the search to maximize human health span and improve quality of life.

Keywords

Aging · Longevity · Nutrient sensing · Cellular processes

8.1 Introduction

Aging is a universal and unavoidable biological process defined by the gradual decline of physiological functioning and increasing liability for disease, leading to death (Saxon et al. 2021). Although it has been a fixed part of life throughout history, recent developments in molecular biology and genetics have fundamentally changed the narrative of aging from one of a purely degenerative process to one of somewhat modifiable and responsive to habituative and genetic factors (Fleck 2022). One of the most exciting advances in the field of biogerontology has been the identification of nutrient-sensing pathways as key facilitators of lifespan and health span regulation (Goel et al. 2025a). Conserved throughout evolution, nutrient-sensing pathways recognize internal and external nutrient availability. These networks of signaling pathways facilitate physiological adaptations to nutrient-sensing organisms and seek to regulate growth, metabolism, stress resilience, and lifespan (Tomtheelnganbee et al. 2022).

The emerging narrative holds that the availability, type of nutrients, and timing of nutrient delivery significantly impact the cellular and systemic processes that determine aging rates (Kassis et al. 2023). This premise has primarily been confirmed through nutritionally controlled studies of model organisms, including yeast, worms, flies, and mice (Goel 2025). Those studies illuminated the ways in which caloric restriction and selective nutrient regulation, such as carbohydrates, dietary fats, and proteins, prolong lifespan and delay the onset of aging-related comorbidities (Cagigas et al. 2025). Based on my experience researching these experimental models, I believe that the understanding of the mechanisms of action of these interventions will depend on the understanding of many of the broadly conserved intercommunicating signaling pathways that are cellular nutrient sensors in metazoans including insulin/IGF-1 signaling, mechanistic target of rapamycin (mTOR), AMP-activated protein kinase (AMPK), and the sirtuins pathways (Bhushan et al. 2025). Individual pathways provide information about particular elements of nutrient status, such as glucose status, amino acid levels, and energy content, respectively. These signal to different adaptive responses that promote metabolic homeostasis and cellular resilience under nutrient scarcity (Tippairote 2025).

The relationship between aging and the nutrient-sensing pathways is complex and multifactorial (Fernandes and Demetriades 2021). For example, downregulation of insulin/IGF-1 signaling and mTOR activity, both of which support anabolic processes during high nutritional conditions, generally corresponds to increased lifespan across all species (Goel et al. 2025b). Conversely, activation of catabolism, such as autophagy, is induced by AMPK and sirtuin signaling when nutrients are scarce, ultimately leading to the removal of damaged cellular components and facilitating a metabolic shift to a longer life (Xu et al. 2023). The availability of nutrients leads to a dynamic equilibrium between anabolic and catabolic states via the nutrient-sensing pathways, which represents a critical determinant of aging trajectories and has led to extensive studies utilizing therapeutic interventions targeting these pathways to drive healthy aging (Addo and Khan 2024).

The evolutionarily conserved mechanisms represented above are shared across so many different species, and thus, importantly, they are ultimately rooted in survival and adaptation (Shumayla et al. 1927). For any organism, including unicellular yeast, complex multicellular organisms like mammals and humans, the capacity to sense nutrient signals and then respond accordingly contributed to reproductive success and survival in a variable environment (Chauhan et al. 2024). It is this evolutionary pressure that has resulted in the conservation of important nutrient-sensing pathways over the common evolutionary history of many species (Cavender-Bares et al. 2022). Given this conservation, the contributions of using such simplified model organisms to major discoveries related to humans are not surprising (Pal et al. 2025a). Using these pathways, they not only impact lifespan in humans, but these pathways also directly relate to the onset of age-related diseases such as type 2 diabetes, cardiovascular disease, neurodegeneration, and cancer, which all share the commonality of metabolic dysfunction (Kopp 2024).

It is therefore vital that we understand how the pathways of nutrient sensing affect aging, both at the molecular and systemic levels, to develop new strategies that can be used to extend health span (Tomtheelnganbee et al. 2022). The abundance of evidence showing that both dietary interventions, pharmacological compounds that modulate nutrient sensing, and lifestyle behavior changes like intermittent fasting inspire the quest for a practical means to slow aging and enhance quality of life (Pal et al. 2025b). As we increasingly understand the complexities between these pathways and how they interact, the possibility of exploiting the biology of the pathways of nutrient sensing to influence longevity becomes an attainable reality, not just a goal (Tomtheelnganbee et al. 2022).

8.2 Major Nutrient-Sensing Pathways in Longevity

Aging and longevity are increasingly recognized not just as stochastic damage accumulation, but as regulated by evolutionarily conserved signaling pathways that sense nutrients when we eat, and even when we do not eat (Watts et al. 2022). These nutrient-sensing pathways are the first molecular switch to understand external and internal nutritional cues, which they integrate in the regulation of cellular growth, metabolism, repair, and stress responses (Rai et al. 2021a). The four major pathways of Insulin/IGF-1 signaling (IIS), mechanistic target of rapamycin (mTOR), AMP-activated protein kinase (AMPK), and sirtuins (SIRT1–7) are interconnected tightly to form a network that balances growth with maintenance and, consequently, is vital in the lifespan of all animals (Pontzer and McGrosky 2022).

The Insulin/IGF-1 signaling (IIS) pathway is by far one of the most described nutrient-sensing pathways concerning aging and longevity (Engel 2023). In conditions of nutrient overload, particularly glucose and other carbohydrates, IIS is used to promote cellular growth, proliferation, and energy storage through anabolic events (Rai et al. 2025). In model organisms including *Caenorhabditis elegans*, *Drosophila melanogaster*, and mice, reduced IIS activities have been directly associated with lifespan extension and increased resistance to stress (Rai et al. 2025). The major downstream components affecting IIS responses are the PI3K/AKT axis and FOXO transcription factor family members that regulate genes pivotal for oxidative stress responses, metabolism, and autophagy (Kma and Baruah 2022). By reducing or inhibiting IIS, the activity of the FOXO proteins and nuclear abundance are promoted, which facilitates a switch in the cellular program from growth to maintenance and repair that can be viewed as programmed for longevity (Rai et al. 2022). While there is no direct evidence in humans, polymorphisms in the IGF-1 and FOXO3 genes have been associated with increased longevity, and suggest that the benefits of modulating IIS and even directly modulating IIS will be evolutionarily conserved (Fischer et al. 2022).

The mTOR pathway is another important nutrient sensor that responds mainly to amino acids, energy status, and growth factors (Hu and Guo 2021). mTOR is located in two complexes, mTORC1 and mTORC2, with mTORC1 being the focus of aging regulation. When nutrients are available, mTORC1 increases protein synthesis, ribosome biogenesis, and inhibits autophagy (Ballesteros-Álvarez and Andersen 2021). There is an association with chronic mTORC1 activation to accelerating aging and age-associated diseases; acute genetic manipulation or pharmacological manipulation of mTOR (rapamycin) has been shown to extend lifespan in many organisms (Rai et al. 2021b). Importantly, mTOR inhibition slows aging, but also improves health span via increasing immune function, reducing inflammation, and autophagic clearance of damaged proteins and organelles (Mannick and Lamming 2023).

AMP-activated protein kinase (AMPK) acts as a key cellular energy sensor. A significant change in the adenine monophosphate / adenosine triphosphate (AMP/ATP) ratio stimulates the allosteric activation of AMPK and signifies a state of energy stress or loss of energy sources (Park et al. 2023). AMPK is essentially responsible for altering cellular metabolism during activation. Following AMPK activation, a transition from anabolic to catabolic metabolism occurs, leading to the stimulation of fatty acid oxidation, glucose uptake, and autophagy while inhibiting lipid and protein synthesis (Zhang et al. 2022). AMPK phosphorylates upstream AMPK regulators that signal inhibition of mTORC1, resulting in protection of energy stores and maintenance of cellular function (Singh et al. 2024a). The role of AMPK in modifying longevity in several experimental animal models has been documented. Pharmacological approaches, including the use of metformin, have demonstrated a potential for modulating energy metabolism to mimic the effects of caloric restriction while extending lifespan (Giacomello and Toniolo 2021).

Sirtuins, and in particular SIRT1 in mammals, are NAD^+-dependent deacetylases that couple cellular energy status with transcriptional regulation and responses to stress (Santos et al. 2023). Sirtuin activity is regulated by levels of NAD^+, which is dependent on the availability of nutrients in the cell. Sirtuins have a vast array of substrates. They can deacetylate histones, regulatory transcription factors such as the FOXO transcription factors, and enzymes (Santos et al. 2023). Sirtuins are known to promote mitochondrial function, DNA repair, and autophagy, while stimulating reduced oxidative stress. SIRT1 also stimulates FOXO transcription factors and PGC-1α simultaneously, leading to an enhanced metabolic state that is conducive to lifespan extension. In the context of the antiaging effects of nutritional interventions such as dietary restrictions and intermittent fasting, sirtuin activity, as well as NAD^+ levels, increases (Duan et al. 2022).

The mechanisms by which these pathways regulate aging are not independent. They are connected to one another in a complex signaling network with cross talk and feedback loops that modulate the cellular response to environmental and nutritional changes or cues(Tabibzadeh 2021). For example, AMPK-rich environments can aim to induce autophagy and mitochondrial biogenesis through sirtuin pathways and AMP-activated protein kinase signaling. Both AMPK and sirtuin pathways inhibit the mTOR pathway. IIS and mTOR signaling represents a generally progrowth signaling state that is incompatible with AMPK and sirtuin signaling. These signals allow a cell to discriminate between anabolic and catabolic states according to nutrient availability, which may influence aging and organismal longevity (Zhou et al. 2021).

These interconnected pathways present more opportunities for understanding the biological basis for aging and point to potential targets for therapeutic interventions designed to extend health span (Guo et al. 2022). Lifestyle changes, dietary changes, or pharmacological agents that can alter the behavior of these nutrient-sensing networks represent an important strategy for healthy aging and age-related diseases, including Alzheimer's disease (Fekete et al. 2022) (Fig. 8.1).

8.3 Nutrient Sensing and Cellular Processes

Nutrient-sensing pathways regulate systemic metabolic responses that determine how long an organism can live, but they also profoundly regulate major intracellular processes related to cellular homeostasis (Tomtheelnganbee et al. 2022). These

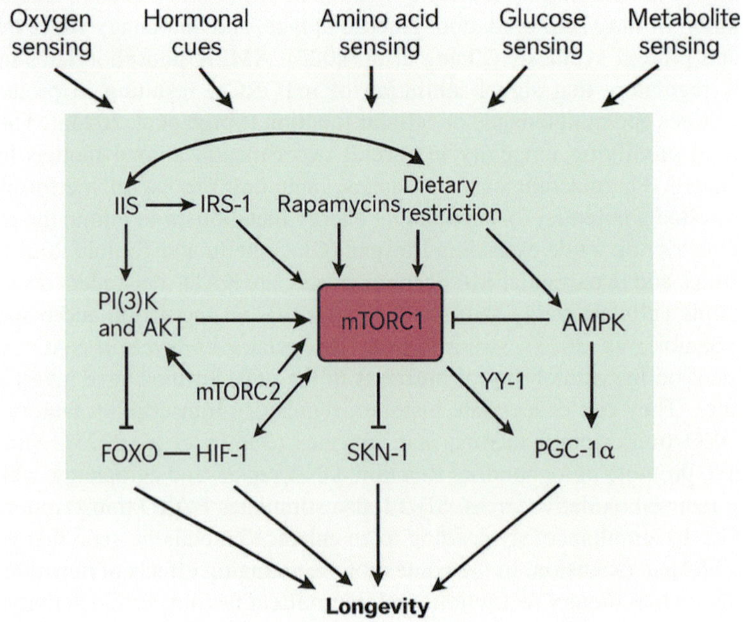

Fig. 8.1 Key nutrient-sensing pathways that regulate longevity and aging processes. This figure illustrates the key nutrient-sensing pathways (insulin/IGF-1 signaling, AMP-activated protein kinase, the mTOR signaling pathway, and sirtuins) and how these pathways regulate lifespan in interconnected ways. In response to energy status and nutrient availability, these pathways regulate downstream targets involved in autophagy, resistance to oxidative stress, mitochondrial functioning, and metabolic regulation. Furthermore, cross talk between these nutrient-sensing pathways organizes cellular maintenance through stresses and lifespan in many organisms

processes include autophagy, proteostasis, mitochondrial function, oxidative stress responses, and immune regulation. Nutrient-sensing pathways integrate information about nutrient availability, energy status, and stress signals to activate or suppress a variety of cellular processes that preserve function and slow aging. Disruptions of these finely regulated processes across the lifespan are common features of aging and contribute to the initiation of multiple chronic diseases (Guo et al. 2022). Thus, understanding the mechanisms by which nutrient sensing regulates these processes will be important to develop approaches and treatments that potentially extend health span (Tomtheelnganbee et al. 2022).

Autophagy is a tightly controlled catabolic pathway that is important for degrading and recycling damaged organelles, misfolded proteins, and other cytoplasmic constituents (Lin et al. 2021). It is a critical quality control mechanism in the cell that prevents toxic cellular debris from accumulating. Nutrient-sensing pathways are key regulators of autophagy. The mechanistic target of rapamycin complex 1 (mTORC1) is a critical inhibitor of autophagy during nutrient-rich conditions, with a prioritized anabolic state (Chen et al. 2024). In nutrient-deprived states, mTORC1 is inhibited, and AMPK and sirtuins are activated, which stabilizes the initiation of autophagy, allowing the cell to reclaim energy and building blocks during cellular stress. Enhanced longevity of several model organisms has been linked with efficient autophagic flux, and defects in autophagy are a contributor to neurodegeneration, metabolic disorders, and cancer. As such, nutrient sensing acts as a switch to turn on, and stabilize, cellular growth versus autophagic maintenance (Sanchez-Garrido and Shenoy 2021).

Proteostasis, or protein homeostasis, includes the regulation of protein synthesis, folding, trafficking, and degradation. Proteostasis is required to maintain cellular function and viability (Ottens et al. 2021). Proteostasis declines with age, leading to the build-up of dysfunctional or aggregated proteins, which are implicated in many diseases associated with aging (e.g., Alzheimer's disease, Parkinson's disease). Nutrient-sensing pathways act on proteostasis in a variety of ways (Kasar et al. 2025). Sirtuins mediate the expression of chaperone proteins and components of proteasomes, which increases cellular ability to fold and degrade proteins correctly. Similarly, AMPK supports both autophagy and proteasomal degradation, both required to remove misfolded proteins. However, chronic activation of mTOR supports protein synthesis while not supporting compensatory clearance, resulting in greater proteostatic load and eventually contributing to cellular aging (Ottens et al. 2021).

Mitochondria—the energy factories of the cell—are strictly regulated by nutrient-sensing pathways, given their important role in energy production and regulation of metabolism (Anania 2022). With aging, there is mitochondrial dysfunction characterized by decreased production of ATP, increased amounts of ROS, and failure of mitochondrial dynamics. Nutrient-sensing pathways, such as AMPK and SIRT1, promote mitochondrial function via enhancement of mitochondrial biogenesis. Both pathways enhance mitochondrial biogenesis via activation of PGC1α, the master regulator of mitochondrial gene expression (Kong et al. 2022). In addition, these pathways promote mitophagy, the selective autophagic degradation of damaged mitochondria, ensuring mitochondrial quality and preventing the propagation of mitochondrial damage. In other words, nutrient excess or impairment of nutrient sensing contributes to the introduction of additional stress and dysfunction to mitochondria, and subsequently promotes cellular senescence and aging (Miwa et al. 2022).

Oxidative stress occurs due to the presence of excessive ROS, which can be harmful in different ways to our cellular components, including DNA, proteins, and lipids (Juan et al. 2021). ROS are byproducts of normal mitochondrial respiration and can be used to transmit signals within a cell at low levels. It is repression of the intracellular redox state through too much ROS accumulation that can damage the protein, DNA, and lipids as described above. Nutrient-sensing pathways are the responsible pathways that can be regulated to maintain the redox state by modulating the response to ROS production, either through downregulation of ROS or upregulation of antioxidant defenses (Huchzermeyer et al. 2022). For example, AMPK and sirtuins are regulated by cellular energy levels and cellular nutrient levels. Both of these pathways act through the FOXO and NRF2 signaling axis and are used by the cell to modulate the antioxidant enzymes, such as SOD and catalase, through transcriptional regulation and posttranslational modification. In addition to antioxidant enzymes, these pathways also produce other metabolites and small molecules that can resist and scavenge oxidative damage. In addition, these pathways help manage ROS levels by having stimulatory effects on both mitochondrial bioenergetics and the moderation of ROS through autophagy pathways. On the flip side, persistent overabundance of nutrients, which leads to hyperactive insulin and insulin-like signaling (IIS) and mTOR regulation will suppress both antioxidant sensing pathways and promote VO2 max reduction and overall enhanced oxidative damage. Given that modulating the intracellular response and nutrient level is important for the homeostasis of redox level, longevity of the cell, and stress resilience, the importance of good nutrient sensing and regulation is extremely important (Tomtheelnganbee et al. 2022) (Table 8.1).

Inflammation, especially the low-grade chronic inflammation referred to as "inflammaging," is a core aspect of aging (Bauer et al. 2021) and is associated with metabolic and immune dysfunction (Bleve et al. 2022). Immunosenescence, which is the gradual deterioration of the immune system associated with aging, contributes to this inflammatory environment. Importantly, cellular nutrient-sensing pathways also play an important role in determining immune responses and inflammatory signaling. For example, activation of the mTOR pathway in immune cells leads to increased production of proinflammatory cytokines and reduced differentiation of memory T cells. Alternatively, AMPK and SIRT1 pathways have been shown to suppress inflammatory pathways by inhibiting NF-κB signaling and promoting anti-inflammatory cytokine signaling (Song et al. 2022). Additionally, caloric restriction and caloric restriction mimetics alter these cellular nutrient-sensing pathways and promote reductions in systemic inflammation and improvements in immune function. Nutrient sensing modulates the proportion of responses that are proinflammatory or anti-inflammatory, ensuring that immune response activity is appropriate and maintains an anti-inflammatory component to alleviate immune decline related to aging (Balamurugan et al. 2024).

Cellular nutrient-sensing pathways are central regulators of a number of cellular processes that ensure homeostasis and resistance to age-related pathology (Guo et al. 2022). In aging models, the ability of nutrient-sensing pathways to regulate autophagy, proteostasis, mitochondrial health, oxidative stability, and immune function ensures amelioration of the pathophysiology of aging and contributes to mechanisms of lifespan extension. In humans, by regulating cellular nutrient-sensing pathways, there is also a mechanistic basis to formulate therapeutic interventions to promote healthy aging (Guo et al. 2022) (Fig. 8.2).

Table 8.1 Nutrient sensing and cellular processes

Nutrient sensor	Key nutrient detected	Associated pathway	Major cellular processes regulated	Physiological/pathological implications	References
mTOR (mechanistic target of rapamycin)	Amino acids, glucose, oxygen, and energy status	mTORC1/ mTORC2 signaling	Cell growth, protein synthesis, autophagy inhibition, lipid metabolism	Dysregulation leads to cancer, obesity, neurodegeneration, and aging	Marafie et al. (2024)
AMPK (AMP-activated protein kinase)	AMP/ATP ratio (energy status)	AMPK signaling pathway	Autophagy activation, lipid oxidation, mitochondrial biogenesis, and inhibition of anabolism	Protects against metabolic disorders, promotes longevity	Ge et al. (2022)
SIRT1 (sirtuin 1)	NAD^+ levels	Sirtuin signaling	DNA repair, mitochondrial function, inflammation control, and apoptosis	Linked to aging, metabolic homeostasis, and neuroprotection	Zhao et al. (2021)
Insulin/IGF-1 receptor	Glucose, insulin	PI3K-AKT-mTOR pathway	Glucose uptake, glycogen synthesis, and cell proliferation	Insulin resistance leads to diabetes, cancer, and accelerated aging	Ramasubbu and Devi Rajeswari (2023)
GCN2 (general control nonderepressible 2)	Amino acid deprivation	Integrated stress response (ISR)	Inhibition of protein synthesis, stress adaptation, and autophagy	Adaptive response to starvation; dysfunction linked to cognitive deficits	Ilyin et al. (2024)
FOXO transcription factors	Indirectly via insulin/IGF-1 and AMPK	FOXO signaling pathway	Antioxidant defense, DNA repair, cell cycle arrest, apoptosis	Involved in lifespan extension and stress resistance	Guan et al. (2025)
ChREBP (carbohydrate-responsive element-binding protein)	Glucose metabolites	Carbohydrate-sensing pathway	Lipogenesis, glycolysis	High activity contributes to hepatic steatosis, obesity	Ramatchandirin et al. (2023)
HIF-1α (hypoxia-inducible factor 1-alpha)	Oxygen levels	HIF pathway	Glycolysis, angiogenesis, erythropoiesis	Crucial in cancer metabolism and hypoxic adaptation	Dzhalilova and Makarova (2021)

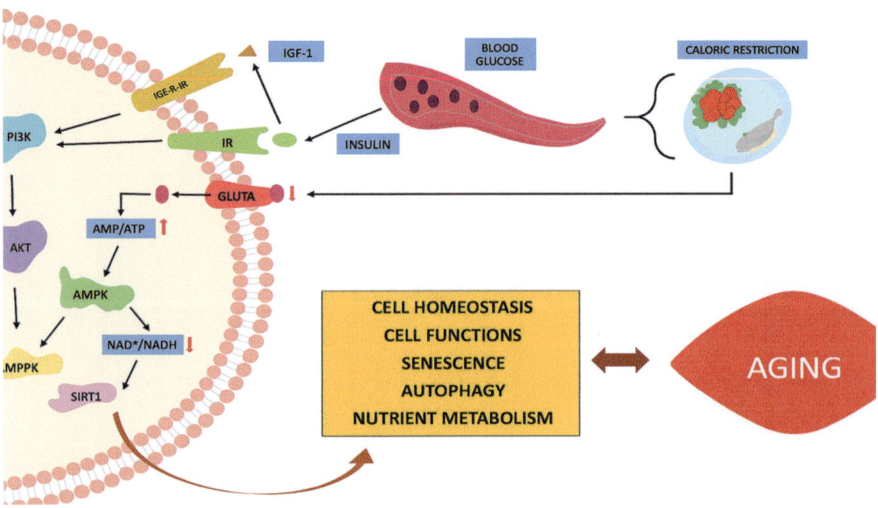

Fig. 8.2 The figure schematically depicts nutrient-sensing pathways and their role in regulating key cellular processes. Each nutrient-sensing mechanism (such as mTOR, AMPK, and the sirtuins) integrates signals pertaining to nutrient availability and directs several vital cellular processes, including autophagy, mitochondrial biogenesis, protein synthesis, and stress response. These pathways work to improve metabolic homeostasis and play vital functions in aging, disease resistance, and energy balance

8.4 Nutrient Restriction and Longevity

Dietary intake has long been a source of curiosity among researchers in the fields of aging and metabolism (Morita et al. 2024). One of the most sturdy and reproducible findings in the field of gerontology is that nutrient restriction, as opposed to malnutrition, can delay aging and increase longevity in biological systems ranging from yeast and worms to primates. Additionally, this expansion of life promotes the idea that aging is not fully programmed into the genetic code, but instead, can be influenced by environmental and lifestyle stressors, like dietary intake (Morita et al. 2024). The important link between the modulated diet and aging is based on how cells and organisms sense nutrients and respond to nutrient levels. Nutritional restriction (of any kind, caloric restriction or restriction of specific macronutrients) invokes conserved cellular processes that favor cellular maintenance and repair over reproduction, thereby promoting health span and delaying age-related disease (Secomandi et al. 2022).

Caloric restriction (CR), commonly described as a 20–40% decrease in total caloric intake without malnutrition, is the most well-researched type of nutrient restriction (Oudmaijer et al. 2024). Caloric restriction was first noted to increase lifespan in animals back in the 1930s in rodents, and more recently it has been shown to have a similar influence in areas such as yeast, flies, worms, and primates. CR achieves this through multiple means, including modulation of key nutrient-sensing pathways such as decreased insulin/IGF-1 signaling, decreased mTOR activity, and increasing AMPK and sirtuins. The reductions in these pathways modulate the cellular environment even further by increasing energy efficiency and also making cells more resilient to stressors, increase DNA repair, and increase autophagic activity (Agbemafle et al. 1927). Caloric restriction also leads to decreased systemic inflammation, improved insulin sensitivity, and lowered oxidative stress, which are all important to the development of diseases of aging and reducing the incidence of chronic inflammatory disease (Jain 2023, 1752. In nonhuman primates, CR has been associated with better cardiovascular health, lower incidences of cancer, and stable cognitive functioning, probably due to the aforementioned pathways. In humans, there are ongoing long-term studies on caloric restriction; however, there is only short-term data on caloric restriction as a whole, but many benefits such as improved metabolic markers, reduction of inflammation, and even improvement in body composition are noticeable even in extreme CR in short durations (Lao et al. 2023). Not only can molecular-level knowledge improve understanding of the mechanisms implicated in aging, but omics approaches also hold promise for identifying effective biomarkers, therapeutic targets, and personalized strategies to promote healthy longevity. This raises consideration for alternative methods of CR that would provide similar benefits without meal-by-meal and day-by-day caloric restriction (Hofer et al. 2022).

Intermittent fasting (IF) and time-restricted feeding (TRF) have transitioned from radical ideas into novel approaches other than continuous caloric restriction (Soliman 2022). IF, which can take the form of alternate-day fasting, the 5:2 diet, and periodic 24-hour fasts, is alternating between periods of eating and fasting (Rai 2021, 1753). TRF restricts eating to a specific number of hours each day, typically 6–10, but doesn't necessarily limit total caloric intake. Both of these feeding strategies represent an exploitation of natural circadian rhythms and metabolic cycles to activate nutrient-sensing pathways and signaling and stress response networks (de Oliveira Melo et al. 2024). Fasting periods activate AMPK and sirtuin activation, suppress mTOR, and stimulate autophagy, just like many of the other molecular changes occurring in CR. Studies in rodents have demonstrated that IF and TRF positively influence insulin sensitivity, reduce fat mass, decrease blood pressure, and extend longevity, among other benefits, even in cases where isolated calorie intake was unchanged. Human studies, although not as abundant, seem to suggest that IF and TRF inform changes in body weight, lipid profiles, reduce inflammation, and improve glycemic control. To conclude, IF and TRF may be preferred interventions in prolonging health span in a real-world context as they are perceived to be more sustainable and culturally acceptable than chronic caloric restriction (Kebbe et al. 2021).

In addition to overall caloric consumption and meal timing, the other important element of the diet relates to what it consists of, that is, the composition, especially the macronutrient ratio, which includes protein, carbohydrate, and fat (Feng 2025). Macronutrient ratio is now emerging as an important regulator of aging that may exert control over lifespan and health span, independent of caloric status. Low-protein, high-carbohydrate diets have been linked with increased lifespan in mice, which is probably due in part to decreased activity of the IGF-1/mTOR pathways. Based on current evidence, specific amino acids, particularly methionine and branched-chain amino acids (BCAAs), have generated considerable interest as modulators of nutrient signaling. For instance, methionine restriction extends lifespan and improves metabolic health by decreasing caspase 9 (a trigger of apoptosis) and lengthening the lifespan of fission yeast and worms. The methionine restriction presumably achieves this through reducing mTOR activity and improving resistance to oxidative and other types of stress (Kitada et al. 2021). Reducing dietary BCAAs has improved insulin sensitivity and led to less fat mass. Conversely, high protein intakes, especially from animal protein, correlated with higher circulating IGF-1 levels and an increased probability of age-related disease in epidemiological studies. As well, fat quality may also be important; unsaturated fat diets (especially omega-3) improve mitochondrial function and decrease inflammation. Diets high in saturated fat promote metabolic dysregulation and accelerated aging (Kitada et al. 2021).

In conclusion, restriction of nutrients, either through caloric restriction, fasting, or modulating macronutrient intake, activates pathways that were conserved through evolution, make cells more resilient, maintain metabolic equilibrium, and decelerate functional decline associated with aging (Moldakozhayev and Gladyshev 2023). These strategies are important and powerful interventions for extending lifespan in experimental subjects and for increasing health span as well as reducing the disease burden in humans. As we learn more about the molecular features of nutrient sensing, we can say with confidence that the quality, quantity, and timing of food intake are strong modifiers of the biology of aging with significant public health and preventative medicine ramifications (Zhang et al. 2023).

8.5 Molecular Targets and Therapeutic Interventions

The complex web of nutrient-sensing pathways that regulate longevity has created exciting possibilities for pharmacological interventions targeting modes of aging and aging-associated diseases (Guo et al. 2022). As the molecular mechanisms of caloric restriction and nutrient sensing are further uncovered, researchers are working to determine important nodes along these pathways that can be stimulated with a specific drug or natural compound needed to replicate the beneficial effects of specific diets without the need for ongoing restriction. The most studied therapeutic targets include the mechanistic target of rapamycin (mTOR), AMPK, and sirtuins. Not only have preclinical models demonstrated remarkable potential for pharmacological modulation of these pathways, and several of those agents are now in clinical testing, but increasing interest is also devoted to using combinations of agents aimed at targeting multiple pathways at once in the hopes of ultimately getting better efficacy and broader protection against the rise and fall of aging's complex processes (Wang et al. 2022).

The mTOR pathway, specifically the mTORC1 complex, is a major regulator of growth, metabolism, and autophagy, making it an attractive target for antiaging therapies (Fu and Wu 2023). Rapamycin was the first pharmacological mTOR inhibitor to extend lifespan in mammals, and it is a macrolide originally developed as an immunosuppressant. Rapamycin extends lifespan in yeast, worms, flies, and mice, and improves multiple health span markers, including immune function, cognitive performance, and cardiovascular health. The antiaging effects of rapamycin are thought to be mainly attributed to decreased protein synthesis and increased autophagy, and decreased cellular senescence (Fu and Wu 2023). Despite these antiaging benefits, chronic use of rapamycin may lead to unwanted consequences related to immunosuppression, glucose intolerance, and dyslipidemia. Some scholars have focused on developing rapalogs (rapamycin analogs) or stones of intermittent dosing, for example, to take rapamycin every other week, to hopefully achieve the desired benefits of extended lifespan, but with the least adverse effects. Furthermore, there remain ongoing efforts to develop mTORC1-specific inhibitors that would not inhibit mTORC2 functions related to insulin signaling and cell survival (Szwed et al. 2021).

AMPK, a cellular energy sensor, is activated when energy sources are low, and has also become an important target in research into longevity (Penugurti et al. 2024). The most famous AMPK activator is metformin, an oral antihyperglycemic drug commonly used in the management of type 2 diabetes, which has become an even more highly publicized potential antiaging therapeutic. Metformin has been linked to improved metabolic health, reduced cancer incidence, and increased lifespan, particularly in animals under conditions of metabolic stress. Metformin appears to act primarily by inhibiting mitochondrial complex I, which leads to increased AMP levels and activation of AMPK (Feng et al. 2022). AMPK activation effects include promoting autophagy, inhibiting mTOR signaling, improving insulin sensitivity, and enhancing oxidative stress resistance. Epidemiologic data suggest that long-term exposure to metformin in humans may lead to lower rates of age-related diseases and mortality. The Targeting Aging with Metformin (TAME) clinical trial is currently underway to test the hypothesis that metformin can delay the onset of age-related disease and potentially provide a new model of gerotherapeutics. Other AMPK activators are also being examined for their potential antiaging properties (Zhao et al. 2024).

Sirtuins are NAD^+-dependent deacetylases that connect energy metabolism to cellular stress responses and longevity programs, with SIRT1 being the most studied (Watroba and Szukiewicz 2021). Sirtuin activation leads to improved mitochondrial health, genomic stability, inflammation resolution, and metabolic protection. Resveratrol, a polyphenol compound prevalent in grapes and red wine, was the first compound identified as a sirtuin activator. In preclinical studies, resveratrol demonstrated the ability to elicit a caloric restriction-like phenotype, improve insulin sensitivity, and provide neuroprotective and cardioprotective effects. However, resveratrol's bioavailability and potency are relatively low, leading researchers to pursue more potent synthetic sirtuin activators. In parallel, there has been continued interest in differing strategies to increase NAD^+ levels in a variety of forms, as well as supplementing with NAD^+ precursors, such as nicotinamide riboside (NR) and nicotinamide mononucleotide (Alegre and Pastore 2023). While NR and NMN increase sirtuin activity, they boost sirtuin activity indirectly by essentially replenishing intracellular NAD^+ pools to enhance metabolic flexibility and cellular repair. Early phase clinical trials have shown promise with respect to improvements in muscle strength, mitochondrial health, and measures of metabolic health; however, the long-term efficacy and safety remain under investigation (Barber et al. 2021).

Acknowledging that aging is governed by multiple interrelated pathways, scientists are increasingly investigating the use of combinatorial and synergistic therapies that can target multiple nutrient-sensing pathways simultaneously (Narayan 2024). For example, utilizing a combination of metformin and rapamycin could produce additive or synergistic effects by activating AMPK while inhibiting mTOR, thereby promoting autophagy and nutrient metabolism in a more drastic way (Deleyto-Seldas and Efeyan 2021). Combining NAD^+ precursor supplementation with sirtuin activators or AMPK agonists may further enhance beneficial effects on inflammation, mitochondrial function, and mitochondrial stress resistance. Even herbal and dietary products targeting multiple pathways (e.g., curcumin, quercetin, etc.) are under investigation for their multifunctional geroprotective benefits. Given the complexity of signaling interactions, careful attention must be made to the dose, timing of administration, and assessment of biomarkers, in an effort to avoid adverse effects such as immune dysfunction or metabolic discord (Chennamadhavuni et al. 2022).

In conclusion, targeting nutrient-sensing pathways is an attractive therapeutic target for aging and age-related diseases (Aiello et al. 2017). Agents that have modulating activity on mTOR, AMPK, and sirtuins are likely to extend lifespan and optimize health span in preclinical trials. While we have a long way to go toward clinical application, investigational therapies are being studied in clinical trials as drug development platforms continue to advance our pharmacokinetic and pharmacodynamic capabilities, indicating we are getting closer to impactful therapies in humans. As we better understand the molecular basis of aging, the future of delivering safe, effective, and patient-tailored combinations of these agents to facilitate healthy longevity in the aging population is very exciting (Andreozzi et al. 2022).

8.6 Model Organisms in Nutrient Sensing and Aging Research

Model organisms in the field of aging biology have provided a powerful platform for the discovery of the basic principles that govern nutrient sensing, and other pathways that promote longevity (Obata and Miura 2024). Model organisms, from unicellular yeast to mammals, possess unique attributes, such as being genetically tractable, the short time frame of the life cycle, as well as conserved metabolic pathways (Soanes et al. 2008). Model organisms have provided relevance to the identification and characterization of molecular participants in nutrient-sensing pathways like insulin/IGF-1 signaling (IIS), target of rapamycin (mTOR), AMPK, and sirtuins, all acknowledged as some of the central regulators of lifespan and age-related diseases today. While many differences between species exist, the remarkable conservation of these pathways over evolution allows for cross-species confirmation of longevity interventions, while providing greater confidence in translating evidence from preclinical models into the research of human aging (Holtze et al. 2021).

Budding yeast, or *Saccharomyces cerevisiae*, was the first experimental organism used to study aging at the cellular level (Eigenfeld et al. 2021). Yeast is a simple, genetically tractable model on which to study the basic biology of nutrient sensing. It has been instrumental in making strides toward understanding the role of TOR signaling and sirtuins in regulating lifespan. Two types of aging exist in yeast: replicative lifespan, measuring the number of daughter cells a mother cell produces, and chronological lifespan, reflecting the aging/survival of a nondividing cell in the stationary phase. Caloric restriction extends both replicative lifespan and chronological lifespan in yeast through reduced TOR signaling activity, enhanced autophagy, and increased stress resistance (Mirisola and Longo 2022). Altering the TOR signaling pathway (i.e., deleting TOR1 or overexpressing the sirtuin gene SIR2) also demonstrated lifespan extension in yeast, pointing to conserved molecular targets pertinent to all eukaryotes. Yeast continues to be a useful organism for high-throughput screening of aging-related genes and compounds (Hong 2024).

Caenorhabditis elegans, a nematode worm, has made a significant impact on our knowledge of the genetics of aging (Chakravarty 2022). *C. elegans* has a life span of only 2–3 weeks and has a fully sequenced genome that is amenable to genetic and pharmacological analysis. There have been a multitude of important findings from studying *C. elegans*, including its discovery of the insulin/IGF-1 receptor homolog, daf-2, and the downstream transcription factor, daf-16 (FOXO), that regulate life span. Mutations in the daf-2 can double the life span of the worm, and the daf-2 effect is abrogated by knocking out daf-16, indicating an important role for IIS in aging. *C. elegans* has also helped to establish that caloric restriction—by bacterial dilution or genetically disrupting feeding— can extend life span via the AMPK, TOR, and sirtuin pathways. *C. elegans* (Laskovs 2024) is also widely used to study mitochondrial function and stress responses, such as heat shock and oxidative stress responses to aging. The use of RNA interference (RNAi) techniques allows functional gene analysis to occur over very short time periods, increasing the effectiveness of *C. elegans* as a model system for understanding complex interactions of nutrient-sensing pathways (Chakravarty 2022).

Drosophila melanogaster (fruit fly) is an intermediate model with more complex tissue organization and physiology than yeast or nematodes, but has allowed us to study how systemic nutrient sensing coordinates with organ-specific processes to modulate aging (Kim et al. 2021). Similar to *C. elegans*, *Drosophila* has conserved IIS and TOR pathways, and genetic perturbations of these pathways have a major impact on lifespan. For example, overexpressing dFOXO in the fat body, which is analogous to mammalian liver and adipose tissue, extends lifespan, and the attenuation of insulin-like peptides generates increased longevity. Caloric restriction in flies also activates AMPK and downregulates mTOR signaling, similar to the mechanisms observed in lower organisms (Livelo et al. 2023). The short lifespan and relatively low cost of maintenance, in combination with the ample array of increasingly sophisticated genetic tools, including the GAL4/UAS system, make *Drosophila* an attractive model to assess the impact of nutrient-modulating signals, intervening diets, or environmental exposures on aging and metabolic health. Studies in flies have also provided insight into circadian rhythms, neuronal aging, and gut microbiota as they relate to longevity regulation (Cheng et al. 2022).

Rodents and higher mammals, and specifically mice, are more complex models with closer physiological, genetic, and metabolic features than humans (Vatashchuk et al. 2022). Mouse models are an important resource for confirming findings derived from simpler organisms and for evaluating the translational potential of antiaging interventions. Caloric restriction in mice strongly extends lifespan and the onset of age-related diseases, such as cancer, diabetes, and neurodegeneration. While transgenic mice models have tested the function of key nutrient-sensing pathways (mTOR, AMPK, sirtuins) to mediate dietary and pharmacological effects on aging (Xu et al. 2023), rapamycin treatment in mice has been shown to directly extend lifespan and health markers, while SIRT1 overexpression in mice increases the biochemical and phenotypic resistances toward metabolic and oxidative stress. Metformin and NAD^+ precursors also show improvements in metabolic health, through the parallel of caloric restriction models. Nonhuman primates and rhesus monkeys specifically can act as transitional models between rodents and humans. Long-term caloric restriction studies in primates show improvements to multiple health span markers, and the evaluation of lifespan is ongoing. These models provide critical information about the systemic effects of nutrient sensing, and evaluate potential side effects, sexual dimorphism, to antiaging therapies (Theodore Armand et al. 2024).

In conclusion, model organisms have been very important for understanding the complex intersection between nutrient sensing and aging (Fernandes and Demetriades 2021). Yeast, flies, worms, fish, and mice have revealed conserved pathways and molecular targets controlling lifespan, efficiency with which nutrients are converted to body mass (growth), and responsiveness to diets that prevent disease. All models possess unique strengths and facilitate a multidimensional assessment of how dietary and genetic factors influence the health span and lifespan of animals (Raubenheimer et al. 2022). The field of research will continue to advance, and it will remain important to apply an integrative approach using multiple model organisms to help us better translate discoveries made in the laboratory into therapeutic approaches to extend healthy aging in humans (Holtze et al. 2021).

8.7 Systems Biology and Omics Approaches

Given the intricate and multilevel nature of nutrient sensing and its role in longevity potential, we must carefully employ analytical frameworks that employ more than biomedical reductionism focused on single genes or single pathways (Vega 2023). Systems biology provides investigative frameworks that integrate high-throughput data generation, computational analysis, and biological meaning, as the large nondiscriminate molecular networks that regulate aging and nutrient response can be discovered. Using omics-based technologies, such as autologous and genomic studies (e.g., transcriptomics, proteomics, and metabolomics), parental and mutated genomes differing environmental influences can be modeled using computational models in order to create a systems biology picture, generating a dynamic view of how organisms sense a nutrient rationale leading to nutrient effects (Patoine et al. 2022). Not only can molecular-level knowledge improve understanding of the mechanisms implicated in aging, but omicsapproaches also hold promise for identifying effective biomarkers, therapeutic targets, and personalized strategies to promote healthy longevity (Wu et al. 2021).

Transcriptomics and proteomics give the initial clues of gene expression and protein behavior that engage nutrient sensing and aging (Wahl and LaRocca 2021). Transcriptomics is an umbrella term for analyses of global mRNA levels under different dietary conditions or genetic manipulations targeting pathways like mTOR, AMPK, and sirtuins. Within these studies comparing gene expression under caloric restriction, intermittent fasting, or with pharmacological interventions such as metformin or rapamycin, conserved gene signatures associated with longer lifespan (Mohammed et al. 2021), increased stress resistance, and metabolic adaptation have been discovered. Indeed, caloric restriction studies often revealed downregulation of inflammatory genes and upregulation of genes, for example, involved in mitochondrial function and autophagy. Advances in single-cell RNA sequencing (scRNA-seq) now allow for tissue-specific and cell-type-resolved transcriptomic analyses, allowing investigation into nutrient-sensing pathways within heterogeneous biological contexts (Megas et al. 2024).

Proteomics, which directly determines the abundance, modifications, and interactions of proteins, is an excellent complement to transcriptomic data because it provides data about the actual functional outcome of gene expression (Sanches et al. 2024). While transcriptomics provides mRNA information, proteomics provides a comprehensive understanding of protein function, especially posttranslational modifications like phosphorylation and acetylation, which are known to be important in nutrient sensing through the regulation of key signaling proteins like AMPK, mTOR, and sirtuins (Tomtheelnganbee et al. 2022). Using quantitative proteomics through mass spectrometry, we can learn about protein networks affected by age or under diminished nutrient resources, and we can even observe changes in proteostasis, mitochondrial function, and responses to oxidative stress. Together, transcriptomics and proteomics represent an extremely powerful approach to identifying and mapping adaptive changes in response to nutrient dynamics and understanding how those changes may shape the ability to live for a long time in mammals from a systems perspective (Stroud and Ratcliff 2025).

Metabolomics and nutrigenomics add to systems-level analyses by taking into account both the biochemical and dietary context of pathways involved in nutrient sensing (Nasar et al. 2024). Metabolomics refers to the large-scale profiling of the small-molecule metabolites that serve as substrates, intermediates, or products of metabolic reactions, including amino acids, lipids, sugars, and signaling molecules, which are influenced by dietary intake and energy metabolism (Ruddick-Collins et al. 2022). By profiling metabolite signatures between long- and short-lived organisms or in the same organism following specific interventions, it is possible to implicate metabolic pathways that are closely associated with the aging process. For example, age-associated changes in NAD^+, ketone bodies, or branched-chain amino acids are often correlated with changes in energy metabolism and redox homeostasis that have an impact on lifespan (Jamerson and Bradshaw 2024).

Nutrigenomics examines personal genetic backgrounds and how they influence individual responses to dietary components, which provide insights about personalized nutrient sensing and aging (Lagoumintzis and Patrinos 2023). Polymorphisms in genes accountable for insulin signaling, sirtuin function, or lipid metabolism could influence the effectiveness of dietary restriction or pharmacological interventions. Nutrigenomics uses genomic and metabolomic data to develop individualized dietary plans that optimize health span and minimize disease risk by each individual's genetics. Nutrigenomics allows precision gerontology, based on an individual's unique molecular profile (Rezaei and Reynoso-Sánchez 2024).

Computational modeling of longevity networks is a vital aspect of systems biology that facilitates the simulation and understanding of complex biological systems as they evolve over space and time (Cohen et al. 2022). Modeling is done by first creating mathematically based models and network diagrams of pathways relating to nutrient sensing and their interactions. From these diagrams, researchers can identify the critical nodes of regulation, feedback mechanisms, and potential points for intervention (Uleman et al. 2024). These models incorporate multiomics data, such as gene expression, protein abundance, and metabolite concentrations, to be able to predict how a biological system will respond when faced with certain stimuli such as genetic mutations, dietary changes, or drug treatments. For instance, dynamic models of mTOR and AMPK signaling have been reported that demonstrate how different patterns/levels of nutrient availability (or pharmacological input) affect autophagy, cellular stress responses, and patterns of energy metabolism (Kumar et al. 2023).

Machine learning or artificial intelligence will provide further insights into computational modeling through identifying patterns in large and complex datasets (multidimensional datasets) and subsequently predicting experiences based on those identified patterns (Hu et al. 2024). Because they can reveal insights into nutrient signals and their relationships with aging phenotypes that may not be apparent through traditional analysis, they can be incredibly useful. Predictive profiles, such as we illustrate here, also aid in drug discovery and/or drug repurposing (Rodriguez et al. 2021). These models can help scientists to screen candidate compounds for those that emulate caloric restriction or minimize other counterproductive effects when combined with therapeutic agents. They could also be implemented to model the heterogeneity in aging trajectories across individuals to account for the influences of genetic variation, exposed environments, and personal lifestyle factors (Ferrucci and Kuchel 2021).

Systems biology methods have transformed our understanding of nutrient sensing and aging by allowing us to see the biological complexity holistically and in a more data-driven way (Marabita et al. 2022). By employing transcriptomics, proteomics, metabolomics, and computational modeling, systems biology promotes the dissection of the multifaceted regulation of longevity and the discovery of new biomarkers and therapies. Each technology discussed above provides the opportunity to move from reductionist to integrative science and will allow for actionable changes to promote healthy aging, thus affecting a personalized, predictive, and preventive approach of healthy aging across heterogeneous populations (Addo and Khan 2024) (Table 8.2).

8.8 Human Studies and Clinical Implications

Although model organisms have paved the way for understanding nutrient sensing and its relationships to longevity, studies in humans are critical for verifying these findings and advancing their translational potential into real-world health interventions (Prasad et al. 2024). Human physiology, lifestyle, environmental exposures, and iatrogenesis introduce a level of complexity and variability that is not found in controlled laboratory studies. Human-centered research in long-lived populations, nutritional epidemiology, biomarker development, and precision medicine studies has provided a useful beginning of understanding how nutrient-sensing pathways operate in humans and how they can be manipulated to optimize healthy aging (Kroemer et al. 2025). This human-focused research emphasizes the translational potential of utilizing nutrient-sensing pathways to promote health and longevity; however, it also highlights the complexities that must be considered when going from bench to bedside (Wang et al. 2024).

Perhaps the most persuasive source of evidence comes from studies of centenarians and long-lived populations (Topol 2025). These individuals can maintain functional independence into older adulthood and either delay or have significantly fewer instances of age-related disease, including cardiovascular disorders, cancer, and neurodegeneration. Studies on different "Blue Zones," regions with an abundance of centenarians near specific geographic coordinates (for example, Okinawa, Japan; Sardinia, Italy; Nicoya, Costa Rica), have revealed some common lifestyle characteristics of these Blue Zones. Blue Zone individuals had a normal caloric intake, a predominantly plant-based diet, an activity or exercise routine, and strong social support. Nutritionally, these populations consume low protein, low sugar, and high-antioxidant, high-polyphenol-based diets that may restrict central nutrient-sensing pathways, including mTOR, AMPK, and the sirtuins (Tomtheelnganbee et al. 2022). Genetic studies of centenarians have also revealed an enrichment in certain alleles associated with favorable effects on lipid metabolism, increased insulin sensitivity, and decreased inflammatory status, further substantiating the working hypothesis that both environmental and genetic factors dictate longevity and health span along common nutrient-responsive molecular circuits (Dabrowska et al. 2022).

The discovery of nutrient-sensing and aging biomarkers has become central to attempts to measure biological aging and predict responsiveness to dietary or pharmacological intervention (Siopis and Porter 2022). Chronological age does not accurately reflect a person's physiological state or disease risk, and looking for molecular biomarkers to track the activity of aging pathways is becoming more

Table 8.2 Table summarizing systems biology and omics approaches

Omics layer	Focus	Key technologies/tools	Insights provided	Applications in research and medicine	References
Genomics	DNA sequence and structure	Whole-genome sequencing (WGS), GWAS, CRISPR	Genetic variants, mutations, SNPs, structural variants	Disease susceptibility, personalized medicine, gene therapy	Khan et al. (2025)
Transcriptomics	RNA expression profiles	RNA-seq, microarrays, qRT-PCR	Gene expression patterns, alternative splicing	Biomarker discovery, cell type identification, disease progression	Azadeh et al. (2023)
Proteomics	Protein abundance, modification, interaction	Mass spectrometry, 2D-GE, protein arrays	Posttranslational modifications, signaling networks	Drug target validation, disease mechanism elucidation	Ogunjobi et al. (2024)
Metabolomics	Small-molecule metabolites	NMR spectroscopy, LC-MS, GC-MS	Metabolic signatures, energy metabolism, nutrient flux	Early diagnostics, metabolic pathway analysis, nutritional studies	Marchev et al. (2021)
Epigenomics	DNA methylation, histone modification, chromatin structure	ChIP-seq, bisulfite sequencing, ATAC-seq	Gene regulation, environmental effects on the genome	Cancer epigenetics, transgenerational inheritance, and developmental biology	Bi et al. (2024)
Microbiomics	Composition and function of microbiota	16S rRNA sequencing, metagenomics	Host-microbe interactions, dysbiosis	Gut-brain axis, probiotics, immune regulation	Kandpal et al. (2022)
Lipidomics	Lipid species and pathways	MS-based lipid profiling	Lipid metabolism and signaling	Cardiometabolic diseases, neurodegenerative disorders	Sarkar et al. (2024)
Glycomics	Glycan structures and interactions	HPLC, lectin arrays, MS	Protein glycosylation, cell-cell communication	Cancer diagnostics, immune modulation	Kandpal et al. (2022)
Interactomics	Molecular interaction networks	Yeast two-hybrid, co-IP, BioID	Protein-protein, protein-DNA, signaling networks	Systems-level drug target discovery, pathway reconstruction	Reitz et al. (2023)
Phenomics	Observable traits and phenotypes	High-throughput imaging, wearable sensors, and AI analytics	Integration of genotype-environment-phenotype	Precision medicine, behavioral phenotyping, clinical stratification	Tanaka (2025)

commonplace. These biomarkers include: circulating insulin, IGF-1, adiponectin, inflammatory cytokines, and metabolites such as NAD+, glucose, and ketone bodies. Recently, it has also been possible to develop composite "aging clocks" that aggregate epigenetic, transcriptomic, proteomic, and metabolomic data to better estimate biological age (Yusri et al. 2024). For example, DNA methylation clocks can track the effects of lifestyle interventions like dietary restriction and fasting on biological aging. Quantifying the dynamics of sirtuin activity, mitochondrial function, and autophagic flux through minimally invasive means is also improving, and allows for the dynamic assessment of nutrient-sensing pathways in real time (Bharti et al. 2024).

Despite the expanding evidence base, several translational challenges remain that limit the clinical utility of nutrient-sensing-based approaches (Pokushalov et al. 2024). Conducting human trials is more challenging in practice due to ethical, operational, and temporal factors; lifetime studies are impractical, and when focusing on health span, can take years of follow-up. In addition, there is considerable interindividual variability in genes, microbiome composition, and follow-on dietary and environmental exposures. What works for one person or population group may not yield a similar result for another, and as a result, more individualized/precision nutrition is needed. In this context [of] nutrient sensing (Liang et al. 2024), the new fields of nutrigenomics and nutrigenetics will become extremely relevant, especially as applied research inevitably will lead to personalized dietary plans based on genetic risk scores, microbiome characteristics, and metabolic indicators to maximize nutrient sensing to promote health span (Singh et al. 2024b).

Personalized strategies benefit from digital health innovations, such as wearable devices and mobile applications, related to dietary intake, activity, sleep, and physiological responses (Nur 2024). Digital health innovations provide real-time feedback, which allows response and cocreation of interventions. In addition, digital health innovations and basic science are paving the way for machine learning and AI to interpret complex health data and even predict future responses to specific diets or drugs. Indeed, this digitized flow of information, coupled with machine learning, is offering a more predictive and preventive model of care (Guo 2022).

Simultaneously, a number of pharmacologic agents with potentially similar effects on nutrient-sensing pathways, like metformin, rapamycin, NAD^+ precursors, and resveratrol (Sorrenti et al. 2022), are being investigated for their ability to mimic caloric restriction and delay decline during aging in clinical studies. The Targeting Aging with Metformin (TAME) trial, for instance, is designed to investigate whether metformin can modulate health span via delay in multiple chronic diseases. Long-term safety, dosing indices, and criteria for patient selection requirements are still being established. Regulatory processes may also need to evolve to address the unique dynamics of therapy development for the aging process itself, which, in most clinical designations, is not yet a treatable condition (Pimenta et al. 2021).

To sum up, human studies are vital to advancing our knowledge of nutrient sensing, and the future clinical applications of nutrient sensing for aging and longevity (Singh et al. 2024b). Research in long-lived populations, biomarker development in clinical trials, and research on personalized nutrition offer us both ideas and tools to support health span. New findings from our model organism studies will invariably need to be translated, which require awareness of the inherent biological complexity, individual variability, and potential ethical ramifications of our research (Vashishat et al. 2024). We believe that a systems-level, personalized, and preventative approach that combines omics data, digital health technologies, and precision therapeutics presents the greatest potential for accessing nutrient-sensing pathways to enhance human health and ultimately, functional longevity (Tomtheelnganbee et al. 2022).

8.9 Emerging Trends and Future Directions

The landscape of nutrient sensing and longevity research is changing rapidly as a result of improvements in molecular biology, systems medicine, artificial intelligence, and ethics (Biswas et al. 2023). With an aging global population and increasing interest in extending health span, growing numbers of researchers are exploring new links between nutrient-sensing pathways and the pathophysiology of age-related diseases, using AI to write algorithms to decipher biological networks, and reviewing substantial ethical considerations surrounding the future of life extension (Izankar et al. 2024). While the scientific landscape is continuing to develop a broader understanding of aging, these trends also open up critical discussions on how societies should prepare for, and ethically grapple with, the potential for prolonged lifespans in the future (Keeler and Bernstein 2021).

One important emerging trend includes exploring nutrient-sensing pathways involved in age-related diseases that share commonality in mechanisms such as impaired metabolism (Kopp 2024), oxidative stress, dysfunction in autophagy, and chronic, low-grade inflammation. These pathways have been studied in common age-related diseases, including cancer, cardiovascular disease, type 2 diabetes, neurodegenerative disease, including Alzheimer's and Parkinson's disease, and chronic inflammation. These diseases have been linked to disrupted cellular metabolism, oxidative stress, impaired autophagy, and chronic (Chang et al. 2022), low-grade inflammation modulated by nutrient-sensing pathways such as mTOR, AMPK, sirtuins, and insulin/IGF-1 signaling, for example, excess activation of mTOR has been implicated in traits of several age-related diseases, including insulin resistance, cell proliferation in malignancy, and neurodegeneration while reduced activity of sirtuins has been linked to mitochondrial dysfunction and cognitive decline. These nutrient-sensing pathways provide potential therapeutic targets that can delay aging and prevent or mitigate age-related diseases. Nutrient-sensing pathways are the focus of investigation in many preclinical and clinical studies related to age-associated diseases that program the homeostasis of our nutrient-sensing systems to deteriorate, including caloric restriction mimetics (e.g., metformin and resveratrol), types of diets (e.g., low-protein or ketogenic diets), and regimes that can be classified as intermittent fasting (Giugliano et al. 2024).

In addition to biological discoveries, AI and machine learning are revolutionizing longevity research by giving researchers the capacity to analyze unprecedented amounts of complex data generated from proteomics (Guo et al. 2025), genomics, transcriptomics, metabolomics, electronic health records, wearable sensors, and clinical trials. AI models can analyze and integrate genomic, transcriptomic, proteomic, and metabolomic data with respect to aging biomarkers, the risk of disease (Ogunjobi et al. 2024), the interaction of biologically relevant nutrients in pathways, and the development and personalization of interventions. Deep learning algorithms can create biological aging clocks, determine individual responses to caloric restriction or other pharmacological agents, and discover novel compounds that target nutrient-sensing pathways. Additionally, AI enables the repurposing of drugs and designing synergistic interventions through predictive modeling of biochemical networks. Digital twins—virtual representations of individual persons are emerging as platforms to model the effects of various lifestyle, diet, and treatment options on the individual trajectory of aging in real-time. The maturation of these technologies will ultimately be critical to making precision aging interventions adaptable, scalable, and clinically actionable (Chen 2024).

While there are significant new existing sciences and technologies to offer, there remains a considerable space for ethical and societal discussions about lifespan extension (Raman et al. 2025). As therapies targeting nutrient-sensing pathways become more accessible and affordable, we must consider more challenging issues in equity, accessibility, and societal preparedness (Mozaffarian et al. 2022). Who gets access to lifespan-extending therapies? Will lifespan extension exacerbate existing health inequalities across existing populations with differing socioeconomic means? Certainly, using engineered aging to generate a mature population with relatively long lifespans will raise additional complications for our healthcare systems, pension systems, employment and retirement arrangements, not to mention the impact upon relationships between generations and families. The psychological aspects involved in delayed aging (identity, purpose, and belonging) merit more attention (Gaviano et al. 2024).

Ethical quandaries can be found within the sphere of clinical research. The development and approval of antiaging interventions twisted with the fact that current regulations exist to address diseases, not aging as a process we can modify (Fletcher 2022). We continue to debate whether aging constitutes a disease, and how one would define appropriate endpoints for a clinical trial investigating health span and longevity. Informed consent will need to address uncertainties in long-term outcomes and the risk of unintended consequences when fundamentally altering biological processes. As resources for longevity researchers become increasingly commercial, issues of ethical scrutiny, transparency, and public engagement will be necessary to achieve responsible innovation (Li et al. 2023).

In summary, nutrient-sensing research in aging is at an important interface involving molecular and digital science and ethical governance (Singh et al. 2024b). By mapping out a more complete picture of how nutrient signals regulate the processes of aging and disease, the combination of AI, personalized medicine, and systems biology will expand the boundaries of longevity science, but we also need to develop inclusive and ethically sound mechanisms to translate these developments equitably to all people. While the future is promising, it is complex—and achieving a longer life must not outpace its quality and access by continuing to work and collaborate across disciplines (Knopp et al. 2023).

8.10 Conclusion

The investigation of nutrient-sensing pathways has dramatically altered our fundamental perspective on aging, health span, and longevity. Aging is no longer simply a passive manifestation or outcome of time; rather, it is an active, comprehensively regulated process of a multitude of signaling pathways of varying complexities that quantitatively respond to nutrients, energy status, and other environmental stimuli. These pathways are now more than theoretical concepts; they have specific biologically defined functions related to growth, metabolism, stress responses, and cellular upkeep; their recent diverse combinations—over the past 4 billion years of biology—and their physiologically control—through the many biological pathways of nutrient sensing—communicate a number of roles that are evolutionary preserved (i.e., the insulin/IGF-1 signaling (IIS) pathway, mTOR, AMPK, and sirtuins) in all living organisms. Importantly, these pathways work in an integrated fashion, often collectively responding to extreme metabolic variation; they cross-regulate each other with feedback loops, which facilitates this harmonious resilience. Modulating these nutrient-sensing pathways has been demonstrated in a number of diverse model organisms to delay aging, promote stress resistance, and/or diminish the incidence of age-related disease, thus informing therapies and interventions aimed at healthy aging.

One major take-home message from this scholarship is that nutrients influence not only a wide range of metabolic measures but also energy balance. Nutrient monitoring pathways govern critical cellular processes, including autophagy, mitochondrial biogenesis, proteostasis, redox homeostasis, and inflammation, which decline with age. From caloric restriction, intermittent fasting, and selected macronutrient intake, nutritional interventions are well-established to increase lifespan and health span across multiple model organisms, and now into humans. Pharmacological compounds, such as rapamycin, metformin, NAD^+-augmenting precursors, and resveratrol, represent dietary mimetics and targeted public health strategies that address nutrient-sensing nodes—and many are currently in earlier-stage and late-stage clinical trials to combat age-related pathologies.

Simultaneously, omics technologies and systems biology are facilitating a level of resolution to aging that has never been seen before. Transcriptomics, proteomics, metabolomics, and nutrigenomics have identified biomarkers and molecular signatures of nutrient-response longevity, while computational modeling and AI are paving a way to map, predict, and manipulate these networks. Human studies involving centenarians, aging biomarkers, and responsive dietary changes further corroborate these conclusions and point to the translational possibilities of nutrient-sensing research. Moving forward into precision gerontology will require that these advances be used to create personalized, preventive, and ethically responsible interventions.

Considered collectively, the pathways of nutrient sensing are thought of as regulators of the biological process of aging and potential therapeutic targets for maintaining health across a lifespan. Pathways of nutrient sensing can affect cellular homeostasis, risk of disease, physiological resilience, and ultimately health across a lifespan. The future directions would be to apply these molecular-based scientific findings into a practical and safe intervention that is supported by reasonable evidence for clinical practice and ethical considerations to ensure the benefits are equitably distributed. By appreciating a holistic perspective, in which molecular biology, lifestyle medicine, digital health, and social value are viewed collectively, we can better utilize the pathways of nutrient sensing for healthy aging and improved well-being in the aging global population.

References

Addo KM, Khan H (2024) Factors affecting healthy aging and its interconnected pathways. Turkish J Healthy Aging Med 1:9–24

Agbemafle W, Jayasinghe V, Bassham DC (1927) Can autophagy enhance crop resilience to environmental stress? Philos Trans B 2025(380):20240245

Aiello A et al (2017) Nutrient sensing pathways as therapeutic targets for healthy ageing. Expert Opin Ther Targets 21(4):371–380

Alegre GFS, Pastore GM (2023) NAD+ precursors nicotinamide mononucleotide (NMN) and nicotinamide riboside (NR): potential dietary contribution to health. Curr Nutr Rep 12(3):445–464

Anania S (2022) Identification of myoferlin as a component of mitochondria-associated membranes and discovery of a new function in calcium transfer in PDAC cell lines

Andreozzi F et al (2022) New perspectives in treating acute myeloid leukemia: driving towards a patient-tailored strategy. Int J Mol Sci 23(7):3887

Azadeh M et al (2023) Integrated high-throughput bioinformatics (microarray, RNA-Seq, and RNA interaction) and qRT-PCR investigation of BMPR1B axis as a potential diagnostic biomarker of Isfahan breast cancer. Adv Biomed Res 12(1):120

Balamurugan BS et al (2024) Micro nutrients as immunomodulators in the ageing population: a focus on inflammation and autoimmunity. Immun Ageing 21(1):88

Ballesteros-Álvarez J, Andersen JK (2021) mTORC2: the other mTOR in autophagy regulation. Aging Cell 20(8):e13431

Barber TM et al (2021) The low-carbohydrate diet: short-term metabolic efficacy versus longer-term limitations. Nutrients 13(4):1187

Bauer ME, Teixeira AL (2021) Neuroinflammation in mood disorders: role of regulatory immune cells. Neuroimmunomodulation. Aug 24;28(3):99–107

Bharti A, Jain U, Chauhan N (2024) From lab to field: Nano-biosensors for real-time plant nutrient tracking. Plant Nano Biology:100079

Bhushan R, Goel F, Singh S (2025) Beyond dopamine: novel therapeutic pathways for Parkinson's disease through receptor signaling. CNS Neurol Disorders-Drug Targets

Bi F, Gao C, Guo H (2024) Epigenetic regulation of cardiovascular diseases induced by behavioral and environmental risk factors: mechanistic, diagnostic, and therapeutic insights. FASEB BioAdvances 6(11):477–502

Biswas A et al (2023) Revolutionizing biological science: the synergy of genomics in health, bioinformatics, agriculture, and artificial intelligence. OMICS J Integr Biol 27(12):550–569

Bleve A et al (2022) Immunosenescence, inflammaging, and frailty: role of myeloid cells in age-related diseases. Clin Rev Allergy Immunol:1–22

Cagigas ML et al (2025) Dietary and pharmacological energy restriction and exercise for healthspan extension. Trends Endocrinol Metabol

Cavender-Bares J et al (2022) Integrating remote sensing with ecology and evolution to advance biodiversity conservation. Nat Ecol Evol 6(5):506–519
Chakravarty B (2022) The evolving role of the Caenorhabditis elegans model as a tool to advance studies in nutrition and health. Nutr Res 106:47–59
Chang K-C et al (2022) The interplay of autophagy and oxidative stress in the pathogenesis and therapy of retinal degenerative diseases. Cell Biosci 12:1–20
Chauhan K, Goel F, Singh S (2024) Apigenin protects melanocytes and improve tyrosinase activity in a hydroquinone induced vitiligo mouse model targeting P38 MAP kinase signaling: histopathology and immunohistochemistry analysis. Naunyn Schmiedeberg's Arch Pharmacol 397(7):4859–4869
Chen L-K (2024) Integrating precision health with multidomain interventions to advance strategies for healthy aging. J Nutr Health Aging 28(10):100354
Chen R et al (2024) Targeting the mTOR-autophagy Axis: unveiling therapeutic potentials in osteoporosis. Biomolecules 14(11):1452
Cheng W-Y, Ho Y-S, Chang RC-C (2022) Linking circadian rhythms to microbiome-gut-brain axis in aging-associated neurodegenerative diseases. Ageing Res Rev 78:101620
Chennamadhavuni A et al (2022) Risk factors and biomarkers for immune-related adverse events: a practical guide to identifying high-risk patients and rechallenging immune checkpoint inhibitors. Front Immunol 13:779691
Cohen AA et al (2022) A complex systems approach to aging biology. Nat Aging 2(7):580–591
Dabrowska A, Kumar J, Rallis C (2022) Nutrient-response pathways in healthspan and lifespan regulation. Cells 11(9):1568
de Oliveira Melo NC et al (2024) Biological rhythms, Chrono-nutrition, and gut microbiota: Epigenomics insights for precision nutrition and metabolic health. Biomolecules 14(5):559
Deleyto-Seldas N, Efeyan A (2021) The mTOR–autophagy axis and the control of metabolism. Front Cell Dev Biol 9:655731
Duan H et al (2022) Dietary strategies with anti-aging potential: dietary patterns and supplements. Food Res Int 158:111501
Dzhalilova DS, Makarova OV (2021) HIF-dependent mechanisms of relationship between hypoxia tolerance and tumor development. Biochem Mosc 86:1163–1180
Eigenfeld M, Kerpes R, Becker T (2021) Understanding the impact of industrial stress conditions on replicative aging in Saccharomyces cerevisiae. Front Fungal Biol 2:665490
Engel MG (2023) Deciphering the role of insulin and IGF-1 in brain aging, Alzheimer's disease, and vascular dementia. Albert Einstein College of Medicine
Fekete M et al (2022) Nutrition strategies promoting healthy aging: from improvement of cardiovascular and brain health to prevention of age-associated diseases. Nutrients 15(1):47
Feng LB (2025) The correlation between different lifestyles and body composition focuses on eating habits, nutritional status, and physical exercise components. Hormones:1–21
Feng J et al (2022) Mitochondria as an important target of metformin: the mechanism of action, toxic and side effects, and new therapeutic applications. Pharmacol Res 177:106114
Fernandes SA, Demetriades C (2021) The multifaceted role of nutrient sensing and mTORC1 signaling in physiology and aging. Frontiers in Aging 2:707372
Ferrucci L, Kuchel GA (2021) Heterogeneity of aging: individual risk factors, mechanisms, patient priorities, and outcomes. J Am Geriatr Soc 69(3):610
Fischer F et al (2022) Evolutionarily conserved transcription factors as regulators of longevity and targets for geroprotection. Physiol Rev
Fleck LM (2022) Precision medicine and distributive justice: wicked problems for democratic deliberation. Oxford University Press
Fletcher D-J (2022) Gerontological hygiene: the role of anti-aging somatechnologies in the abolition of old age. Macquarie University
Fu W, Wu G (2023) Targeting mTOR for anti-aging and anti-cancer therapy. Molecules 28(7):3157
Gaviano L et al (2024) Definitions of ageing according to the perspective of the psychology of ageing: a scoping review. Geriatrics 9(5):107

Ge Y et al (2022) Role of AMPK-mediated pathways in autophagy and aging. Biochimie 195:100–113
Giacomello E, Toniolo L (2021) The potential of calorie restriction and calorie restriction mimetics in delaying aging: focus on experimental models. Nutrients 13(7):2346
Giugliano F et al (2024) Fast mimicking diets and other innovative nutritional interventions to treat patients with breast cancer. In: Breast cancer research and treatment: innovative concepts. Springer, pp 199–218
Goel F (2025) Exploring the therapeutic role of Moringa oleifera in neurodegeneration: antioxidant, anti-inflammatory, and neuroprotective mechanisms. Inflammopharmacology:1–17
Goel F, Dobhal V, Singh S (2025a) Protective effects of diacerein against quinolinic acid–induced Huntington's disease–like symptoms in adult zebrafish by targeting GSK-3β signalling. Naunyn Schmiedeberg's Arch Pharmacol:1–23
Goel F, Kumar D, Sharma A (2025b) Impact of corticoid receptors on Alzheimer's disease: a neuroendocrine perspective. Inflammopharmacology:1–16
Guan G, Chen Y, Dong Y (2025) Unraveling the AMPK-SIRT1-FOXO pathway: the in-depth analysis and breakthrough prospects of oxidative stress-induced diseases. Antioxidants 14(1):70
Guo Y (2022) A new paradigm of "real-time" stroke risk prediction and integrated care management in the digital health era: innovations using machine learning and artificial intelligence approaches. Thromb Haemost 122(01):005–007
Guo J et al (2022) Aging and aging-related diseases: from molecular mechanisms to interventions and treatments. Signal Transduct Target Ther 7(1):391
Guo T, Steen JA, Mann M (2025) Mass-spectrometry-based proteomics: from single cells to clinical applications. Nature 638(8052):901–911
Hofer SJ et al (2022) The ups and downs of caloric restriction and fasting: from molecular effects to clinical application. EMBO Mol Med 14(1):e14418
Holtze S et al (2021) Alternative animal models of aging research. Front Mol Biosci 8:660959
Hong J (2024) A high-throughput screening platform for longevity gene and Anti-aging drug discovery. University of California, San Francisco
Hu X, Guo F (2021) Amino acid sensing in metabolic homeostasis and health. Endocr Rev 42(1):56–76
Hu X et al (2024) Deep learning algorithm-enabled sediment characterization techniques to determination of water saturation for tight gas carbonate reservoirs in Bohai Bay basin, China. Sci Rep 14(1):12179
Huchzermeyer B et al (2022) Metabolic pathway of natural antioxidants, antioxidant enzymes and ROS providence. Antioxidants 11(4):761
Ilyin N, Nikitin V, Kalueff A (2024) The role of the integrated stress response (ISR) in neuropsychiatric disorders. J Evol Biochem Physiol 60(6):2215–2240
Izankar SV, Kumar P, Waghmare G (2024) Evolution of artificial intelligence in biotechnology: from discovery to ethical and beyond. In: AIP conference proceedings. AIP Publishing
Jamerson LE, Bradshaw PC (2024) The roles of white adipose tissue and liver NADPH in dietary restriction-induced longevity. Antioxidants 13(7):820
Juan CA et al (2021) The chemistry of reactive oxygen species (ROS) revisited: outlining their role in biological macromolecules (DNA, lipids and proteins) and induced pathologies. Int J Mol Sci 22(9):4642
Kandpal M et al (2022) Dysbiosis of gut microbiota from the perspective of the gut–brain axis: role in the provocation of neurological disorders. Meta 12(11):1064
Kasar GN et al (2025) Proteostasis in aging: mechanistic insights and therapeutic opportunities. Aging Pathobiol Ther 7(1):25–43
Kassis A et al (2023) Nutritional and lifestyle management of the aging journey: a narrative review. Front Nutr 9:1087505
Kebbe M et al (2021) Beyond weight loss: current perspectives on the impact of calorie restriction on healthspan and lifespan. Expert Rev Endocrinol Metabol 16(3):95–108
Keeler LW, Bernstein MJ (2021) The future of aging in smart environments: four scenarios of the United States in 2050. Futures 133:102830

Khan A et al (2025) Genomic medicine and personalized treatment: a narrative review. Ann Med Surg:10.1097

Kim SK et al (2021) Discovering signaling mechanisms governing metabolism and metabolic diseases with Drosophila. Cell Metab 33(7):1279–1292

Kitada M et al (2021) Effect of methionine restriction on aging: its relationship to oxidative stress. Biomedicine 9(2):130

Kma L, Baruah TJ (2022) The interplay of ROS and the PI3K/Akt pathway in autophagy regulation. Biotechnol Appl Biochem 69(1):248–264

Knopp MI et al (2023) AI-enabled medical education: threads of change, promising futures, and risky realities across four potential future worlds. JMIR Med Educ 9:e50373

Kong S, Cai B, Nie Q (2022) PGC-1α affects skeletal muscle and adipose tissue development by regulating mitochondrial biogenesis. Mol Gen Genomics 297(3):621–633

Kopp W (2024) Aging and "age-related" diseases-what is the relation? Aging Dis 16(3):1316

Kroemer G et al (2025) From geroscience to precision geromedicine: understanding and managing aging. Cell 188(8):2043–2062

Kumar RN et al (2023) Unveiling the cardioprotective power: liquid chromatography–mass spectrometry (LC–MS)-analyzed Neolamarckia cadamba (Roxb.) Bosser leaf ethanolic extract against myocardial infarction in rats and in silico support analysis. Plants 12(21):3722

Lagoumintzis G, Patrinos GP (2023) Triangulating nutrigenomics, metabolomics and microbiomics toward personalized nutrition and healthy living. Hum Genomics 17(1):109

Lao B-N et al (2023) Time-restricted feeding's effect on overweight and obese patients with chronic kidney disease stages 3-4: a prospective non-randomized control pilot study. Front Endocrinol 14:1096093

Laskovs M (2024) Investigating the metabolic effects of RAS/MAPK pathway inhibition during ageing. Aston University

Li W et al (2023) The making of responsible innovation and technology: an overview and framework. Smart Cities 6(4):1996–2034

Liang Y et al (2024) AI nutritionist: intelligent software as the next generation pioneer of precision nutrition. Comput Biol Med:108711

Lin PW, Chu ML, Liu HS (2021) Autophagy and metabolism. Kaohsiung J Med Sci 37(1):12–19

Livelo C et al (2023) Time-restricted feeding promotes muscle function through purine cycle and AMPK signaling in drosophila obesity models. Nat Commun 14(1):949

Mannick JB, Lamming DW (2023) Targeting the biology of aging with mTOR inhibitors. Nature Aging 3(6):642–660

Marabita F et al (2022) Multiomics and digital monitoring during lifestyle changes reveal independent dimensions of human biology and health. Cell systems 13(3):241–255. e7

Marafie SK, Al-Mulla F, Abubaker J (2024) mTOR: its critical role in metabolic diseases, cancer, and the aging process. Int J Mol Sci 25(11):6141

Marchev AS et al (2021) Metabolomics and health: from nutritional crops and plant-based pharmaceuticals to profiling of human biofluids. Cell Mol Life Sci 78(19–20):6487–6503

Megas S et al (2024) Spatial transcriptomics of the respiratory system. Annu Rev Physiol 87

Mirisola MG, Longo VD (2022) Yeast chronological lifespan: longevity regulatory genes and mechanisms. Cells 11(10):1714

Miwa S et al (2022) Mitochondrial dysfunction in cell senescence and aging. J Clin Invest 132(13)

Mohammed I et al (2021) A critical review of the evidence that metformin is a putative anti-aging drug that enhances healthspan and extends lifespan. Front Endocrinol 12:718942

Moldakozhayev A, Gladyshev VN (2023) Metabolism, homeostasis, and aging. Trends Endocrinol Metabol 34(3):158–169

Morita S et al (2024) Diets and leisure activities are associated with curiosity. PLoS One 19(12):e0314384

Mozaffarian D et al (2022) A food is medicine approach to achieve nutrition security and improve health. Nat Med 28(11):2238–2240

Narayan S (2024) Leveraging a systems Geroscience approach to identify combinatorial drug therapies for Alzheimer's disease. Albert Einstein College of Medicine

Nasar S et al (2024) OMICS-based knowledge for achieving food and nutritional security. In: OMICs-based techniques for global food security, pp 67–90

Nur S (2024) The role of digital health technologies and sensors in revolutionizing wearable health monitoring systems. Int J Innov Res Comput Sci Technol 12(6):69–80

Obata F, Miura M (2024) Regulatory mechanisms of aging through the nutritional and metabolic control of amino acid Signaling in model organisms. Annu Rev Genet 58

Ogunjobi TT et al (2024) Bioinformatics applications in chronic diseases: a comprehensive review of genomic, transcriptomics, proteomic, metabolomics, and machine learning approaches. Medinfo

Ottens F, Franz A, Hoppe T (2021) Build-UPS and break-downs: metabolism impacts on proteostasis and aging. Cell Death Differ 28(2):505–521

Oudmaijer C et al (2024) Clinical implications of nutritional interventions reducing calories, a systematic scoping review. Clin Nutr ESPEN

Pal A, Goel F, Garg VK (2025a) From genetics to function: the role of ABCA12 in autism neurobiology. J Mol Neurosci 75(2):1–13

Pal A, Goel F, Garg VK (2025b) Zebrafish as a tool for autism research: unraveling the roles of Shank3, Cntnap2, Neuroligin3, and Arid1b in synaptic and behavioral abnormalities. Neurogenetics 26(1):1–17

Park J-M, Lee D-H, Kim D-H (2023) Redefining the role of AMPK in autophagy and the energy stress response. Nat Commun 14(1):2994

Patoine G et al (2022) Drivers and trends of global soil microbial carbon over two decades. Nat Commun 13(1):4195

Penugurti V et al (2024) AMPK: the energy sensor at the crossroads of aging and cancer. In: Seminars in cancer biology. Elsevier

Pimenta C et al (2021) Advanced therapies and regulatory framework in different areas of the globe: past, present, and future. Clin Ther 43(5):e103–e138

Pokushalov E et al (2024) Biomarker-guided dietary supplementation: a narrative review of precision in personalized nutrition. Nutrients 16(23):4033

Pontzer H, McGrosky A (2022) Balancing growth, reproduction, maintenance, and activity in evolved energy economies. Curr Biol 32(12):R709–R719

Prasad P et al (2024) Fostering healthy longevity through regenerative and precision medicine: biodiversity's vital role and equitable benefit sharing. In: Biodiversity and business. Springer, pp 563–594

Rai SN et al (2021a) Promising drug targets and associated therapeutic interventions in Parkinson's disease. Neural Regen Res 16(9):1730–1739

Rai SN et al (2021b) Therapeutic potential of vital transcription factors in Alzheimer's and Parkinson's disease with particular emphasis on transcription factor EB-mediated autophagy. Front Neurosci 15:777347

Rai SN et al (2022) Exploring the paradox of COVID-19 in neurological complications with emphasis on Parkinson's and Alzheimer's disease. Oxidative Med Cell Longev 2022(1):3012778

Rai SN et al (2025) Autophagy modulating potential of Mucuna pruriens in Parkinson's disease. Curr Tradit Med

Raman R et al (2025) Navigating artificial general intelligence development: societal, technological, ethical, and brain-inspired pathways. Sci Rep 15(1):1–22

Ramasubbu K, Devi Rajeswari V (2023) Impairment of insulin signaling pathway PI3K/Akt/mTOR and insulin resistance induced AGEs on diabetes mellitus and neurodegenerative diseases: a perspective review. Mol Cell Biochem 478(6):1307–1324

Ramatchandirin B, Pearah A, He L (2023) Regulation of liver glucose and lipid metabolism by transcriptional factors and coactivators. Life 13(2):515

Raubenheimer D et al (2022) An integrative approach to dietary balance across the life course. Iscience 25(5)

Reitz CJ, Kuzmanov U, Gramolini AO (2023) Multi-omic analyses and network biology in cardiovascular

Rezaei MS, Reynoso-Sánchez LF (2024) Nutrigenomics and personalized nutrition in the context of aging
Rodriguez S et al (2021) Machine learning identifies candidates for drug repurposing in Alzheimer's disease. Nat Commun 12(1):1033
Ruddick-Collins LC et al (2022) Timing of daily calorie loading affects appetite and hunger responses without changes in energy metabolism in healthy subjects with obesity. Cell Metab 34(10):1472–1485. e6
Sanches PHG et al (2024) Integrating molecular perspectives: strategies for comprehensive multi-omics integrative data analysis and machine learning applications in transcriptomics, proteomics, and metabolomics. Biology 13(11):848
Sanchez-Garrido J, Shenoy AR (2021) Regulation and repurposing of nutrient sensing and autophagy in innate immunity. Autophagy 17(7):1571–1591
Santos L et al (2023) Sirtuins: the NAD+-dependent multifaceted modulators of inflammation. Antioxid Redox Signal 39(16):1185–1208
Sarkar S et al (2024) Clinical advances in analytical profiling of signature lipids: implications for severe non-communicable and neurodegenerative diseases. Metabolomics 20(2):37
Saxon SV et al (2021) Physical change and aging: a guide for helping professions. Springer Publishing Company
Secomandi L et al (2022) The role of cellular senescence in female reproductive aging and the potential for senotherapeutic interventions. Hum Reprod Update 28(2):172–189
Shumayla et al (1927) The promise of resurrection plants in enhancing crop tolerance to water scarcity. Philos Trans B 2025(380):20240231
Singh S, Rai SN, Singh SK (2024a) Synaptic plasticity in neurodegenerative disorders. CRC Press
Singh VK et al (2024b) Precision nutrition-based strategy for management of human diseases and healthy aging: current progress and challenges forward. Front Nutr 11:1427608
Siopis G, Porter J (2022) Contribution of biological age–predictive biomarkers to nutrition research: a systematic review of the current evidence and implications for future research and clinical practice. Adv Nutr 13(5):1930–1946
Soanes DM et al (2008) Comparative genome analysis of filamentous fungi reveals gene family expansions associated with fungal pathogenesis. PLoS One 3(6):e2300
Soliman GA (2022) Intermittent fasting and time-restricted eating role in dietary interventions and precision nutrition. Front Public Health 10:1017254
Song Y, Wu Z, Zhao P (2022) The protective effects of activating Sirt1/NF-κB pathway for neurological disorders. Rev Neurosci 33(4):427–438
Sorrenti V et al (2022) Immunomodulatory and antiaging mechanisms of resveratrol, rapamycin, and metformin: focus on mTOR and AMPK signaling networks. Pharmaceuticals 15(8):912
Stroud JT, Ratcliff WC (2025) Long-term studies provide unique insights into evolution. Nature 639(8055):589–601
Szwed A, Kim E, Jacinto E (2021) Regulation and metabolic functions of mTORC1 and mTORC2. Physiol Rev 101(3):1371–1426
Tabibzadeh S (2021) Signaling pathways and effectors of aging. Growth 3:53
Tanaka M (2025) From serendipity to precision: integrating AI, multi-omics, and human-specific models for personalized neuropsychiatric care. Biomedicine 13(1):167
Theodore Armand TP, Kim H-C, Kim J-I (2024) Digital anti-aging healthcare: an overview of the applications of digital Technologies in Diet Management. J Personal Med 14(3):254
Tippairote T et al (2025) The metabolic cost of resilience: bioenergetic trade-offs in stress adaptation, aging, and chronic disease
Tomtheelnganbee E, Sah P, Sharma R (2022) Mitochondrial function and nutrient sensing pathways in ageing: enhancing longevity through dietary interventions. Biogerontology 23(6):657–680
Topol E (2025) Super agers: an evidence-based approach to longevity. Simon and Schuster
Uleman JF et al (2024) Triangulation for causal loop diagrams: constructing biopsychosocial models using group model building, literature review, and causal discovery. NPJ Complexity 1(1):19
Vashishat A et al (2024) Alternatives of animal models for biomedical research: a comprehensive review of modern approaches. Stem Cell Rev Rep 20(4):881–899

Vatashchuk MV et al (2022) Metabolic syndrome: lessons from rodent and drosophila models. Biomed Res Int 2022(1):5850507
Vega RA (2023) Nested ecologies: a multilayered ethnography of functional medicine. University of Texas Press
Wahl D, LaRocca TJ (2021) Transcriptomic effects of healthspan-promoting dietary interventions: current evidence and future directions. Front Nutr 8:712129
Wang Y et al (2022) Immunosenescence, aging and successful aging. Front Immunol 13:942796
Wang T et al (2024) From bench to bedside: a review of clinical trialsin drug discovery and development. arXiv preprint arXiv:2412.09378
Watroba M, Szukiewicz D (2021) Sirtuins at the service of healthy longevity. Front Physiol 12:724506
Watts AG et al (2022) The physiological control of eating: signals, neurons, and networks. Physiol Rev 102(2):689–813
Wu L et al (2021) Integrated multi-omics for novel aging biomarkers and antiaging targets. Biomolecules 12(1):39
Xu W et al (2023) Targeting AMPK signaling by polyphenols: a novel strategy for tackling aging. Food Funct 14(1):56–73
Yusri K et al (2024) Towards healthy longevity: comprehensive insights from molecular targets and biomarkers to biological clocks. Int J Mol Sci 25(12):6793
Zhang S et al (2022) The regulation, function, and role of lipophagy, a form of selective autophagy, in metabolic disorders. Cell Death Dis 13(2):132
Zhang K et al (2023) Metabolic diseases and healthy aging: identifying environmental and behavioral risk factors and promoting public health. Front Public Health 11:1253506
Zhao Y et al (2021) NAD+ improves cognitive function and reduces neuroinflammation by ameliorating mitochondrial damage and decreasing ROS production in chronic cerebral hypoperfusion models through Sirt1/PGC-1α pathway. J Neuroinflammation 18:1–16
Zhao Z et al (2024) Phytochemicals targeting Alzheimer's disease via the AMP-activated protein kinase pathway, effects, and mechanisms of action. Biomed Pharmacother 173:116373
Zhou Y, Hu G, Wang MC (2021) Host and microbiota metabolic signals in aging and longevity. Nat Chem Biol 17(10):1027–1036

The Gut-Brain Axis (GBA): Implications for Brain Longevity

9

Priyanka Gautam, Ranjeet Kumar Vishwakarma, and Abhishek Pathak

Abstract

The Gut-Brain Axis (GBA) is an evolving communication process using immunological, hormonal, metabolic, and neurological routes to connect our gut and brain. Central to this axis is the gut microbiota (GM), a diverse microbial community essential in shaping how our brain develops, functions, and ages. This chapter delves into how these gut microbes influence memory, mood, and mental clarity by producing key molecules like short-chain fatty acids (SCFA), neurotransmitter building blocks, and immune signals. When this microbial balance is disturbed, a state known as dysbiosis, it can contribute to conditions such as Alzheimer's disease (AD), amyotrophic lateral sclerosis (ALS), Parkinson's disease (PD), as well as anxiety and depression. The chapter also highlights how everyday choices like diet, exercise, sleep, and stress management can positively influence our gut microbes and, in turn, support brain health. Popular nutritional strategies, including the Mediterranean, ketogenic, and MIND diets, are explored alongside lifestyle practices that enhance resilience. It also examines emerging therapies such as prebiotics, probiotics, synbiotics, fecal microbiota, and transplantation (FMT) and advanced tools like Clustered Regularly Interspaced Short Palindromic Repeats (CRISPR) and artificial intelligence (AI) that offer exciting new ways to tailor treatments to individuals. Ultimately, by understanding and supporting the gut-brain connection, we may unlock powerful ways to maintain mental sharpness, emotional well-being, and cognitive vitality throughout life.

P. Gautam · A. Pathak (✉)
Department of Neurology, Institute of Medical Sciences, Banaras Hindu University, Varanasi, India

R. K. Vishwakarma
Department of Physiology, Institute of Medical Sciences, Banaras Hindu University, Varanasi, India

A. Kumar Singh, S. Nand Rai (eds.), *Rejuvenating the Brain: Nutraceuticals, Autophagy, and Longevity*, Nutritional Neurosciences,
https://doi.org/10.1007/978-981-95-2790-8_9

Keywords

Gut-brain axis (GBA) · Gut microbiota (GM) · Dysbiosis · Neurodegenerative diseases · Emerging therapies

Learning Objectives

- Explain the gut-brain axis and its communication via neural, hormonal, immune, and microbial pathways.
- Describe how gut microbiota regulates neurochemicals like serotonin and GABA and their impact on mood and memory.
- Discuss how lifestyle factors (diet, exercise, sleep) influence the gut microbiota and brain aging.
- Identify emerging microbiome-based therapies including CRISPR, AI, and personalized probiotics.

9.1 Introduction

9.1.1 Overview of the Gut-Brain Axis (GBA)

Microorganisms have been integral to human life since its origins. Over millions of years, humans and microbes have coevolved to form a mutualistic relationship, where the body offers shelter and nutrients to these microbes, and in return, the microbiota actively supports the host's health during both stable and diseased states (Caballero-Flores et al. 2023). The most diversified microbial community in the human body is found in the gut. It is a complex ecology with viruses, fungus, bacteria, and even some archaea that coexist in delicate balance (Hoegenauer et al. 2022). The human gut is home to about 2000 different kinds of bacteria, and the gut microbiota (GM) is thought to have almost 150 times the number of genes found in the human genome (Qin et al. 2010). The GM can have an impact on almost every aspect of human biology, particularly development, longevity, disease, and health maintenance, thanks to its extensive genetic and metabolic capabilities (Lynch and Pedersen 2016; Loh et al. 2024).

The immune system is significantly shaped by the gut microbiota since birth. Microbial colonization starts soon after birth and allows babies to start producing vital vitamins and amino acids by the time they are 4 months old (Loh et al. 2024; Donald and Finlay 2023; Bäckhed et al. 2015). The microbiome usually stabilizes into an adult-like condition by the time a child is 3–6 years old, and this state lasts throughout adulthood (Bäckhed et al. 2015; Yatsunenko et al. 2012; Roswall et al. 2021). Key functions of the adult microbiota include dietary nutrient extraction (Jumpertz et al. 2011), immune and autoimmune regulation (Qin et al. 2010; Schluter et al. 2020), maintenance of intestinal barrier (Ghosh et al. 2021), cholesterol metabolism and bile acids (BAs) metabolism (Yao et al. 2020; Collins et al.

2023), production of antimicrobial peptides (Ma et al. 2022), and drug metabolism (Collins et al. 2023; Zimmermann et al. 2019). According to recent research, the human GM significantly influences the plasma metabolome and may even be more important than genetics (Collins et al. 2023; Diener et al. 2022). Dysbiosis has been identified as one of the 12 revised signs of aging, which is of interest and highlights the significance of the microbiome (López-Otín et al. 2023).

Over the last 30 years, growing scientific evidence has unveiled a remarkable link between the GI tract and the brain, an intricate communication network now widely known as Gut-Brain Axis (GBA). This bidirectional communication system is heavily influenced by the GM and plays a crucial role in shaping brain function and development (Hattori and Yamashiro 2021). Diet, in particular, affects microbial composition and metabolic output, which in turn impact the brain through neural, immune, and hormonal pathways (Deverman and Patterson 2009).

The GBA is structured around both central and peripheral components. The brain, including the cerebral cortex, limbic system, and hypothalamic-pituitary axis (HPA), forms the central hub. Particularly, the limbic system uses information from areas such as the hippocampus to evaluate behavioral and emotional cues (Perlman 2002). The autonomic, enteric, and sympathetic nervous systems are the means by which peripheral communication takes place (Udit and Gautron 2013; Forsythe et al. 2014). The brain is connected to the intestine wall's enteric nervous system (ENS) by the vagus nerve, dorsal root ganglia, and nodose ganglia (Udit and Gautron 2013). Other pathways such as the HPA and sympathetic system, are also integral to this network (Hattori and Yamashiro 2021).

9.1.2 Historical Perspectives on the Gut-Brain Connection

Ayurveda (Lewandowska-Pietruszka et al. 2022) and Traditional Chinese Medicine are two ancient medicinal traditions that consistently emphasize the significance of gut health in preserving mental and physical well-being, proving that the gut and brain are interconnected (Steer 2019; Lewandowska-Pietruszka et al. 2022). In the nineteenth century, the anatomical discovery of the vagus nerve strengthened the insights of physiological links between the gut and brain. Yet, it was not until the late twentieth and early twenty-first centuries, fueled by advances in microbiome research, that scientist truly began to appreciate how gut microbes influence brain chemistry and behavior.

Compelling evidence emerged from germ-free animal studies demonstrating behavioral and neurochemical alterations in the absence of gut microbiota (Udit and Gautron 2013; Lewandowska-Pietruszka et al. 2022; Parrish et al. 2008). Though the gut and brain are physically separate, multiple communication routes bridge the two, including immune modulation, vagus nerve signaling, ENS activity, neuroendocrine responses, and the circulatory delivery of microbial metabolites and hormones (Morais et al. 2021; Liu et al. 2022). Dopamine, gamma-aminobutyric acid (GABA), serotonin (Yano et al. 2015), and other important neurotransmitters are known to be produced by or influenced by gut microorganisms (Loh et al. 2024; Yano et al. 2015; Strandwitz et al. 2019).

While early work in this area often relied on simplified animal models, recent advances in sequencing and culture-independent techniques have allowed scientists to move beyond correlation and explore the underlying mechanisms of microbe–host interactions (Yano et al. 2015). Both preclinical and clinical research now support the role of GM in modulating behaviors related to social interaction, mood, motivation, and physical performance (Strandwitz et al. 2019; Fontana et al. 2023).

9.1.3 Gut-Brain Harmony: A Foundation for Healthy Aging

As the global population continues to age, the need to preserve brain function and promote cognitive longevity has never been more pressing. Among the most fascinating and quickly developing fields of study tackling this problem is the GBA (Fung et al. 2017). Recent findings suggest that keeping the gut microbiome balanced and healthy may help protect the brain from age-related memory and thinking problems. It does so by reducing systemic inflammation and supporting neuroplasticity which is the brain's extraordinary capacity to rearrange and adapt throughout life (Foster and McVey Neufeld 2013).

On the other hand, disturbances in the gut microbiome, known as dysbiosis, have been linked to a range of neurodegenerative diseases, including AD and PD, as well as mental health disorders such as depression and anxiety (Loh et al. 2024; Foster and McVey Neufeld 2013; Ferreiro et al. 2023; Huang et al. 2023). These associations underscore the significance of the GBA as a critical interface between gut health and brain function.

Encouragingly, interventions targeting the GBA, through dietary strategies, lifestyle changes, and precision therapeutics, have shown potential to enhance brain health and preserve cognitive function in old age. Animal model studies further reinforce this potential, revealing that alterations in the GM can directly impact the development of neurodegenerative diseases, largely by modulating microglial activity (Seo et al. 2023). Since microglial activation and chronic neuroinflammation are key features in many neurodegenerative conditions, the microbiota-GBA axis offers a promising, actionable target for slowing or preventing disease progression (Cook and Prinz 2022).

This chapter delves into the complex biological mechanisms that underpin the GBA, its pivotal role in promoting brain longevity, and emerging interventions that aim to leverage this powerful connection for healthier aging.

9.2 Understanding the GBA

9.2.1 Anatomy and Physiology of the GBA

The GBA is a sophisticated communication network that integrates the central nervous system (CNS), ENS, autonomic nervous system (ANS), HPA axis, the immune system, and GM. These interconnected systems synchronize digestive, immune, metabolic, and emotional processes.

9.2.1.1 The ENS: The "Second Brain"

The ENS, sometimes referred to as the "second brain," is made up of 200–600 million neurons arranged in a network of glial cells and ganglia that are dispersed throughout the GI system, from the anal sphincter to the esophagus (Annahazi and Schemann 2020). It also connects with accessory digestive organs like the liver, pancreas, gallbladder, and biliary tract. The ENS is made up of two key plexuses, the myenteric and submucosal, which regulate muscle movement, epithelial function, blood flow, and immune interactions. Originating mostly from vagal neural crest cells, the ENS also draws aids from sacral neural crest cells and Schwann cell precursors (Annahazi and Schemann 2020; Stamp 2017).

Although the ENS operates semi-independently, it maintains close communication with the CNS via the GBA (Furness and Stebbing 2018; Mayer 2011). Parasympathetic nerves promote motility and secretion, while sympathetic nerves inhibit these functions. The ENS serves as a critical interface for both sensory (afferent) and motor (efferent) signals. Intestinofugal neurons further extend ENS influence by linking it with sympathetic ganglia and other organs (Furness and Stebbing 2018). Hormonal pathways, such as the HPA and sympatho-adrenal axes, allow the CNS to regulate ENS function. Enteroendocrine cells (EECs), which come in over 20 types, act as sensors that communicate with the CNS through hormones and nerve signals, helping control digestion and appetite (Annahazi and Schemann 2020).

The ENS is essential for GI regulation and even impacts CNS activity. Its absence, as seen in conditions like Hirschsprung's disease, is fatal despite intact external innervation. The myenteric plexus oversees motility, while the submucosal plexus manages secretion and absorption. ENS neurons are multifunctional and share neurotransmitter systems with the CNS, though glycine acts excitatorily in the ENS. Notably, emerging studies suggest that the ENS may possess rudimentary forms of learning and memory, with implications for understanding gut-based disorders (Furness et al. 2014; Schemann et al. 2020).

9.2.1.2 The Vagus Nerve: A Critical Communication Pathway

The vagus nerve, the tenth cranial nerve, plays a central role in maintaining brain-body communication. It regulates vital autonomic processes such as digestion, respiration, and heart rate (Ma et al. 2025). The longest cranial nerve ranges from the brainstem to the abdomen, innervating organs such as the heart, lungs, liver, pancreas, and GI tract. It is a mixed nerve, with 20% efferent and 80% afferent fibers, forming a key part of the parasympathetic nervous system within the ANS, which maintains homeostasis alongside the sympathetic nervous system. Spanning from the brainstem to the abdomen, the vagus nerve innervates organs like the heart, lungs, liver, pancreas, and GI tract. Approximately 80% of its fibers are afferent, transmitting sensory information, such as gut distension, nutrients, and hormones like cholecystokinin (CCK), ghrelin, and leptin, from the gut to the nucleus tractus solitarius (NTS) in the brainstem. The remaining 20% are efferent fibers, which regulate autonomic functions, including digestive motility, immune modulation, and inflammatory responses (Ma et al. 2025; Oakes et al. 2016).

Through the cholinergic anti-inflammatory pathway, the vagus nerve influences immune cells such as macrophages, thereby modulating the release of cytokines during gut stress or infection (Hattori and Yamashiro 2021; Parrish et al. 2008). For example, microbial components like lipopolysaccharides (LPS) can breach a compromised intestinal barrier, enter systemic circulation, and trigger immune activation through pathogen-recognition receptors, contributing to systemic inflammation and neuroimmune signaling (Forsythe et al. 2014).

Additionally, vagal signaling influences the HPA axis, modulating the release of cortisol and neuropeptides that impact stress, mood, and inflammation. Specialized gut sensory cells called neuropods have recently been identified as rapid communicators with the brain via vagal circuits. These findings position the vagus nerve as a central mediator of gut-derived neuroactive signals, systemic immune responses, and cognitive and emotional regulation, reinforcing its therapeutic relevance in disorders of the brain and body (Liu and Zhu 2018; Mörkl et al. 2023; Colzato et al. 2017).

9.2.1.3 Role of the GM in GBA Signaling

The GM plays a pivotal role in GBA communication, influencing both local intestinal functions and central processes through metabolic, neuroendocrine, and immune pathways (Carabotti et al. 2015). One of the earliest clues to this interaction came from observations that oral antibiotics improved hepatic encephalopathy symptoms (Mayer et al. 2014).

Apart from being linked to mental health issues like anxiety and depression, dysbiosis may also be a factor in neurodevelopmental disorders like autism (Simrén et al. 2013). It also contributes to functional GI disorders like irritable bowel syndrome (IBS), which is denoted by altered motility, hypersensitivity, and immune dysfunction. Microbiota instability, post-infectious IBS, and small intestinal bacterial overgrowth are common features. Both probiotics and non-systemic antibiotics have shown promise in alleviating symptoms (Mayer and Tillisch 2011). Notably, microbiota from IBS patients can induce hypersensitivity in germ-free rats, reinforcing the idea that IBS is a disorder deeply rooted in the microbiome-GBA interface (Simrén et al. 2013; Mayer and Tillisch 2011).

9.2.2 Key Mechanisms of Communication

The gut and brain maintain a complex and continuous conversation via neural, endocrine, immune, and microbial signaling pathways.

9.2.2.1 Neural Pathways: Vagus Nerve and Spinal Cord

The GBA links emotional and cognitive centers of the brain with the ENS and gut physiology. The vagus nerve is particularly critical, with the majority of its fibers relaying gut-derived signals back to the brain. It also plays a regulatory role in the HPA axis, which governs stress responses via hormones like Corticotropin-releasing factor (CRF), Adrenocorticotropic Hormone (ACTH), and cortisol (Loh et al. 2024; Carabotti et al. 2015; Howland 2014). The vagus nerve and HPA axis coordinate the brain's influence on intestinal immune cells, neurons, smooth muscle cells, and enterochromaffin cells, all of which interact with GM. Studies suggest that microbiota affect anxiety and depression-like behaviors, as germ-free animals show exaggerated stress responses with increased ACTH and cortisol levels (Pariante and Lightman 2008).

Sensory endings of the vagus nerve detect signals such as gut stretch, nutrients, and hormones including CCK, ghrelin, and leptin, relaying these to the brain (Badman and Flier 2005). These hormones control satiety, digestion, and regulating energy by influencing the brainstem's solitary tract nucleus.

- *CCK* is a gut hormone and neurotransmitter that suppresses appetite and slows gastric emptying (Sayegh and Ritter 2000). It also influences brain regions such as the cortex, hypothalamus, and brainstem. Its effects are largely mediated through vagal pathways, as shown by reduced neuronal activation following vagotomy in animal models (Little et al. 2005; Klok et al. 2007). Elevated CCK levels have been associated with anxiety, making it a useful model for studying anxiety-related disorders.
- *Ghrelin*, a stomach-derived hormone, stimulates appetite by suppressing vagal afferent activity. Its levels increase during a fast and decrease following a meal. In both rats and humans, ghrelin increases food intake, effects nullified by vagotomy in animals, emphasizing its role in hunger regulation and links to obesity and insulin resistance (Klok et al. 2007).
- *Leptin*, a key metabolic hormone, signals satiety by acting on vagal afferents and working alongside CCK to suppress appetite. However, in obesity, leptin resistance diminishes this regulatory effect. The gut also serves as a sensory organ, where GPCRs, including bitter taste receptors, modulate appetite by promoting CCK release and altering ghrelin levels, thereby influencing vagal signaling and hunger (Badman and Flier 2005; Tanaka et al. 2008).

9.2.2.2 Endocrine Pathways: Hormones and Gut-Derived Peptides

Enteroendocrine cells (EECs) in the GI lining respond to mechanical and nutritional stimuli by releasing a diverse array of hormones. These hormones (Table 9.1) influence digestion, glucose regulation, and brain function (Annahazi and Schemann 2020; Habib et al. 2012). They act either locally or systemically, affecting organs like the pancreas, liver, and brain (Klok et al. 2007; Gribble and Reimann 2019; Beumer et al. 2018).

9.2.2.3 Key Gut Hormones Include

- *Serotonin* - regulates intestinal motility and appetite.
- *GLP-1 and GIP* - enhance insulin secretion and promote satiety.
- *Ghrelin* - stimulates hunger and gastric motility.
- *PYY and CCK* - reduce appetite and slow digestion.

Table 9.1 Key gut-derived hormones

Hormone	Producing cell type	Major sites of secretion	Physiological stimuli for secretion	Main physiological effects
Serotonin	Enterochromaffin cells	Duodenum, jejunum, ileum, colon, rectum	Luminal nutrients; mechanical and irritant stimulation of the intestine	Gastrointestinal: Regulates intestinal motility and fluid secretion; visceral pain perception Adipose tissue: Fat volume regulation Brain: Appetite control
Somatostatin	D cells	Gastric fundus, duodenum	Nutrient detection by D cells; feedback from other gut hormones	Gastrointestinal: Inhibits other gut hormones; reduces stomach acid secretion
Motilin	M cells	Duodenum, jejunum	Low luminal pH; presence of fat and bile acids	Gastrointestinal: Enhances gut motility
Growth differentiation factor 15 (GDF15)	Enterocytes	Colon, small intestine	Cell stress pathways	Brain: Appetite regulation; implicated in nausea from chemotherapy, cancer cachexia, and hyperemesis gravidarum
Ghrelin	X cells	Gastric fundus	Fasting	Brain: Stimulates appetite Pituitary: Triggers growth hormone release Gastrointestinal: Increases gastric emptying and motility
Glucagon-like peptide 1 (GLP-1)	L cells	Jejunum, ileum, colon, rectum	Nutrients (carbohydrates, proteins, fats); bile acids	Pancreas: Stimulates insulin, inhibits glucagon Brain: Promotes satiety Gastrointestinal: Slows gastric emptying
Glucose-dependent insulinotropic peptide (GIP)	K cells	Duodenum, jejunum	Nutrients (carbohydrates, proteins, fats)	Pancreas: Enhances insulin and glucagon secretion Adipose tissue: Regulates lipid metabolism Brain: Promotes satiety GI tract: Slows motility
Insulin-like peptide 5 (INSL5)	L cells	Colon, rectum	Unknown (co-secreted with GLP-1 and PYY)	Brain: Stimulates appetite Gastrointestinal: Increases colonic propulsion
Peptide YY (PYY)	L cells	Ileum, colon, rectum	Nutrients (carbohydrates, proteins, fats); bile acids	Brain: Induces satiety Gastrointestinal: Slows gastric emptying and intestinal transit

Gastrin	G cells	Gastric antrum	Protein presence; gastric distension	Gastrointestinal: Increases gastric acid secretion and motility
Fibroblast growth factor 19 (*FGF15 in mice*)	Enterocytes	Terminal ileum	Bile acids	Liver: Suppresses bile acid synthesis; promotes glycogen storage; inhibits gluconeogenesis
Guanylin and uroguanylin	Enterocytes, goblet, Paneth cells	Small intestine, colon	Uncertain; possibly neurotransmitters and inflammatory mediators	Gastrointestinal: Regulates chloride and fluid secretion by epithelial cells
Cholecystokinin (CCK)	I cell	Duodenum, jejunum	Nutrients (fats, proteins)	Brain: Induces satiety Gastrointestinal: Slows gastric emptying; stimulates gallbladder contraction and pancreatic enzyme release
Secretin	S cells	Duodenum, jejunum	Low luminal pH; protein content	Gastrointestinal: Reduces gastric acid secretion and motility; increases pancreatic bicarbonate production
Neurotensin	N cells	Jejunum, ileum, proximal colon	Lipid presence	Gastrointestinal: Facilitates lipid absorption Brain: Regulates appetite

Source: Stimuli, and Physiological Functions Across the Gut-Brain Axis
This table summarizes major hormones produced in the gastrointestinal (GI) tract, detailing their cellular origins, primary secretion sites, physiological triggers, and systemic effects on gastrointestinal function, metabolism, and brain signaling. It spotlights the integrative role of gut-derived hormones in controlling appetite, digestion, nutrient sensing, motility, and inter-organ communication within the gut-brain axis (GBA).

9.2.2.4 Immune Pathways: Cytokines and Inflammation

Cytokines such as IL-1β, IL-6, and Tumor Necrosis Factor (TNF-α) play central roles in inflammation and pain.

- IL-1β, released by immune and sensory cells during injury or inflammation, heightens pain sensitivity by boosting substance P and prostaglandin E2 levels. Blocking its receptor can help reduce pain (Sweitzer et al. 2001; Copray et al. 2001; Zhang and An 2007).
- IL-6 plays a key role in nerve repair and neuronal activation. While elevated IL-6 levels are linked to increased pain sensitivity, blocking its receptors can hinder nerve recovery (Zhang and An 2007).
- TNF-α promotes inflammatory and neuropathic pain via Tumor Necrosis Factor Receptor (TNFR1 and TNFR2) by activating Necrosis Factor (NF-κB) and stress signaling pathways. Its inhibition alleviates hyperalgesia and lowers IL-1β levels, whereas its administration triggers Wallerian degeneration and pain-related behaviors (Zhang and An 2007; Boka et al. 1994; Wagner and Myers 1996).

9.2.2.5 Microbial Metabolites: SCFAs and Neurotransmitters

GM produces short-chain fatty acids (SCFAs) during the fermentation of dietary fiber, primarily acetate, propionate, and butyrate (Miller and Wolin 1996; Luu et al. 2019; Gautam et al. 2025a). They are absorbed by monocarboxylate and sodium-dependent transporters, used by colonocytes, or transported to the liver, with only minimal amounts reaching systemic circulation (Miller and Wolin 1996; Cummings et al. 1987).

SCFAs enhance gut barrier function, stimulate mucus production, and suppress inflammation. They act through G-protein coupled receptors (GPCR) free fatty acid receptor (FFAR2, FFAR3, GPR109a) and serve as histone deacetylase (HDAC) inhibitors, altering gene expression in immune and neural cells (Lewis et al. 2010; Mohajeri et al. 2018; Stilling et al. 2016).

Butyrate enhances immune regulation by promoting T regulatory (Treg) cells and suppressing inflammation, with therapeutic relevance for conditions like inflammatory bowel disease (IBD) (Zhang and An 2007). In addition, SCFAs influence metabolism by modulating energy balance, liver function, appetite, and sleep (Szentirmai et al. 2019). While their impact on brain health is increasingly recognized, the precise neurological mechanisms remain under active investigation.

9.2.3 The Role of the Gut Microbiome

9.2.3.1 Composition and Diversity of GM

The human gut hosts trillions of microbes, outnumbering human cells by at least 10 to 1. These include bacteria, viruses, fungi, and archaea. Among them, *Firmicutes* and *Bacteroidetes* dominate, while *Proteobacteria, Actinobacteria,* and others are present in smaller proportions (Qin et al. 2010; Eckburg et al. 2005).

Gut microbial diversity is shaped by factors like genetics, diet, age, antibiotic use, and geography (Eckburg et al. 2005). Despite individual differences, certain enterotypes reflect shared microbial patterns (Arumugam et al. 2011). The gut microbiome's adaptability makes it a vital player in health maintenance and disease risk.

9.2.3.2 Metabolites and Brain Function

Microbes produce a range of metabolites, especially SCFAs and tryptophan derivatives, that influence inflammation, gut motility, and blood-brain barrier (BBB) integrity (Miller and Wolin 1996; Leyrolle et al. 2023; Gao et al. 2020). Some bacteria can also synthesize neurotransmitters or their precursors, including GABA, dopamine, and noradrenaline, which affect mood, cognition, and behavior (Strandwitz et al. 2019).

9.2.3.3 Dysbiosis and Its Consequences for Brain Health

An imbalanced GM has been related with a range of neurological and psychiatric disorders, depression, anxiety, autism, Parkinson's disease (PD), and Alzheimer's disease (AD) among them (Góralczyk-Bińkowska et al. 2022; Gulas et al. 2018). Dysbiosis may increase gut permeability and inflammation, disrupt metabolite production, and trigger neuroimmune responses. In PD, for example, dysbiosis involves a decrease in beneficial bacteria and a rise in pathogenic species (Zhu et al. 2022). Correcting these imbalances could be key to developing microbiome-targeted therapies for brain health (Góralczyk-Bińkowska et al. 2022).

9.3 The GBA and Brain Health

The GBA is a vital communication network that helps maintain brain health by linking the gastrointestinal system with the central nervous system. When this intricate system is disrupted, it can contribute to a variety of neurological and psychiatric disorders, including neuroinflammation, neurotransmitter imbalances, mood disorders, and age-related cognitive decline.

9.3.1 Neuroinflammation and the GBA

A prevalent characteristic of neurodegenerative diseases is neuroinflammation, which is mainly caused by glial activation and the release of proinflammatory cytokines. Dysbiosis induced endotoxemia can activate the kynurenine pathway, producing neurotoxic metabolites like quinolinic acid that amplify inflammation in the brain.

9.3.1.1 Chronic Inflammation and Neurodegenerative Diseases

Conditions such as AD (Foster and McVey Neufeld 2013; Ferreiro et al. 2023), PD, and multiple sclerosis (MS) (Huang et al. 2023) are characterized by persistent low-grade inflammation. Activated microglia release cytokines like TNF-α, IL-1β, and IL-6, which can damage neurons, impair synapses, and contribute to cognitive decline (Parrish et al. 2008; Copray et al. 2001; Wagner and Myers 1996). Systemic inflammation, often originating in the gut, may weaken the BBB, allowing peripheral immune cells to infiltrate the brain and exacerbate neuroinflammation.

9.3.1.2 GM as a Regulator of Neuroinflammation

Gut microbes influence neuroinflammation by modulating gut permeability, immune responses, and metabolite production. Dysbiosis can result in increased permeability, allowing bacterial products such as LPS to enter circulation and activate immune pathways via toll-like receptors. This promotes cytokine release and systemic inflammation. In contrast, beneficial bacteria produce SCFAs, like butyrate, that reinforce the gut barrier and reduce inflammation by modulating microglial activity (Cummings et al. 1987; Oleskin and Shenderov 2016). The vagus nerve also contributes by transmitting anti-inflammatory signals. Approaches like probiotics, prebiotics, and dietary changes offer promising strategies to restore microbial balance and reduce neuroinflammation (Aoun et al. 2020).

9.3.2 Neurotransmitters and the Gut

The GM and the ENS play significant roles in the synthesis and modulation of neurotransmitters crucial for brain function and emotional regulation.

9.3.2.1 Microbial Production of Serotonin, Dopamine, and GABA

Gut microbes can influence neurotransmitter activity by producing neurotransmitters directly or modulating their precursors. While neurotransmitters themselves often cannot cross the BBB, amino acid precursors like tryptophan and tyrosine can, where they are converted into serotonin and dopamine, respectively (Yano et al. 2015; Strandwitz et al. 2019; Góralczyk-Bińkowska et al. 2022).

- *Dopamine*, the most abundant catecholamine neurotransmitter in the brain, plays a central role in regulating movement, mood, and reward pathways. It is predominantly generated by dopamine-releasing neurons in the ventral tegmental region and substantia nigra within the central nervous system. Dysregulation of the dopaminergic system is a hallmark of several neurological disorders, including PD and schizophrenia (McCutcheon et al. 2019). Dopamine is produced from the amino acid tyrosine, a dietary precursor that crosses the BBB to support central synthesis.

  Interestingly, dopamine is not limited to brain production. Over 50% of the body's dopamine is synthesized in the gastrointestinal tract, where it also serves key regulatory functions. In the gut, dopamine and its receptors are widely expressed and influence processes such as gastric secretion, intestinal motility, and mucosal blood flow (Zhu et al. 2022). Certain gut-residing bacteria, particularly *Staphylococcus* species, have been found to produce dopamine by converting L-3,4-dihydroxy-phenylalanine (L-DOPA), an intermediate form of tyrosine, into dopamine via the enzyme staphylococcal aromatic amino acid decarboxylase (SadA) (McCutcheon et al. 2019). These findings underscore the gut's significant contribution to systemic dopamine levels and emphasize the GM role in modulating neurochemical signaling both locally and systemically.

- *Serotonin*, a key neurotransmitter involved in regulating mood, cognition, and gastrointestinal function, is primarily synthesized in the CNS by serotonergic neurons located in the raphe nuclei. Disruptions in serotonin signaling have been strongly associated with mental health conditions such as depression and anxiety (Helton and Lohoff 2015).

  Surprisingly, only a small portion of the body's serotonin is produced in the brain. Nearly 90% is synthesized peripherally, primarily by enterochromaffin cells in the intestinal epithelium. Although serotonin itself cannot cross the BBB, its precursor, tryptophan, can enter the CNS, where it is converted into serotonin. In the gut, dietary tryptophan is absorbed and utilized by enterochromaffin cells for serotonin synthesis. This process is intricately regulated by gut microbes via the kynurenine synthesis pathway, which controls tryptophan availability and metabolism (Erickson et al. 2012; Bailey and Cryan 2017).

  Spore-forming gut bacteria, especially from the *Clostridia*, further influence serotonin levels by upregulating tryptophan hydroxylase 1 (TPH1), the rate-limiting enzyme in serotonin biosynthesis within the colon. Bacterial metabolites act as signaling molecules that stimulate this enzymatic activity (Yano et al. 2015). Additionally, certain strains of Staphylococcus have been shown to directly synthesize serotonin by expressing the enzyme aromatic amino acid decarboxylase (SadA), which converts 5-hydroxytryptophan (5-HTP) into serotonin.

- *GABA* is the central nervous system's main inhibitory neurotransmitter, and it is essential for controlling the excitability of neurons and numerous physiological and metabolic processes. In the brain, GABA is synthesized by GABAergic neurons through the enzymatic conversion of glutamate, catalyzed by glutamic acid decarboxylase (Strandwitz et al. 2019). Beyond the CNS, intriguing discoveries have revealed that certain gut bacteria are also capable of producing GABA. This was first observed unexpectedly during a co-culture study, which isolated a member of the *Ruminococcaceae* family, strain KLE1738, from the human gut. This strain was found to rely on Bacteroides fragilis for growth, which produced GABA as its primary metabolic output and growth factor (Lee et al. 2019). Subsequent studies confirmed that other genera, including *Parabacteroides, Eubacterium, and Bifidobacterium*, also possess the ability to synthesize GABA, a finding further supported by transcriptomic analyses of human fecal samples.

  Although GABA cannot cross the BBB, gut-derived GABA may influence central physiology via local ENS signaling or through vagus nerve pathways (see Sect. 9.2.2 for vagal mechanisms) (Strandwitz et al. 2019; Frost et al. 2014) (Table 9.2 and Fig 9.1).

Table 9.2 Microbial and cellular contributions to neurotransmitter production in the gut-brain axis (GBA)

Neurotransmitters	Precursors	Gut microbiota	Intestinal cells	Functions
Dopamine McCutcheon et al. (2019)	Tyrosine L-DOPA	*Staphylococcus*	–	Regulates gut motility and secretion; affects central dopamine signaling
Serotonin Yano et al. (2015)	5-HTP Tryptophan	*Staphylococcus Clostridial species*	Enterochromaffin cells	Stimulates gut motility; modulates mood and cognition
GABA Strandwitz et al. (2019), Lee et al. (2019)	Acetate	*Bifidobacterium Bacteroides fragilis Parabacteroides Eubacterium*	Myenteric neurons Mucosal endocrine-like cells	Modulates enteric signaling, intestinal motility.
Acetylcholine Parrish et al. (2008)	Choline	*Lactobacillus plantarum Bacillus acetylcholini Bacillus subtilis Escherichia coli Staphylococcus aureus*	Myenteric neurons	Regulates GI motility and secretion
Phenylethylamine Sejbuk et al. (2024)	Phenylalanine	*Staphylococcus*	–	Neuromodulatory functions
Glutamate Lee et al. (2019)	Acetate	*Lactobacillus plantarum Bacteroides vulgatus Campylobacter jejuni*	Enteroendocrine cells	Connects gut signals to the brain via vagus nerve
Tyramine Miller and Wolin (1996)	Tyrosine	*Staphylococcus Providencia*	–	Precursor to octopamine
Norepinephrine Habib et al. (2012)	Tyrosine	–	–	Regulates stress response, energy balance
Tryptamine Strandwitz et al. (2019)	Tryptophan	*Staphylococcus Ruminococcus gnavus Clostridium sporogenes*	–	Trigger serotonin release, promotes gut motility

This table outlines key neurotransmitters synthesized in the gut, their biochemical precursors, associated gut microbiota, relevant intestinal cell types, and primary physiological functions. It highlights the intricate interplay between microbial metabolism and host signaling pathways that influence gut motility, immune regulation, and central nervous system activity, emphasizing the gut's role as a neuroactive organ

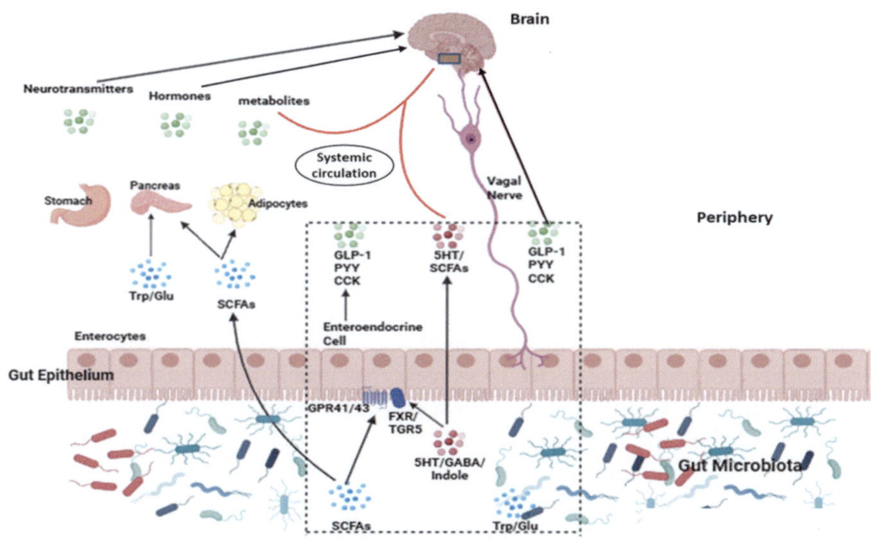

Fig. 9.1 Bidirectional communication between the gut microbiota and brain via the GBA. This schematic illustrates the multifaceted pathways by which gut microbiota influence brain function through neural, hormonal, immune, and metabolic signaling. Microbial metabolites such as SCFAs, neurotransmitters (e.g., 5-HT, GABA), and precursors (Trp/Glu) interact with gut epithelial and enteroendocrine cells, activating receptors like GPR41/43 and FXR/TGR5. These signals are transmitted to the brain via the vagus nerve and systemic circulation. Hormones including GLP-1, PYY, and CCK, as well as microbial metabolites, modulate central and peripheral physiology, highlighting the gut's integral role in neuroendocrine and behavioral regulation

9.3.3 Stress, Anxiety, and Depression

9.3.3.1 Microbiota and Stress Response

Psychological stress disrupts GM through neuroendocrine and immune mechanisms. It can induce gut inflammation, enhance the proliferation of pathogenic bacteria, and increase intestinal permeability (Yang et al. 2017). Other animal and human studies have demonstrated that acute and chronic stress rapidly alter microbial composition. Notably, catecholamines released during stress have been shown to significantly accelerate microbial growth, up to 10,000-fold, in certain bacterial species (Freestone et al. 2002). Chronic stress in humans is associated with diminished microbial diversity and heightened inflammatory markers (Chen et al. 2018). Elevated intestinal permeability under stress conditions facilitates the translocation of bacteria or bacterial products into systemic circulation, contributing to systemic inflammation. This phenomenon has been observed in individuals exhibiting elevated cortisol levels and in those experiencing interpersonal conflict (e.g., hostile couples). Moreover, stress-induced activation of mast cells further compromises intestinal barrier integrity, collectively promoting dysbiosis, systemic inflammation, and gut dysfunction (Vanuytsel et al. 2014).

9.3.3.2 GBA Dysregulation and Microbiota-Targeted Interventions in Mood Disorders

Dysregulation of the GBA has been increasingly associated with mood disorders such as anxiety and depression. These conditions are often accompanied by reduced levels of beneficial microbial genera, such as *Lactobacillus* and *Bifidobacterium*, and an increased abundance of potentially pathogenic species. The GM influences stress responses via vagal afferent signaling and modulates the synthesis of key neurotransmitters including serotonin, GABA, and dopamine. A neuroimaging study involving 40 women demonstrated that specific gut microbial profiles were associated with differential brain activation patterns in response to emotional stimuli, underscoring the role of gut microbes in affective regulation (Tillisch et al. 2017). Randomized controlled trials (RCT) suggest that probiotic supplementation may enhance stress resilience, improve ANS balance, promote sleep quality, and lower circulating cortisol levels (Tillisch et al. 2013). Additionally, fecal microbiota transplantation (FMT) from individuals with depression into rodents has been shown to induce depressive-like behaviors, supporting a causal link between gut dysbiosis and altered mood (Kelly et al. 2016). In line with this, a large UK cohort study found that repeated antibiotic exposure was significantly associated with an increased risk of anxiety and depression, with individuals receiving multiple courses exhibiting up to a 50% higher risk. While probiotics may help mitigate these effects, clinical findings remain mixed and context-dependent (Aoun et al. 2020; Tillisch et al. 2013).

Building on these observations, modulating GM through dietary interventions, probiotics, and prebiotics has emerged as a promising strategy for managing mood disorders. According to the International Scientific Association for Probiotics and Prebiotics (ISAPP), prebiotics are substrates that are specifically used by good bacteria, whereas probiotics are live microorganisms that provide health advantages to the host (Loh et al. 2024; Hill et al. 2014). Advances in sequencing technologies have broadened the scope of probiotics beyond traditional strains like Lactobacillus and Bifidobacterium, encompassing next-generation probiotics (NGPs) such as *Akkermansia muciniphila* and *Faecalibacterium prausnitzii* (Hitch et al. 2022).

Prebiotics such as inulin and galacto oligosaccharides (GOS) support the growth of these beneficial microbes and enhance the production of their health-promoting metabolites (Hill et al. 2014; Hitch et al. 2022). In models of neurodegenerative disease, both probiotics and prebiotics have demonstrated the ability to reinforce intestinal and BBB integrity, reduce systemic and neuroinflammation, and restore neurotransmitter balance factors that collectively improve cognitive and emotional functioning (Loh et al. 2024).

9.3.4 GM, Cognitive Function, and Neurodegeneration in Aging

The GBA plays a central role in preserving cognitive function across the lifespan, and its dysregulation has been increasingly implicated in age-related cognitive decline and neurodegenerative diseases. Aging induces notable alterations in GM composition, characterized by reduced microbial diversity and a relative increase in pro-inflammatory bacterial taxa (Zhang and An 2007). These changes can trigger chronic low-grade inflammation, increased gut permeability, and dysregulation of immune signaling. Importantly, alterations in the secretion of SCFAs, notably acetate, propionate, and butyrate can compromise the integrity of both the intestinal lining and the BBB, facilitating neuroinflammatory processes that are associated with impairments in short-term memory and verbal learning (Luu et al. 2019; Cummings et al. 1987; Erickson et al. 2012).

The GM also exerts profound influence on neurotransmitter systems that govern cognitive and emotional regulation. Through interactions with the ENS and modulation of metabolic precursors, the microbiota helps maintain the balance between excitatory neurotransmitters (glutamate, dopamine, acetylcholine) and inhibitory ones (GABA, serotonin). Aging-associated microbial shifts can disrupt these homeostatic mechanisms, contributing to neurochemical imbalances commonly observed in disorders such as AD, PD, anxiety, and depression (Foster and McVey Neufeld 2013; Ferreiro et al. 2023; Zhu et al. 2022). Moreover, SCFAs produced by gut bacteria play an essential role in neuronal health by modulating intracellular signaling cascades, supporting neurotrophic factor expression, and influencing hypothalamic and limbic system activity. A decline in SCFA-producing bacteria with age may reduce acetylcholine and serotonin synthesis, thereby impairing synaptic plasticity and cognitive performance (Frost et al. 2014; Nankova et al. 2014; Dalile et al. 2019).

Understanding the multifaceted role of the GM in cognitive aging provides a promising avenue for therapeutic innovation. By targeting microbiota-mediated pathways, such as neuroinflammation, neurotransmitter biosynthesis, and SCFA production, emerging interventions may offer novel strategies for the prevention or management of neurodegenerative diseases.

9.3.4.1 The Role of the GBA in Age-Related Neurodegenerative Diseases (NDD): AD, PD, ALS

Building on the role of GM in cognitive aging, emerging evidence underscores the GBA as a key mediator in the development and progression of major age-related neurodegenerative diseases, including AD, PD, and Amyotrophic Lateral Sclerosis (ALS). These conditions share pathophysiological features such as neuroinflammation, oxidative stress, protein misfolding, and synaptic dysfunction, processes that are increasingly linked to GM dysbiosis (Gautam et al. 2021; Gautam et al. 2025b). Disruption of intestinal homeostasis may exacerbate CNS pathology through increased gut permeability, systemic inflammation, altered microbial metabolites, and impaired BBB integrity.

9.3.4.1.1 Alzheimer's Disease (AD)

The primary cause of dementia, AD, is typified by intracellular neurofibrillary tangles made of hyperphosphorylated tau protein and extracellular amyloid beta (Aβ) plaques (Kowalski and Mulak 2019). Gut dysbiosis in AD promotes intestinal barrier dysfunction and systemic inflammation, which can compromise the BBB and intensify neuroinflammatory responses (Ferreiro et al. 2023). Aβ, although known as a hallmark pathological protein, also functions as an antimicrobial peptide; cross-seeding between bacterial amyloids and host Aβ may facilitate misfolding and accumulation, activating microglia and accelerating neurodegeneration (Ho et al. 2018). Notably, aging, the most significant risk factor for AD, is associated with progressive gut microbial alterations. Yet, studies of healthy centenarians reveal preserved gut microbial diversity and enrichment in beneficial commensals. Transplantation of fecal microbiota from young donors (yFMT) into aged mice has been shown to restore youthful microglial gene expression, particularly genes such as DNAX-activating protein of 12 kDa (DAP12) Tyrosine kinase binding protein (*TYROBP*) and *C1q*, which regulate microglial activation and synaptic pruning both central to AD pathology (Kowalski and Mulak 2019; Dejanovic et al. 2022). These findings emphasize the gut microbiome's potential to influence neuroinflammatory and neurodegenerative processes and support gut-targeted therapeutic strategies in AD.

9.3.4.1.2 Parkinson's Disease (PD)

PD, the second most common neurodegenerative disorder, is marked by progressive motor deficits and intracellular aggregation of α-synuclein. The gut has emerged as an early site of pathology, with studies linking gastrointestinal dysfunction to prodromal PD symptoms. Gut dysbiosis increases intestinal permeability and systemic inflammation, which may facilitate the misfolding and propagation of α-synuclein (Hirayama and Ohno 2021). Certain microbes, such as *Akkermansia muciniphila*, a bacterium linked to PD, implicated in PD, induce mitochondrial calcium overload and promote α-synuclein aggregation within enteroendocrine cells (Amorim Neto et al. 2022).

Further, dysbiosis permits translocation of LPS and environmental neurotoxins, exacerbating neurodegenerative cascades. *Proteus mirabilis*, significantly enriched in PD mouse models, has been shown to trigger dopaminergic neuron death and motor impairment, reinforcing the microbiota-neurodegeneration link (Choi et al. 2018). Additionally, *Helicobacter pylori* may aggravate motor symptoms by depleting dopamine levels and impairing the absorption of levodopa, a frontline PD therapy. The vagus nerve likely acts as a conduit for gut-to-brain α-synuclein transmission. As microbial composition directly impacts α-synuclein pathology, interventions targeting gut dysbiosis may offer promising disease-modifying avenues for PD.

9.3.4.1.3 Amyotrophic Lateral Sclerosis (ALS)

ALS is a fatal neurodegenerative condition characterized by progressive motor neuron degeneration. Though historically considered a CNS restricted disease, recent findings suggest that GBA dysfunction may influence ALS onset and progression (Kaul et al. 2024). In SOD1G93A, transgenic mouse models of ALS, gut dysbiosis, intestinal barrier disruption, and increased systemic inflammation have been observed. Correspondingly, stool microbiota analyses from ALS patients reveal diminished microbial diversity and significant shifts in the *Firmicutes* and *Bacteroidetes* (Zhang et al. 2017). A decreased *Firmicutes/Bacteroidetes* (F/B) *ratio* coupled with lower *Ruminococcus* correlates with heightened intestinal inflammation and disease severity (Boddy et al. 2021). Interestingly, microbial metabolites altered in disease states are also present in other biofluids, indicating their systemic influence beyond the gut (Gautam et al. 2021; Gautam et al. 2025a, b; Gray et al. 2015). These data suggest that microbial dysbiosis may contribute to a pro-inflammatory milieu that facilitates neurodegeneration in ALS.

9.4 Strategies for Enhancing Brain Longevity Through the GBA

As we've explored in earlier sections, disruptions in the GBA are tightly linked to cognitive decline and the development of NDD like AD, PD, and ALS. What if this same gut-brain connection could also help us preserve brain health and slow down aging? Recent research suggests that it can, especially when we focus on diet, lifestyle, and targeted microbiota interventions. These modifiable factors hold powerful potential to protect cognitive function, promote resilience to stress, and support brain longevity well into older age.

9.4.1 Diet and Nutrition: Foundations for a Resilient Brain

Our diet does not just feed us, it feeds our microbiota, which in turn shapes how our brain functions. The gut microbiome assists in the digestion of complex carbohydrates, synthesizes essential vitamins such as K and B, and produces SCFAs that regulate immune and neural functions (Mills et al. 2019). In fact, experiments have shown that transplanting GM from obese individuals into germ-free mice (GFM) can actually induce weight gain, proving that microbial communities can carry metabolic traits (Vendrik et al. 2020).

In diseases like ALS, where maintaining body weight is critical, there is growing interest in using diet to modulate the microbiome and potentially slow disease progression. Longitudinal studies, such as those by Di Gioia et al., show that patients with ALS experience a gradual loss of microbial diversity partly due to disease-related changes in activity, diet, and feeding methods (Di Gioia et al. 2020). However, it remains unclear whether these microbiome changes are driving the disease or are simply a consequence of its progression.

9.4.1.1 Mediterranean Diet (MD), Ketogenic Diet (KD), and Other Nutritional Approaches

Among the best-studied dietary strategies for brain health is the Mediterranean diet (Medi), known for its high content of antioxidants, polyphenols, and healthy fats. Rich in olive oil, leafy greens, fatty fish, and nuts, this diet helps fight oxidative stress and inflammation, two major culprits in neurodegenerative disease (Kowalski and Mulak 2019; Di Gioia et al. 2020). Nutrients like Docosahexaenoic acid (DHA), found in fish, reduce tau buildup and promote neurogenesis, while monounsaturated fats (MUFAs) from olive oil support autophagy, a key process for clearing cellular waste like beta-amyloid (Yao et al. 2020; Ferreiro et al. 2023; Zhang and An 2007; Ho et al. 2018).

Another approach is the KD, which shifts the brain's energy source from glucose to ketones, particularly helpful in AD, where glucose metabolism falters. By supplying a steady stream of ketone bodies like β-hydroxybutyrate (BHB), the KD helps fuel brain cells and may also rebalance the microbiota (Kossoff et al. 2009; Cardoso et al. 2016).

The Mediterranean-DASH Intervention for Neurodegenerative Delay (MIND) diet combines the Medi and Dietary Approaches to Stop Hypertension (DASH) diets, emphasizing brain-protective foods while limiting sugar, red meat, and processed items. It is linked to reduced AD risk and better cognitive function by targeting vascular health and inflammation (Strandwitz et al. 2019; Cardoso et al. 2016; Livingston et al. 2020). Studies suggest that up to 40% of dementia cases could be preventable through such lifestyle strategies (Livingston et al. 2020; Hosking et al. 2019; Chiavaroli et al. 2019). Additionally, time-restricted feeding and intermittent fasting show promise by enhancing autophagy and reducing tau and beta-amyloid accumulation (Loh et al. 2024).

9.4.1.2 Polyphenols and Fermented Foods: Natural Allies for Gut and Brain Health

Polyphenols are bioactive plant compounds known for their antioxidant, anti-inflammatory, and neuroprotective properties (Sejbuk et al. 2024). Their efficacy depends on bioavailability, which is influenced by food source, processing, and GM metabolism. Once metabolized, polyphenols promote gut barrier integrity, stimulate SCFA production, and regulate immune responses (Dalile et al. 2019; Lippolis et al. 2023).

Fermented foods, such as yogurt, kefir, and kimchi, further support microbiota diversity and mucosal immunity. Studies show these foods can increase beneficial taxa like *Faecalibacterium prausnitzii*, *Lachnospiraceae*, and *Ruminococcaceae*, even in individuals with health conditions (Taylor et al. 2020; Le Roy et al. 2022; Yamamoto et al. 2018). However, more longitudinal studies are needed to confirm their long-term cognitive benefits and mechanistic effects (Fig. 9.2).

9.4.2 Lifestyle Factors: Enhancing the GBA Through Daily Habits

While nutrition lays the foundation for a healthy GBA, lifestyle factors such as physical activity, sleep, circadian rhythm regulation, and stress management also play essential roles in maintaining brain function and longevity. These behaviors collectively shape the gut microbiome, reduce inflammation, and support neural resilience.

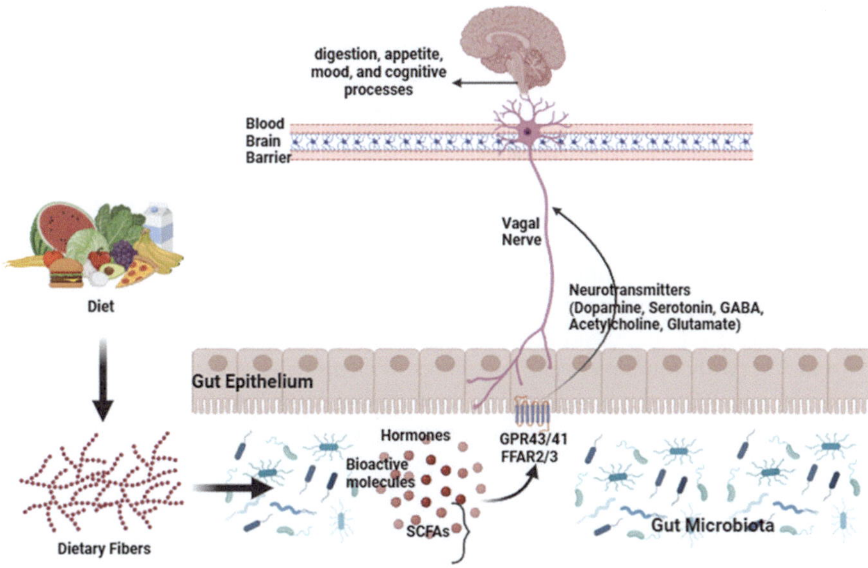

Fig. 9.2 Dietary influence on gut microbiota–brain communication via the GBA. This figure illustrates how dietary intake, particularly dietary fibers, modulates the gut microbiota to produce SCFAs and other bioactive molecules. These metabolites interact with gut epithelial cells and activate receptors such as GPR41/43 (FFAR3/2), leading to the release of gut-derived hormones and neurotransmitters (e.g., dopamine, serotonin, GABA, acetylcholine, glutamate). Signals are transmitted to the brain via the vagus nerve and systemic circulation, impacting digestion, appetite, mood, and cognitive functions across the BBB

9.4.2.1 Exercise: A Natural Modulator of the Microbiome and the Mind

Regular physical activity is one of the most effective non-pharmacological strategies to enhance both metabolic and mental health. Exercise increases microbial diversity and enriches populations of beneficial gut bacteria that produce SCFAs, such as butyrate, which have neuroprotective and anti-inflammatory effects (Dalile et al. 2019; Kim et al. 2020).

Different types of exercise influence the microbiota uniquely. Moderate endurance and resistance training are associated with improved microbiome stability, while high-intensity or prolonged exertion may temporarily increase intestinal permeability and inflammation (Sweitzer et al. 2001; Copray et al. 2001). Athletes tend to have distinct microbial profiles, with higher SCFA production, which may contribute to better energy metabolism and reduced disease risk (Wegierska et al. 2022).

In older adults, greater muscle strength correlates with healthier GM profiles, an important finding given the link between sarcopenia, frailty, and cognitive decline (Cummings et al. 1987; Oleskin and Shenderov 2016; Frost et al. 2014). Exercise also enhances hippocampal neurogenesis and cognitive performance, particularly in memory and attention domains, reinforcing its role in GBA regulation.

9.4.2.2 Impact of Sleep on Microbiome and Inflammation

Sleep plays a vital role in human health, underpinning growth, cognitive development, immune function, and metabolic regulation (Matenchuk et al. 2020). Sleep influences and is influenced by gut microbiota (Matenchuk et al. 2020). Poor sleep quality impairs microbial diversity and raises inflammation, contributing to metabolic dysregulation. Microbial metabolites, in turn, modulate neurotransmitters and immune responses, reinforcing a bidirectional gut-brain-sleep loop (Badman and Flier 2005; Kozyrskyj et al. 2009).

A central mechanism linking sleep and immune regulation involves the release of pro-inflammatory cytokines. Sleep deprivation increases pro-inflammatory cytokines such as IL-1β, IL-6, and TNF-α, and elevates C-reactive protein (CRP). These cytokines act as both markers and regulators of sleep-wake cycles, promoting compensatory non-REM sleep in response to stress but contributing to chronic inflammation when persistently elevated (Deverman and Patterson 2009; Parrish et al. 2008; Zhang and An 2007). Prolonged sleep restriction further exacerbates immune activation, upregulating cytokines such as IL-1β and IL-17, both of which are associated with chronic inflammation and cognitive dysfunction (Szentirmai et al. 2019; Ho et al. 2018; Matenchuk et al. 2020).

9.4.2.3 Cytokines and the Sleep Cycle: A Feedback Mechanism

Interestingly, several pro-inflammatory cytokines are not only markers of sleep disruption but also active regulators of sleep itself. IL-1β, in particular, is a well-characterized sleep-promoting molecule. Its receptor accessory protein plays a crucial role in the initiation of non-rapid eye movement (NREM) sleep following sleep deprivation (van Leeuwen et al. 2009). This dynamics highlight the bidirectional interplay between immune activation and sleep behavior (Matenchuk et al. 2020).

9.4.2.4 Circadian Rhythms and Microbial Oscillations

Closely linked to sleep, circadian rhythms regulate physiological processes across a 24-h cycle. Disruption of circadian rhythms—due to irregular sleep schedules, shift work, or jet lag, has been associated with increased risk of obesity, insulin resistance, and neuroinflammation (Badman and Flier 2005; Klok et al. 2007).

Up to 60% of gut microbes display circadian rhythmicity. Key bacterial genera, including *Parabacteroides*, *Lachnospira*, *Roseburia*, *Veillonella*, and *Ruminococcus* demonstrate time-dependent abundance patterns (Hitch et al. 2022). Disruptions in host circadian rhythms, through jet lag, shift work, or erratic sleep patterns, reduce microbial diversity, increase the *F/B* ratio like *Christensenellaceae* and *Lactobacillus* (Klok et al. 2007), and impair microbial metabolite production. Experimental models confirm that circadian misalignment leads to dysbiosis, inflammation, and cognitive decline (Matenchuk et al. 2020). These findings underscore the importance of a stable sleep-wake cycle not only for brain health but also for maintaining microbial balance and metabolic integrity.

9.4.2.5 Stress and Mindfulness: Calming the Gut-Brain Conversation

Stress and mindfulness practices modulate parasympathetic tone, likely through vagus nerve activity—thus impacting gut microbial composition, inflammation, and brain signaling (see Sect. 9.2.2) (Hattori and Yamashiro 2021; Matenchuk et al. 2020; Hu 2003).

Mindfulness-based stress reduction (MBSR) programs, rooted in contemplative practices, teach techniques such as breathing meditation, body scans, and gentle movement to help individuals respond more thoughtfully to stressors. Different interventions have demonstrated positive effects in both clinical and non-clinical populations, improving emotional regulation, reducing anxiety, and enhancing psychological resilience (Vanuytsel et al. 2014; Kabat-Zinn 1982).

9.4.3 GM Modulation: Tools for Enhancing Cognitive and Emotional Health

As research continues to uncover the deep interconnection between the gut and brain, targeted modulation of the GM has emerged as a promising strategy for promoting mental well-being and cognitive longevity. Probiotics, prebiotics, synbiotics, and even FMT are examples of nutritional therapies that try to minimize neuroinflammation, improve gut-brain connection, and restore or maintain microbial balance.

9.4.3.1 Probiotics, Prebiotics, and Synbiotics: Nourishing the Microbial Ecosystem

Prebiotics are indigestible food ingredients that specifically nourish good gut flora, particularly *Lactobacillus* and *Bifidobacterium* species. Found in dietary fibers, inulin, and oligosaccharides, prebiotics improve digestion, modulate immunity, and enhance microbial diversity, key factors in maintaining emotional and cognitive resilience (Aoun et al. 2020; Hill et al. 2014).

Probiotics are live bacteria that, when given in sufficient quantities, improve intestinal barrier function, produce neuroactive chemicals, and balance the GM. Increasingly, studies have shown that certain probiotic strains (*lactobacilli* and *bifidobacterium*) reduce symptoms of anxiety, depression, and stress, likely by modulating the GBA through inflammatory, hormonal, and neural pathways (Foster and McVey Neufeld 2013; Pariante and Lightman 2008; Lewis et al. 2010; Aoun et al. 2020; Tillisch et al. 2013).

Synbiotics combine prebiotics and probiotics into synergistic formulations that enhance the survival and efficacy of probiotic strains. These formulations have shown potential in treating conditions such as IBS, attention-deficit/hyperactivity disorder (ADHD), chronic fatigue syndrome, and mood disorders, while also improving cognitive function and overall quality of life (Liu and Zhu 2018; Simrén et al. 2013; Oleskin and Shenderov 2016; Ronald et al. 2008).

A well-balanced microbiota not only supports digestion and immune health but also contributes to neurotransmitter synthesis, SCFA production, and inflammation control, all of which are vital for brain longevity in modern, fast-paced lifestyles (Hattori and Yamashiro 2021; Aoun et al. 2020).

9.4.3.2 Fecal Microbiota Transplantation (FMT): Resetting the Gut Ecosystem

While nutritional approaches offer gradual modulation of the gut ecosystem, more direct interventions like FMT, the transfer of stool from a healthy donor to a recipient, offers a more direct approach to restoring microbial balance. While it has proven highly effective in treating recurrent *Clostridioides difficile* infections, FMT is now being explored for its potential to treat neurological and psychiatric conditions (Sanzone et al. 2024; Zhang et al. 2012).

Beyond its gastrointestinal benefits, FMT may help recalibrate the GBA. Early studies suggest that FMT can improve cognition and mood. A more intensive intervention, FMT, is covered in depth in Sect. 9.4.4, highlighting its emerging use in NDDs. Advances such as Microbial Ecosystem Therapeutics-2 (MET-2), a standardized, oral capsule-based microbiome therapy, are being developed for psychiatric disorders like Generalized Anxiety Disorder (GAD) and Major Depressive Disorder (MDD) (Chinna Meyyappan and Milev 2020). These therapies aim to combine precision with safety, selecting and cultivating gut bacteria from healthy donors for targeted intervention.

Despite its potential, FMT faces important challenges, including standardization, donor screening, safety regulation, and understanding long-term effects. More clinical research is needed to fully define its role in neurodegenerative and neuropsychiatric disease treatment.

9.4.3.3 Personalized Interventions: Microbiome Medicine for the Future

Modern lifestyles, characterized by high consumption of processed foods, low fiber intake, physical inactivity, and widespread antibiotic use, have significantly impacted the human microbiome. These changes are linked to increased rates of cognitive decline, mood disorders, and cardiometabolic disease (Aoun et al. 2020; Hill et al. 2014; Hu 2003; Patangia et al. 2022).

Targeted, personalized approaches to GM modulation may offer a path forward. A clinical trial using BugSpeaks®, a microbiome sequencing platform, demonstrated that personalized nutrition based on individual microbiota profiles could improve metabolic markers in people with hyperglycemia and hyperlipidemia (Kallapura et al. 2024). Such precision strategies could be expanded to enhance cognitive health and reduce neurodegeneration risk.

By customizing dietary and microbial interventions to an individual's microbiome composition, researchers hope to optimize gut-brain communication and develop highly tailored treatments for cognitive and emotional well-being.

9.4.4 Emerging Therapies and Interventions: Reimagining the Gut-Brain Connection

As evidence continues to mount on the pivotal role of the GBA in NDDs, emerging therapies are exploring new frontiers in microbiota modulation. Beyond conventional prebiotics, probiotics, and synbiotics, novel interventions, from precision microbial engineering to AI-driven dietary strategies, aim to reshape brain health through the gut. These approaches offer hope for slowing neuroinflammation, preserving cognition, and improving quality of life in aging populations.

9.4.4.1 Integrating Nutrition, Lifestyle, and Therapeutics

The foundation of brain longevity still rests on modifiable lifestyle factors. As elaborated in Sect. 9.4.1, the MD's polyphenol-rich and fiber-dense components support gut-brain integrity by modulating inflammation and microbial metabolites such as SCFAs and trimethylamine N-oxide (TMAO) (Kowalski and Mulak 2019; Hosking et al. 2019; De Filippis et al. 2016). Dietary polyphenols like resveratrol, covered previously, continue to gain attention as neuroprotective agents acting through gut microbial metabolism and antioxidant signaling (Seo et al. 2023; Vanuytsel et al. 2014; Lippolis et al. 2023).

Prebiotics and probiotics support gut and brain health by modulating microbial composition and enhancing barrier function. Synbiotics may further amplify these effects by combining the benefits of both components (Aoun et al. 2020). In animal studies, synbiotic formulations have preserved dopaminergic neurons in PD models and reduced amyloid burden in AD models (Huang et al. 2023; Westfall et al. 2019). Lifestyle interventions such as the MIND diet, already outlined in Sect. 9.4.1, synergize with microbial therapies to support brain aging (Hirayama and Ohno 2021; Kossoff et al. 2009).

KD, discussed earlier, show promise in promoting mitochondrial and microbial resilience, particularly in conditions with impaired glucose metabolism (Loh et al. 2024; Cunningham et al. 2021). Lifestyle interventions, when combined with medical therapy, form the foundation of an integrative strategy to support brain aging.

9.4.4.2 Fecal Microbiota Transplantation (FMT): A Restorative Reset

FMT continues to gain attention for its broader therapeutic potential. While widely used for recurrent C. difficile infections, FMT is being studied as a tool to restore microbial equilibrium in NDDs. In AD mouse models, FMT from wild-type donors reduced amyloid pathology, improved cognitive function, and modulated inflammatory signaling through intestinal macrophages (Ho et al. 2018). Similarly, small pilot studies in PD patients have demonstrated improvements in both motor and non-motor symptoms following FMT (Huang et al. 2023; Choi et al. 2018; Park et al. 2022).

Although promising, FMT is not without limitations. Standardization of donor screening, delivery methods, and long-term safety protocols remain active areas of investigation. As microbiome-based therapies evolve, the need for rigorously controlled human trials grows more urgent.

9.4.4.3 Prebiotics and Their Role in NDDs

Prebiotics are indigestible dietary components that selectively nourish beneficial gut microbes, primarily in the large intestine (Aoun et al. 2020). They include fructans (e.g., inulin and fructo-oligosaccharides (FOS)), galacto-oligosaccharides (GOS), resistant starch (RS), and non-carbohydrate oligosaccharides (Foster and McVey Neufeld 2013; Sayegh and Ritter 2000).

Prebiotics undergo fermentation in the gut, producing SCFAs that lower intestinal pH, promote beneficial bacteria like *Lactobacilli* and *Bifidobacteria*, and inhibit pathogens (Oleskin and Shenderov 2016; Dalile et al. 2019). SCFAs also enter the bloodstream, influencing systemic health, including brain function. RS helps maintain bacterial diversity, improve lipid metabolism, and support intestinal function. In murine models, RS from legumes enhanced SCFA production, reduced bile acids, and mitigated obesity-related metabolic disorders (Badman and Flier 2005; Aoun et al. 2020).

FOS, found in fruits and vegetables, promotes GM stability. In AD models, FOS improved neurotransmitter synthesis, reduced neuronal apoptosis, and alleviated cognitive impairment via GM regulation and the GLP-1 pathway (Sayegh and Ritter 2000). Another study showed that FOS from *Morinda officinalis* enhanced learning and memory in AD rats by reducing oxidative stress and inflammation, further supporting gut-brain interaction in neuroprotection.

9.4.4.4 Probiotics in Cognitive Health

Probiotics are live, non-pathogenic microorganisms that provide health benefits when consumed adequately. Common strains include *Bifidobacterium*, *Lactobacillus*, *Bacillus*, and *Streptococcus* (Kechagia et al. 2013). They support gut health by regulating microbial populations, preventing pathogen colonization, reinforcing the intestinal barrier, and producing SCFAs with neuroprotective and anti-inflammatory effects. Additionally, probiotics can cross the BBB, influencing CNS immune function and neurotransmitter synthesis (Erickson et al. 2012; Hill et al. 2014; Dalile et al. 2019).

Probiotic formulations like SLAB51 and *L. acidophilus* (discussed in Sect. 9.4.3) have shown efficacy in modulating neuroinflammation and supporting neurotransmitter homeostasis. While decreasing *Campylobacterales*, leading to higher SCFA production and reduced Aβ aggregates in an AD murine model (Bonfili et al. 2020). It also restored neuronal proteolytic pathways, reduced oxidative stress via the Sirtuin 1 (SIRT1) pathway, and improved glucose metabolism (Ho et al. 2018; Bonfili et al. 2018).

Clinical research shows that supplementing AD patients with *Lactobacillus acidophilus*, *Lactobacillus casei*, *Bifidobacterium bifidum*, and *Lactobacillus fermentum* improves cognitive function as well as metabolic markers including triglyceride levels and insulin sensitivity (Klok et al. 2007). Long-term supplementation with *L. paracasei* K71 also prevented age-related cognitive decline by increasing the hippocampus's brain-derived neurotrophic factor (BDNF) (Corpuz et al. 2018).

Probiotic-4, containing *Lactobacillus*, *Bifidobacterium*, and *Typhimurium*, improved gut and BBB integrity in an ALS model, reducing inflammation and inhibiting the Toll-like receptor (TLR4) and NF-κB inflammatory pathways (Boddy et al. 2021; van Leeuwen et al. 2009; Kechagia et al. 2013).

9.4.4.5 Synbiotics and Their Synergistic Effects

Synbiotics are formulations combining prebiotics and probiotics to enhance their viability and efficacy. Studies suggest synbiotics are more effective than prebiotics or probiotics alone in modulating GM and promoting health benefits (Aoun et al. 2020). A novel synbiotic containing polymannuronic acid and *Lacticaseibacillus rhamnosus GG* demonstrated neuroprotection in PD mice by preserving dopaminergic neurons and improving motor function (Huang et al. 2023). Another formulation, combining *Lactobacillus plantarum*, *Lactobacillus fermentum*, and *Bifidobacterium longum*, along with a polyphenol-rich prebiotic, delayed AD onset in a *Drosophila* model (Westfall et al. 2019).

A randomized controlled trial investigating a 12-week synbiotic supplementation in PD patients reported improved antioxidant capacity, reduced depression symptoms, and enhanced cognitive function and quality of life (Hirayama and Ohno 2021; Kossoff et al. 2009).

9.4.4.6 Fecal Microbiota Transplantation (FMT) in NDDs

FMT restores a healthy gut microbiome by transferring prescreened donor feces. It has been extensively studied for treating IBD and metabolic disorders, and there is growing interest in its potential for treating neurodegenerative disease (NDD) (Vendrik et al. 2020; Park et al. 2022). In an AD mouse model, FMT from conventionally raised AD mice worsened cerebral Aβ pathology, whereas FMT from wild-type mice did not (Ho et al. 2018). Another study discovered that while intestinal macrophage activity and systemic inflammation were modulated, repeated FMT from wild-type mice decreased amyloid plaque formation and cognitive impairment in AD-like pathology with amyloid and neurofibrillary tangles (ADLPAPT) mice. FMT has also shown promise in PD. A study involving 11 PD patients found that FMT reconstructed GM and improved both motor and non-motor symptoms (Huang et al. 2023; Choi et al. 2018). However, challenges such as donor selection, stool availability, potential adverse effects, and regulatory hurdles remain. Standardized protocols and further clinical research are needed to confirm its safety and efficacy. GM modulation through diet, prebiotics, probiotics, synbiotics, and FMT offers promising therapeutic avenues for NDDs.

9.4.4.7 Synthetic Biology and CRISPR: Engineering the Microbiome for Precision Health

Synthetic biology offers exciting new tools to customize and enhance the gut microbiome. By modifying bacteria through gene editing technologies such as CRISPR, researchers can selectively enhance probiotic strains or eliminate harmful microbes. For instance, CRISPR-Cas3 has been used to engineer Bifidobacterium and Lactobacillus to support immune function and gut health (Bonfili et al. 2020; Bonfili et al. 2018).

CRISPR is also being repurposed as an antimicrobial alternative to antibiotics. Engineered bacteriophages carrying CRISPR payloads can target pathogens like C. difficile with precision, preserving the rest of the microbial ecosystem (Corpuz et al. 2018). While these technologies are in early stages, they hold significant promise for treating infections, inflammation, and perhaps neurodegeneration by editing the gut microbiome at the genetic level.

9.4.4.7.1 Personalized Probiotics

Synthetic biology enables the development of probiotics tailored to an individual's microbiome. By addressing specific dysbiosis or deficiencies, personalized probiotics offer targeted interventions that align with unique microbiota compositions, optimizing gut health and overall well-being (Zhou et al. 2020).

9.4.4.7.2 Engineered Biosensors

Genetically modified microbes can act as biosensors to detect environmental chemicals, pathogens, or gut health conditions. These sensors can produce observable signals, such as fluorescence, in response to specific stimuli, enabling real-time infection monitoring and early diagnosis (Saifi et al. 2022).

9.4.4.7.3 Targeted Drug Delivery

Engineered bacteria can be programmed to deliver therapeutic compounds precisely within the body, releasing their cargo in response to environmental triggers (Patangia et al. 2022). This targeted approach enhances treatment efficacy while minimizing side effects, representing a major advancement in microbial therapeutics.

Overall, synthetic biology revolutionizes microbiota engineering, offering novel strategies for improving health through engineered microbial systems.

9.4.4.8 AI-Driven Personalization: Toward Precision Microbiome Medicine

Artificial intelligence (AI) is revolutionizing microbiome research by integrating complex multi-omics data to identify disease biomarkers and predict therapeutic responses. Machine learning (ML) models like random forests and deep learning architectures such as convolutional neural networks (CNNs) have shown high accuracy in classifying disease states based on microbial signatures (Ma et al. 2022; Cummings et al. 2016; Strandwitz 2018).

In clinical settings, AI enables the development of personalized diets and microbial interventions. By analyzing GM profiles, platforms like BugSpeaks® can generate individualized nutrition plans to improve metabolic and cognitive outcomes (Chinna Meyyappan and Milev 2020). These tools may soon guide microbiota-targeted therapies for NDDs, adapting treatments to each patient's microbial and genetic makeup.

Despite their promise, AI and synthetic biology approaches face hurdles, including data standardization, privacy concerns, and a limited understanding of many gut microbes. Nevertheless, as research deepens, these technologies are poised to enhance the precision, efficacy, and scalability of gut-brain therapeutics.

9.4.4.9 Dietary Interventions

Diet plays a key role in neuroprotection by reducing inflammation and oxidative stress and supporting the GBA. The Mediterranean diet, rich in antioxidants and polyphenols, is linked to lower cognitive decline. Curcumin, resveratrol, and flavonoids offer anti-inflammatory benefits, while KD improve mitochondrial function (Cunningham et al. 2021). Probiotics, prebiotics, and essential nutrients like vitamin D, B vitamins, and magnesium support cognitive health, potentially mitigating neurodegeneration in conditions like AD and PD (Zhu et al. 2022; Kowalski and Mulak 2019; Bonfili et al. 2020).

9.4.4.10 Lifestyle Modifications

Lifestyle factors like exercise, sleep quality, and cognitive engagement are vital for brain health (Lin et al. 2024). Physical activity boosts neuroplasticity, reduces inflammation, and increases BDNF, aiding neuronal survival. Exercise benefits AD and PD patients by improving motor and cognitive function (Zhu et al. 2022; Kowalski and Mulak 2019). Sleep disturbances linked to neurodegeneration can be managed with good sleep hygiene and melatonin (Lin et al. 2024). Mental stimulation strengthens synaptic plasticity, delaying dementia. Stress management techniques like meditation and yoga help reduce neuroinflammation and oxidative stress, supporting neurological health.

9.4.4.11 Medical and Pharmacological Interventions

Current medical treatments focus on symptom management and neuroprotection (Cummings et al. 2016). Cholinesterase inhibitors (e.g., donepezil, rivastigmine) and N-methyl-D-aspartate receptor (NMDAR) antagonists (e.g., memantine) provide symptomatic relief in AD by enhancing neurotransmitter activity (Kowalski and Mulak 2019; Cummings et al. 2016; Strandwitz 2018). Dopaminergic therapies, such as levodopa and dopamine agonists, remain the cornerstone of PD management (Hirayama and Ohno 2021). Nonetheless, non-pharmacological methods such as transcranial magnetic stimulation (TMS) and deep brain stimulation (DBS) provide supplementary advantages in modifying brain networks. Advances in stem and gene therapy are promising for neuronal regeneration and targeted disease modification. Additionally, anti-amyloid and anti-tau monoclonal antibodies are being explored for their potential to slow AD progression (Ashique et al. 2024).

9.4.4.12 Integrative Approach and Future Directions

Combining dietary, lifestyle, and medical interventions offers a holistic strategy for managing neurodegenerative diseases. Personalized medicine, leveraging genetic and metabolic profiling, can optimize treatment efficacy. The GBA remains a key area of exploration, as GM modulation through probiotics and dietary interventions may provide neuroprotection. Ongoing research into neuroinflammation, mitochondrial dysfunction, and novel therapeutic targets will further refine integrative treatment strategies. Ultimately, a multidisciplinary approach, incorporating neurology, nutrition, exercise science, and behavioral therapy, provides the best opportunity to improve quality of life and slow neurodegenerative progression.

9.5 Challenges and Future Directions

The GBA has emerged as a powerful framework for understanding and potentially mitigating cognitive decline and neurodegenerative diseases. However, translating this promising science into consistent clinical practice remains challenging. From methodological limitations in microbiome studies to the complexity of personalized interventions, the path forward requires innovation, collaboration, and a systems-level approach.

9.5.1 Research Challenges: Complexity, Variability, and Standardization

Despite a growing body of evidence linking GM to cognitive health, significant research challenges remain. The microbiome's intrinsic complexity and variety rank among the most urgent problems. Individual differences in microbial makeup can be attributed to age, medication use, diet, environment, and genetics. This variability complicates the identification of disease-specific microbial markers and limits the generalizability of study findings.

Furthermore, most microbiome studies rely on relative abundance data, which are sensitive to technical variations in sample collection, sequencing platforms, and data processing pipelines. These inconsistencies hinder reproducibility across studies (Hitch et al. 2022; Nankova et al. 2014). Quantitative microbiome profiling, combining absolute microbial counts with relative data—has been proposed as a more accurate method to assess intervention outcomes and microbial shifts.

Standardization of methodologies, including sample processing protocols, sequencing techniques, and analytical frameworks, is essential to ensure consistency and comparability across studies. Longitudinal, multi-center human trials with diverse populations is also needed to assess causal relationships between gut dysbiosis and neurodegenerative outcomes.

9.5.2 Clinical Translation: From Mechanisms to Therapies

While preclinical and early clinical trials have shown promising results, translating microbiota-based strategies into standardized treatments faces several hurdles. Probiotics and synbiotics vary widely in formulation, dosage, and delivery method, often lacking consensus on strain-specific effects or therapeutic targets (Oleskin and Shenderov 2016; Kelly et al. 2016). Regulatory frameworks for microbiota-based therapies, such as FMT and microbial consortia, are still evolving, raising concerns around safety, quality control, and long-term effects.

FMT, though effective in some cases, poses logistical challenges, including donor selection, potential pathogen transfer, and variable outcomes across recipients. Its application in neurological diseases is still experimental and requires rigorous validation through placebo-controlled trials (Sanzone et al. 2024). Likewise, personalized nutrition and microbiome-based diagnostics—though highly promising, require integration into clinical workflows, digital infrastructure, and patient education platforms to become widely scalable.

9.5.3 Ethical and Regulatory Considerations

Emerging technologies like CRISPR-based microbiome editing and AI-driven therapeutic personalization raise important ethical and regulatory questions. The ability to engineer microbial genomes or design phage-based antimicrobials offers not only unprecedented precision but also demands strict oversight to prevent misuse, off-target effects, and unintended ecological consequences (Bonfili et al. 2020; Bonfili et al. 2018; Corpuz et al. 2018).

As microbiome data become a key part of health records, issues of privacy, data security, and equitable access must also be addressed. Incorporating ethical frameworks into microbiome research and therapeutic design is critical to ensure that these innovations benefit all populations, particularly those historically underrepresented in clinical trials.

9.5.4 Future Directions: Toward Integrative, Personalized Brain Health

The future of microbiome-based brain health lies in integration, of disciplines, data types, and therapeutic approaches. Combining dietary strategies, exercise, sleep optimization, and stress management with targeted microbial interventions may offer synergistic benefits for brain longevity.

Advances in multi-omics technologies (genomics, transcriptomics, metabolomics, proteomics) are enabling deeper insights into host-microbe interactions. These tools will allow researchers to unravel the mechanisms by which specific microbial pathways influence brain signaling, inflammation, and neuroplasticity. Precision approaches using AI and ML can help identify individualized patterns of dysbiosis and predict which interventions will be most effective for a given patient (Ma et al. 2022; Cummings et al. 2016; Strandwitz 2018).

Collaboration across neuroscience, microbiology, nutrition, data science, and ethics will be essential to unlock the full potential of the GBA in promoting brain longevity. As research continues to evolve, the next decade may witness the emergence of truly personalized, gut-centered strategies to prevent or slow cognitive decline and enhance quality of life in aging populations.

9.5.4.1 Gaps in Understanding Causality vs. Correlation

A significant challenge in GBA research is distinguishing between causality and correlation. Many studies have identified associations between GM composition and brain health outcomes, but establishing causal relationships is more complex.

9.5.4.1.1 How Diet and Therapy Can Make a Difference

Research highlights GM as a key factor in early-life brain health programming, alongside genetics and environmental influences (Tillisch et al. 2017). Animal and human trials examine dietary interventions, including probiotics, prebiotics, synbiotics, postbiotics, omega-3 fatty acids, and polyphenols. High-fiber diets enhance SCFA production, mitigating maternal-obesity-related cognitive and social deficits (Dalile et al. 2019; Cunningham et al. 2021; Sydenham et al. 2012). FMT also shows promise in modulating gut-brain communication and hedonic food intake (Vendrik et al. 2020), offering potential therapeutic strategies for brain health and disease management.

9.5.4.1.2 Regulation of Stress, Mood, and Anxiety

The GM is closely associated with stress and mood-related disorders (Liu and Zhu 2018; Vanuytsel et al. 2014). Clinical studies indicate that probiotic and prebiotic supplementation may reduce stress (Cunningham et al. 2021). For instance, supplementation with *Lactobacillus rhamnosus* has been linked to reduced anxiety and depression-like behaviors in animal models (Romijn and Rucklidge 2015). However, clinical trials have yielded mixed results, with some demonstrating improvements in mood and anxiety disorders, particularly in patients with IBD. In contrast, others found limited effects due to population heterogeneity and variations in cognitive testing methods.

9.5.4.1.3 Implications for Autism Spectrum Disorder (ASD)

Microbiota modulation has been implicated in ASD, with studies showing that transplantation of microbiota from ASD-diagnosed individuals induces ASD-like behaviors in animal models (Mayer et al. 2014). Conversely, clinical trials suggest that antibiotics, prebiotics, probiotics, and FMT may improve social behavior, reduce anxiety, and mitigate hyperactivity and defiance behaviors in ASD patients (Aoun et al. 2020; Zhang et al. 2012; Patangia et al. 2022). Children with ASD are also four times more likely to experience gastrointestinal symptoms, highlighting the gut-brain connection in neurodevelopmental disorders (Mayer et al. 2014).

9.5.4.1.4 Learning and Memory

Emerging research suggests that GM plays a crucial role in learning and memory development (Ma et al. 2022). Animal studies indicate that alterations in GM composition affect visual-spatial learning and memory performance. Despite the paucity of human research, some results link microbial diversity to infant cognitive development. Innovative treatments for learning and memory disorders may result from a more integrated research approach that takes into account microbiota in addition to other biological systems (Fontana et al. 2023; Asnicar et al. 2024).

9.5.4.1.5 Cognitive Performance and Age-Related Disorders

Microbial diversity appears to be linked to healthy aging. FMT has been shown to restore age-related immune deficiencies in mice, while transplants from aged to young mice negatively impact central nervous system functions (Carabotti et al.

2015; Vendrik et al. 2020; Chakrabarti et al. 2022). These findings suggest that a youthful GM may help maintain cognitive function in later life.

Neurological research also points to a connection between GM and NDD such as PD and AD (Zhu et al. 2022; Cummings et al. 2016). Studies in transgenic mouse models suggest a causal relationship between GM alterations, protein aggregation, and cognitive decline (McCutcheon et al. 2019; Helton and Lohoff 2015; Erickson et al. 2012). More research is necessary to confirm these findings and explore their therapeutic potential.

9.5.4.1.6 Knowledge with Potential

The role of microbiota in brain disorders remains an area of active investigation. While some therapeutic interventions show promise, much remains to be understood regarding the physiological mechanisms underlying these associations. Positive indications exist for interventions in neurodevelopmental and age-related disorders (Chakrabarti et al. 2022), as well as for stress and anxiety management. The challenge is to bridge knowledge gaps and advance research toward practical applications.

9.5.4.2 Shortfalls and Challenges: Bottlenecks to Progress

9.5.4.2.1 Resolving Knowledge Gaps with Strict Approaches

The microbiota, GBA, demands a multidisciplinary approach, as research is largely observational and reliant on animal models. Human RCTs are limited, and individual microbiota variability complicates findings. Integrated studies are needed to optimize microbiota-targeted brain health interventions.

9.5.4.2.2 Understanding Changes and Mechanisms

The definition of a healthy GM remains unclear, though alterations are linked to various disorders (Klok et al. 2007; Kim et al. 2024). Large-scale studies suggest microbiota uniqueness increases with age, potentially informing healthy aging (Corpuz et al. 2018). Biomarkers for disease prediction and treatment efficacy are lacking. While probiotics, prebiotics, and dietary interventions show promise for mental health, inconsistencies in study design limit conclusions (Sejbuk et al. 2024; Xiao and Zhao 2024). Key knowledge gaps include the immunological roles of microbes in neurodevelopmental and neuropsychiatric disorders, microbiota-regulated neurotransmitter mapping, microbial by-products' impact on brain function, and specific microbes' roles in early brain development. Rigorous clinical trials are needed to establish microbial signatures relevant to neurological health (Chakrabarti et al. 2022).

9.5.4.2.3 Few and Varied Clinical Studies

Human intervention studies remain limited, with early research relying on in vitro and animal models. Recent clinical trials suffer from small sample sizes, short durations, and non-standardized methodologies, complicating meta-analyses (Chen et al. 2018; Singh et al. 2014). The lack of disease-specific biomarkers, inadequate behavioral assessments, and poor cohort stratification further constrain findings. Meta-analyses suggest psychobiotic benefits for anxiety, schizophrenia, and cognition, but confounding factors affect outcomes (Romijn and Rucklidge 2015; Singh et al. 2014). Future research should explore psychotropic drug interactions with microbiota-based therapies. Advancing the field requires interdisciplinary collaboration, integrating microbiology, neuroscience, immunology, and clinical research to overcome limitations and establish effective microbiota-targeted interventions.

9.5.4.3 Beyond Theories: Proven Therapeutic and Nutritional Approaches

9.5.4.3.1 Practical Proposals for Advancing Microbiota–Gut–Brain Axis Research

A holistic, multi-system approach is necessary to move beyond correlations and establish causation. Scientists have documented numerous associations between GM and brain disorders, but the next step involves identifying the mechanisms underlying these relationships. Understanding how dietary habits, lifestyle factors, and genetic predispositions influence microbiota composition is crucial for developing targeted interventions.

9.5.4.3.2 More Targeted, Gold-Standard Clinical Studies with Reproducible Results

Future clinical studies should prioritize

- Standardized methodologies to facilitate reproducibility and meta-analyses.
- Randomized, placebo-controlled, double-blind study designs.
- Compliance with ethical guidelines, including the Declaration of Helsinki and Good Clinical Practice (GCP) standards.
- Large-scale trials with diverse populations to account for individual variability.

Conducting rigorous and repeatable studies is essential for building a solid evidence base and translating microbiota research into viable therapeutic strategies.

9.5.5 Ethical and Practical Considerations

Translating GBA research into clinical practice raises several ethical and practical considerations that must be addressed.

9.5.5.1 Ethical Issues in Microbiome Manipulation

9.5.5.1.1 Personal Identity

Traditionally, it was straightforward, but scientific advancements have made identity a multifaceted term. "I" now encompasses an individual's moral agency, consciousness, genotype, phenotype, and microbiome, which consists of organisms that are both part of and distinct from the human body (Gordon et al. 2006). Different contexts emphasize different aspects of identity, ethically, not all humans are moral persons, as responsibility for actions is age-dependent. In disease contexts, identity can be tied to one's body or microbiome, as the latter affects health and varies between individuals. The uniqueness of each person's microbiome complicates definitions of normalcy (Gordon et al. 2006). Integrating the microbiome into our understanding of personhood requires reconsidering identity and its ethical implications (Rhodes 2016).

9.5.5.1.2 Privacy

Privacy has traditionally been linked to physical boundaries, such as one's body, home, or private thoughts. Confidentiality, distinct from privacy, is a professional obligation to safeguard sensitive information within specific domains, such as medicine or law (McGuire et al. 2008). In medical contexts, patients trust that their information will remain confidential within healthcare teams. However, microbiome research presents new privacy challenges. Since 95% of fecal matter consists of microbiota, a person's microbiome can reveal personal details, including their environment and social interactions (Vendrik et al. 2020; Sanzone et al. 2024; Park et al. 2022). This raises questions about law enforcement access to microbiome data, akin to fingerprint collection (McGuire et al. 2008). Should microbiome information be protected as private, or should its use in forensic investigations be permissible?

9.5.5.1.3 Property

Philosophical and legal debates about ownership extend to biological samples (Rhodes 2016). Does an individual own their microbiome simply because their body hosts it? Governments mandate bodily interventions like vaccinations and military service, raising questions about compulsory microbiome contributions to research or public health. If microbiome studies yield commercially valuable findings, determining who profits remains contentious. While legal disputes concerning microbiome ownership have yet to arise, past rulings on genetic samples provide a framework for future debates (Zhang et al. 2012).

9.5.5.1.4 Human Subject Research

Key milestones like the Nuremberg Code and the Helsinki Declaration shaped the evolution of research ethics, emerging alongside civil rights movements that emphasized individual autonomy (GeoH 1949). Foundational documents such as the Belmont Report and the Common Rule prioritize informed consent and privacy, reinforcing the ethical obligation to respect research participants (Rhodes 2016; McGuire et al. 2008; Department of Health, Education, and Welfare, National Commission for the Protection of Human Subjects of Biomedical and Behavioral Research 2014). Traditionally, human subject research relies on voluntary and informed participation, ensuring individuals knowingly accept associated risks. Ethical discourse, including debates from the Human Genome Project, has largely focused on safeguarding autonomy and individual rights. However, this focus often neglects an equally important issue: the responsibilities of research participants. As genomic science advances, researchers increasingly study broad interactions, such as microbiome-genome links in diabetes and inflammatory bowel disorders (Diener et al. 2022; Simrén et al. 2013; Hu 2003; Li and Peng 2019). These investigations require large-scale biobank participation, raising ethical questions about collective responsibility. Public funding already supports genetic and microbiome research, meaning taxpayers indirectly contribute. This prompts critical inquiries: Should individuals have a moral duty to donate biological samples? Does benefiting from scientific advancements without participating constitute free riding? Could refusal to contribute create an ethical injustice?

Addressing these concerns requires reevaluating the balance between individual autonomy and societal responsibility. Scientific progress depends on broad participation, yet ethical frameworks have primarily focused on protecting participants rather than considering their obligations. A deeper examination of individual responsibility within the wider social contract of research is essential for advancing genomic and microbiome studies while maintaining ethical integrity.

9.5.5.1.5 Public Health

Public health measures, including disease surveillance and outbreak control, rely on data collection (Rhodes 2016; Department of Health, Education, and Welfare, National Commission for the Protection of Human Subjects of Biomedical and Behavioral Research 2014). Since the microbiome plays a role in infectious diseases, monitoring it is essential. However, obtaining microbiome data in emergencies may bypass traditional informed consent protocols. Unlike individual-focused clinical research, microbiome studies involving prebiotics, probiotics, and phages introduce public risks. Altering microbial populations could have unintended consequences, as beneficial microbes for some individuals may harm others (McCarty et al. 2005). Given the transmissibility of microbes, ethical considerations extend beyond individuals to the broader community. Research on obesity and asthma must account for potential public health risks.

9.5.5.1.6 Biobanks

Medical advancements depend on biobanks and human research samples. For example, personalized cancer treatment requires studying diverse tumor samples (McCarty et al. 2005). The increasing reliance on biobanks raises governance issues: Who decides sample access? What information should be shared? Which research projects should benefit? Establishing ethical frameworks for biobank management is crucial for balancing scientific progress with individual rights and societal responsibilities (Gordon et al. 2006). In conclusion, microbiome research intersects with key ethical, legal, and social concerns, including identity, privacy, ownership, research participation, public health, and biobank governance. As our understanding of the microbiome expands, so must our ethical frameworks to address these emerging challenges.

9.5.5.2 Accessibility and Equity in Gut-Brain Health Interventions

9.5.5.2.1 Microbiome-Based Public Health Interventions: Promise and Challenges

Most microbiome-based interventions target individuals, such as fecal transplants for *C. difficile* infections (Taylor et al. 2020; Park et al. 2022). However, large-scale microbiome modifications like vitamin D fortification in milk could provide widespread health benefits. Potential strategies include adding probiotics to food or water to prevent diseases like *H. pylori*-induced stomach cancer or respiratory infections (Cunningham et al. 2021). Probiotics could also support mental health, particularly in at-risk populations, by mitigating the effects of stress and trauma.

Microbiome-based approaches may also aid in disaster response (Qin et al. 2010; Fontana et al. 2023). For example, after the Walkerton water contamination tragedy, affected individuals faced long-term gastrointestinal issues, raising the possibility that probiotics could help mitigate such consequences in future outbreaks. Despite these benefits, large-scale microbiome interventions present ethical and practical concerns. Public health strategies that bypass individual choice must justify their necessity by demonstrating that societal benefits outweigh the potential loss of autonomy.

Additionally, the unintended effects of microbiome modifications require careful assessment. Eliminating *H. pylori*, for instance, may increase susceptibility to conditions like asthma or allergies (Caballero-Flores et al. 2023; Kozyrskyj et al. 2009). Interactions between multiple microbiome interventions must be thoroughly studied to prevent unforeseen complications. Regulatory oversight and transparent policies are crucial for ensuring safety and maintaining public trust. As microbiome science advances, ethical considerations must guide implementation to balance innovation with potential risks. Careful evaluation of benefits, risks, and public acceptance will be crucial for integrating microbiome interventions into public health strategies responsibly.

9.5.6 Future Research Directions

9.5.6.1 Longitudinal Studies on GM and Brain Aging

Longitudinal studies are essential for understanding how GM changes and its relationship to brain aging. These studies help identify links between microbiome composition, cognitive decline, neuroinflammation, and other aging markers. Aging alters gut function and structure, affecting bacterial survival and diversity. Reduced bile secretion, weakened mucosal lining, slower gastric emptying, and decreased gut motility contribute to longer whole-gut transit times and constipation (Yatsunenko et al. 2012; López-Otín et al. 2023; Cardoso et al. 2016; Corpuz et al. 2018; Bitar et al. 2011). Furthermore, as we age, our guts become more permeable, which makes it possible for bacterial endotoxins to get into our circulation and cause long-term, low-grade inflammation (Zhang and An 2007). Appetite and digestion also change, with older adults experiencing delayed gastric emptying, reduced

ANS function, and increased digestive discomfort. Conditions such as colonic diverticula and functional constipation become more prevalent, affecting up to 40% of those over 85. However, diagnosing gastrointestinal disorders in older adults is challenging due to decreased pain sensitivity and incomplete symptom reporting (Bitar et al. 2011).

GM composition shifts significantly with age. Prolonged transit time lowers microbial diversity and increases protein fermentation in the colon, which can negatively impact health. *H. pylori*, a bacterium linked to gastritis and ulcers, become more common in later life (Kozyrskyj et al. 2009). Despite these trends, microbiome changes are not inevitable, as some individuals maintain stable GM across their lifespan. However, studies show centenarians tend to have a more pro-inflammatory microbiota, supporting the concept of inflamm-aging, in which chronic low-grade inflammation accelerates aging (Erickson et al. 2012).

GM may serve as an indicator of biological age. Researchers have developed a "microbiomic clock" that predicts age within 6 years based on the composition of gut microbes. Certain bacterial species, particularly those in the *Bifidobacterium* genus, are linked to younger biological ages and lower levels of depression, anxiety, and inflammation (Foster and McVey Neufeld 2013; Mörkl et al. 2023; Zhang and An 2007; Helton and Lohoff 2015; Chen et al. 2018). Some microbiota species associated with youthfulness also produce butyrate, a SCFA essential for gut-brain communication. A study in Japan identified key microbiome transitions at ages 3, 33, 42, 77, and 94, suggesting that gut microbial composition shifts at specific life stages (Odamaki et al. 2016).

While natural microbiome shifts occur, external factors such as medical treatments, antibiotics, and proton pump inhibitors can significantly alter GM (Patangia et al. 2022; Department of Health, Education, and Welfare, National Commission for the Protection of Human Subjects of Biomedical and Behavioral Research 2014). Aging-related microbial changes have been linked to infections, colorectal cancer, cardiovascular disease, and frailty (Chiavaroli et al. 2019; van Leeuwen et al. 2009). Importantly, microbiota composition correlates more strongly with frailty and functionality than chronological age. Frail individuals exhibit distinct GM patterns characterized by lower diversity and an overrepresentation of harmful bacterial genera that weaken the gut barrier and promote systemic inflammation (Zimmermann et al. 2019; Strandwitz 2018). This cycle of gut dysfunction accelerates physical decline, reinforcing the critical role of microbiota in aging and overall health.

Understanding how GM influences aging opens possibilities for targeted interventions. Strategies to maintain a healthy gut include dietary modifications, the use of probiotics, and the consumption of prebiotics to preserve microbial diversity and reduce inflammation. As research advances, microbiome-based approaches may become valuable tools for promoting longevity and mitigating aging-related diseases.

9.5.6.2 Development of Biomarkers for GBA Health

Biomarkers play a crucial role in diagnosing and monitoring GBA health, yet mental health disorders lack specific biomarkers that can reliably differentiate conditions (Zapata and Quagliarello 2015). Many microbial biosensors currently detect general signals linked to multiple diseases, such as indole-producing bacteria, which are associated with inflammatory bowel disease, metabolic disorders, neurodegeneration, and cancer (Loh et al. 2024; Collins et al. 2023; Seo et al. 2023; Yang et al. 2017). Similarly, overlapping symptoms among mental health conditions make microbiome-based differentiation difficult, highlighting the need for more precise biomarkers or strategies that detect multiple biomarkers simultaneously.

One approach to improving specificity is integrating multiple biological signals. Studies have shown that sensing multiple cues enhances the accuracy of disease detection. For example, *Salmonella Typhimurium* is engineered to detect hypoxia and lactate and improve tumor specificity (Udit and Gautron 2013). Similarly, living cells sensing up to three signals have been used for accurate cancer detection. Researchers are exploring how such integrated sensing mechanisms could aid mental health diagnostics (Liu et al. 2022). Current advances in protein-based biosensors and transcriptional logic gates, which process multiple biological inputs, demonstrate the potential for refining disease detection in this field (Chien et al. 2021).

Bacteria naturally possess sophisticated two-component systems (TCSs) for processing multiple signals. These kinase networks allow for signal crosstalk, increasing their ability to integrate diverse inputs (Beumer et al. 2018; Chen et al. 2018; Westfall et al. 2019). Some kinases feature tandem sensors with multiple sensory domains, enabling them to function as logic gates. Both natural and synthetic tandem sensors have demonstrated signal integration capabilities, though their design principles remain poorly understood. Advances in ML-based protein structure prediction and engineering may facilitate the development of more effective biosensors for mental health disorders.

Beyond existing biosensors for indole metabolites, inflammation, pH, and oxygen, biosensors are urgently needed to target other mental health-related biomarkers (Chien et al. 2021). For instance, microbial biosensors for kynurenine metabolites, important in depression, remain undeveloped. SCFAs such as acetate, butyrate, and propionate are another promising biomarker group, as they influence the central nervous system via orphan GPCRs and HDACs, both of which are implicated in depression and other mental health conditions (Forsythe et al. 2014; Miller and Wolin 1996). Developing bacterial biosensors for SCFAs could enhance understanding of GBA dysregulation and improve mental health diagnostics (Roswall et al. 2021; Hattori and Yamashiro 2021; Chien et al. 2021).

The potential of bacterial TCSs as biosensors is further demonstrated by histidine kinases that detect I3A and tetrathionate. Gut microbes constantly respond to fluctuating metabolite levels, making them an untapped resource for biosensor development. By analyzing the vast array of TCSs in gut bacterial genomes, researchers may discover novel sensory components for constructing microbiome-based diagnostic tools. This approach offers a promising avenue for advancing GBA research and improving the precision of mental health diagnostics.

9.5.6.3 Exploration of Microbial Therapeutics for Brain Longevity

Exploring microbial therapeutics for brain longevity is a fascinating and rapidly evolving field. Researchers are delving into how the gut microbiome, a complex ecosystem of microorganisms, influences brain health and aging (Yatsunenko et al. 2012; Cardoso et al. 2016; Le Roy et al. 2022; Marraffini 2015). The GBA, a bidirectional communication network, is pivotal in this interaction. Future research directions include:

1. *Probiotics and Prebiotics*: Investigating how specific strains of beneficial bacteria and dietary fibers can enhance cognitive functions and delay neurodegenerative processes (Aoun et al. 2020; Hill et al. 2014; Cunningham et al. 2021).
2. *FMT*: Exploring its potential to restore GM balance and improve brain health in aging populations (Chen et al. 2018; Vendrik et al. 2020; Sanzone et al. 2024; Park et al. 2022).
3. *Synthetic Biology*: Developing engineered microbes tailored to produce neuroprotective compounds or modulate brain-related pathways (Cubillos-Ruiz et al. 2021).
4. *Dietary Interventions*: Studying the impact of personalized diets on GM composition and their subsequent effects on brain longevity (Sejbuk et al. 2024; De Filippis et al. 2016).
5. *Phage Therapy*: Utilizing bacteriophages to target harmful gut bacteria that may negatively affect brain health (Sahu et al. 2024).

These approaches aim to harness the gut microbiome's metabolic and signaling capabilities to promote brain health and longevity.

9.6 Conclusion

Exploring the microbiome has transformed our understanding of human biology, health, and disease. As research advances, the microbiome is increasingly recognized as a critical component of identity, privacy, ownership, public health, and ethical considerations in human subject research. This chapter has examined the microbiome's profound implications, highlighting its relevance in various domains, from individual health interventions to large-scale public health strategies.

A key takeaway is that the microbiome challenges traditional notions of personal identity. We used to believe our identity was defined by our own genes and choices but science now shows who we are is also shaped by trillions of microbes living inside us. These microbes interact with our body and brain in dynamic ways, influencing metabolism, immunity, and even behavior. This fluid relationship challenges fixed ideas of normalcy and individuality, prompting a reexamination of the ethical and legal boundaries that guide microbiome research and its clinical applications.

Privacy concerns surrounding microbiome data are increasingly relevant as technological advancements enable forensic applications and environmental tracking. Unlike genetic data, microbiome profiles can provide insights into an individual's diet, lifestyle, and geographic location, raising ethical concerns about consent and data protection. The potential use of microbiome data in law enforcement, health monitoring, and personalized medicine requires careful regulation to balance innovation with individual rights.

Similarly, the concept of ownership remains unresolved in microbiome research. While individuals may assume ownership of their microbiota, legal and ethical precedents regarding genetic samples suggest that broader societal benefits and corporate interests may challenge this assumption. The commercial potential of microbiome-based therapies and probiotics further complicates intellectual property and profit distribution discussions. As scientific discoveries progress, clear legal frameworks will be necessary to ensure fair and ethical management of microbiome-derived products.

The ethical considerations of human subject research are particularly pressing as microbiome studies increasingly rely on biobank contributions and large-scale sample repositories. While informed consent remains a cornerstone of research ethics, the potential societal benefits of microbiome research raise questions about moral obligations to contribute biological samples. Public funding supports many microbiome studies, leading to debates about whether non-participation constitutes free-riding or an injustice. Addressing these ethical concerns will require balancing individual autonomy and collective responsibility.

Public health applications of microbiome research also present significant challenges and opportunities. Interventions such as probiotic fortification of food and water could provide widespread health benefits akin to historical public health measures like vitamin D fortification. However, these initiatives raise ethical concerns regarding individual consent, unintended health consequences, and potential disparities in access. The transmissible nature of microbes means that microbiome-based interventions must consider community-wide implications, requiring transparent policies and regulatory oversight.

Furthermore, the role of the microbiome in aging, mental health, and chronic diseases underscores its importance in medical research. The GBA highlights the intricate connections between microbiota and neurological function, suggesting potential biomarkers and therapeutic targets for depression and neurodegeneration. Developing microbial biosensors and two-component systems offers promising advancements in diagnostics and treatment, but ethical considerations must guide their implementation.

As microbiome research expands, it is imperative to integrate ethical, legal, and social considerations into scientific progress. The complex interplay between the microbiome, identity, privacy, ownership, and public health necessitates ongoing discourse and regulatory evolution. Ensuring that microbiome research aligns with ethical principles will maximize its benefits while safeguarding individual rights and societal well-being. Ultimately, a multidisciplinary approach that combines scientific innovation with ethical responsibility will be essential in shaping the future of microbiome research and its impact on human health.

9.7 Chapter Summary

This chapter explores the fascinating two-way communication system between our gut and brain, known as the *GBA*. At the heart of this connection lies the *GM,* the trillions of microbes in our digestive tract that help regulate everything from mood and memory to immune responses and brain aging.

The chapter breaks down how these microbes produce important compounds like *SCFAs* and *neurotransmitters* (like serotonin and GABA) that directly influence how our brain feels and functions. It shows how imbalances in the microbiome, called *dysbiosis*, are linked to mental health conditions (e.g., anxiety, depression) and *neurodegenerative diseases* such as *AD, PD, MS, and ALS*.

Importantly, it highlights how *diet, exercise, sleep, and stress* affect the gut, and through it, the brain. Popular diets like the *Mediterranean, ketogenic, and MIND diets* are shown to support a healthy microbiome and cognitive function. The chapter also explores emerging therapies including *probiotics, prebiotics, synbiotics*, and even *FMT,* plus cutting-edge tools like *CRISPR* and *AI-driven personalization* for tailoring treatments to individual microbiomes.

The overall message? Taking care of our gut, through lifestyle, nutrition, and emerging therapies, could be one of the most powerful ways to *protect our brain, boost mood, and stay mentally sharp as we age.*

References

Amorim Neto DP, Bosque BP, Pereira de Godoy JV, Rodrigues PV, Meneses DD, Tostes K et al (2022) Akkermansia muciniphila induces mitochondrial calcium overload and α -synuclein aggr7egation in an enteroendocrine cell line. iScience 25(3):103908

Annahazi A, Schemann M (2020) The enteric nervous system: "a little brain in the gut". Neuroforum 26(1):31–42

Aoun A, Darwish F, Hamod N (2020) The influence of the gut microbiome on obesity in adults and the role of probiotics, prebiotics, and synbiotics for weight loss. Prev Nutr Food Sci 25(2):113–123

Arumugam M, Raes J, Pelletier E, Le Paslier D, Yamada T, Mende DR et al (2011) Enterotypes of the human gut microbiome. Nature 473(7346):174–180

Ashique S, Mohanto S, Ahmed MG, Mishra N, Garg A, Chellappan DK et al (2024) Gut-brain axis: a cutting-edge approach to target neurological disorders and potential synbiotic application. Heliyon 10(13):e34092

Asnicar F, Thomas AM, Passerini A, Waldron L, Segata N (2024) Machine learning for microbiologists. Nat Rev Microbiol 22(4):191–205

Bäckhed F, Roswall J, Peng Y, Feng Q, Jia H, Kovatcheva-Datchary P et al (2015) Dynamics and stabilization of the human gut microbiome during the first year of life. Cell Host Microbe 17(5):690–703

Badman MK, Flier JS (2005) The gut and energy balance: visceral allies in the obesity wars. Science 307(5717):1909–1914

Bailey MT, Cryan JF (2017) The microbiome as a key regulator of brain, behavior and immunity: commentary on the 2017 named series. Brain Behav Immun 66:18–22

Beumer J, Artegiani B, Post Y, Reimann F, Gribble F, Nguyen TN et al (2018) Enteroendocrine cells switch hormone expression along the crypt-to-villus BMP signalling gradient. Nat Cell Biol 20(8):909–916

Bitar K, Greenwood-Van Meerveld B, Saad R, Wiley JW (2011) Aging and gastrointestinal neuromuscular function: insights from within and outside the gut. Neurogastroenterol Motil 23(6):490–501

Boddy SL, Giovannelli I, Sassani M, Cooper-Knock J, Snyder MP, Segal E et al (2021) The gut microbiome: a key player in the complexity of amyotrophic lateral sclerosis (ALS). BMC Med 19(1):13

Boka G, Anglade P, Wallach D, Javoy-Agid F, Agid Y, Hirsch EC (1994) Immunocytochemical analysis of tumor necrosis factor and its receptors in Parkinson's disease. Neurosci Lett 172(1–2):151–154

Bonfili L, Cecarini V, Cuccioloni M, Angeletti M, Berardi S, Scarpona S et al (2018) SLAB51 probiotic formulation activates SIRT1 pathway promoting antioxidant and neuroprotective effects in an AD mouse model. Mol Neurobiol 55(10):7987–8000

Bonfili L, Cecarini V, Gogoi O, Berardi S, Scarpona S, Angeletti M et al (2020) Gut microbiota manipulation through probiotics oral administration restores glucose homeostasis in a mouse model of Alzheimer's disease. Neurobiol Aging 87:35–43

Caballero-Flores G, Pickard JM, Núñez G (2023) Microbiota-mediated colonization resistance: mechanisms and regulation. Nat Rev Microbiol 21(6):347–360

Carabotti M, Scirocco A, Maselli MA, Severi C (2015) The gut-brain axis: interactions between enteric microbiota, central and enteric nervous systems. Ann Gastroenterol 28(2):203–209

Cardoso C, Afonso C, Bandarra NM (2016) Dietary DHA and health: cognitive function ageing. Nutr Res Rev 29(2):281–294

Chakrabarti A, Geurts L, Hoyles L, Iozzo P, Kraneveld AD, La Fata G et al (2022) The microbiota-gut-brain axis: pathways to better brain health. Perspectives on what we know, what we need to investigate and how to put knowledge into practice. Cell Mol Life Sci CMLS 79(2):80

Chen Z, Li J, Gui S, Zhou C, Chen J, Yang C et al (2018) Comparative metaproteomics analysis shows altered fecal microbiota signatures in patients with major depressive disorder. Neuroreport 29(5):417–425

Chiavaroli L, Viguiliouk E, Nishi SK, Blanco Mejia S, Rahelić D, Kahleová H et al (2019) DASH dietary pattern and cardiometabolic outcomes: an umbrella review of systematic reviews and meta-analyses. Nutrients 11(2):338

Chien T, Harimoto T, Kepecs B, Gray K, Coker C, Hou N et al (2021) Enhancing the tropism of bacteria via genetically programmed biosensors. Nat Biomed Eng 6(1):94–104

Chinna Meyyappan A, Milev R (2020) The safety, efficacy, and tolerability of microbial ecosystem Therapeutic-2 in people with major depression and/or generalized anxiety disorder: protocol for a phase 1, open-label study. JMIR Res Protoc 9(6):e17223

Choi JG, Kim N, Ju IG, Eo H, Lim SM, Jang SE et al (2018) Oral administration of Proteus mirabilis damages dopaminergic neurons and motor functions in mice. Sci Rep 8(1):1275

Collins SL, Stine JG, Bisanz JE, Okafor CD, Patterson AD (2023) Bile acids and the gut microbiota: metabolic interactions and impacts on disease. Nat Rev Microbiol 21(4):236–247

Colzato LS, Sellaro R, Beste C (2017) Darwin revisited: the vagus nerve is a causal element in controlling recognition of other's emotions. Cortex J Devoted Study Nerv Syst Behav 92:95–102

Cook J, Prinz M (2022) Regulation of microglial physiology by the microbiota. Gut Microbes 14(1):2125739

Copray JC, Mantingh I, Brouwer N, Biber K, Küst BM, Liem RS et al (2001) Expression of interleukin-1 beta in rat dorsal root ganglia. J Neuroimmunol 118(2):203–211

Corpuz HM, Ichikawa S, Arimura M, Mihara T, Kumagai T, Mitani T et al (2018) Long-term diet supplementation with lactobacillus paracasei K71 prevents age-related cognitive decline in senescence-accelerated mouse prone 8. Nutrients 10(6):762

Cubillos-Ruiz A, Guo T, Sokolovska A, Miller PF, Collins JJ, Lu TK et al (2021) Engineering living therapeutics with synthetic biology. Nat Rev Drug Discov 20(12):941–960

Cummings JH, Pomare EW, Branch WJ, Naylor CP, Macfarlane GT (1987) Short chain fatty acids in human large intestine, portal, hepatic and venous blood. Gut 28(10):1221–1227

Cummings J, Aisen PS, DuBois B, Frölich L, Jack CR, Jones RW et al (2016) Drug development in Alzheimer's disease: the path to 2025. Alzheimers Res Ther 8:39

Cunningham M, Azcarate-Peril MA, Barnard A, Benoit V, Grimaldi R, Guyonnet D et al (2021) Shaping the future of probiotics and prebiotics. Trends Microbiol 29(8):667–685

Dalile B, Van Oudenhove L, Vervliet B, Verbeke K (2019) The role of short-chain fatty acids in microbiota–gut–brain communication. Nat Rev Gastroenterol Hepatol 16(8):461–478

De Filippis F, Pellegrini N, Vannini L, Jeffery IB, La Storia A, Laghi L et al (2016) High-level adherence to a Mediterranean diet beneficially impacts the gut microbiota and associated metabolome. Gut 65(11):1812–1821

Dejanovic B, Wu T, Tsai MC, Graykowski D, Gandham VD, Rose CM et al (2022) Complement C1q-dependent excitatory and inhibitory synapse elimination by astrocytes and microglia in Alzheimer's disease mouse models. Nat Aging 2(9):837–850

Department of Health, Education, and Welfare, National Commission for the Protection of Human Subjects of Biomedical and Behavioral Research (2014) The Belmont report. Ethical principles and guidelines for the protection of human subjects of research. J Am Coll Dent 81(3):4–13

Deverman BE, Patterson PH (2009) Cytokines and CNS development. Neuron 64(1):61–78

Di Gioia D, Bozzi Cionci N, Baffoni L, Amoruso A, Pane M, Mogna L et al (2020) A prospective longitudinal study on the microbiota composition in amyotrophic lateral sclerosis. BMC Med 18(1):153

Diener C, Dai CL, Wilmanski T, Baloni P, Smith B, Rappaport N et al (2022) Genome-microbiome interplay provides insight into the determinants of the human blood metabolome. Nat Metab 4(11):1560–1572

Donald K, Finlay BB (2023) Early-life interactions between the microbiota and immune system: impact on immune system development and atopic disease. Nat Rev Immunol 23(11):735–748

Eckburg PB, Bik EM, Bernstein CN, Purdom E, Dethlefsen L, Sargent M et al (2005) Diversity of the human intestinal microbial flora. Science 308(5728):1635–1638

Erickson MA, Dohi K, Banks WA (2012) Neuroinflammation: a common pathway in CNS diseases as mediated at the blood-brain barrier. Neuroimmunomodulation 19(2):121–130

Ferreiro AL, Choi J, Ryou J, Newcomer EP, Thompson R, Bollinger RM et al (2023) Gut microbiome composition may be an indicator of preclinical Alzheimer's disease. Sci Transl Med 15(700):eabo2984

Fontana F, Longhi G, Tarracchini C, Mancabelli L, Lugli GA, Alessandri G et al (2023) The human gut microbiome of athletes: metagenomic and metabolic insights. Microbiome 11(1):27

Forsythe P, Bienenstock J, Kunze WA (2014) Vagal pathways for microbiome-brain-gut axis communication. In: Lyte M, Cryan JF (eds) Microbial endocrinology: the microbiota-gut-brain axis in health and disease [internet]. Springer, New York, pp 115–133. (Advances in Experimental Medicine and Biology; vol. 817). Available from: https://link.springer.com/10.1007/978-1-4939-0897-4_5. [Cited 2023 May 12]

Foster JA, McVey Neufeld KA (2013) Gut-brain axis: how the microbiome influences anxiety and depression. Trends Neurosci 36(5):305–312

Freestone PP, Williams PH, Haigh RD, Maggs AF, Neal CP, Lyte M (2002) Growth stimulation of intestinal commensal Escherichia coli by catecholamines: a possible contributory factor in trauma-induced sepsis. Shock Augusta Ga 18(5):465–470

Frost G, Sleeth ML, Sahuri-Arisoylu M, Lizarbe B, Cerdan S, Brody L et al (2014) The short-chain fatty acid acetate reduces appetite via a central homeostatic mechanism. Nat Commun 5(1):3611

Fung TC, Olson CA, Hsiao EY (2017) Interactions between the microbiota, immune and nervous systems in health and disease. Nat Neurosci 20(2):145–155

Furness JB, Stebbing MJ (2018) The first brain: species comparisons and evolutionary implications for the enteric and central nervous systems. Neurogastroenterol Motil 30(2). https://doi.org/10.1111/nmo.13234

Furness JB, Callaghan BP, Rivera LR, Cho HJ (2014) The enteric nervous system and gastrointestinal innervation: integrated local and central control. Adv Exp Med Biol 817:39–71

Gao K, Mu CL, Farzi A, Zhu WY (2020) Tryptophan metabolism: a link between the gut microbiota and brain. Adv Nutr Bethesda Md 11(3):709–723

Gautam P, Jogi MK, Pathak A (2021) Novel therapeutic targets in amyotrophic lateral sclerosis. In: Rai SN (ed) Recent advances in the treatment of neurodegenerative disorders [internet]. Bentham Science Publishers, pp 160–177. Available from: https://www.eurekaselect.com/node/192975. [Cited 2025 Jun 14]
Gautam P, Yadav R, Vishwakarma RK, Shekhar S, Pathak A, Singh C (2025a) An integrative analysis of metagenomic and metabolomic profiling reveals gut microbiome dysbiosis and metabolic alterations in ALS: potential biomarkers and therapeutic insights. ACS Chem Neurosci 16:acschemneuro.5c00254
Gautam P, Yadav R, Vishwakarma RK, Pathak A, Singh C (2025b) Metabolic dysregulation in amyotrophic lateral sclerosis: insights from 1H NMR-based metabolomics in a tertiary care center in India. Metab Brain Dis 40(5):196
GeoH S (1949) Doctors of infamy; the story of the Nazi medical crimes. Yale J Biol Med 21(4):362
Ghosh S, Whitley CS, Haribabu B, Jala VR (2021) Regulation of intestinal barrier function by microbial metabolites. Cell Mol Gastroenterol Hepatol 11(5):1463–1482
Góralczyk-Bińkowska A, Szmajda-Krygier D, Kozłowska E (2022) The microbiota-gut-brain axis in psychiatric disorders. Int J Mol Sci 23(19):11245
Gordon JI, Ley RE, Wilson R, Mardis E, Xu J, Fraser CM, et al. Extending our view of self: the human gut microbiome initiative (HGMI). 2006; Available from: https://www.researchgate.net/publication/228348902_Extending_our_view_of_self_the_Human_Gut_Microbiome_Initiative_HGMI. [Cited 2025 Mar 26]
Gray E, Larkin JR, Claridge TDW, Talbot K, Sibson NR, Turner MR (2015) The longitudinal cerebrospinal fluid metabolomic profile of amyotrophic lateral sclerosis. Amyotroph Lateral Scler Front Degener 16(7–8):456–463
Gribble FM, Reimann F (2019) Function and mechanisms of enteroendocrine cells and gut hormones in metabolism. Nat Rev Endocrinol 15(4):226–237
Gulas E, Wysiadecki G, Strzelecki D, Gawlik-Kotelnicka O, Polguj M (2018) Can microbiology affect psychiatry? A link between gut microbiota and psychiatric disorders. Psychiatr Pol 52(6):1023–1039
Habib AM, Richards P, Cairns LS, Rogers GJ, Bannon CAM, Parker HE et al (2012) Overlap of endocrine hormone expression in the mouse intestine revealed by transcriptional profiling and flow cytometry. Endocrinology 153(7):3054–3065
Hattori N, Yamashiro Y (2021) The gut-brain axis. Ann Nutr Metab 77(Suppl. 2):1–3
Helton SG, Lohoff FW (2015) Serotonin pathway polymorphisms and the treatment of major depressive disorder and anxiety disorders. Pharmacogenomics 16(5):541–553
Hill C, Guarner F, Reid G, Gibson GR, Merenstein DJ, Pot B et al (2014) Expert consensus document. The international scientific association for probiotics and prebiotics consensus statement on the scope and appropriate use of the term probiotic. Nat Rev Gastroenterol Hepatol 11(8):506–514
Hirayama M, Ohno K (2021) Parkinson's disease and gut microbiota. Ann Nutr Metab 77(Suppl. 2):28–35
Hitch TCA, Hall LJ, Walsh SK, Leventhal GE, Slack E, de Wouters T et al (2022) Microbiome-based interventions to modulate gut ecology and the immune system. Mucosal Immunol 15(6):1095–1113
Ho L, Ono K, Tsuji M, Mazzola P, Singh R, Pasinetti GM (2018) Protective roles of intestinal microbiota derived short chain fatty acids in Alzheimer's disease-type beta-amyloid neuropathological mechanisms. Expert Rev Neurother 18(1):83–90
Hoegenauer C, Hammer HF, Mahnert A, Moissl-Eichinger C (2022) Methanogenic archaea in the human gastrointestinal tract. Nat Rev Gastroenterol Hepatol 19(12):805–813
Hosking DE, Eramudugolla R, Cherbuin N, Anstey KJ (2019) MIND not Mediterranean diet related to 12-year incidence of cognitive impairment in an Australian longitudinal cohort study. Alzheimers Dement J Alzheimers Assoc 15(4):581–589
Howland RH (2014) Vagus nerve stimulation. Curr Behav Neurosci Rep 1(2):64–73
Hu FB (2003) Sedentary lifestyle and risk of obesity and type 2 diabetes. Lipids 38(2):103–108

Huang B, Chau SWH, Liu Y, Chan JWY, Wang J, Ma SL et al (2023) Gut microbiome dysbiosis across early Parkinson's disease, REM sleep behavior disorder and their first-degree relatives. Nat Commun 14(1):2501

Jumpertz R, Le DS, Turnbaugh PJ, Trinidad C, Bogardus C, Gordon JI et al (2011) Energy-balance studies reveal associations between gut microbes, caloric load, and nutrient absorption in humans. Am J Clin Nutr 94(1):58–65

Kabat-Zinn J (1982) An outpatient program in behavioral medicine for chronic pain patients based on the practice of mindfulness meditation: theoretical considerations and preliminary results. Gen Hosp Psychiatry 4(1):33–47

Kallapura G, Prakash AS, Sankaran K, Manjappa P, Chaudhary P, Ambhore S et al (2024) Microbiota based personalized nutrition improves hyperglycaemia and hypertension parameters and reduces inflammation: a prospective, open label, controlled, randomized, comparative, proof of concept study. PeerJ 12:e17583

Kaul M, Mukherjee D, Weiner HL, Cox LM (2024) Gut microbiota immune cross-talk in amyotrophic lateral sclerosis. Neurother J Am Soc Exp Neurother 21(6):e00469

Kechagia M, Basoulis D, Konstantopoulou S, Dimitriadi D, Gyftopoulou K, Skarmoutsou N et al (2013) Health benefits of probiotics: a review. ISRN Nutr 2013:481651

Kelly JR, Borre Y, O'Brien C, Patterson E, El Aidy S, Deane J et al (2016) Transferring the blues: depression-associated gut microbiota induces neurobehavioural changes in the rat. J Psychiatr Res 82:109–118

Kim MH, Yun KE, Kim J, Park E, Chang Y, Ryu S et al (2020) Gut microbiota and metabolic health among overweight and obese individuals. Sci Rep 10(1):19417

Kim B, Song A, Son A, Shin Y (2024) Gut microbiota and epigenetic choreography: implications for human health: a review. Medicine (Baltimore) 103(29):e39051

Klok MD, Jakobsdottir S, Drent ML (2007) The role of leptin and ghrelin in the regulation of food intake and body weight in humans: a review. Obes Rev 8(1):21–34

Kossoff EH, Zupec-Kania BA, Rho JM (2009) Ketogenic diets: an update for child neurologists. J Child Neurol 24(8):979–988

Kowalski K, Mulak A (2019) Brain-gut-microbiota axis in Alzheimer's disease. J Neurogastroenterol Motil 25(1):48–60

Kozyrskyj AL, Kendall GE, Zubrick SR, Newnham JP, Sly PD (2009) Frequent nocturnal awakening in early life is associated with nonatopic asthma in children. Eur Respir J 34(6):1288–1295

Le Roy CI, Kurilshikov A, Leeming ER, Visconti A, Bowyer RCE, Menni C et al (2022) Yoghurt consumption is associated with changes in the composition of the human gut microbiome and metabolome. BMC Microbiol 22(1):39

Lee SE, Lee Y, Lee GH (2019) The regulation of glutamic acid decarboxylases in GABA neurotransmission in the brain. Arch Pharm Res 42(12):1031–1039

van Leeuwen WMA, Lehto M, Karisola P, Lindholm H, Luukkonen R, Sallinen M et al (2009) Sleep restriction increases the risk of developing cardiovascular diseases by augmenting proinflammatory responses through IL-17 and CRP. PLoS One 4(2):e4589

Lewandowska-Pietruszka Z, Figlerowicz M, Mazur-Melewska K (2022) The history of the intestinal microbiota and the gut-brain axis. Pathog Basel Switz 11(12):1540

Lewis K, Lutgendorff F, Phan V, Söderholm JD, Sherman PM, McKay DM (2010) Enhanced translocation of bacteria across metabolically stressed epithelia is reduced by butyrate. Inflamm Bowel Dis 16(7):1138–1148

Leyrolle Q, Prado-Perez L, Layé S (2023) The gut-derived metabolites as mediators of the effect of healthy nutrition on the brain. Front Nutr 10:1155533

Li Y, Peng N (2019) Endogenous CRISPR-Cas system-based genome editing and antimicrobials: review and prospects. Front Microbiol 10:2471

Lin Z, Jiang T, Chen M, Ji X, Wang Y (2024) Gut microbiota and sleep: interaction mechanisms and therapeutic prospects. Open Life Sci 19(1):20220910

Lippolis T, Cofano M, Caponio GR, De Nunzio V, Notarnicola M (2023) Bioaccessibility and bioavailability of diet polyphenols and their modulation of gut microbiota. Int J Mol Sci 24(4):3813

Little TJ, Horowitz M, Feinle-Bisset C (2005) Role of cholecystokinin in appetite control and body weight regulation. Obes Rev 6(4):297–306

Liu L, Zhu G (2018) Gut-brain axis and mood disorder. Front Psych 9:223

Liu L, Huh JR, Shah K (2022) Microbiota and the gut-brain-axis: implications for new therapeutic design in the CNS. EBioMedicine 77:103908

Livingston G, Huntley J, Sommerlad A, Ames D, Ballard C, Banerjee S et al (2020) Dementia prevention, intervention, and care: 2020 report of the lancet commission. Lancet Lond Engl 396(10248):413–446

Loh JS, Mak WQ, Tan LKS, Ng CX, Chan HH, Yeow SH et al (2024) Microbiota-gut-brain axis and its therapeutic applications in neurodegenerative diseases. Signal Transduct Target Ther 9(1):37

López-Otín C, Blasco MA, Partridge L, Serrano M, Kroemer G (2023) Hallmarks of aging: an expanding universe. Cell 186(2):243–278

Luu M, Pautz S, Kohl V, Singh R, Romero R, Lucas S et al (2019) The short-chain fatty acid pentanoate suppresses autoimmunity by modulating the metabolic-epigenetic crosstalk in lymphocytes. Nat Commun 10(1):760

Lynch SV, Pedersen O (2016) The human intestinal microbiome in health and disease. N Engl J Med 375(24):2369–2379

Ma Y, Guo Z, Xia B, Zhang Y, Liu X, Yu Y et al (2022) Identification of antimicrobial peptides from the human gut microbiome using deep learning. Nat Biotechnol 40(6):921–931

Ma L, Wang HB, Hashimoto K (2025) The vagus nerve: an old but new player in brain-body communication. Brain Behav Immun 124:28–39

Marraffini LA (2015) CRISPR-Cas immunity in prokaryotes. Nature 526(7571):55–61

Matenchuk BA, Mandhane PJ, Kozyrskyj AL (2020) Sleep, circadian rhythm, and gut microbiota. Sleep Med Rev 53:101340

Mayer EA (2011) Gut feelings: the emerging biology of gut-brain communication. Nat Rev Neurosci 12(8):453–466

Mayer EA, Tillisch K (2011) The brain-gut axis in abdominal pain syndromes. Annu Rev Med 62:381–396

Mayer EA, Padua D, Tillisch K (2014) Altered brain-gut axis in autism: comorbidity or causative mechanisms? BioEssays News Rev Mol Cell Dev Biol 36(10):933–939

McCarty CA, Wilke RA, Giampietro PF, Wesbrook SD, Caldwell MD (2005) Marshfield clinic personalized medicine research project (PMRP): design, methods and recruitment for a large population-based biobank. Pers Med 2(1):49–79

McCutcheon RA, Abi-Dargham A, Howes OD (2019) Schizophrenia, dopamine and the striatum: from biology to symptoms. Trends Neurosci 42(3):205–220

McGuire AL, Fisher R, Cusenza P, Hudson K, Rothstein MA, McGraw D et al (2008) Confidentiality, privacy, and security of genetic and genomic test information in electronic health records: points to consider. Genet Med 10(7):495–499

Miller TL, Wolin MJ (1996) Pathways of acetate, propionate, and butyrate formation by the human fecal microbial flora. Appl Environ Microbiol 62(5):1589–1592

Mills S, Stanton C, Lane JA, Smith GJ, Ross RP (2019) Precision nutrition and the microbiome, part I: current state of the science. Nutrients 11(4):923

Mohajeri MH, Brummer RJM, Rastall RA, Weersma RK, Harmsen HJM, Faas M et al (2018) The role of the microbiome for human health: from basic science to clinical applications. Eur J Nutr 57(S1):1–14

Morais LH, Schreiber HL, Mazmanian SK (2021) The gut microbiota-brain axis in behaviour and brain disorders. Nat Rev Microbiol 19(4):241–255

Mörkl S, Butler MI, Wagner-Skacel J (2023) Gut-brain-crosstalk- the vagus nerve and the microbiota-gut-brain axis in depression. A narrative review. J Affect Disord Rep 13:100607

Nankova BB, Agarwal R, MacFabe DF, La Gamma EF (2014) Enteric bacterial metabolites propionic and butyric acid modulate gene expression, including CREB-dependent catecholaminergic neurotransmission, in PC12 cells – possible relevance to autism spectrum disorders. Tsuji Y, editor. PLoS ONE 9(8):e103740

Oakes PC, Fisahn C, Iwanaga J, DiLorenzo D, Oskouian RJ, Tubbs RS (2016) A history of the autonomic nervous system: part I: from Galen to Bichat. Childs Nerv Syst ChNS Off J Int Soc Pediatr Neurosurg 32(12):2303–2308

Odamaki T, Kato K, Sugahara H, Hashikura N, Takahashi S, Zhong XJ et al (2016) Age-related changes in gut microbiota composition from newborn to centenarian: a cross-sectional study. BMC Microbiol 16(1):90

Oleskin AV, Shenderov BA (2016) Neuromodulatory effects and targets of the SCFAs and gasotransmitters produced by the human symbiotic microbiota. Microb Ecol Health Dis [Internet] 27(0). Available from: http://www.microbecolhealthdis.net/index.php/mehd/article/view/30971. [Cited 2025 Mar 26]

Pariante CM, Lightman SL (2008) The HPA axis in major depression: classical theories and new developments. Trends Neurosci 31(9):464–468

Park SH, Lee JH, Kim JS, Kim TJ, Shin J, Im JH et al (2022) Fecal microbiota transplantation can improve cognition in patients with cognitive decline and Clostridioides difficile infection. Aging 14(16):6449–6466

Parrish WR, Rosas-Ballina M, Gallowitsch-Puerta M, Ochani M, Ochani K, Yang LH et al (2008) Modulation of TNF release by choline requires alpha7 subunit nicotinic acetylcholine receptor-mediated signaling. Mol Med Camb Mass 14(9–10):567–574

Patangia DV, Anthony Ryan C, Dempsey E, Paul Ross R, Stanton C (2022) Impact of antibiotics on the human microbiome and consequences for host health. MicrobiologyOpen 11(1):e1260

Perlman JM (2002) Cognitive and behavioral deficits in premature graduates of intensive care. Clin Perinatol 29(4):779–797

Qin J, Li R, Raes J, Arumugam M, Burgdorf KS, Manichanh C et al (2010) A human gut microbial gene catalogue established by metagenomic sequencing. Nature 464(7285):59–65

Rhodes R (2016) Ethical issues in microbiome research and medicine. BMC Med 14(1):156

Romijn AR, Rucklidge JJ (2015) Systematic review of evidence to support the theory of psychobiotics. Nutr Rev 73(10):675–693

Ronald A, Simonoff E, Kuntsi J, Asherson P, Plomin R (2008) Evidence for overlapping genetic influences on autistic and ADHD behaviours in a community twin sample. J Child Psychol Psychiatry 49(5):535–542

Roswall J, Olsson LM, Kovatcheva-Datchary P, Nilsson S, Tremaroli V, Simon MC et al (2021) Developmental trajectory of the healthy human gut microbiota during the first 5 years of life. Cell Host Microbe 29(5):765–776.e3

Sahu M, Vishwakarma RK, Kumar D, Gangwar M, Parmar K, Nath G (2024) Synergistic effects of phage and antibiotic combinations against pathogenic bacteria. In: Gangwar M, Nath G (eds) Emerging paradigms for antibiotic-resistant infections: beyond the pill [internet]. Springer Nature Singapore, pp 435–447. Available from: https://link.springer.com/10.1007/978-981-97-5272-0_20. [Cited 2025 Mar 22]

Saifi S, Ravi V, Sharma S, Swaminathan A, Chauhan NS, Pandey R (2022) SARS-CoV-2 VOCs, mutational diversity and clinical outcome: are they modulating drug efficacy by altered binding strength? Genomics 114(5):110466

Sanzone J, Life M, Reiss D, May D, Hartley B, Spiddle P et al (2024) Uses of fecal microbiota transplantation in neurodegenerative disease: a scoping review. Cureus 16(6):e62265

Sayegh AI, Ritter RC (2000) Vagus nerve participates in CCK-induced Fos expression in hindbrain but not myenteric plexus. Brain Res 878(1–2):155–162

Schemann M, Frieling T, Enck P (2020) To learn, to remember, to forget-how smart is the gut? Acta Physiol Oxf Engl 228(1):e13296

Schluter J, Peled JU, Taylor BP, Markey KA, Smith M, Taur Y et al (2020) The gut microbiota is associated with immune cell dynamics in humans. Nature 588(7837):303–307

Sejbuk M, Mirończuk-Chodakowska I, Karav S, Witkowska AM (2024) Dietary polyphenols, food processing and gut microbiome: recent findings on bioavailability, bioactivity, and gut microbiome interplay. Antioxid Basel Switz 13(10):1220

Seo DO, O'Donnell D, Jain N, Ulrich JD, Herz J, Li Y et al (2023) ApoE isoform- and microbiota-dependent progression of neurodegeneration in a mouse model of tauopathy. Science 379(6628):eadd1236

Simrén M, Barbara G, Flint HJ, Spiegel BMR, Spiller RC, Vanner S et al (2013) Intestinal microbiota in functional bowel disorders: a Rome foundation report. Gut 62(1):159–176

Singh B, Parsaik AK, Mielke MM, Erwin PJ, Knopman DS, Petersen RC et al (2014) Association of mediterranean diet with mild cognitive impairment and Alzheimer's disease: a systematic review and meta-analysis. J Alzheimers Dis JAD 39(2):271–282

Stamp LA (2017) Cell therapy for GI motility disorders: comparison of cell sources and proposed steps for treating Hirschsprung disease. Am J Physiol Gastrointest Liver Physiol 312(4):G348–G354

Steer E (2019) A cross comparison between ayurvedic etiology of major depressive disorder and bidirectional effect of gut dysregulation. J Ayurveda Integr Med 10(1):59–66

Stilling RM, Van De Wouw M, Clarke G, Stanton C, Dinan TG, Cryan JF (2016) The neuropharmacology of butyrate: the bread and butter of the microbiota-gut-brain axis? Neurochem Int 99:110–132

Strandwitz P (2018) Neurotransmitter modulation by the gut microbiota. Brain Res 1693(Pt B):128–133

Strandwitz P, Kim KH, Terekhova D, Liu JK, Sharma A, Levering J et al (2019) GABA-modulating bacteria of the human gut microbiota. Nat Microbiol 4(3):396–403

Sweitzer S, Martin D, DeLeo JA (2001) Intrathecal interleukin-1 receptor antagonist in combination with soluble tumor necrosis factor receptor exhibits an anti-allodynic action in a rat model of neuropathic pain. Neuroscience 103(2):529–539

Sydenham E, Dangour AD, Lim WS (2012) Omega 3 fatty acid for the prevention of cognitive decline and dementia. Cochrane Database Syst Rev 2012(6):CD005379

Szentirmai É, Millican NS, Massie AR, Kapás L (2019) Butyrate, a metabolite of intestinal bacteria, enhances sleep. Sci Rep 9(1):7035

Tanaka T, Katsuma S, Adachi T, Koshimizu T-a, Hirasawa A, Tsujimoto G (2008) Free fatty acids induce cholecystokinin secretion through GPR120. Naunyn Schmiedebergs Arch Pharmacol 377(4–6):523–527

Taylor BC, Lejzerowicz F, Poirel M, Shaffer JP, Jiang L, Aksenov A et al (2020) Consumption of fermented foods is associated with systematic differences in the gut microbiome and metabolome. mSystems 5(2):e00901–e00919

Tillisch K, Labus J, Kilpatrick L, Jiang Z, Stains J, Ebrat B et al (2013) Consumption of fermented milk product with probiotic modulates brain activity. Gastroenterology 144(7):1394–1401, 1401.e1–4

Tillisch K, Mayer EA, Gupta A, Gill Z, Brazeilles R, Le Nevé B et al (2017) Brain structure and response to emotional stimuli as related to gut microbial profiles in healthy women. Psychosom Med 79(8):905–913

Udit S, Gautron L (2013) Molecular anatomy of the gut-brain axis revealed with transgenic technologies: implications in metabolic research. Front Neurosci 7:134

Vanuytsel T, van Wanrooy S, Vanheel H, Vanormelingen C, Verschueren S, Houben E et al (2014) Psychological stress and corticotropin-releasing hormone increase intestinal permeability in humans by a mast cell-dependent mechanism. Gut 63(8):1293–1299

Vendrik KEW, Ooijevaar RE, de Jong PRC, Laman JD, van Oosten BW, van Hilten JJ et al (2020) Fecal microbiota transplantation in neurological disorders. Front Cell Infect Microbiol 10:98

Wagner R, Myers RR (1996) Endoneurial injection of TNF-alpha produces neuropathic pain behaviors. Neuroreport 7(18):2897–2901

Wegierska AE, Charitos IA, Topi S, Potenza MA, Montagnani M, Santacroce L (2022) The connection between physical exercise and gut microbiota: implications for competitive sports athletes. Sports Med Auckl NZ 52(10):2355–2369

Westfall S, Lomis N, Prakash S (2019) A novel synbiotic delays Alzheimer's disease onset via combinatorial gut-brain-axis signaling in Drosophila melanogaster. PLoS One 14(4):e0214985

Xiao L, Zhao F (2024) Exploring the frontier of microbiome biomarker discovery with artificial intelligence. Natl Sci Rev 11(11):nwae325

Yamamoto K, Kushida M, Tsuduki T (2018) The effect of dietary lipid on gut microbiota in a senescence-accelerated prone mouse model (SAMP8). Biogerontology 19(5):367–383

Yang T, Ahmari N, Schmidt JT, Redler T, Arocha R, Pacholec K et al (2017) Shifts in the gut microbiota composition due to depleted bone marrow beta adrenergic signaling are associated with suppressed inflammatory transcriptional networks in the mouse colon. Front Physiol 8:220

Yano JM, Yu K, Donaldson GP, Shastri GG, Ann P, Ma L et al (2015) Indigenous bacteria from the gut microbiota regulate host serotonin biosynthesis. Cell 161(2):264–276

Yao H, Fan C, Fan X, Lu Y, Wang Y, Wang R et al (2020) Effects of gut microbiota on leptin expression and body weight are lessened by high-fat diet in mice. Br J Nutr 124(4):396–406

Yatsunenko T, Rey FE, Manary MJ, Trehan I, Dominguez-Bello MG, Contreras M et al (2012) Human gut microbiome viewed across age and geography. Nature 486(7402):222–227

Zapata HJ, Quagliarello VJ (2015) The microbiota and microbiome in aging: potential implications in health and age-related diseases. J Am Geriatr Soc 63(4):776–781

Zhang JM, An J (2007) Cytokines, inflammation, and pain. Int Anesthesiol Clin 45(2):27–37

Zhang F, Luo W, Shi Y, Fan Z, Ji G (2012) Should we standardize the 1,700-year-old fecal microbiota transplantation? Am J Gastroenterol 107(11):1755; author reply p. 1755–1756

Zhang YG, Wu S, Yi J, Xia Y, Jin D, Zhou J et al (2017) Target intestinal microbiota to alleviate disease progression in amyotrophic lateral sclerosis. Clin Ther 39(2):322–336

Zhou Z, Chen X, Sheng H, Shen X, Sun X, Yan Y et al (2020) Engineering probiotics as living diagnostics and therapeutics for improving human health. Microb Cell Factories 19(1):56

Zhu M, Liu X, Ye Y, Yan X, Cheng Y, Zhao L et al (2022) Gut microbiota: a novel therapeutic target for Parkinson's disease. Front Immunol 13:937555

Zimmermann M, Zimmermann-Kogadeeva M, Wegmann R, Goodman AL (2019) Mapping human microbiome drug metabolism by gut bacteria and their genes. Nature 570(7762):462–467

Neurogenesis and Brain Plasticity: Nutraceutical Enhancers

10

Krishnendu Adhikary, Riya Sarkar, Baisakhi Mukherjee, Ishani Sadhukhan, and Adrija Gupta

Abstract

Neurogenesis and brain plasticity are essential processes underlying cognition and adaptability throughout life. In contrast to previous speculations, adult neurogenesis is ongoing, particularly in the hippocampus, a brain area of interest in learning and memory. Various factors like aging, stress, and neurodegenerative disease can perturb these processes. Nutraceuticals—biologically active food-derived molecules—have been identified as potential candidates to trigger synaptic plasticity and neurogenesis. Omega-3 fatty acids, flavonoids, curcumin, and phosphatidylserine are some molecules with neuroprotection by reducing oxidative stress, modulating inflammation, and promoting synaptic growth. These bioactive molecules account for cognitive health by modulating neurotransmitter systems, regulating neuronal survival, and inducing long-term potentiation. Lifestyle treatments like exercise, stress management, and nutritional balance also optimize neurogenesis and brain plasticity. While promising outcomes have been noted, optimal dosages, bioavailability, and long-term effec-

K. Adhikary (✉) · B. Mukherjee
Department of Medical Laboratory Technology, Paramedical College Durgapur, Durgapur, West Bengal, India

R. Sarkar
Department of Medical Lab Technology, Dr. B. C. Roy Academy of Professional Courses, Durgapur, West Bengal, India

I. Sadhukhan
Department of Food Processing, Indian Institute of Engineering Science and Technology, Shibpur, Howrah, West Bengal, India

A. Gupta
Department of Microbiology, University of Calcutta, Ballygunge Circular Road, Kolkata, West Bengal, India

A. Kumar Singh, S. Nand Rai (eds.), *Rejuvenating the Brain: Nutraceuticals, Autophagy, and Longevity*, Nutritional Neurosciences,
https://doi.org/10.1007/978-981-95-2790-8_10

tiveness of nutraceutical therapy need to be determined. Standardized guidelines must be drawn by further studies to ensure application to cognitive health practice. The use of nutraceuticals in inducing neurogenesis and brain plasticity and the evidence supporting that they are an organic and holistic remedy for the prevention of cognitive dysfunction and neurodegenerative disease are discussed in this chapter. With the combination of a healthy lifestyle and nutraceuticals, individuals can optimize brain function, enable memory retention, and achieve cognitive resilience throughout life. Specific molecular mechanisms and clinical applications need to be delineated in future studies to ensure the therapeutic action of these molecules in neurology.

Keywords

Neurogenesis · Brain Plasticity · Nutraceuticals · Cognitive Health · Neuroprotection

10.1 Introduction

The human brain is a dynamic one with a vast potential for adaptation, reorganization, and the production of new neurons—a phenomenon called neurogenesis (Gage 2019). The human brain was previously believed to produce new neurons only at birth; however, recent evidence implies that the adult brain can also produce new neurons, which occur in the hippocampus, an area implicated in memory and learning (Boldrini et al. 2018). The process is intrinsically linked with brain plasticity, a feature of the brain to adapt its form and function as per experience, learning, and environmental effect (Mateos-Aparicio and Rodríguez-Moreno 2019). Neurogenesis and brain plasticity are significant in cognitive function, emotional processing, and neurological resilience. The processes are vulnerable to aging, stress, neurodegenerative disease, and harmful living (Kumar et al. 2022). The past decade has seen increased concern with the potential role of nutraceuticals—biologically active food-derived molecules—in the preservation of neurological health (Mohajeri et al. 2021). Nutraceuticals are a wide range of substances from polyphenols, omega-3 fatty acids, flavonoids, to vitamins, all of which have been proven to be neuroprotective (Parletta et al. 2019). The substances manifest their protective actions by initiating a multitude of mechanisms from the inhibition of oxidative stress, modulation of inflammation, to the induction of synaptic plasticity (Freeman et al. 2020). For instance, omega-3 fatty acids of fish oil have been demonstrated to trigger neurogenesis and synaptic plasticity and thus enhance cognitive function and the prevention of neurodegenerative diseases such as Alzheimer's (Dyall 2015). Likewise, flavonoids found in berries and dark chocolate were also found to improve memory as well as neuron preservation from oxidative damage (Spencer 2018).

With growing neurologic disorders, i.e., Alzheimer's disease, Parkinson's disease, and depression, there is a changing need to find safe and effective interventions in brain health promotion (Wu et al. 2021). The traditional pharmacological treatments are accompanied by catastrophic side effects and are unable to treat the underlying mechanisms of neurologic dysfunction. In contrast, nutraceuticals offer a holistic and natural alternative by enhancing neurogenesis, brain plasticity, and inhibition of neuroinflammation (Singh et al. 2020). Incorporation of nutraceuticals into a daily routine can potentially yield a preventive effect, retarding the rate of onset of cognitive impairment and improving mental well-being (Bjørklund et al. 2022).

Furthermore, evidence from the literature reveals that the synergistic action of nutraceuticals with other life interventions, e.g., physical exercise, cognitive training, and stress management, can augment their effectiveness in further preserving brain function (Fernandes et al. 2020). Exercise, for example, has been shown to increase the expression of brain-derived neurotrophic factor (BDNF), the central protein that boosts neurogenesis and synaptic plasticity (Szuhany et al. 2015). In the same vein, meditation and mindfulness have been linked to cognitive flexibility and greater cortical thickness, providing the synergistic benefit of holistic living in maintaining brain health (Tang et al. 2015). This paper presents the complex interaction among neurogenesis, brain plasticity, and the consumption of nutraceuticals in the maintenance of optimal neurological function. By reviewing the scientific basis underlying these interactions, this discussion aims to highlight the promise of nutraceuticals as an effective intervention for cognitive enhancement and neuroprotection against neurodegenerative diseases.

10.2 Understanding Neurogenesis and Brain Plasticity

10.2.1 Definitions and Mechanisms of Neurogenesis

Neurogenesis is how neural stem cells (NSCs) become new neurons that are involved in nervous system development and repair. It's a trait common in almost all animals, with the exception of placozoans and sponges (Porifera). Examples of the kinds of NSCs are neuroepithelial cells (NECs), radial glial cells (RGCs), basal progenitors (BPs), intermediate neuronal precursors (INPs), and astrocytes in the subventricular and subgranular zones that are involved in neurogenesis. Although this activity is highest in embryonic stages when the whole neural network could be established, it still takes place in some parts of the adult brain in other species (Garcıa and Alvarez-Buylla 2002). After neurons are produced, they never divide again, and most survive an organism's entire life unless attacked by disease or injury. The best example is the hippocampus, a brain region responsible for memory and orientation, consisting of at least 2 eurons of various subtypes (Bergmann et al. 2015). In spite of all research carried out, the precise processes involved in neurogenesis and the staggering number of neurons are among science's unsolved mysteries.

10.2.1.1 Neural Stem and Progenitor Cells

Neural progenitor cells (NPCs) are a particular subtype of NSCs with the ability to develop into neurons or glial cells, such as oligodendrocytes and astrocytes (Aigner et al. 2018). As opposed to the broader differentiation capacity of NSCs, NPCs are more undifferentiated in their fate. NSCs possess multipotent properties in the sense that they can give rise to the major cell types that form the brain, including glial cells and neurons (Wexler et al. 2009).

10.2.1.2 Mechanism of Neurogenesis

The process of neurogenesis involves the generation of neurons from NSCs, contributing to the growth and maintenance of the nervous system (Fig. 10.1). Notably, this process does not occur in sponges or placozoans, making them exceptions in the animal kingdom. Several key NSC types, including NECs, RGCs, BPs, INPs, and astrocytes in the subventricular and subgranular zones, are responsible for neuronal generation (Tartt et al. 2018). While neurogenesis is crucial for the formation of neural circuits during embryonic development, it also continues in specific brain regions during adulthood in various species. Once generated, neurons generally do not undergo further division and can persist for the entirety of an organism's life, except in cases of disease or neurological disorders (Kang et al. 2016).

Molecular and genetic factors primarily control the rate of neurogenesis and the kind of neuron formed (generally, excitatory or inhibitory). The Notch signaling pathway is one of these factors, and the regulation of the Notch pathway has been connected to numerous genes. As axons and dendrites grow, neurons do not instantly create neural circuits (Broadbent et al. 2009). Rather, in order to mature and eventually produce brain circuitry, young neurons must first migrate great distances to reach their final destinations. For instance, the cerebral cortex is formed by the

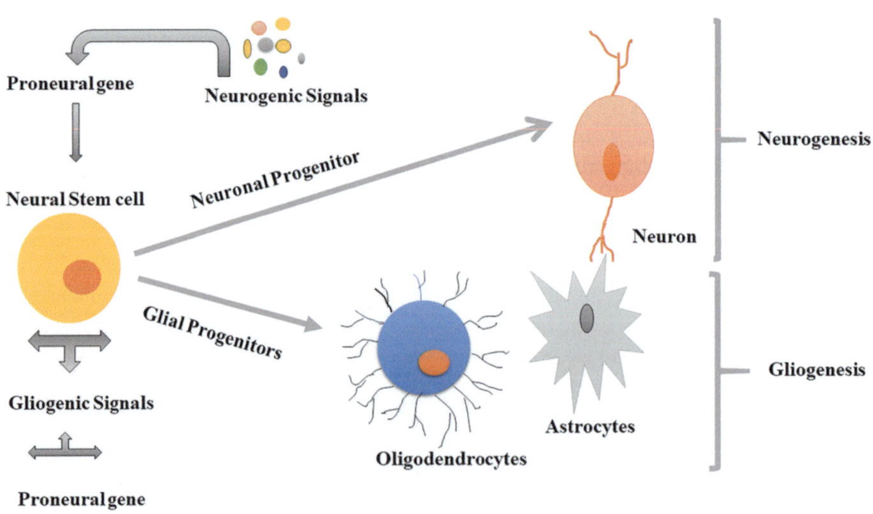

Fig. 10.1 Different factors affecting neurogenesis and gliogenesis

accumulation of neurons that migrate radially from the ventricular zone to the cortical plate. Therefore, the "neurogenic niche" that their parent stem cells occupy is the precise tissue compartment where neurons are generated (Andreu et al. 2010).

10.2.2 Phases of Neurogenesis

(a) *Proliferation*:

Neurogenesis is initiated by the proliferation of NSCs and NPCs. Cells divide by mitosis in regions like the subgranular zone (SGZ) of the hippocampus or the subventricular zone (SVZ) lining the lateral ventricles. The daughter cells formed at this stage can either remain progenitor cells (self-renewal) or differentiate into a new neural precursor.

(b) *Differentiation*:

Transcription factors such as NeuroD1, Mash1, Ngn1, and Ascl1 direct progenitors to differentiate into neurons instead of glial cells, whereas molecular signals such as Notch, Wnt, BMPs, and Sonic Hedgehog (Shh) regulate the process by which NPCs mature into immature neurons after proliferation (Aigner et al. 2012).

(c) *Migration*:

Then, the newly formed neurons migrate to their final positions. For example, these neurons migrate to the granule cell layer (GCL) of the dentate gyrus of the hippocampus. Immature neuron migration is tightly controlled by signaling pathways, including those involving chemokines, guidance molecules, and cell adhesion molecules (Faigle and Song 2013).

(d) *Maturation*:

This includes becoming integrated into pre-existing brain pathways, axonal and dendritic elongation, and the development of synapses (Tchorz et al. 2014). These developing neurons form effective connections with existing neurons in the brain through the process of synaptic plasticity which is essential in processes such as memory consolidation and learning.

10.2.3 The Role of Neurogenesis in Cognitive Function

10.2.3.1 Role in Memory Resolution

Regarding neurogenesis function, a somewhat different viewpoint focuses on what new neurons encode rather than how they encode. The other main function that the DG is thought to have, in addition to its pattern separation function, is conjunctive encoding. Conjunctive encoding is essentially the process by which the DG (as well as the downstream hippocampus) integrates both spatial and nonspatial information into memories (Schouten et al. 2014).

10.2.3.2 A Neurogenesis Pattern Separation Function Is Supported by the Following

Although the computational mechanisms underlying a link between neurogenesis and pattern separation remain incompletely understood, behavioral studies utilizing various neurogenesis knockdown techniques have demonstrated a robust correlation between neurogenesis and spatial discriminating tests (Innamorati et al. 2012). The two-choice discrimination test, fear context discrimination, and the radial-arm maze are some of the spatial discrimination tasks that have been employed.

10.2.3.3 Adult Neurogenesis Function in Temporal Context Encoding

A related possible function for new neurons offers a fresh viewpoint on DG and hippocampus memory formation, even if the pattern separation function mentioned above has a lengthy history in the DG (Alkass et al. 2013). According to one theory, by preventing fresh and old memories from interfering with one another, teaching learning to young neurons can enhance pattern separation. However, when the process is seen from a temporal viewpoint, the direct pattern separation role of new neurons becomes most apparent.

10.2.4 Plasticity Across the Lifespan: From Development to Aging

- *Neuro-Plasticity*

The brain's ability to adapt is one of its fundamental characteristics. The malleability of neural circuitry and connectivity is referred to by neuroscientists as "plasticity." Cellular physical changes show up as circuit-level modifications in neuronal firing patterns, and it is these circuit-level modifications that enable learning, memory, and adaptation to shifting environmental and physiological situations (Hurley et al. 2012). These cells may receive and send electrical activity on a millisecond scale.

10.2.5 Factors Influencing Neurogenesis and Brain Plasticity

- A synapse is formed when a neuron joins another neuron. A few single positive impulses are typically insufficient to trigger a neuron's response; instead, a neuron must receive numerous positive impulses almost simultaneously in order for it to fire (Mathews et al.). Additionally, the cumulative excitatory signals must be greater than the cumulative inhibitory signals in order to fire (Table 10.1).

Table 10.1 Factors influencing neurogenesis and brain plasticity

Nutrients	Source	Impact on brain function
Fatty acids that are omega-3 (like docosahexaenoic acid)	Walnuts, butternuts, krill, chia, kiwi fruit, flax seeds, and salmon	Reduction of cognitive decline in the elderly; Foundation for mood disorder treatment; enhancement of cognitive function in rodents with traumatic brain damage; Reduction of cognitive decline in an Alzheimer's disease mouse model.
Flavonoids	Citrus fruits, wine (especially red wine), dark chocolate, green tea, and ginkgo trees	Exercise-induced cognitive enhancement in rodents. Cognitive function improvement in the elderly.
Calcium, selenium	Calcium is found in milk and coral. Foods high in selenium include nuts, cereals, meat, fish, and eggs.	In older adults, a higher serum calcium level is linked to a quicker rate of cognitive deterioration. Humans with lifetime low selenium levels have poorer cognitive performance.
Iron	Fish, poultry, lentils, beans, and red meat	Iron therapy helps young women's cognitive function return to normal.
Copper	Brazil nuts, blackstrap molasses, cocoa, black pepper, oysters, and beef or lamb liver	Low copper plasma levels in Alzheimer's disease patients are associated with cognitive deterioration.
Vitamin B complex	Several natural sources. Vitamin B12 cannot be obtained from plant-based products.	Memory performance in women of all ages is improved by taking supplements of vitamin B6, vitamin B12, or folate; in rats given a diet lacking in choline, vitamin B12 reduces cognitive impairment.
Vitamin E	Red palm oil, avocado, peanuts, almonds, olives, and asparagus	Reduction of cognitive deterioration in the elderly and amelioration of cognitive impairment following brain trauma in rodents,
Vitamin D	Fortified foodstuffs, mushrooms, fish liver, fatty fish, milk, soy milk,	Crucial for maintaining older people's cognitive abilities.

- Almost all kids pick up their native tongue with ease, using proper grammar and pronunciation. The process has become time-consuming and essentially restricted. From a neurological standpoint, this suggests that the brain has a window of opportunity to become particularly proficient in language, after which the window closes and learning a new language becomes extremely challenging. This window of opportunity is known as the "sensitive" or "critical" phase.
- The proper innervation of muscles by nerves requires spontaneous bursts of depolarization in the spinal cord; if the frequency of these depolarizations is changed, axons will innervate the incorrect muscles. In this instance, spontaneous activity is believed to affect axon guidance by means of the previously described chemical pathways (Mateo et al. 2013).

10.3 Nutraceuticals: A New Frontier in Brain Health

10.3.1 What Are Nutraceuticals?

The term "nutraceutical," coined in 1989 by Dr. Stephen De Felice, Chairman of the Foundation for Innovation in Medicine, combines the concepts of nutrition and pharmaceuticals. It refers to nutritional supplements marketed to treat or prevent diseases, although it lacks a formal regulatory definition. Essentially, a nutraceutical can be any substance classified as food or part of food that provides medical or health benefits, such as disease prevention and treatment.

Nutraceuticals encompass a broad range of products, including isolated nutrients, dietary supplements, genetically modified foods, herbal products, and processed foods like cereals, soups, and beverages. The term "dietary supplements" is often used interchangeably with nutraceuticals (Souyoul et al. 2018). According to the Dietary Supplement Health and Education Act (DSHEA), a dietary supplement is defined as a product meant to enhance the diet, containing one or more ingredients such as vitamins, minerals, herbs, amino acids, or other dietary substances. These products are considered to enhance the day-by-day addition of individual nutrients and are reduced in liquids, tablets, capsules, or pills. They are not meant to replace conventional food or serve as the sole component of a meal, and they must be labeled as "dietary supplements." This category also includes items like approved new drugs, certified antibiotics, or licensed biologics that were sold as dietary supplements or food before their official approval or certification. The difference between nutraceuticals and dietary supplements is subtle. While dietary supplements aim to enhance overall nutrition, nutraceuticals are marketed not only to supplement the diet but also to prevent or treat diseases or disorders. Furthermore, nutraceuticals are often positioned as conventional food or as the primary component of a meal.

Nutraceuticals can be classified in several ways (Table 10.2):

1. By chemical composition: including phenols, alkaloids, fibers, fatty acids, minerals, terpenes, etc.
2. Traditional vs. Nontraditional: such as fruits, vegetables, or fortified juices.
3. By type of food: like vitamins, minerals, or dairy products.

There has long been concern about the potential benefits and side effects of nutraceuticals. Research suggests that nutraceuticals can boost immunity, with compounds such as flavonoids, green tea, and quercetin (found in onions) showing positive effects. They may also support recovery from degenerative diseases, including substances like vitamin E, creatine, and turmerone. Beyond these benefits, nutraceuticals can help prevent or manage chronic conditions, such as heart disease, through the use of antioxidants, fiber, minerals, vitamins, and flavonoids.

In addition, some nutraceuticals play a role in addressing obesity, with soy foods, flavonoids, and green tea extracts proving particularly effective. Certain nutraceuticals, including green tea, dietary fibers, and antioxidants, have also shown

Table 10.2 Category of nutraceuticals

Category	Examples	Benefits
Probiotics	*Lactobacillus*, *Bifidobacterium*	Gut health, immune system support
Vitamins	Vit C, D, E	Antioxidants, immunity boosting
Minerals	Ca, iron, zinc	Bone health, energy metabolism
Herbal supplement	Turmeric, ginger, *Ginkgo biloba*	Anti-inflammatory properties, antioxidant properties
Omega-3 fatty acids	EPA, DHA	Heart health, brain function boosting, anti-inflammatory properties
Antioxidants	Coenzyme, polyphenols	Reduce oxidative stress, protects from cell damage
Digestive enzymes	Amylase, lipase, protease	Reduce IBS symptoms, promotes protein digestion
Prebiotics	Inulin, fructooligosaccharides	Maintains gut health, promotes the number of gut health bacteria

antidiabetic properties. Moreover, compounds like lycopene, soy foods, and saponins from vegetables like spinach, tomatoes, and potatoes have demonstrated potential anticancer effects.

10.3.2 Mechanisms of Action in the Brain

The brain's neurotransmitter system functions through a complex network of chemical messengers, and nutraceuticals are crucial in influencing this system. For instance, 5-HTP, derived from Griffonia simplicifolia seeds, has been transformed into serotonin by the enzyme aromatic L-amino acid decarboxylase. This process is important for mood regulation and psychological well-being and mood regulation. Additionally, nutraceuticals have changed the accuracy of neurotransmitter receptors, as illustrated with *Mucuna pruriens*, which includes L-DOPA. This compound changes into dopamine in the brain, enhancing motivation and increasing cognitive processing.

However, some nutraceuticals also delay monoamine oxidase (MAO) enzymes, which helps manage neurotransmitter levels in synaptic clefts and improves a certain neurochemical environment. For instance, Lion's Mane mushroom has the capacity to increase neuroplasticity by inducing Brain-Derived Neurotrophic Factor (BDNF). This begins a chain of molecular interactions that facilitate neuronal growth and synapse formation by gene expression control and protein synthesis. Such modifications are important on a biological level. Neurons develop new axons and dendrites, leading to a complex branching structure. Nerve growth factor (NGF) describes neurite growth in interaction with its receptor, tropomyosin receptor kinase (TrkA). This process may be represented as the brain changing itself, with new neural pathways cutting through the dense forest of information and continually altering in response to new learning.

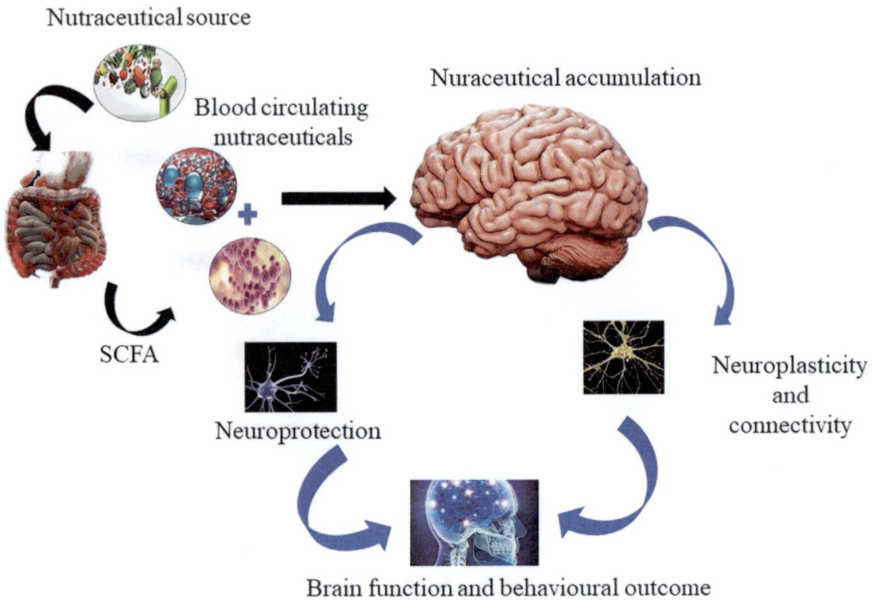

Fig. 10.2 Nutraceuticals in brain function

The brain is constantly subjected to oxidative stress, with free radicals compromising cell integrity. Nutraceuticals function as sophisticated molecular protectors, going beyond just neutralizing these free radicals. Curcumin, produced from turmeric, is an excellent example of this protective action. Curcumin induces a defensive response in the brain by activating the nuclear factor erythroid 2-related factor 2 (Nrf2), a major regulator of cellular protection. This activation causes the body to boost antioxidant response components and create defense enzymes like superoxide dismutase, which aggressively seek out and neutralize dangerous free radicals.

Furthermore, nutraceutical antioxidants and phosphatidylserine promote cellular protection and communication, notably in the brain (Fig. 10.2). Antioxidants protect mitochondria from lipid peroxidation, allowing the electron transport chain and oxidative phosphorylation to operate properly, both of which are necessary for nerve cell health. Phosphatidylserine is essential for strengthening cellular communication, increasing membrane fluidity, and promoting better interactions between proteins and signaling molecules. This enhancement in cellular dynamics allows for greater receptor function and more efficient signal transmission, including improved G-protein coupling and calcium signaling. Finally, these nutraceuticals promote a more sensitive and efficient neural communication system, which improves the brain's ability to process and transfer information.

10.3.3 The Growing Interest in Brain-Boosting Supplements

When investigating nutraceuticals as potential stroke therapeutics, it is vital to understand their ability to reduce stroke risk, increase ischemic tolerance, and alleviate post-stroke symptoms. However, it is vital to stress the need for doing thorough preclinical research in order to fully understand their advantages and limitations in clinical settings. For example, a study on Asparagus racemosus, a plant used in traditional medicine, stresses the importance of standardizing herbal formulations and determining appropriate dosages and contraindications before proceeding with clinical trials in neuromedicine. Similarly, phytochemicals derived from Azadirachta indica, another plant with long-standing therapeutic usage, show promise in treating neurological disorders. However, warn that additional research into potential neurotoxicity is needed before completely assessing their therapeutic efficacy. There is growing interest in combining nutraceuticals with nanocarriers to improve bioavailability and blood-brain barrier bridging. However, such combinations can disrupt metabolic pathways and transport systems, needing careful consideration.

10.4 Key Nutraceuticals in Enhancing Neurogenesis

10.4.1 Omega-3 Fatty Acids and Their Effects on the Brain

Omega-3 fatty acids are vital for brain health, especially because they supply docosahexaenoic acid (DHA), which is required for neural tissue formation and function. Insufficient omega-3 consumption can lower DHA levels while raising omega-6 fatty acids in the brain. This imbalance may have a deleterious influence on neurogenesis, neurotransmitter function, and cognitive functions including learning and visual processing. The current Western diet frequently lacks appropriate supplies of omega-3 s, such as alpha-linolenic acid from plant oils and DHA from seafood, which can worsen these problems. Maternal DHA levels are important for babies' early brain development. The DHA level of neonates and breastfed babies is directly influenced by the mother's diet, and it can vary greatly. Research has indicated that low maternal DHA levels are linked to a greater risk of poor brain development in offspring. On the other hand, research suggests that providing adequate DHA consumption during pregnancy may promote improved neurological and visual development in babies. As a result, good maternal nutrition is critical for delivering enough DHA to the growing fetus and newborn, with long-term consequences for brain development. Individuals' omega-3 dietary needs can also be influenced by genetic variances in fatty acid metabolism. Because DHA is an important component of synaptic membranes, particularly in the form of phosphatidylserine (PS), maintaining a healthy balance of omega-3 and omega-6 fatty acids is critical during pregnancy, nursing, and early infancy. Interestingly, despite consuming little dietary DHA, herbivores retain high DHA levels in their cerebral tissues. DHA and PS concentrations increase as the brain ages, but phosphatidylcholine (PC) levels decrease. The fusion of lipid transport vesicles with the plasma membrane promotes lipid synthesis, which is required for neural membrane development. This mechanism, like neurotransmitter release, requires SNARE proteins and is regulated by omega-3 and omega-6 fatty acid supply, including DHA (Bourre 2004).

10.4.2 Polyphenols and Antioxidants

Polyphenols and other antioxidants are believed to protect cellular components from oxidative damage by neutralizing free radicals. Cells primarily respond to polyphenols through direct interactions with receptors or enzymes involved in signal transduction, potentially altering the cell's redox status and triggering a cascade of redox-dependent reactions. Polyphenols exhibit both antioxidant and prooxidant properties, which can lead to varying effects on cellular physiological processes (Scalbert et al. 2005). Polyphenols can promote cell survival as antioxidants, while as prooxidants, they may induce apoptosis and inhibit tumor growth. However, the biological effects of polyphenols likely extend beyond their role in oxidative stress. A notable example is the interaction of soy isoflavones with estrogen receptors, which impacts endocrine function. This interaction may help explain how isoflavones can prevent bone resorption in postmenopausal women.

10.4.3 Curcumin Neuroprotective Properties

Curcumin, a naturally occurring polyphenol extracted from the rhizome of *Curcuma longa*, has gained significant attention for its potential therapeutic role in treating various neurodegenerative diseases. According to research, it may be beneficial in ailments such as Parkinson's Disease (PD), Alzheimer's Disease (AD), Huntington's Disease (HD), Multiple Sclerosis (MS), Amyotrophic Lateral Sclerosis (ALS), and prion disorders. Curcumin's potent anti-inflammatory, antioxidant, anticancer, immunomodulatory, neuroprotective, and antibacterial effects have sparked worldwide attention. For millennia, it has been an essential component of traditional medicinal and nutritional practices in India, China, and other Asian nations. Despite its potential pharmacological benefits, curcumin's clinical use is limited due to a number of difficulties. These include its low water solubility, instability in the circulation, fast metabolism, limited bioavailability, and difficulties crossing the blood-brain barrier (BBB) (Cole et al. 2007). Researchers have investigated numerous nanocarrier technologies to overcome these restrictions and improve brain delivery. Advanced approaches for improving curcumin absorption and stability include liposomes, micelles, dendrimers, cubosome nanoparticles, polymer nanoparticles, and solid lipid nanoparticles. Functionalizing these nanocarriers with brain-targeting ligands improves their capacity to reach neural tissues, boosting bioavailability while reducing negative effects (Genchi et al. 2024).

10.4.4 Phosphatidylserine and Brain Cell Communication

Phosphatidylserine is the most acidic phospholipid found in human membranes, accounting for 2–20% of the total phospholipid content in adult human plasma and intracellular membranes. Myelin in a healthy brain includes a high quantity of phosphatidylserine, which grows twice in gray matter from birth to age 80 (Glade and

Smith 2015). Phosphatidyl serine is a structural component of the endoplasmic reticulum, nuclear envelopes, Golgi apparatus, inner plasma membrane leaflets, outer mitochondrial membranes, and myelin (Svennerholm 1968). Approximately 20–30% of phosphatidyl serine in human gray matter resides as 1-stearoyl-2-docos ahexaenoyl-sn-glycero-3-phosphoserine (Kim et al. 2014). The presence of docosahexaenoic acid (DHA) in neuronal phosphatidyl serine is vital for its function, as a reduction in DHA levels in the cortex is associated with the progression of mild cognitive impairment to Alzheimer's disease. Thus, incorporating phosphatidyl serine into human membranes depends on the availability of both phosphatidyl serine and DHA. Exogenous phosphatidyl serine, in doses of 300–800 mg daily, is efficiently absorbed by humans, crosses the blood-brain barrier, and can safely slow, stop, or reverse biochemical changes and structural damage in nerve cells (Choi et al. 2018). It also supports various cognitive functions, including short-term memory formation, long-term memory consolidation, memory retrieval, learning and recalling information, attention, reasoning, problem-solving, language skills, communication, motor functions, rapid reactions, and reflexes.

10.5 Nutraceuticals in Promoting Brain Plasticity

10.5.1 The Concept of Brain Plasticity

There are two types of plasticity: structural and functional. Structural plasticity refers to changes in brain architecture such as dendritic branching and synapse rearrangement. In contrast, functional plasticity enables other brain regions to compensate for lost functions or take on new tasks. Experience-dependent plasticity research examines how environmental events impact brain circuit development. Furthermore, long-term potentiation (LTP) and synaptic pruning are key plasticity processes that promote cognitive function and contribute to neurodevelopmental disorders (Bliss and Lømo 1973). Neuroplasticity is especially important in neurorehabilitation, since clinical recovery in stroke patients is frequently achieved with targeted therapies that encourage cortical reconfiguration. Furthermore, advances in neuroimaging methods, such as functional MRI, have helped us better understand how plasticity works throughout life (Pascual-Leone et al. 2005). Ongoing research into brain plasticity holds the potential to open new therapeutic avenues for treating neurological and psychiatric disorders.

10.5.2 How Nutraceuticals Affect Synaptic Plasticity

Synaptic plasticity, the strength of synapses to increase or decrease over time through usage, is essential for learning and memory. Nutraceuticals, bioactive compounds found in foodstuffs, have also been reported to modulate synaptic plasticity via multiple molecular pathways (Table 10.3) (Gómez-Pinilla 2008).

Table 10.3 Some important nutraceuticals and their mechanism of action

Nutraceutical	Mechanism of Action
Polyphenols	Induce synaptic plasticity by enhancing the expression of brain-derived neurotrophic factor (BDNF) and triggering cascading signals such as ERK-CREB.
Docosahexaenoic acid (DHA)	Preserves cell membrane fluidity and promotes synaptic transmission by modulating AMPA and NMDA receptor activity (Cao et al. 2009).
Curcumin	Neuroprotective by suppressing oxidative stress and neuroinflammation and promoting long-term potentiation (LTP).
Resveratrol	Upregulates sirtuins and AMP-activated protein kinase (AMPK) pathways, raising synaptic plasticity and cognition (Vauzour 2017).
Probiotics	Improve synaptic function & neurotransmission (Kim et al. 2021).

Nutraceuticals significantly protect synaptic integrity, prevent neurodegeneration, and augment intellectual functions. Despite this, proper dose and bioavailability of the compounds must be researched utilizing clinical studies to make them useful in the therapy.

10.5.3 The Impact of Nutraceuticals on Neurotransmitter Systems

Nutraceuticals, or bioactive dietary components with medicinal characteristics, have received attention for their ability to alter neurotransmitter systems. These substances have an impact on cognitive function, mood, and neuroprotection via influencing neurotransmitters including serotonin, dopamine, gamma-aminobutyric acid (GABA), and acetylcholine. Green tea and berries, for example, contain polyphenols that stimulate dopamine and serotonin pathways, boosting cognitive function and lowering depressive symptoms. Omega-3 fatty acids, particularly those found in fish oil, are essential for the formation and maintenance of neuronal membranes, as well as modulating serotonin and dopamine activity, which is good for those suffering from mood disorders (Bazinet and Layé 2014). L-tryptophan and L-theanine serve as precursors in the synthesis of serotonin and GABA, modulating stress and relaxation responses (Jenkins et al. 2016). Additionally, vitamins B6, B9, and B12 are required for neurotransmitter metabolism, and deficits have been associated with cognitive impairment and mood problems (Selhub et al. 2010). New research reveals that flavonoids such as resveratrol and quercetin control cholinergic neurotransmission and have neuroprotective properties against neurodegeneration. While nutraceuticals promise neuroprotection, their long-term efficacy and bioavailability can only be confirmed through clinical trials. Understanding these compounds' molecular mechanisms of action could lead to new therapeutic strategies for treating neuropsychiatric diseases.

10.5.4 Supporting Learning and Memory Through Natural Supplements

Cognitive function, learning, and memory are directly relevant to daily life and academic performance. As a result, there has been growing interest in natural supplements as potential cognitive function enhancers. Several bioactive compounds derived from plants, fungi, and marine sources have been identified for their positive effects on memory recall and neuroprotection. *Bacopa monnieri*, a well-studied Ayurvedic herb, is renowned for its nootropic effects. *Bacopa monnieri* appears to improve synaptic transmission, boost resilience to oxidative stress, and stimulate long-term memory formation (Calabrese et al. 2008). Similarly, *Ginkgo biloba*, a traditional Chinese medicine staple, boosts brain circulation and has been found to help prevent cognitive loss, especially in elderly persons (Gauthier and Schlaefke 2014). Omega-3 fatty acids, notably docosahexaenoic acid (DHA) found in fish oil, are another highly researched supplement. DHA is important for synaptic plasticity and neurogenesis, which are both required for learning and memory (Dyall 2015). Curcumin, a bioactive phytochemical contained in turmeric, is a neuroprotective drug with antioxidant and anti-inflammatory properties that may improve cognitive performance (Small et al. 2018). While further study is needed to identify the best doses and long-term benefits of these natural supplements, existing studies indicate that taking these bioactive substances may improve cognitive functions and memory recall.

10.6 Clinical Studies and Evidence-Based Applications

10.6.1 Reviewing Clinical Trials Involving Nutraceuticals

Nutraceuticals, including functional foods and supplements, have gained wide popularity in clinical studies for their ability to enhance health. Despite the fact that nutraceuticals are being handled as food, and not as a drug, this implies implications for study design and regulatory purposes. Despite this, their disease prevention and treatment promises have welcomed mass-scale clinical trials.

Nutraceutical clinical trials usually adhere to traditional randomized controlled trial (RCT) designs but present some peculiar challenges. Bioavailability may be unreliable because of variability in extraction, processing, and metabolism, so uniform dosing is challenging (Daliu et al. 2019). The leniency of the regulatory agencies most often leads to variable study results. The placebo effect poses another issue, as natural products tend to be psychologically linked with health effects. Trial lengths are, at times, too brief to determine long-term effectiveness. Furthermore, patient heterogeneity, as well as lifestyle issues including diet and exercise, makes it difficult to interpret findings (Cencic and Chingwaru 2010).

Despite these challenges, several nutraceuticals have demonstrated notable therapeutic value. Omega-3 fatty acids help reduce triglycerides and inflammation, polyphenols like those found in green tea and curcumin improve metabolic syndrome and insulin sensitivity, and probiotics support gastrointestinal health, benefiting conditions such as IBS and IBD (Ford et al. 2018). However, more rigorous and high-quality trials are needed to confirm their efficacy. In the future, nutrigenomics and personalized nutrition research may enable the development of nutraceutical treatments that are more precisely tailored for improved therapeutic outcomes (Grosso et al. 2020).

10.6.2 Efficacy and Safety Profiles of Popular Brain Enhancers

The quest for brain enhancement has led to research on various cognitive enhancers, each with varying degrees of efficacy and safety. Central nervous system stimulants such as amphetamines and methylphenidate can boost working memory and attention but carry risks of tachycardia, addiction, and require close medical supervision (Massie et al. 2017). Non-invasive brain stimulation (NIBS) techniques, including transcranial magnetic stimulation (TMS), have been studied for conditions like Alzheimer's disease and mild cognitive impairment. While some studies have suggested cognitive enhancement, results are not consistent and need further research to ascertain safety and long-term effects (Wu et al. 2025). Omega-3 fatty acids and *Ginkgo biloba* are also studied for possible cognitive-enhancing effects. *Ginkgo biloba* is studied for its use as a memory enhancer, but efficacy is questionable, and it carries the risk of uncontrolled bleeding. Omega-3 fatty acids have been linked with improved brain activity, but not enough evidence exists to support recommending that they provide definite cognitive enhancement in healthy humans. Another compound being studied for its neuroprotective effects, particularly the maintenance of mitochondrial function, is methylene blue. However, stricter clinical trials must be undertaken to validate its efficacy and safety. Since there is risk involved, self-treatment with cognitive enhancers is not encouraged, and medical consultation is necessary to best balance risks and benefits.

10.6.3 Future Directions and Emerging Research

The brain plasticity supplement market is also evolving rapidly with the development of nutrigenomics and neuroscience. Future research will explore how bioactive molecules influence synaptic plasticity, neurogenesis, and cognitive resilience. Polyphenols, omega-3 fatty acids, and neuropeptides play core functions in the regulation of neurotrophic factors like BDNF, which play a critical role in synaptic reorganization and learning (Ansari et al. 2023).

The gut-brain axis has been receiving a lot of attention, and prebiotics and probiotics are on the horizon to modulate cognition by the action of microbiota-derived metabolites. Emerging advances in precision nutrition and metabolomics will allow for personalized nutraceutical therapies based on the genetic and epigenetic profiling of a person. Nanotechnology-mediated drug delivery systems also have the possibility of enhancing bioavailability and crossing the blood-brain barrier (Singh et al. 2022). Combination of transcranial stimulation and cognitive training with nutraceuticals can be beneficial for brain injury rehabilitation. Yet problems exist, e.g., dosing standardization, safety, and long-term safety, as well as clinical effectiveness. It is necessary to remember drug synergies and chemical interactions when applying nutraceuticals efficiently and safely within the clinical environment.

10.7 Lifestyle Factors Supporting Neurogenesis and Brain Plasticity

Neurogenesis and brain plasticity are key features of repairing and enhancing cognitive function over a lifetime. Neurogenesis refers to the production of new neurons, while brain plasticity (or neuroplasticity) is the capacity of the brain to adapt and form new neural connections due to learning, experience, or damage. There are lifestyle aspects that may induce neurogenesis and brain plasticity like:

10.7.1 Exercise and Physical Activity

Exercise and physical activity are essential in provoking neurogenesis (creation of new neurons) and brain plasticity (brain ability to adapt and develop new neural circuits) (Maharjan et al. 2020). Greater insight into the effects of exercise on the brain is as follows:

Aerobic Exercise and Neurogenesis: Aerobic exercise—exercise that enhances heart rates and cardiovascular health—possesses the best-studied advantages to neurogenesis. Aerobic exercise includes running, cycling, swimming, and brisk walking (Déry et al. 2013).

10.7.1.1 Benefits of Aerobic Exercise to Neurogenesis

- *BDGF (Brain-Derived Neurotrophic Factor)*: Aerobic training enhances the release of BDNF, a protein essential for neurogenesis, for survival of the neuron, and for differentiation, particularly in the hippocampus, a brain area that has a function in memory and learning (Déry et al. 2013).
- *Improved Blood Flow*: Aerobic exercise enhances blood flow, delivering greater oxygen and nutrients to the brain and hence enables new neurons to be created and the brain to be healthy.
- *Hippocampal Neurogenesis*: A consistent aerobic exercise has been reported to enhance the creation of new hippocampal neurons, enhancing memory, learning, and cognitive functioning (Déry et al. 2013).
- *Stress Reduction*: Aerobic exercise decreases cortisol levels (stress hormone), preventing unwanted activity in the brain and providing neurogenesis benefits.
- *Strengthening and Training the Brain*: While aerobic exercise is globally acclaimed for increasing neurogenesis, there are long-term brain benefits with resistance and strength training (Maharjan et al. 2020).

10.7.1.1.1 Importance of Strength Training and Brain Function

Cognitive Improvement There is evidence that strength training can improve cognitive functions such as attention, memory, and executive functions by triggering Brain-Derived Neurotrophic Factor (BDNF) and superior brain plasticity (Herold et al. 2019).

Hippocampal Neurogenesis Resistance training has also been associated with enhanced neurogenesis, particularly in regions that are responsible for spatial memory and locomotion.

Increased Brain Structure Strength training has been demonstrated to enhance gray matter density in the prefrontal cortex, a region of the brain responsible for decision-making, planning, and working memory (Herold et al. 2019).

High-Intensity Interval Training (HIIT) HIIT is performed through alternating between short intervals of high-intensity exercise and quick recovery time. This type of exercise is extremely effective at enhancing cardiovascular health and has been proven to have strong effects on brain health (Lei et al. 2024).

10.7.1.1.2 Importance of High-Intensity Interval Training (HIIT)

- *Enhanced Mental Function*: Research has demonstrated that HIIT enhances memory, learning, and cognitive function, as the severity of the exercise triggers the growth factors such as BDNF to be released (Lei et al. 2024).
- *Enhanced Neuroplasticity*: HIIT's short periods of high-intensity exercise tell the brain to adapt itself to novel experiences, hence enhancing neuroplasticity (Lei et al. 2024).
- *Hormonal Benefits*: HIIT augments the secretion of growth hormones, which help promote neuronal regeneration and enhance neuroplasticity.

Exercise and Aging Brain: Exercise is quite possibly the most important advantage for the possible avoidance of age-related cognitive impairment.

Importance of Exercise to the Aging Brain:

- *Preventing Brain Shrinkage*: Exercise retards the process of aging in the brain, particularly protecting the hippocampus from shrinkage, which is usually linked to age-related loss of memory among elderly people (Maharjan et al. 2020).
- *Boosting Cognitive Resilience*: Consistent physical exercise enhances the ability of the brain to adjust and reorganize, and make it resilient to age-related change and neurodegenerative diseases like Alzheimer's.

10.7.2 Sleep and Its Role in Brain Health

Sleep plays a crucial role in maintaining normal brain functioning, influencing cognitive processes, mental health, and overall brain health. During sleep, the brain undergoes important processes that support learning, memory consolidation, and brain plasticity (Tanaka and Shirakawa 2004). Here's an overview of how sleep contributes to brain health:

10.7.2.1 Memory Consolidation

One of sleep's primary functions is consolidating memories.

1. During sleep, particularly in deep sleep stages (slow-wave sleep), the brain processes and strengthens new information learned during the day, transferring memories from short-term to long-term storage.
2. The hippocampus, a key brain region for memory, plays a central role in this process. Sleep stabilizes and solidifies memories, making them more easily accessible when needed (Tanaka and Shirakawa 2004).
3. In deeper stages of NREM sleep, especially slow-wave sleep, the brain replays and reorganizes information, moving it from the hippocampus to the neocortex for long-term storage. This process integrates new knowledge and experiences into existing memory networks.
4. REM sleep, in particular, is crucial for emotional and procedural memory (such as motor skills). During REM sleep, the brain strengthens and fine-tunes memories, which enhances memory retention and recall (Tanaka and Shirakawa 2004).

10.7.2.2 Brain Detoxification and Waste Removal

Sleep is essential for clearing toxins and waste products that build up in the brain during the day.

1. The glymphatic system, which is most active during sleep, helps remove metabolic waste, including beta-amyloid plaques linked to neurodegenerative diseases like Alzheimer's (Cook 2009).
2. During deep NREM sleep, the brain's interstitial fluid increases, enhancing the removal of these toxic proteins and other waste products, thus preventing potential impairments to brain function.

10.7.2.3 Neurogenesis and Brain Plasticity

Sleep plays a key role in neurogenesis (the formation of new neurons) and brain plasticity (the brain's ability to adapt and form new neural connections).

1. Deep sleep has been shown to support neurogenesis in the hippocampus, a region critical for learning, memory, and emotional regulation, thus promoting cognitive function and emotional well-being (Sailor et al. 2017).
2. Sleep supports synaptic plasticity, which enables the brain to modify the strength of synapses between neurons depending on experience. Learning and memory rely on this process, and sleep optimizes the brain's capability to refine such neural connections (Sailor et al. 2017).

10.7.2.4 Emotional and Mental Health

Adequate and comfortable sleep is essential to emotional regulation and psychological well-being.

1. Sleep allows for emotion regulation through the regulation of normal levels of neurotransmitters like serotonin and dopamine. Disturbed sleep tends to be related to irritability, mood swings, and heightened susceptibility to stress and anxiety (Ganguly et al. 2024).
2. During sleep, the brain and body repair stress-induced damage, diffusing the impact of stress hormones like cortisol. Good sleep on a regular basis can prevent chronic stress, which, if not prevented, can impair mental functioning, increase brain inflammation, and suppress neurogenesis (Bandyopadhyay et al. 2025; Tanaka and Shirakawa 2004).
3. Mental illnesses such as depression and anxiety also have robust relationships with sleep disorders. Loss of sleep exacerbates them, but good sleep improves emotional control and overall mental health (Tanaka and Shirakawa 2004).

10.7.3 Stress Management and Its Effect on Neurogenesis

Good stress management is important in reducing the damaging effects of chronic stress on the brain, promoting neurogenesis, and protecting the hippocampus from destruction. Various techniques have been shown to reduce stress and promote brain health (Adhikary et al. 2024a; Mirescu and Gould 2006):

10.7.3.1 Mindfulness and Meditation

Interventions such as mindfulness and meditation are successful in lowering stress levels and promoting brain health.

10.7.3.2 Effect on Cortisol Levels

Mindfulness meditation has been found to decrease cortisol levels, successfully lowering stress and averting the damaging effects of long-term cortisol exposure.

Promoting Neurogenesis

Consistent meditation has been associated with higher gray matter density in brain areas involved in memory, emotional regulation, and learning, like the hippocampus. This is to say that mindfulness exercises may stimulate neurogenesis and improve cognition.

Stress Reduction and Emotional Reslience

Meditation generates emotional resilience through the power to conquer stress and remain calm when confronted with challenging situations. Emotional regulation is the potential to shield the brain from chronic stress's damaging effects and supply neuroplasticity (Mirescu and Gould 2006).

Physical Exercise
The regular exercise is one of the biggest weapons against stress and has gigantic significance for neurogenesis.

Cortisol Regulation via Exercise Exercise regulates cortisol levels, minimizing the stress impact on the brain. Chronic stress has been said to be lowered through aerobic exercise and increased secretion of Brain-Derived Neurotrophic Factor (BDNF), a neurotrophic factor that promotes neurogenesis (Adhikary et al., 2024b).

Hippocampus and Neurogenesis Exercise has been shown to increase neurogenesis in the hippocampus, which reverses the adverse effects of chronic stress. It also enhances cognition and mood, building emotional resilience (Yau et al. 2011).

Stress Relief through Movement Physical Exercise Stimulates the Release of Endorphins, which Elevate Mood and Promote Relaxation. This Helps Reduce the Body's Overall Stress Response, Enabling the Brain to Function more Efficiently

Adequate Sleep Sleep Is Critical for Stress Recovery and Neurogenesis

- *Sleep and Cortisol Regulation*: Proper, restorative sleep helps regulate cortisol levels, keeping them within a healthy range. Sleep deprivation, on the other hand, increases cortisol production and exacerbates stress, impeding neurogenesis (Guzman-Marin and McGinty 2006).
- *Sleep and Brain Detoxification*: During sleep, the brain undergoes detoxification through the *glymphatic system*, removing waste products and harmful proteins that accumulate during the day. This process helps protect brain cells and supports neurogenesis.
- *REM Sleep and Emotional Regulation*: REM sleep plays a key role in processing emotions and consolidating memories. It helps the brain process emotional experiences, which can reduce the impact of stress and prevent emotional overload (Guzman-Marin and McGinty 2006).

Social Support and Connection Strong social support networks can buffer the brain from the harmful effects of stress.

Social Interaction and Stress Buffering Positive social interactions reduce the impact of stress and increase emotional resilience. Engaging in supportive relationships releases *oxytocin*, a hormone that promotes feelings of calm and reduces cortisol levels.

- *Group Activities and Neurogenesis*: Socializing through activities such as group exercise or shared hobbies has been shown to enhance brain function and promote neuroplasticity. These interactions foster positive emotions and reduce the brain's stress response.

Healthy Nutrition: Diet plays a crucial role in managing stress and supporting brain health.

- *Anti-inflammatory Diets*: Diets rich in antioxidants, omega-3 fatty acids, and other anti-inflammatory nutrients can help reduce the inflammatory effects of stress on the brain. Nutrients like *omega-3s* (found in fish and flaxseeds) are known to support neurogenesis and improve mood (Wilson 2020).
- *Stabilizing Blood Sugar Levels*: Stable blood sugar levels help prevent the spikes in cortisol that can occur with blood sugar fluctuations. A balanced diet, rich in whole grains, fruits, and vegetables, helps stabilize energy levels and reduce stress.

10.7.4 Cognitive Training and Brain Challenges

Cognitive training, also known as brain training, refers to activities or exercises designed to enhance mental functions such as memory, attention, problem-solving, reasoning, and executive functions. These exercises enhance the brain's plasticity—the capacity to reorganize by forming new neural connections in response to learning, experiences, and changes in the environment. Engaging in cognitive challenges not only promotes brain health but also encourages neurogenesis (the creation of new neurons) and improves brain function (Wilson 2020). Cognitive training involves using specific exercises or tasks to improve mental functions that decline with age, stress, or injury. It aims to enhance several cognitive areas:

- *Memory*: This involves exercises that improve both short-term and long-term memory. Memory tasks can include recalling lists of words, names, or numbers, and practicing the retention of new information (Adhikary et al., 2024c).
- *Attention*: Activities that require sustained concentration, like solving puzzles or participating in games that demand focus, improve the brain's ability to concentrate and filter distractions (Mirescu and Gould 2006).
- *Problem-Solving*: Cognitive exercises often involve puzzles, strategic games, and tasks that require creative thinking, helping to improve the brain's problem-solving abilities.
- *Executive Functioning*: This encompasses higher-level cognitive functions like decision-making, reasoning, and planning. Games or tasks that require multitasking or organizing help enhance executive functioning skills (Mirescu and Gould 2006).
- *Processing Speed*: Cognitive exercises designed to improve reaction times, such as time-based challenges, improve the brain's ability to process and respond to information quickly.

10.8 Conclusion and Future Scopes

Neurogenesis and brain plasticity are essential processes that enable the brain to adapt, learn, and recover throughout life. Neurogenesis, the creation of new neurons—especially in the hippocampus—and brain plasticity, which refers to the brain's ability to reorganize neural connections, play a vital role in cognitive health, memory, and overall brain function. Nutraceuticals have shown promising potential in supporting and enhancing these processes. Compounds such as omega-3 fatty acids, curcumin, flavonoids, *Bacopa monnieri*, *Ginkgo biloba*, and Lion's Mane mushroom have demonstrated neuroprotective effects, promoting neurogenesis, improving brain plasticity, and possibly enhancing cognitive performance. By reducing oxidative stress and inflammation, these nutraceuticals help create an optimal environment for neural growth and plasticity. While current research is promising, more large-scale and rigorous studies are needed to fully understand the long-term effects, ideal dosages, and safety profiles of these compounds. Additionally, individual responses to nutraceuticals may vary, and their benefits are likely to be most effective when combined with a healthy lifestyle, including proper nutrition, regular physical activity, mental stimulation, and stress management. Nutraceuticals offer a promising approach to enhancing brain health, supporting neurogenesis, and improving brain plasticity, especially as we age. They should be considered as part of a holistic strategy for cognitive well-being, in conjunction with other lifestyle factors, and under the guidance of healthcare professionals.

References

Adhikary K, Banerjee P, Barman S, Bandyopadhyay B, Bagchi D (2024a) Nutritional aspects, chemistry profile, extraction techniques of lemongrass essential oil and it's physiological benefits. J Am Nutr Assoc 43(2):183–200. https://doi.org/10.1080/27697061.2023.2245435

Adhikary K, Mohanty S, Bandyopadhyay B, Maiti R, Bhattacharya K, Karak P (2024b) β-Amyloid peptide modulates peripheral immune responses and neuroinflammation in rats. Biomol Concepts 15(1). https://doi.org/10.1515/bmc-2022-0042

Adhikary K, Barman S, Banerjee P, Mondal P, Dam B, Misra A, Chandra Mandal N, Mukherjee D, Bandyopadhyay B, Maiti R (2024c) Characterization and antimicrobial activity of essential oils extracted from lemongrass (*Cymbopogon flexuosus*) using microwave-assisted hydro distillation. Zeitschrift fur Naturforschung. C, J Biosci. https://doi.org/10.1515/znc-2024-0095. Advance online publication

Aigner S, Singer T, Bracko O et al (2012) The regulation of adult hippocampus neurogenesis by IGF2 is shown by gene expression profiling of neural stem cells and their neuronal offspring. J Neurosci. 32(19), 6535–6547

Aigner L, Kempermann G, Gauge FH et al (2018) Adult neurogenesis in humans: data and unanswered issues. Cell Stem Cell 23(1):25–30. https://doi.org/10.1016/j.stem.2018

Alkass K, Bernard S, Salehpour M, Huttner HB, Bergmann O, Spalding KL et al (2013) Adult human hippocampus neurogenesis dynamics. Cell. 153(6), 1219–1227

Andreu Z, Suh H, Mira H et al (2010) The adult hippocampus's neural stem cells' quiescence and long-term activity are controlled by signalling via BMPR-IA. Cell Stem Cell. 7(1), 78–89

Ansari F, Neshat M, Pourjafar H, Jafari SM, Samakkhah SA, Mirzakhani E (2023) The role of probiotics and prebiotics in modulating of the gut-brain axis. Front Nutr 10:1173660

Bandyopadhyay U, Sen D, Ahuja D, Mahapatra SP, Biswas D, Maiti R, Chakraborty S, Hazra A, Parua S, Basak AK, Das A, Paul N, Purkait MP, Syamal AK, Dey R, Bhattacharya K, Adhikary K, Bhattacharjee A (2025) Interplay of calcium, vitamin D, and parathormone in the milieu of infections and immunity: reassessed in the context of COVID-19. J Steroid Biochem Mol Biol 245:106624. https://doi.org/10.1016/j.jsbmb.2024.106624

Bazinet RP, Layé S (2014) Polyunsaturated fatty acids and their metabolites in brain function and disease. Nat Rev Neurosci 15(12):771–785

Bergmann O, Spalding KL, Frisén J (2015) Human adult neurogenesis. Cold Spring Harb Perspect Biol 7:a018994

Bjørklund G, Tinkov AA, Hosnedlová B, Kizek R, Ajsuvakova OP, Chirumbolo S, Skalny AV (2022) The role of nutrients in neurodevelopmental disorders. Acta Neurobiol Exp 82(1):1–17

Bliss TV, Lømo T (1973) Long-lasting potentiation of synaptic transmission in the dentate area of the anaesthetized rabbit following stimulation of the perforant path. J Physiol 232(2):331–356

Boldrini M, Fulmore CA, Tartt AN, Simeon LR, Pavlova I, Poposka V et al (2018) Human hippocampal neurogenesis persists throughout aging. Cell Stem Cell 22(4):589–599.e5

Bourre JM (2004) Roles of unsaturated fatty acids (especially omega-3 fatty acids) in the brain at various ages and during ageing. J Nutr 8:163–174

Broadbent NJ, Jessberger S, Clark RE et al (2009) Adult rats' spatial and object recognition memory is affected when adult neurogenesis specific to the dentate gyrus is knocked down. Learning & Memory, 16(2), 79–83

Calabrese C, Gregory WL, Leo M, Kraemer D, Bone K, Oken B (2008) Effects of a standardized Bacopa monnieri extract on cognitive performance, anxiety, and depression in the elderly: a randomized, double-blind, placebo-controlled trial. J Altern Complement Med 14(6):707–713

Cao D, Kevala K, Kim J, Moon HS, Jun SB, Lovinger D, Kim HY (2009) Docosahexaenoic acid promotes hippocampal neuronal development and synaptic function. J Neurochem 111(2):510–521

Cencic A, Chingwaru W (2010) The role of functional foods, nutraceuticals, and food supplements in intestinal health. Nutrients 2(6):611–625

Choi J, Yin T, Shinozaki K, Lampe JW, Stevens JF, Becker LB, Kim J (2018) Comprehensive analysis of phospholipids in the brain, heart, kidney, and liver: brain phospholipids are least enriched with polyunsaturated fatty acids. Mol Cell Biochem 442:187–201

Cole GM, Teter B, Frautschy SA (2007) Neuroprotective effects of curcumin. Adv Exp Med Biol 595:197–212

Cook MS (2009) The brain wash: a powerful, all-natural program to protect your brain against Alzheimer's, chronic fatigue syndrome, depression, Parkinson's, and other diseases. Wiley

Daliu P, Santini A, Novellino E (2019) From pharmaceuticals to nutraceuticals: bridging disease prevention and management. Expert Rev Clin Pharmacol 12(1):1–7

Déry N, Pilgrim M, Gibala M, Gillen J, Wojtowicz JM, MacQueen G, Becker S (2013) Adult hippocampal neurogenesis reduces memory interference in humans: opposing effects of aerobic exercise and depression. Front Neurosci 7:66

Dyall SC (2015) Long-chain omega-3 fatty acids and the brain: a review of the independent and shared effects of EPA, DPA, and DHA. Front Aging Neurosci 7:52

Faigle R, Song H (2013) Signalling pathways controlling adult brain stem cells and neurogenesis. BiochimBiophys Acta. 1830(2), 2435–2448

Fernandes J, Arida RM, Gomez-Pinilla F (2020) Physical exercise as an epigenetic modulator of brain plasticity and cognition. Neurosci Biobehav Rev 115:25–36

Ford AC, Harris LA, Lacy BE, Quigley EM, Moayyedi P (2018) Systematic review with meta-analysis: the efficacy of prebiotics, probiotics, synbiotics and antibiotics in irritable bowel syndrome. Aliment Pharmacol Ther 48(10):1044–1060

Freeman LR, Krey JF, Jicha GA, Keller JN (2020) Nutritional interventions in aging and age-associated diseases. Adv Nutr 11(5):1223–1241

Gage FH (2019) Adult neurogenesis in mammals. Science 364(6436):827–830

Ganguly K, Adhikary K, Acharjee A, Acharjee P, Trigun SK, Mutlaq AS, Ashique S, Yasmin S, Alshahrani AM, Ansari MY (2024) Biological significance and pathophysiological role of

matrix metalloproteinases in the central nervous system. Int J Biol Macromol 280(Pt 3):135967. Advance online publication. https://doi.org/10.1016/j.ijbiomac.2024.135967
García-Verdugo JM, Alvarez-Buylla A (2002) Adult subventricular zone neurogenesis. J Neurosci. 20021;22(3):629–34
Gauthier S, Schlaefke S (2014) Efficacy and tolerability of Ginkgo biloba extract EGb 761® in dementia: a systematic review and meta-analysis of randomized placebo-controlled trials. Clin Interv Aging 9:2065–2077
Genchi G, Lauria G, Catalano A, Carocci A, Sinicropi MS (2024) Neuroprotective effects of curcumin in neurodegenerative diseases. Foods 13(11):1774
Glade MJ, Smith K (2015) Phosphatidylserine and the human brain. Nutrition 31(6):781–786
Gómez-Pinilla F (2008) Brain foods: the effects of nutrients on brain function. Nat Rev Neurosci 9(7):568–578
Grosso G, Mateo A, Rangelov N, Buzeti T, Birt C (2020) Nutrition in the context of the sustainable development goals. Eur J Pub Health 30(Supplement_1):i19–i23
Guzman-Marin R, McGinty D (2006) Sleep deprivation suppresses adult neurogenesis: clues to the role of sleep in brain plasticity. Sleep Biol Rhythms 4(1):27–34
Herold F, Törpel A, Schega L, Müller NG (2019) Functional and/or structural brain changes in response to resistance exercises and resistance training lead to cognitive improvements–a systematic review. Eur Rev Aging Phys Act 16:1–33
Hurley RA, Kays JL, Taber KH (2012) Neuroplasticity and mental wellness in the dynamic brain. J Neuropsychiatry Clin Neurosci. 24(2), 118–124
Innamorati M, Pompili M, Serafini G et al (2012) The part microRNAs play in suicidal behaviour, severe emotional disorders, and synaptic plasticity. Neurosci Res. 74(2), 157–166
Jenkins TA, Nguyen JC, Polglaze KE, Bertrand PP (2016) Influence of tryptophan and serotonin on mood and cognition with a possible role of the gut-brain axis. Nutrients 8(1):56
Kang E, Wen Z, Song H et al (2016) Psychiatric illnesses and adult neurogenesis. Harb Perspect Biol Cold Spring. 8(9), a019026
Kim HY, Huang BX, Spector AA (2014) Phosphatidylserine in the brain: metabolism and function. Prog Lipid Res 56:1–18
Kim CS, Cha J, Sim M, Jung S, Chun WY, Baik HW, Shin DM (2021) Probiotic supplementation improves cognitive function and mood with changes in gut microbiota in community-dwelling older adults: a randomized, double-blind, placebo-controlled, multicenter trial. J Gerontol A Biol Sci Med Sci 76(1):32–40
Kumar A, Sidhu J, Goyal A, Tsao JW (2022) Alzheimer disease. StatPearls Publishing
Lei Z, Mozaffaritabar S, Kawamura T, Koike A, Kolonics A, Kéringer J et al (2024) The effects of long-term lactate and high-intensity interval training (HIIT) on brain neuroplasticity of aged mice. Heliyon 10(2):e24421
Maharjan R, Bustamante LD, Ghattas KN, Ilyas S, Al-Refai R, Khan S (2020) Role of lifestyle in neuroplasticity and neurogenesis in an aging brain. Cureus 12(9):e10639
Massie CF, Yamga EM, Boot BP. (2017). Neuroenhancement: A call for better evidence on safety and efficacy. In: Maslen H, Savulescu J, Sahakian BJ, editors. Rethinking Cognitive Enhancement. Oxford: Oxford University Press; p. 57–68
Mateo JL, Zhou B, Andersen J, Achimastou A, Martynoga B, Urban N, et al (2013) An important role for NFIX in neural stem cell quiescence is shown by epigenomic enhancer annotation. Genes Dev 27(16), 1769–1786
Mateos-Aparicio P, Rodríguez-Moreno A (2019) The impact of studying brain plasticity. Front Cell Neurosci 13:66
Mirescu C, Gould E (2006) Stress and adult neurogenesis. Hippocampus 16(3):233–238
Mohajeri MH, Troesch B, Weber P (2021) Inadequate supply of vitamins and DHA in the elderly: implications for brain aging and Alzheimer's disease. Nutrition 82:111053
Parletta N, Milte CM, Meyer BJ (2019) Nutritional modulation of cognitive function and mental health. J Nutr Biochem 70:1–10
Pascual-Leone A, Amedi A, Fregni F, Merabet LB (2005) The plastic human brain cortex. Annu Rev Neurosci 28(1):377–401

Sailor KA, Schinder AF, Lledo PM (2017) Adult neurogenesis beyond the niche: its potential for driving brain plasticity. Curr Opin Neurobiol 42:111–117
Scalbert A, Johnson IT, Saltmarsh M (2005) Polyphenols: antioxidants and beyond. Am J Clin Nutr 81(1):215S–217S
Schouten M, Fitzsimons CP, van Bodegraven E et al (2014) Implications of adult brain stem cell epigenetic regulation for Alzheimer's disease. Neurodegener Mol. 9(1), 25
Selhub J, Troen A, Rosenberg IH (2010) B vitamins and the aging brain. Nutr Rev 68(suppl_2):S112–S118
Singh A, Kukreti R, Saso L, Kukreti S (2020) Oxidative stress: a key modulator in neurodegenerative diseases. Mol Neurobiol 57(7):2850–2867
Singh AR, Desu PK, Nakkala RK, Kondi V, Devi S, Alam MS et al (2022) Nanotechnology-based approaches applied to nutraceuticals. Drug Deliv Transl Res:12(3);1–15
Small GW, Siddarth P, Li Z, Miller KJ, Ercoli L, Emerson ND et al (2018) Memory and brain amyloid and tau effects of a bioavailable form of curcumin in non-demented adults: a double-blind, placebo-controlled 18-month trial. Am J Geriatr Psychiatry 26(3):266–277
Souyoul SA, Saussy KP, Lupo MP (2018) Nutraceuticals: a review. Dermatol Ther 8:5–16
Spencer JP (2018) Flavonoids and brain health: multiple effects underpinned by common mechanisms. Genes Nutr 13:19
Svennerholm L (1968) Distribution and fatty acid composition of phosphoglycerides in normal human brain. J Lipid Res 9(5):570–579
Szuhany KL, Bugatti M, Otto MW (2015) A meta-analytic review of the effects of exercise on brain-derived neurotrophic factor. J Psychiatr Res 60:56–64
Tanaka H, Shirakawa S (2004) Sleep health, lifestyle and mental health in the Japanese elderly: ensuring sleep to promote a healthy brain and mind. J Psychosom Res 56(5):465–477
Tang YY, Hölzel BK, Posner MI (2015) The neuroscience of mindfulness meditation. Nat Rev Neurosci 16(4):213–225
Tartt AN, Fulmore CA, Boldrini M et al (2018) As people age, their hippocampus continues to grow new neurones. Cell Stem Cell. 22(4), 589–599.e5
Tchorz JS, Barz M, Giachino C et al (2014) In the adult hippocampus, GABA inhibits neurogenesis via GABAB receptors. Progress. 141(1), 83–90
Vauzour D (2017) Polyphenols and brain health. Ocl 24(2):A202
Wexler EM, Paucer A, Kornblum HI, Palmer TD, Geschwind DH (2009) Endogenous Wnt signalling preserves neural progenitor cell potency. Stem Cells. 27(5), 1130–1141
Wilson K (2020) How to build a healthy brain: reduce stress, anxiety and depression and future-proof your brain. Hachette
Wu Y, Zhang C, Rong S, Shi Y (2021) Brain plasticity and neurodegenerative diseases: targeting neurogenesis. Neural Regen Res 16(9):1723–1731
Wu M, Song W, Wang X, Teng L, Li J, Zhang J et al (2025) Efficacy of non-invasive brain stimulation interventions on cognitive impairment: an umbrella review of meta-analyses of randomized controlled trials. J Neuroeng Rehabil 22(1):22
Yau SY, Lau BWM, So KF (2011) Adult hippocampal neurogenesis: a possible way how physical exercise counteracts stress. Cell Transplant 20(1):99–111

Sirtuins, Autophagy, and Brain Longevity

11

Shradha Bisht, Vishakha Jaiswal, and Swati Prakash

11.1 Introduction

The aging brain represents one of the most complex challenges in modern medicine, involving intricate interactions between cellular senescence, metabolic dysregulation, and impaired repair mechanisms. Emerging research highlights sirtuins and autophagy as central regulators of brain longevity, operating through interconnected pathways that mitigate age-related cognitive decline. This chapter explores their roles in maintaining neuronal health and combating neurodegenerative processes.

11.1.1 The Aging Brain: A Complex Challenge

The aging brain undergoes significant changes that affect both its structure and function. These changes can lead to cognitive decline, which is a major concern for older adults. Approximately 40% of individuals aged 65+ experience memory deficits, while 14–16% develop mild cognitive impairment (MCI) or dementia (Brito et al. 2023). Unlike Alzheimer's disease characterized by neuronal death, normal aging primarily involves synaptic degeneration and inefficient recruitment of brain regions during cognitive tasks. Motor-associated areas, such as the

S. Bisht
Uttaranchal Institute of Pharmaceutical Sciences, Uttaranchal University, Dehradun, Uttarakhand, India
e-mail: shradhabisht@uumail.com

V. Jaiswal · S. Prakash (✉)
Amity Institute of Pharmacy, Amity University, Lucknow, Uttar Pradesh, India
e-mail: vjaiswal@lko.amity.edu; sprakash@lko.amity.edu

A. Kumar Singh, S. Nand Rai (eds.), *Rejuvenating the Brain: Nutraceuticals, Autophagy, and Longevity*, Nutritional Neurosciences,
https://doi.org/10.1007/978-981-95-2790-8_11

cerebellum, also exhibit hyperactivity in older adults, reflecting compensatory mechanisms that often fail to improve outcomes (Brito et al. 2023).

As we age, the number of synapses, the connections between neurons, decreases. This reduction in synaptic density contributes to impaired communication between neurons, affecting memory and learning. White matter, which consists of myelinated nerve fibers, plays a crucial role in transmitting signals across the brain. Its reduction can slow down neural communication, impacting cognitive speed and efficiency.

Older brains often exhibit different patterns of neuronal activation compared to younger brains. For example, older adults might show increased activity in certain brain regions as a compensatory mechanism to maintain cognitive performance. Even areas like the cerebellum, traditionally associated with motor coordination, show hyperactivity in older adults. This hyperactivity can be part of the brain's attempt to compensate for age-related declines in other cognitive functions.

Cognitive decline with aging can range from mild cognitive impairment (MCI) to full-blown dementia. Alzheimer's disease is a prominent example of neurodegenerative pathology, characterized by the accumulation of amyloid plaques and tau tangles, leading to neuronal death and severe cognitive impairment.

11.1.1.1 Cellular Senescence and Dysfunction

Cellular senescence is a state where cells cease to divide but remain metabolically active. Senescent cells accumulate with age and contribute to aging and age-related diseases through several mechanisms (Zhang et al. 2022):

Senescent cells produce a senescence-associated secretory phenotype (SASP), which includes pro-inflammatory cytokines. This chronic inflammation can damage surrounding tissues and promote aging (Saito et al. 2024). These cells can inhibit the proliferation of nearby cells, impairing tissue repair and regeneration. Senescent cells have been implicated in neurodegenerative diseases like Alzheimer's and Parkinson's. They may contribute to the progression of these diseases by promoting inflammation and oxidative stress. Experiments where senescent cells are transplanted into young animals have shown that these cells can induce age-related phenotypes, such as decreased physical function and increased inflammation. This demonstrates the causal role of senescent cells in driving aging processes (Franco et al. 2025).

11.1.2 Sirtuins and Autophagy: Key Players in Cellular Health

Sirtuins are a family of NAD^+-dependent deacetylases that play critical roles in maintaining cellular health (Yu et al. 2024). Sirtuins and autophagy are key players in maintaining cellular health, particularly in the context of aging and age-related diseases. The primary sirtuin isoforms involved in cellular health are:

SIRT1: Involved in metabolism, stress resistance, and DNA repair. It promotes neuronal survival by enhancing mitochondrial function and reducing oxidative stress. SIRT1 also regulates autophagy by deacetylating key autophagy-related proteins like LC3 and Atg5, enhancing autophagosome formation (Ng and Tang 2013; Baeken 2024).

SIRT3: Localized in mitochondria, SIRT3 regulates mitochondrial biogenesis and function, protecting against oxidative damage. It activates mitophagy, a specific form of autophagy targeting damaged mitochondria, thereby maintaining mitochondrial integrity (Wan et al. 2022).

SIRT6: Known for maintaining genomic stability and telomere integrity, SIRT6 helps delay cellular senescence. It also stabilizes lysosomal membranes, ensuring efficient autophagic degradation (Baeken 2024; Wu et al. 2022).

Autophagy is a cellular recycling process that degrades damaged organelles and protein aggregates (Fig. 11.1). Autophagy helps clear toxic protein aggregates, such as those found in neurodegenerative diseases like Alzheimer's and Parkinson's. This process is essential for preventing neurodegeneration and maintaining neuronal health (Lee 2019; Baeken 2024). Enhanced autophagy has been linked to increased lifespan in model organisms, suggesting its importance in maintaining cellular health over time. Autophagy supports cellular longevity by removing damaged cellular components and recycling essential nutrients (Sun et al. 2023).

11.1.2.1 Interplay Between Sirtuins and Autophagy

The interplay between sirtuins and autophagy is bidirectional and crucial for maintaining cellular health. Sirtuins, particularly SIRT1 and SIRT6, enhance autophagy by deacetylating autophagy-related proteins and regulating gene expression of autophagy genes. SIRT3 promotes mitophagy, ensuring mitochondrial quality (Ng and Tang 2013). Autophagy maintains NAD^+ levels by recycling damaged cellular components, which is essential for sirtuin function. It also removes acetyltransferases that inhibit sirtuins via hyperacetylation, ensuring sirtuin activity remains optimal (Yu et al. 2024).

Understanding the interplay between sirtuins and autophagy has significant therapeutic potential for treating age-related diseases. Enhancing sirtuin activity or autophagy can improve cellular resilience and mitigate neurodegenerative diseases. However, challenges remain in targeting these pathways effectively without off-target effects (Lee 2019; Wu et al. 2022).

Sirtuins and autophagy are central to maintaining cellular health by regulating metabolism, stress responses, and cellular clearance that is critical for preventing age-related diseases and promoting longevity. Further research into these mechanisms could lead to novel therapeutic strategies for enhancing cellular health and combating neurodegeneration.

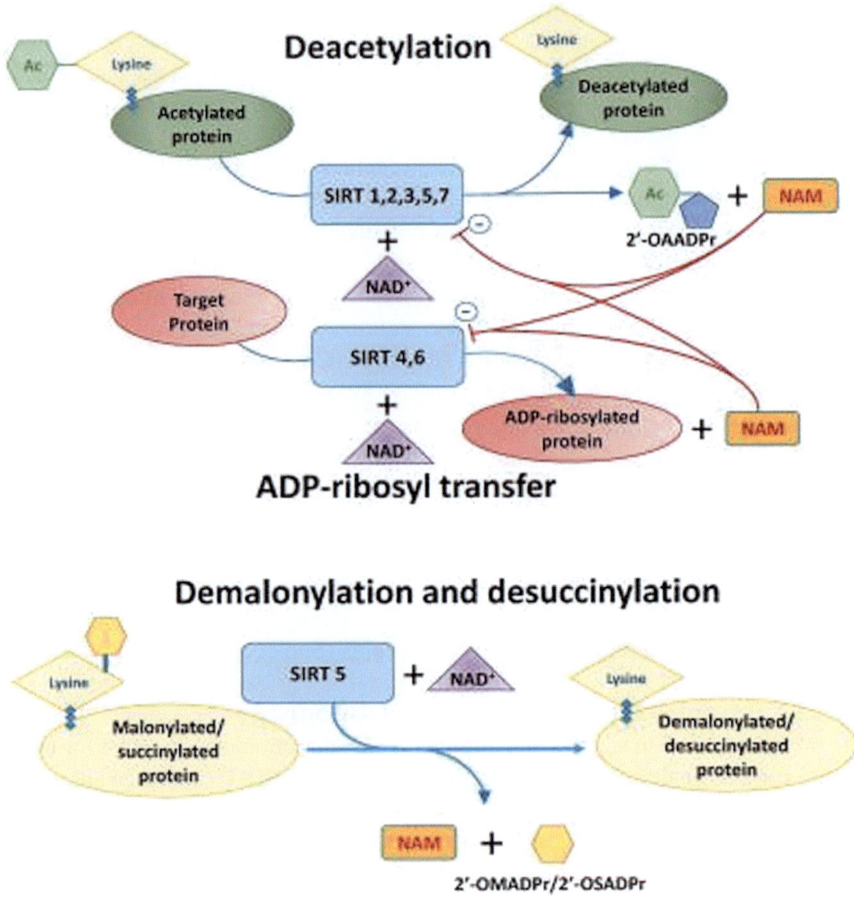

Fig. 11.1 Sirtuins (SIRTs) are enzymes that rely on NAD+ as a co-substrate for their activities, primarily producing nicotinamide (NAM). SIRTs 1, 2, 3, 5, and 7 are involved in deacetylation, removing acetyl groups from lysine residues on target proteins while generating NAM and 2′-O-acetyl-ADP-ribose (2′-OAADPr). SIRT4 functions exclusively as an ADP-ribosyl transferase, transferring ADP-ribose groups from NAD+ to proteins. SIRT6 shares this ADP-ribosyl transfer activity and also exhibits weak deacetylation. Meanwhile, SIRT5 uses NAD+ in demalonylation and desuccinylation processes, producing 2′-O-malonyl-ADP-ribose (2′-OMADPr) and 2′-O-succinyl-ADP-ribose (2′-OSADPr). Overall, while most sirtuins focus on deacetylation, SIRT4 and SIRT6 have specialized roles in ADP ribosylation. (Modification based on Morigi et al. (2018), Kratz et al. (2021))

11.1.3 Synergistic Relationship Between Sirtuins and Autophagy

Sirtuins and autophagy interact through several molecular pathways to enhance cellular resilience:

SIRT1 and Autophagy

SIRT1 plays a crucial role in regulating autophagy by deacetylating and activating key autophagy regulators. This includes deacetylating LC3 and ATG proteins, which are essential for autophagosome formation and maturation. By enhancing autophagic flux, SIRT1 promotes the clearance of damaged cellular components, such as protein aggregates and dysfunctional organelles (Baeken 2024; Kim et al. 2022).

SIRT1 forms molecular complexes with essential components of the autophagy machinery, including Atg5 and Atg7 (Xu and Wan 2023). It also regulates upstream pathways, such as the AMPK/mTOR axis, which are critical for autophagy initiation (Sadria and Layton 2021). Enhanced autophagy helps protect cells from environmental stresses, including nutrient deprivation and oxidative stress (Liang et al. 2020).

SIRT3 and Mitophagy

SIRT3 is primarily localized in mitochondria and plays a pivotal role in activating mitophagy, a specific form of autophagy targeting damaged mitochondria. This process is crucial for reducing oxidative stress and maintaining mitochondrial function, which is vital for cellular energy metabolism (Liang et al. 2020).

SIRT3 interacts with the PINK1/Parkin pathway to selectively remove damaged mitochondria, thereby preventing the accumulation of reactive oxygen species (ROS) (Liang et al. 2020; Huang et al. 2022). Mitophagy helps maintain mitochondrial integrity, which is essential for preventing neurodegenerative diseases and metabolic disorders (Dharmalingam et al. 2023).

SIRT6 and Lysosomal Function

SIRT6 is involved in maintaining lysosomal function, which is critical for the efficient degradation of cellular waste, including senescent cell debris. By stabilizing lysosomal membranes, SIRT6 ensures that autophagosomes can fuse with lysosomes effectively, facilitating the breakdown of damaged cellular components (Son et al. 2021).

SIRT6 regulates the expression of genes involved in lysosomal biogenesis and function, ensuring that lysosomes remain competent for autophagic degradation (Son et al. 2021). Efficient lysosomal function is essential for preventing the accumulation of toxic cellular waste, which can lead to inflammation and cellular senescence.

11.1.4 Impact on Brain Longevity

The synergy between sirtuins and autophagy plays a crucial role in supporting brain health through several key mechanisms:

1. *Recycling Damaged Mitochondria: Mitophagy*

Mitophagy is a specific form of autophagy that targets damaged mitochondria, helping maintain mitochondrial integrity. This process is vital for neuronal energy metabolism, as mitochondria are the primary energy-producing structures within cells. SIRT3, a mitochondrial sirtuin, activates mitophagy by interacting with the PINK1/Parkin pathway, ensuring that dysfunctional mitochondria are removed and replaced with healthy ones. This prevents the accumulation of reactive oxygen species (ROS) and maintains efficient energy production, which is essential for neuronal survival and function (Fernandez et al. 2024).

2. *Reducing SASP-Mediated Inflammation*: *Clearing Senescent Cells*

Senescent cells, which accumulate with age, produce a senescence-associated secretory phenotype (SASP) that includes pro-inflammatory cytokines. This chronic inflammation can damage surrounding tissues and contribute to aging and age-related diseases (Tchkonia et al. 2013). The synergy between sirtuins and autophagy helps clear senescent cells, thereby reducing SASP-mediated inflammation. SIRT1, for example, can suppress inflammatory pathways by deacetylating NF-κB, a key transcription factor involved in inflammation (Chandramowlishwaran et al. 2020). Additionally, autophagy can degrade damaged cellular components, including those from senescent cells, further reducing inflammation.

3. *Preserving Synaptic Plasticity*

Synaptic plasticity is the ability of synapses to strengthen or weaken over time, in response to increases or decreases in their activity. It is crucial for learning and memory. The synergy between sirtuins and autophagy supports synaptic plasticity by maintaining synaptic function and neuronal connectivity (Stampanoni Bassi et al. 2019). SIRT1 enhances synaptic plasticity by regulating genes involved in synaptic function and modulating microRNA expression that affects dendritic branching and synaptic density (Fernandez et al. 2024). Autophagy helps clear protein aggregates and damaged organelles that can impair synaptic function, ensuring that neurons remain healthy and functional (Sun et al. 2023). Enhancing sirtuin activity or autophagy could mitigate neurodegenerative diseases by improving mitochondrial function, reducing inflammation, and preserving synaptic plasticity (Fernandez et al. 2024).

The synergy between sirtuins and autophagy is essential for maintaining brain health by recycling damaged mitochondria, reducing SASP-mediated inflammation, and preserving synaptic plasticity. Understanding these mechanisms can lead to novel therapeutic strategies for combating neurodegenerative diseases and promoting healthy aging (Fernandez et al. 2024). Targeting specific sirtuin isoforms and avoiding off-target effects remain significant challenges in developing effective therapies (Maissan et al. 2021).

11.1.4.1 Chapter Overview: Exploring Their Impact on Brain Longevity

The chapter "Sirtuins, Autophagy, and Brain Longevity" explores the intricate interplay between sirtuins and autophagy in maintaining brain health and combating age-related cognitive decline. Sirtuins, a family of NAD^+-dependent enzymes, regulate key cellular processes such as stress resistance, DNA repair, and mitochondrial function. Specific isoforms like SIRT1, SIRT3, and SIRT6 play pivotal roles in promoting neuronal survival, enhancing mitochondrial quality control, and stabilizing lysosomal function. Meanwhile, autophagy acts as a cellular recycling mechanism that clears damaged organelles and toxic protein aggregates, preventing neurodegenerative diseases. The synergistic relationship between sirtuins and autophagy ensures cellular resilience by maintaining energy metabolism, reducing inflammation, and preserving synaptic plasticity—key factors for brain longevity.

The chapter also highlights the therapeutic potential of targeting these pathways to mitigate neurodegenerative diseases. Caloric restriction, exercise, and natural compounds like resveratrol have been shown to enhance sirtuin activity and autophagy, improving cellular health and lifespan. Sirtuins regulate autophagy by deacetylating proteins involved in autophagosome formation, while autophagy maintains NAD^+ levels essential for sirtuin function. This bidirectional relationship is crucial for recycling damaged mitochondria through mitophagy, reducing inflammation caused by senescent cells, and preserving synaptic connectivity. By understanding these mechanisms, researchers aim to develop novel strategies to enhance brain health and combat aging-related disorders effectively.

11.2 Sirtuins: Guardians of Cellular Health

Sirtuins, a family of NAD^+-dependent deacetylases and ADP-ribosyltransferases, are critical regulators of cellular homeostasis, linking metabolic states to stress adaptation and longevity. Their roles span mitochondrial quality control, DNA repair, inflammation modulation, and neuroprotection, making them central players in combating age-related cognitive decline and neurodegenerative diseases. Below, we expand on their mechanisms, therapeutic potential, and implications for brain health, supported by insights from recent research.

11.2.1 Sirtuins: A Family of Proteins with Diverse Functions

Mammals express seven sirtuin isoforms (SIRT1–7), each localized to specific cellular compartments:

Nuclear sirtuins: SIRT1, SIRT6, and SIRT7 regulate chromatin remodeling, DNA repair, and ribosomal biogenesis. SIRT6 stabilizes telomeres and suppresses age-related gene expression, F2 metabolic signals to cell cycle progression (Lee 2019).

These enzymes rely on NAD^+ as a cofactor, making them sensitive to cellular energy status. Declining NAD^+ levels with age reduce sirtuin activity, accelerating mitochondrial dysfunction and oxidative stress (Xu et al. 2023; Smith et al. 2008).

11.2.2 Sirtuin Activation: Caloric Restriction, Exercise, and Natural Compounds

Sirtuin activity is enhanced by several factors that increase NAD^+ availability and activate key signaling pathways:

1. *Caloric Restriction*: Caloric restriction increases NAD^+ availability, which boosts SIRT1 and SIRT3 activity. This dietary intervention mimics stress resilience and extends lifespan in model organisms by activating sirtuins and inducing autophagy. The rise in NAD^+ levels during CR is crucial for enhancing sirtuin activity, as sirtuins require NAD^+ as a cofactor for their enzymatic functions (Kahmini et al. 2022).
2. *Exercise*: Exercise activates AMPK, a key kinase that elevates NAD^+ levels and promotes mitochondrial biogenesis via SIRT3. AMPK activation enhances SIRT1 activity, which in turn enhances synaptic plasticity by regulating genes involved in synaptic function and modulating microRNA expression (Kahmini et al. 2022). Exercise-induced SIRT1 activation supports neuronal health and cognitive function by maintaining mitochondrial integrity and reducing oxidative stress.
3. *Natural Compounds*: Natural compounds like resveratrol, fisetin, and quercetin are known to activate sirtuins. Resveratrol, a SIRT1 activator, reduces amyloid-β plaques in Alzheimer's models by enhancing autophagy and promoting non-amyloidogenic APP processing. Fisetin and quercetin synergize with caloric restriction to amplify sirtuin-mediated autophagy, further supporting cellular clearance and stress resilience (Xu et al. 2023; Iside et al. 2020).

Pharmacological NAD^+ Boosters Pharmacological NAD^+ boosters, such as nicotinamide mononucleotide (NMN), are being explored in clinical trials for age-related cognitive disorders. These compounds aim to restore NAD^+ levels, which decline with age, thereby enhancing sirtuin activity and autophagy. NMN supplementation has shown promise in improving cognitive function by boosting SIRT1 and autophagic flux (Iside et al. 2020).

11.2.3 Sirtuins and Cellular Repair Mechanisms

11.2.3.1 DNA Repair

Sirtuins play a crucial role in DNA repair mechanisms, contributing to genomic stability and cellular health. Sirtuins are involved in DNA repair at multiple levels, including chromatin remodeling, recruitment of DNA repair machinery, and direct modulation of repair proteins.

1. Chromatin Remodeling: Sirtuins, particularly SIRT1 and SIRT6, modify chromatin structure to facilitate access to DNA damage sites. This remodeling is essential for the recruitment of repair proteins and the initiation of DNA repair processes like non-homologous end joining (NHEJ) and homologous recombination (HR) (Vazquez et al. 2017; Bosch-Presegué and Vaquero 2015).
2. Recruitment of DNA Repair Machinery: Sirtuins interact with key components of the DNA damage response (DDR) pathway. For example, SIRT6 stabilizes DNA-PKcs, a critical enzyme in NHEJ, ensuring efficient repair of double-strand breaks (DSBs) (Vazquez et al. 2017; Van Meter et al. 2011).
3. Direct Modulation of Repair Proteins: SIRT1 and SIRT6 can directly deacetylate and regulate the activity of proteins involved in DNA repair. SIRT1 interacts with proteins from various DNA repair pathways, including HR and NHEJ, facilitating their recruitment to damaged DNA sites (Bosch-Presegué and Vaquero 2015; Alves-Fernandes and Jasiulionis 2019). Specific Roles of Sirtuin Isoforms are-

   SIRT1: Plays a crucial role in maintaining genomic stability by regulating DDR signaling through PP4, which helps dephosphorylate γH2AX, a marker of DNA damage (Alves-Fernandes and Jasiulionis 2019; Rasti et al. 2023). SIRT1 also relocalizes to DSB sites to facilitate repair (Van Meter et al. 2011).

   SIRT6: Involved in NHEJ by stabilizing DNA-PKcs and promoting mono-ADP ribosylation of PARP1, which enhances DNA repair under oxidative stress conditions (Van Meter et al. 2011). SIRT6 also helps restore chromatin structure after repair.

   SIRT7: Essential for the repair of DSBs by deacetylating histone H3 at lysine 18, influencing the recruitment of 53BP1 to DNA damage sites (Vazquez et al. 2017).

Understanding the role of sirtuins in DNA repair can lead to novel therapeutic strategies for diseases associated with genomic instability, such as cancer and neurodegenerative disorders. Enhancing sirtuin activity might improve DNA repair efficiency and mitigate age-related pathologies.

11.2.3.2 Mitochondrial Function

Sirtuins play a crucial role in maintaining mitochondrial quality and controlling inflammation, which are essential for cellular health and preventing age-related diseases. Sirtuins maintain mitochondrial quality through two primary mechanisms:

11.2.3.3 Mitophagy

SIRT1 and SIRT3: Both sirtuins activate the PINK1/Parkin pathway, which is essential for mitophagy. This process selectively removes damaged mitochondria, preventing the accumulation of reactive oxygen species (ROS) and maintaining mitochondrial integrity. SIRT3, localized in mitochondria, is particularly important for this function. Mitophagy ensures efficient energy production and reduces oxidative stress, which is critical for neuronal survival and function (Sack and Finkel 2012; Lombard et al. 2011; Zhang et al. 2020).

11.2.3.4 Metabolic Flexibility

SIRT4 regulates glutamate dehydrogenase (GDH), optimizing energy production during nutrient stress. This metabolic flexibility is crucial for maintaining cellular homeostasis under varying conditions. By modulating GDH, SIRT4 helps balance energy metabolism, ensuring that cells can adapt to changes in nutrient availability (Wang and Wei 2020).

11.2.3.5 Inflammation Control

Sirtuins also play a significant role in controlling inflammation.

SIRT1 and NF-κB Signaling: SIRT1 suppresses NF-κB signaling by deacetylating p65, a subunit of NF-κB. This reduces the production of pro-inflammatory cytokines like TNF-α and IL-6. By inhibiting NF-κB, SIRT1 decreases chronic inflammation, which is associated with various age-related diseases (Kitada et al. 2019; Zhang et al. 2022).

SIRT3 and NLRP3 Inflammasome: SIRT3 inhibits the activation of the NLRP3 inflammasome in microglia. This inflammasome is a key mediator of neuroinflammation. By attenuating NLRP3 inflammasome activation, SIRT3 reduces neuroinflammation, which is crucial for preventing neurodegenerative diseases (Wu et al. 2022).

The main biochemical processes that take place in the human body with the participation of SIRTs are shown in Fig. 11.2.

11.2.4 Neuroprotection: Countering Neurodegenerative Diseases

Sirtuins play a crucial role in neuroprotection by mitigating the pathology of neurodegenerative diseases such as Alzheimer's (AD) and Parkinson's (PD).

11.2.4.1 Amyloid-β Clearance in Alzheimer's Disease

SIRT1 promotes non-amyloidogenic APP processing via α-secretase activation, reducing amyloid-β plaque formation. This process involves the cleavage of amyloid precursor protein (APP) in a way that does not lead to the formation of amyloid-β peptides, which are toxic to neurons and contribute to AD pathology.

SIRT1 enhances the activity of α-secretase, which cleaves APP into soluble APPα (sAPPα), a non-toxic fragment. This reduces the production of amyloid-β peptides, which are the main components of amyloid plaques in AD brains. By reducing amyloid-β burden, SIRT1 helps prevent neuronal damage and cognitive decline associated with AD. Studies have shown that SIRT1 activation can decrease amyloid-β levels by promoting non-amyloidogenic APP processing, offering a

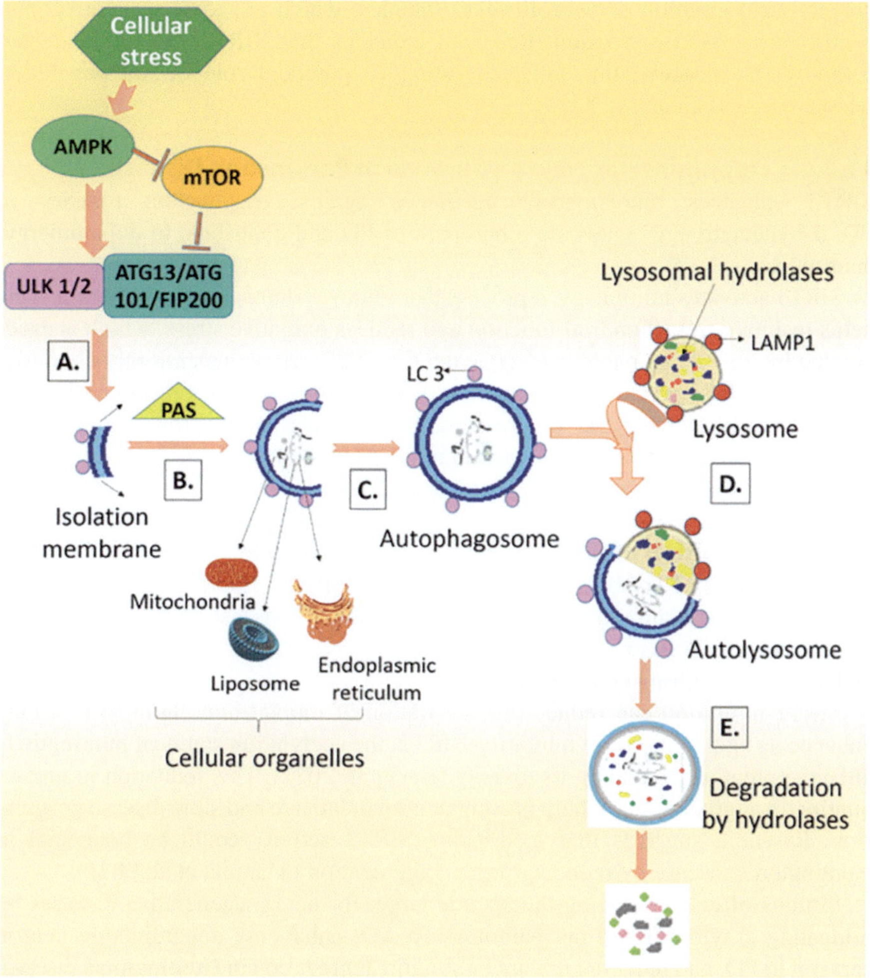

Fig. 11.2 Process of autophagy (**a**). Initiation phase (**b**). Elongation phase (**c**). Maturation phase (**d**). Docking and fusion phase (**e**). Degradation phase (Naseer et al. 2021)

potential therapeutic strategy for AD (Herskovits and Guarente 2013; Fernando and Wijayasinghe 2021).

11.2.4.2 Tau Regulation in Alzheimer's Disease

SIRT2 deacetylates tau, preventing neurofibrillary tangle aggregation in AD models. Tau protein hyperphosphorylation and aggregation are key features of AD pathology.

SIRT2 deacetylates tau protein, which can reduce its aggregation and promote its clearance. Acetylation of tau can enhance its stability and aggregation, contributing to neurofibrillary tangle formation. By reducing tau acetylation, SIRT2 helps prevent the formation of neurofibrillary tangles, which are toxic to neurons and contribute to AD progression. Research indicates that SIRT2 can modulate tau pathology by deacetylating tau, suggesting its potential role in AD prevention (Herskovits and Guarente 2013).

11.2.4.3 Dopaminergic Neuron Survival in Parkinson's Disease

SIRT3 enhances mitochondrial resilience against α-synuclein toxicity in PD. α-Synuclein aggregates are a hallmark of PD and contribute to dopaminergic neuron loss.

SIRT3 activates mitophagy, a process that removes damaged mitochondria. This helps maintain mitochondrial function and reduces oxidative stress, which is exacerbated by α-synuclein aggregates (Li and Cai 2022). By enhancing mitochondrial resilience, SIRT3 protects dopaminergic neurons from α-synuclein-induced toxicity, potentially slowing PD progression. Studies have demonstrated that SIRT3 plays a protective role in PD by maintaining mitochondrial integrity and reducing oxidative damage (Shi et al. 2017; Patel et al. 2024).

11.2.4.4 Inhibition of SIRT2 in Huntington's Disease

Inhibition of SIRT2 shows neuroprotective benefits in Huntington's disease by reducing mutant huntingtin aggregation. Huntington's disease is characterized by the expansion of CAG repeats in the huntingtin gene, leading to protein misfolding and aggregation (Tong et al. 2024).

SIRT2 inhibition can reduce the acetylation of mutant huntingtin, which may enhance its aggregation. By inhibiting SIRT2, the acetylation status of huntingtin is altered, potentially reducing its toxicity (Bai et al. 2022). This reduction in mutant huntingtin aggregation can help prevent neuronal damage and slow disease progression. Research suggests that modulating SIRT2 activity could be beneficial in Huntington's disease by reducing protein aggregation (Manjula et al. 2021).

Sirtuins offer a promising therapeutic target for neurodegenerative diseases by mitigating amyloid-β and tau pathology in AD, enhancing dopaminergic neuron survival in PD, and reducing mutant huntingtin aggregation in Huntington's disease. Understanding these mechanisms can lead to novel therapeutic strategies for combating neurodegenerative disorders.

11.2.5 Cognitive Enhancement: Synaptic Plasticity and Memory

Sirtuins play a pivotal role in cognitive function by regulating synaptic plasticity, neurogenesis, and epigenetic stability, supported by experimental evidence and clinical insights:

1. Synaptic Plasticity

SIRT1 enhances synaptic plasticity by inhibiting miR-134, a microRNA that suppresses the transcription factor CREB (cAMP response element-binding protein). SIRT1 forms a complex with the transcription factor YY1 to suppress miR-134 expression. When SIRT1 is deficient, miR-134 levels rise, leading to translational inhibition of CREB and BDNF (brain-derived neurotrophic factor), impairing synaptic plasticity and memory (Gao et al. 2010; Michán et al. 2010).

Overexpression of miR-134 in hippocampal neurons mimics SIRT1 deficiency, causing defective long-term potentiation (LTP) and impaired contextual fear conditioning (Gao et al. 2010). Conversely, SIRT1 activation restores synaptic plasticity and memory formation. Resveratrol, a SIRT1 activator, restores synaptic plasticity in aging models by enhancing CREB/BDNF signaling (Donmez 2013).

2. Neurogenesis

SIRT1 promotes neurogenesis by upregulating BDNF, which supports neuronal survival and differentiation in the hippocampal dentate gyrus. SIRT1 deficiency reduces dendritic branching, complexity, and hippocampal gene expression linked to synaptic function. BDNF upregulation by SIRT1 enhances neurogenesis, critical for learning and memory (Michán et al. 2010).

SIRT1 knockout (KO) mice exhibit cognitive deficits, while SIRT1 overexpression preserves synaptic plasticity and memory (Michán et al. 2010). NAD^+ boosters like NMN improve cognitive performance in early-stage Alzheimer's trials by enhancing SIRT1 activity and BDNF levels (Donmez 2013).

3. Epigenetic Regulation

SIRT6 stabilizes heterochromatin, preserving genomic integrity in aging neurons. SIRT6 co-localizes with heterochromatin protein HP1β, maintaining telomere integrity and silencing repetitive DNA elements. It deacetylates histone H3K9 and H3K56, ensuring chromatin stability and preventing age-related epigenetic drift (Korotkov et al. 2021).

SIRT6 deficiency accelerates genomic instability, while its overexpression enhances DNA repair and delays cellular senescence. SIRT6's role in heterochromatin maintenance makes it a potential target for combating age-related cognitive decline (Korotkov et al. 2021).

11.2.5.1 Therapeutic Implications and Challenges (Yu et al. 2024)

While sirtuin activators (e.g., STACs, NAD^+ boosters) show promise, several challenges persist:

Tissue Specificity: Mitochondrial sirtuins (SIRT3–5) require targeted delivery to avoid off-target effects. For example, SIRT3 regulates SOD2 in cardiac cells, but systemic activation may disrupt other tissues.

Dosage and Timing: Excessive SIRT1 activation may promote cancer by deacetylating oncogenic targets. Precision in dosing is critical to balance neuroprotection and oncogenic risks.

Clinical Validation: Despite preclinical success in Alzheimer's and Parkinson's models, human trials for sirtuin-based therapies (e.g., resveratrol, NMN) show mixed results. For example, NMN improves cognition in early trials but lacks robust lifespan extension data (Donmez 2013).

Sirtuins bridge metabolic signaling, stress adaptation, and cognitive resilience, offering multifaceted strategies to combat brain aging. Their interplay with autophagy, mitochondrial quality control, and epigenetic regulation underscores their therapeutic potential. However, translating these insights into effective treatments requires optimizing NAD^+-boosting strategies to enhance sirtuin activity without off-target effects, clarifying isoform-specific mechanisms, particularly SIRT6's role in heterochromatin stability (Korotkov et al. 2021) and SIRT1's dual roles in neuroprotection and cancer, and advancing clinical trials to validate sirtuin modulators in neurodegenerative diseases (Donmez 2013). Future research must address tissue-specific delivery, dosage precision, and mechanistic nuances to harness sirtuins' full neuroprotective potential.

11.3 Autophagy: The Cellular Housekeeper

Autophagy is a conserved cellular recycling process critical for maintaining homeostasis by degrading damaged organelles, protein aggregates, and pathogens. It acts as a dynamic quality control system, ensuring cellular survival under stress and preventing age-related diseases. Dysregulated autophagy is implicated in neurodegenerative disorders, cancer, and metabolic diseases, underscoring its importance in health and disease (Kroemer et al. 2010).

11.3.1 What Is Autophagy? The Cellular Recycling Process

Autophagy is a highly regulated cellular recycling process critical for maintaining homeostasis. Below is a detailed explanation of its three primary steps, molecular mechanisms, and activation triggers:

1. *Formation of Autophagosomes*

Autophagosomes are double-membraned vesicles that engulf cytoplasmic cargo (e.g., damaged mitochondria, protein aggregates). This process begins with the activation of the ULK1 kinase complex (ULK1, ATG13, FIP200), which initiates autophagosome nucleation. The PI3K complex (Beclin-1, VPS34, ATG14) then generates phosphatidylinositol 3-phosphate (PI3P) to recruit ATG proteins (e.g., LC3/ATG8) to the phagophore membrane. LC3 is lipidated (LC3-II) to expand the phagophore into a sealed autophagosome (Suzuki and Ohsumi 2007). Key molecules involved in autophagosome formation are:

(a) ULK1/ATG1: Initiates autophagosome formation.
(b) Beclin-1: Regulates phagophore nucleation.
(c) LC3/ATG8: Marks autophagosome membranes for cargo engulfment.

2. *Fusion with Lysosomes*

Autophagosomes fuse with lysosomes to form autolysosomes, where acidic hydrolases (e.g., cathepsins) degrade the cargo. This fusion requires Rab GTPases (Rab7) mediate autophagosome-lysosome docking and SNARE proteins (STX17, SNAP29, VAMP8) enabling membrane fusion and lysosomal membrane protein (LAMP-2) for lysosome maturation.The acidic lysosomal environment (pH ~4.5) activates hydrolytic enzymes, breaking down proteins, lipids, and organelles (Ke 2024).

3. *Recycling of Breakdown Products*

Degraded cargo (e.g., amino acids, fatty acids) is released into the cytoplasm via lysosomal transporters (e.g., LAMP-2) and reused for energy production (amino acids enter the TCA cycle), biosynthesis (Fatty acids rebuild membranes), and stress adaptation (recycling sustains survival during nutrient deprivation (Lahiri et al. 2019). In this process, key role players are LAMP-2 that facilitates lysosomal efflux and transcription factor promoting lysosomal biogenesis (TFEB)

11.3.1.1 Activation of Autophagy

Autophagy is triggered by cellular stressors. In nutrient deprivation, AMPK activates ULK1 by phosphorylating mTORC1 (a nutrient sensor), releasing its inhibition on autophagy. ROS activate FOXO transcription factors, upregulating autophagy genes (e.g., *ATG5*, *LC3*). Pathogen sensors (e.g., TLRs) induce autophagy to eliminate intracellular microbes (e.g., *Mycobacterium tuberculosis*). Autophagy is a dynamic process enabling cells to adapt to stress by recycling damaged components. Its dysregulation is linked to neurodegenerative diseases (e.g., Alzheimer's) and cancer. Understanding its molecular steps offers therapeutic insights for enhancing cellular resilience (Table 11.1).

Table 11.1 Types of autophagy and their key features

Type	Mechanism	Key features
Macroautophagy	Bulk degradation via autophagosomes	Non-selective under starvation; selective for organelles (e.g., mitophagy) (Anding and Baehrecke 2017)
Microautophagy	Lysosomal membrane invagination engulfs cargo	Direct degradation without autophagosomes; less selective (Hashemzaei et al. 2017)
Chaperone-mediated autophagy (CMA)	Hsc70 chaperones deliver proteins to lysosomes	Highly selective; requires a pentapeptide motif (KFERQ) on substrates (Orenstein and Cuervo 2010)
Crinophagy	Lysosomal degradation of secretory granules	Regulates hormone secretion; less studied (Szenci et al. 2023)
Mitophagy	a subtype of macroautophagy, selectively removes damaged mitochondria via PINK1/Parkin signaling	Preventing oxidative stress (Jin and Youle 2012)

11.3.2 Autophagy and Cellular Clearance: Removing Damaged Proteins and Organelles

Autophagy is a vital cellular process that ensures the removal of damaged or dysfunctional cellular components, including proteins and organelles. This process is crucial for maintaining cellular homeostasis and preventing the accumulation of toxic substances that can lead to disease.

Protein Aggregates Mechanism: Autophagy clears toxic protein aggregates, such as β-amyloid in Alzheimer's disease and α-synuclein in Parkinson's disease. This is achieved through the formation of autophagosomes that engulf these aggregates and deliver them to lysosomes for degradation. Proteins like LC3-II and p62/SQSTM1 are essential for autophagosome formation and the selective degradation of ubiquitinated protein aggregates. Defective clearance of protein aggregates contributes to neurodegenerative diseases by allowing these toxic substances to accumulate and damage neurons (Aman et al. 2021).

11.3.2.1 Organelle Turnover

11.3.2.1.1 Mitochondria (Mitophagy)

PINK1/Parkin Pathway: Mitophagy is a specialized form of autophagy that targets damaged mitochondria. The PINK1/Parkin pathway is activated when mitochondria lose membrane potential, leading to the ubiquitination of mitochondrial proteins and their engulfment by autophagosomes.

Mitophagy prevents the accumulation of reactive oxygen species (ROS) and maintains mitochondrial function, which is critical for energy metabolism and neuronal survival (Glick et al. 2010; Khandia et al. 2019).

11.3.2.1.2 ER (Reticulophagy)

Unfolded Protein Response (UPR): Reticulophagy involves the degradation of stressed ER fragments during the unfolded protein response. This process helps maintain ER homeostasis by removing damaged ER components. Dysregulation of reticulophagy can lead to ER stress, contributing to various diseases, including neurodegenerative disorders (Khandia et al. 2019).

11.3.2.2 Pathogen Defense

Xenophagy: Autophagy also plays a role in pathogen defense by targeting intracellular pathogens (e.g., Salmonella) for degradation. Galectin-8 is involved in recognizing and targeting damaged pathogen-containing vacuoles for autophagic clearance. This mechanism helps protect cells from infection by eliminating pathogens and preventing their replication (Van Limbergen et al. 2009).

11.3.3 Autophagy and Cellular Stress Response: Adapting to Changing Conditions

Autophagy is a critical stress response mechanism that helps cells adapt to adverse conditions such as nutrient deprivation, hypoxia, and oxidative stress.

11.3.3.1 Nutrient Deprivation

AMPK/mTORC1 Axis: During starvation, AMPK is activated, which in turn activates ULK1 by phosphorylation. This leads to the initiation of autophagy, allowing cells to recycle macromolecules and sustain energy balance. Autophagy provides essential nutrients by breaking down non-essential cellular components, ensuring cell survival during nutrient scarcity (Glick et al. 2010; Khandia et al. 2019).

11.3.3.2 Hypoxia

HIF-1α Activation: Under hypoxic conditions, HIF-1α is stabilized and upregulates genes involved in mitophagy (e.g., BNIP3/NIX). This helps reduce ROS production by eliminating dysfunctional mitochondria. Mitophagy mitigates oxidative stress, supporting cell survival under low oxygen conditions (Khandia et al. 2019; Aman et al. 2021).

11.3.3.3 Oxidative Stress

ROS Clearance: Autophagy removes ROS-damaged cellular components, supported by sirtuin activation (e.g., SIRT1). SIRT1 enhances autophagy by deacetylating transcription factors like FOXO, which upregulate autophagy-related genes. Efficient clearance of ROS-damaged components prevents oxidative damage and maintains cellular health (Glick et al. 2010; Aman et al. 2021).

11.3.4 Autophagy and Neuroprotection: Preventing Neurodegeneration

Autophagy plays a crucial role in neuroprotection by clearing toxic protein aggregates, maintaining mitochondrial quality, and regulating inflammation.

1. *Aggregate Clearance*

Tau and Huntingtin Degradation: Autophagy clears hyperphosphorylated tau (Alzheimer's) and mutant huntingtin (Huntington's) via selective receptors like p62, reducing neurotoxicity. This process helps prevent neurodegeneration by removing toxic protein aggregates that can damage neurons (Aman et al. 2021; Levine and Klionsky 2017).

2. *Mitochondrial Quality Control*

Mitophagy in Dopaminergic Neurons: SIRT3 enhances PINK1/Parkin-mediated mitophagy, preventing α-synuclein-induced toxicity in Parkinson's disease by removing damaged mitochondria. Maintaining mitochondrial function is essential for preventing oxidative stress and neurodegeneration (Glick et al. 2010; Khandia et al. 2019).

3. *Inflammation Regulation*

NLRP3 Inflammasome Suppression: Autophagy degrades damaged mitochondria, reducing mtROS that activate the NLRP3 inflammasome in microglia. This suppresses neuroinflammation and protects neurons. Chronic inflammation contributes to neurodegenerative diseases; thus, autophagy's role in inflammation regulation is critical for neuroprotection (Khandia et al. 2019; Aman et al. 2021).

11.3.4.1 Therapeutic Challenges

Trehalose Controversy: Initially thought to induce autophagy, trehalose blocks autophagosome-lysosome fusion in vitro. However, it reduces neuropathology in vivo, possibly through indirect mechanisms like modulating gut microbiota.

Drug Development: Rapamycin analogs enhance autophagy but risk immunosuppression. Targeting autophagy pathways requires precision to avoid off-target effects (Van Limbergen et al. 2009).

Autophagy is a multifaceted process essential for cellular clearance, stress adaptation, and neuroprotection. Its dysregulation contributes to neurodegenerative diseases and metabolic disorders. Understanding autophagy's mechanisms and implications can lead to novel therapeutic strategies for enhancing cellular resilience and preventing age-related diseases. However, challenges remain in targeting autophagy pathways effectively without off-target effects.

11.3.4.2 Challenges in Therapeutic Targeting

Autophagy inducers Rapamycin analogs (e.g., everolimus) enhance clearance but risk immunosuppression.

Specificity Selective activation of mitophagy or CMA is needed to avoid off-target effects (Baeken 2024; Lee et al. 2018).

Autophagy is a versatile process essential for cellular resilience, adapting to stress, and preventing neurodegeneration. While its role in clearing toxic aggregates and organelles is well-established, therapeutic strategies must navigate complexities like flux regulation and tissue specificity. Advances in understanding selective autophagy pathways, such as mitophagy and CMA, offer promise for treating neurodegenerative diseases, though controversies around compounds like trehalose highlight the need for deeper mechanistic insights (Anding and Baehrecke 2017; Lee et al. 2018).

11.4 The Sirtuin-Autophagy Connection

11.4.1 Crosstalk Between Sirtuins and Autophagy: A Complex Interplay

The relationship between sirtuins and autophagy is intricate and bidirectional. Both processes are activated in response to cellular stress, such as nutrient deprivation or oxidative damage, and they work together to ensure cellular survival and homeostasis.

11.4.1.1 Shared Metabolic Sensors

NAD^+ Dependency Sirtuins require NAD^+ as a cofactor for their enzymatic activity. Autophagy helps maintain NAD^+ levels by recycling damaged organelles and proteins, which can be used to synthesize NAD^+ precursors.

Stress-Responsive Coordination During stress, AMPK (AMP-activated protein kinase) is activated, which in turn activates autophagy and increases NAD^+ levels, thereby enhancing sirtuin activity.

11.4.2 Regulation of Autophagy by Sirtuins

The regulation of autophagy by sirtuins involves a complex interplay of mechanisms that enhance autophagic processes, contributing to cellular homeostasis and longevity. Sirtuins, particularly SIRT1 and SIRT6, play significant roles in promoting autophagy through various pathways.

11.4.2.1 Mechanisms of Sirtuin-Mediated Autophagy Regulation

Gene Expression Regulation: Sirtuins regulate autophagy by modulating the transcription of autophagy-related genes. SIRT1 plays a central role by interacting with transcription factors such as FOXO, p53, and NF-κB, deacetylating them to influence their activity. For example:

Under nutrient deprivation, SIRT1 levels increase, promoting autophagy by upregulating genes like *LC3* and *Atg5* (Lee 2019). SIRT1 also represses mTOR signaling, a key inhibitor of autophagy, by deacetylating components of the mTOR pathway, thereby enhancing autophagic flux. In *Sirt1*-deficient cells, autophagy is impaired, but reintroducing functional SIRT1 restores autophagic activity, highlighting its transcriptional role (Lee 2019). SIRT1 deacetylates FOXO proteins, which activate stress resistance genes, including those involved in autophagy (Lee 2019), and forms a feedback loop with E2F1 to balance apoptosis and repair during DNA damage, indirectly influencing autophagy-related gene expression (Yang et al. 2022).

Post-Translational Modifications: Sirtuins directly deacetylate autophagy-related proteins to enhance their activity. SIRT1 deacetylates Atg5, Atg7, and LC3, promoting autophagosome formation. Deacetylation of LC3 facilitates its lipidation (conversion from LC3-I to LC3-II), a critical step in autophagosome maturation (Lee 2019). SIRT1 also deacetylates FoxO1, which upregulates autophagy genes under oxidative stress (Ou et al. 2014). Pharmacological inhibition of SIRT1 increases acetylation of these proteins, impairing autophagy. Deacetylation by SIRT1 enhances the enzymatic activity of autophagy machinery components. For instance, deacetylated Atg7 shows increased interaction with other Atg proteins, improving autophagosome assembly (Lee 2019).

Mitochondrial Quality Control (Mitophagy): SIRT3 is critical for mitochondrial homeostasis via the PINK1/Parkin pathway. SIRT3 deacetylates PINK1, enabling its stabilization on damaged mitochondria. This activates Parkin, which ubiquitinates mitochondrial proteins to recruit autophagosomes (Wan et al. 2022).

In osteoarthritis models, SIRT3 deficiency leads to hyperacetylated PINK1, defective mitophagy, and mitochondrial dysfunction, accelerating cartilage degeneration (Cao et al. 2024). Mitophagy prevents ROS accumulation and maintains energy metabolism, protecting against diseases like Parkinson's and cardiovascular disorders (Cen et al. 2021). PINK1-Parkin axis: SIRT3-mediated deacetylation ensures efficient clearance of damaged mitochondria, critical for neuronal and cardiac cell survival.

Lysosomal Function: SIRT6 contributes to autophagic degradation by stabilizing lysosomal function. Prior studies suggest SIRT6 maintains lysosomal membrane integrity, ensuring efficient fusion of autophagosomes with lysosomes (Iachettini et al. 2018). SIRT6 also stabilizes heterochromatin, preserving genomic integrity in aging neurons, which indirectly supports lysosomal gene expression.

Supporting evidence suggests that SIRT1 and SIRT3 are shown to enhance lysosomal activity in other contexts (e.g., SIRT1 interacts with mitofusin-2 to improve mitochondrial-lysosomal coordination in liver cells) (Lambona et al. 2024).

Understanding how sirtuins regulate autophagy has therapeutic potential for treating age-related diseases. Enhancing sirtuin activity can improve autophagy, potentially mitigating neurodegenerative disorders and metabolic diseases. However, challenges remain in targeting specific sirtuin isoforms and avoiding off-target effects.

The regulation of autophagy by sirtuins is a multifaceted process that supports cellular health by enhancing autophagic flux, maintaining mitochondrial integrity, and modulating gene expression. Further research into these mechanisms could lead to novel therapeutic strategies for promoting longevity and combating age-related diseases.

11.4.3 Regulation of Sirtuins by Autophagy

The regulation of sirtuins by autophagy is a crucial aspect of cellular homeostasis, particularly in maintaining sirtuin activity and ensuring proper cellular function. Autophagy supports sirtuin activity through several mechanisms:

Mitochondrial Quality Control: Mitophagy Mitophagy, a specific form of autophagy, targets and removes damaged mitochondria. This process is vital for preventing the accumulation of reactive oxygen species (ROS) that can inactivate sirtuins. Sirtuins, especially SIRT3, play a role in mitophagy by interacting with proteins like PINK1 and Parkin, which are essential for the selective degradation of dysfunctional mitochondria (Wan et al. 2022). By maintaining mitochondrial integrity, mitophagy ensures that sirtuins remain active and functional.

NAD^+ Recycling Autophagy helps recycle damaged cellular components, including organelles and proteins. This recycling process provides precursors for NAD^+ biosynthesis, which is essential for sirtuin activity. NAD^+ serves as a cofactor for sirtuin enzymes, and its availability directly influences their ability to deacetylate target proteins. By maintaining NAD^+ levels, autophagy indirectly supports sirtuin function, ensuring that these enzymes can continue to regulate various cellular processes (Baeken 2024; Wan et al. 2022).

Protein Homeostasis Autophagy also contributes to protein homeostasis by degrading acetyltransferases that inhibit sirtuins through hyperacetylation. Acetyltransferases can modify sirtuin substrates, reducing their activity. By removing these enzymes, autophagy helps maintain sirtuin activity, ensuring that they can effectively regulate their target proteins. This mechanism is crucial for maintaining cellular health and preventing the accumulation of dysfunctional proteins that could lead to disease (Wu et al. 2022).

11.4.3.1 Mechanistic Insights

The interplay between autophagy and sirtuins involves complex signaling pathways. For instance, autophagy can regulate the activity of upstream regulators of sirtuins, such as AMPK and mTOR, which are central to energy metabolism and autophagy induction (Tang et al. 2024). AMPK activation can enhance NAD^+ levels, supporting sirtuin activity, while mTOR inhibition promotes autophagy, further sustaining sirtuin function through NAD^+ recycling and mitochondrial quality control (Wan et al. 2022; Tang et al. 2024).

11.4.3.2 Clinical Implications

Understanding how autophagy regulates sirtuin activity has significant implications for developing therapeutic strategies against age-related diseases. Enhancing autophagy can improve sirtuin function, potentially mitigating neurodegenerative diseases and metabolic disorders. NAD^+ boosters, such as NMN, are being explored for their ability to enhance both autophagy and sirtuin activity, offering a promising avenue for treating age-related cognitive decline (Wan et al. 2022).

The regulation of sirtuins by autophagy is a critical mechanism for maintaining cellular health and preventing age-related diseases. By ensuring mitochondrial quality, recycling NAD^+, and maintaining protein homeostasis, autophagy plays a vital role in sustaining sirtuin activity. Further research into this interplay could lead to novel therapeutic strategies for enhancing sirtuin function and promoting healthy aging.

11.4.4 Importance for Brain Health

The synergy between sirtuins and autophagy plays a pivotal role in maintaining brain health by addressing several critical aspects:

Clearing Neurotoxic Aggregates SIRT1 and Tau: SIRT1 deacetylates tau, reducing its aggregation and neurotoxicity. This process is crucial in Alzheimer's disease, where tau tangles contribute to neuronal damage (Fujita and Yamashita 2018; Fernandez et al. 2024).

APP Processing: SIRT1 promotes non-amyloidogenic APP processing, reducing the formation of amyloid-β plaques, a hallmark of Alzheimer's pathology (Fernandez et al. 2024).

Autophagy and α-Synuclein: Autophagy degrades α-synuclein fibrils, which are associated with Parkinson's disease. Efficient clearance of these aggregates helps prevent neurodegeneration (Chojdak-Łukasiewicz et al. 2022).

Mitochondrial Resilience SIRT3 and Mitophagy: SIRT3 activates mitophagy, selectively removing damaged mitochondria to prevent oxidative stress and maintain neuronal energy metabolism. This process is vital for preserving mitochondrial function and preventing neurodegenerative diseases (Duan 2013).

Consequences of Dysfunction: Loss of SIRT1 or autophagy leads to mitochondrial dysfunction, a common feature in aging brains. This dysfunction contributes to neurodegenerative diseases by impairing energy production and increasing oxidative damage (Chojdak-Łukasiewicz et al. 2022).

Anti-inflammatory Effects SIRT1 deacetylates NF-κB, suppressing its activity and reducing neuroinflammation. Chronic inflammation is a significant contributor to neurodegenerative diseases, making SIRT1's role crucial for maintaining a healthy brain environment.

Autophagy inhibits the activation of the NLRP3 inflammasome in microglia, further reducing neuroinflammation. This mechanism helps protect neurons from inflammatory damage (Chojdak-Łukasiewicz et al. 2022).

NAD^+ Boosters compounds like NMN enhance SIRT1 activity and autophagy, showing promise in early Alzheimer's trials. By boosting NAD^+ levels, these supplements support both sirtuin and autophagy pathways, potentially mitigating age-related cognitive decline (Kahmini et al. 2022).

Caloric Restriction and Exercise: Both interventions amplify the synergy between sirtuins and autophagy, extending cognitive health span. They increase NAD^+ levels and activate AMPK, which enhances autophagy and sirtuin activity.

11.4.4.1 Therapeutic Implications

NAD^+ supplementation restores sirtuin activity and autophagic flux, offering a potential therapeutic strategy for age-related cognitive decline (Fernandez et al. 2024).

Sirtuin-activating compounds (STACs) like resveratrol and SRT2104 are being explored as treatments for neurodegenerative diseases. These compounds aim to enhance SIRT1 activity, though their efficacy and safety require further evaluation.

Tissue-specific targeting (e.g., SIRT3 in mitochondria) and avoiding off-target effects, particularly in cancer, are significant challenges in developing effective therapies (Wu et al. 2022).

The interplay between sirtuins and autophagy is essential for brain health, offering a multifaceted approach to combating neurodegenerative diseases. By enhancing the clearance of neurotoxic aggregates, maintaining mitochondrial resilience, reducing inflammation, and countering age-related decline, this synergy presents a promising therapeutic target for preserving cognitive function and extending health span. However, further research is needed to overcome the challenges associated with targeting these pathways effectively. These studies underscore the critical role of the sirtuin-autophagy axis in brain health and its potential as a therapeutic target for neurodegenerative diseases.

11.5 Sirtuins, Autophagy, and Brain Longevity

Sirtuins and autophagy are two critical biological processes that play significant roles in maintaining brain health and longevity. Sirtuins, a family of NAD++-dependent deacetylases, are involved in various cellular functions including metabolism, DNA repair, and inflammation regulation (Sah et al. 2025). Autophagy, a self-digestive process, helps in the clearance of damaged organelles and proteins, thereby promoting cellular homeostasis (Sah et al. 2025).

Sirtuins and Brain Health

Sirtuins, particularly SIRT1 and SIRT3, have been identified as key neuroprotective agents. SIRT1 is known for its role in inhibiting mTOR signaling, which is crucial for promoting neurite outgrowth and neuronal survival (Lee 2019; Watroba and Szukiewicz 2021). Additionally, SIRT1 helps in reducing the accumulation of neurotoxic proteins such as beta-amyloid, a hallmark of Alzheimer's disease (Lee 2019). SIRT3, localized in mitochondria, exerts antioxidative effects and protects mitochondrial function, contributing to brain health (Watroba and Szukiewicz 2021; Duan 2013).

Autophagy and Brain Longevity

Autophagy is essential for maintaining brain integrity by clearing damaged organelles and protein aggregates. Sirtuins, particularly SIRT1 and SIRT6, have been shown to regulate autophagy by modulating gene expression and post-translational modifications of autophagy-related proteins (Marius 2023). This regulation helps in promoting longevity and protecting against neurodegenerative diseases (Lee 2019; Marius 2023).

Interplay Between Sirtuins and Autophagy
The interplay between sirtuins and autophagy is complex and multifaceted. Sirtuins can induce autophagy by inhibiting mTOR signaling, a negative regulator of autophagy (Lee 2019). This induction of autophagy contributes to the clearance of damaged cellular components, thereby enhancing cellular resilience and longevity (Marius 2023; Aman et al. 2021). Furthermore, autophagy can also influence sirtuin activity by maintaining cellular NAD++ levels, which are essential for sirtuin function (Sah et al. 2025).

Impact on Brain Longevity
Studies have shown that overexpression of SIRT1 in the brain can extend lifespan and delay aging in mice (Yalcin 2018). Moreover, sirtuin-induced autophagy has protective roles in the brain, mitigating neurodegenerative diseases such as Alzheimer's and Parkinson's (Lee 2019; Yalcin 2018). The regulation of autophagy by sirtuins ensures the removal of toxic protein aggregates, which is crucial for maintaining brain health and promoting longevity (Aman et al. 2021; Marius 2023).

Sirtuins and autophagy are critical for maintaining brain health and longevity. Their interplay is essential for regulating cellular homeostasis, preventing neurodegeneration, and promoting lifespan extension. Further research into the mechanisms of sirtuin-mediated autophagy could provide valuable insights into developing therapeutic strategies for age-related brain disorders.

11.5.1 The Impact of Aging on Sirtuin and Autophagy Function

Aging is characterized by a decline in various cellular processes, including sirtuin and autophagy function. Sirtuins, particularly SIRT1, are NAD+^++-dependent deacetylases that play critical roles in metabolism, DNA repair, and stress resistance. Autophagy, a self-digestive process, is essential for removing damaged organelles and proteins, maintaining cellular homeostasis.

11.5.1.1 Impact of Aging on Sirtuin Function

1. *Decline in NAD+ Levels*: Aging is associated with a decrease in NAD+ levels, which are essential for sirtuin activity. This decline impairs sirtuin function, contributing to age-related diseases (Wan et al. 2022; Lee 2019).

NAD+ Availability and Its Impact on Sirtuin Function
NAD+ (Nicotinamide adenine dinucleotide) is a crucial co-substrate for sirtuin enzymes, which play significant roles in cellular metabolism, DNA repair, and stress resistance. The availability of NAD+ directly influences sirtuin activity, and its decline with age contributes to age-related functional decline and disease pathogenesis.

Decline in NAD+ Levels with Age

NAD+ Biosynthesis: NAD+ levels decrease with age due to impaired biosynthesis, primarily mediated by nicotinamide phosphoribosyltransferase (NAMPT). This decline affects multiple organs, including the pancreas, adipose tissue, skeletal muscle, liver, and brain (Imai and Guarente 2014).

Impact on Sirtuin Activity: Reduced NAD+ availability impairs sirtuin function, leading to increased protein acetylation and disruption of cellular processes. This is particularly significant for SIRT1, which regulates metabolism, mitochondrial function, and stress resistance (Imai and Guarente 2016; Johnson and Imai 2018).

Cellular Consequences of Reduced NAD+ Availability

Mitochondrial Dysfunction: Decreased NAD+ levels disrupt nuclear-mitochondrial communication, impairing mitochondrial biogenesis and function. This leads to metabolic dysregulation and increased oxidative stress, contributing to aging and age-related diseases (Imai and Guarente 2016; Roh and Kim 2020).

Increased DNA Damage: Impaired sirtuin activity due to low NAD+ levels results in increased DNA damage, as sirtuins play a role in maintaining genomic stability (Palmer and Vaccarezza 2021).

2. *Increased Protein Acetylation*: With reduced sirtuin activity, there is an accumulation of acetylated proteins, which can disrupt cellular processes and contribute to aging (Wan et al. 2022). Increased protein acetylation is a hallmark of aging, contributing to the decline in cellular function and the development of age-related diseases. This phenomenon is closely linked to the activity of sirtuins, which are NAD+ dependent deacetylases responsible for removing acetyl groups from proteins.

With age, NAD+ levels decrease, impairing sirtuin activity and leading to increased protein acetylation. This is particularly significant for SIRT1, which regulates key cellular processes such as metabolism and stress resistance. Acetylation of mitochondrial proteins can disrupt their function, contributing to mitochondrial dysfunction, a common feature of aging (Kane and Sinclair 2018). This includes the acetylation of enzymes involved in the tricarboxylic acid (TCA) cycle and electron transport chain (ETC), which are crucial for energy metabolism.

Increased acetylation of proteins like PGC-1α, NF-κB, and SOD2 affects their activity and contributes to age-related decline in cellular functions. For example, acetylated PGC-1α has reduced activity, impairing mitochondrial biogenesis and function.

Altered protein acetylation patterns contribute to neurodegenerative diseases by impairing neuronal homeostasis and promoting the accumulation of pathophysiological proteins such as tau (Kabir et al. 2023).

11.5.1.2 Impact of Aging on Autophagy Function

1. *Autophagic Flux Reduction*: Autophagic flux refers to the complete process of autophagy, from the formation of autophagosomes to their degradation by lysosomes. Aging leads to a decrease in autophagic flux, which can lead to the accumulation of damaged organelles and proteins, potentially causing cellular dysfunction and death. (He et al. 2013). The following are some key points related to autophagic flux reduction:

Causes of Autophagic Flux Reduction

Endoplasmic Reticulum (ER) Stress: Elevated ER stress, often due to lipid overload, can impair autophagic flux. This is observed in conditions like non-alcoholic fatty liver disease (NAFLD), where prolonged exposure to saturated fatty acids like palmitic acid (PA) disrupts autophagy, leading to cell death (Gonzalez-Rodriguez et al. 2014).

Oxidative Stress: Oxidative stress can also diminish autophagic flux, contributing to cellular dysfunction. This is proposed as a factor in progressive cellular failure (Porter et al. 2013).

Inhibitors of Autophagy: Certain compounds, such as bafilomycin A1, can inhibit autophagosome-lysosome fusion, thereby blocking autophagic flux. This is often used experimentally to assess autophagic activity (Liebl et al. 2022; Loos et al. 2014).

Impaired autophagy leads to the accumulation of damaged organelles and proteins, which can cause cellular injury and death (Gonzalez-Rodriguez et al. 2014; Porter et al. 2013). In conditions like NAFLD, reduced autophagic flux contributes to disease progression by promoting liver injury (Gonzalez-Rodriguez et al. 2014). Restoring autophagic flux is considered a potential therapeutic strategy to mitigate cellular damage and disease progression (Gonzalez-Rodriguez et al. 2014).

Autophagic flux can be measured by assessing the accumulation of autophagic markers like LC3-II and p62 in the presence of inhibitors like bafilomycin A1. This method helps quantify autophagic activity and identify compounds that modulate autophagy (Loos et al. 2014; Bensalem et al. 2021).

2. *Mitochondrial Dysfunction*: Mitochondrial dysfunction, often seen in aging, disrupts mitophagy, a specialized form of autophagy crucial for maintaining mitochondrial quality (Wan et al. 2022).

Mitochondrial dysfunction is closely linked to autophagic flux, particularly through the process of mitophagy, which is the selective degradation of damaged mitochondria by autophagy. Mitochondrial dysfunction affects autophagic flux and its broader implications:

11.5.1.3 Mitochondrial Dysfunction and Autophagy

Mitochondrial Outputs Regulating Autophagy Mitochondria influence autophagy through several outputs, including ATP production and reactive oxygen species (ROS). Mitochondrial dysfunction can alter these outputs, impacting autophagic pathways. For instance, loss of mitochondrial ATP production can induce autophagy via mTOR/AMPK signaling (Rambold and Lippincott-Schwartz 2011; Lee et al. 2012).

Mitophagy and Mitochondrial Quality Control Mitophagy is crucial for maintaining mitochondrial health by eliminating dysfunctional mitochondria. This process is essential for cellular homeostasis and can be triggered by mitochondrial dysfunction, such as loss of membrane potential (Youn et al. 2021; Li et al. 2022).

Impact on Autophagic Flux Mitochondrial dysfunction can disrupt autophagic flux. For example, mitochondrial depolarization and reduced oxygen consumption have been linked to decreased autophagosome turnover, suggesting that mitochondrial health is critical for efficient autophagy (Rambold and Lippincott-Schwartz 2011).

Impaired mitophagy due to mitochondrial dysfunction contributes to the progression of various diseases, including neurodegenerative disorders and cardiovascular diseases (Li et al. 2022; Lee et al. 2012). While autophagy can be a survival mechanism, insufficient or dysfunctional autophagy in response to mitochondrial stress may lead to cell death (Lee et al. 2012).

Restoring mitochondrial function and enhancing mitophagy are potential therapeutic strategies to improve autophagic flux and mitigate disease progression. Understanding the interplay between mitochondria and autophagy is crucial for developing targeted interventions in conditions where mitochondrial dysfunction plays a key role.

11.5.2 Sirtuins and Autophagy in Age-Related Cognitive Decline

Age-related cognitive decline is a complex process influenced by multiple factors, including impaired sirtuin function and autophagy. Sirtuins, particularly SIRT1, play a crucial role in maintaining cognitive health by regulating synaptic plasticity, epigenetic modifications, and autophagy. Age-related cognitive decline is influenced by both sirtuin and autophagy dysfunction:

Reduced sirtuin activity leads to increased neuroinflammation, which contributes to cognitive decline. SIRT1, in particular, suppresses NF-κB-mediated inflammatory pathways (He et al. 2013). Impaired autophagy allows the accumulation of neurotoxic protein aggregates, further exacerbating cognitive decline (He et al. 2013).

Sirtuins, particularly SIRT1, play a crucial role in cognitive function by enhancing synaptic plasticity, influencing epigenetic regulation, and modulating oxidative stress pathways that affect mitochondrial function. Here, we explore the mechanisms by which sirtuins contribute to cognitive health.

Enhancement of Synaptic Plasticity Long-Term Potentiation (LTP): SIRT1 is essential for LTP, a form of synaptic plasticity associated with learning and memory. Studies in SIRT1 knockout mice have shown impaired LTP in the hippocampus, a region critical for memory formation (Michán et al. 2010; Fagerli et al. 2022).

Regulation of CREB and miR-134 SIRT1 modulates synaptic plasticity by regulating CREB, a transcription factor crucial for memory. It does so by blocking miR-134, which normally inhibits CREB translation, thereby maintaining CREB activity and supporting synaptic plasticity (Fagerli et al. 2022; Lee et al. 2025; Gao et al. 2010).

Epigenetic Regulation

1. *Histone Deacetylation*: Sirtuins, including SIRT1, deacetylate histones, altering chromatin structure and influencing gene expression related to cognitive processes. This epigenetic regulation is essential for maintaining cognitive function and adapting to environmental changes (Gräff and Tsai 2013; Fagerli et al. 2022).
2. *Gene Expression Modulation*: SIRT1 modulates the expression of genes involved in synaptic function, lipid metabolism, and myelination. Its absence leads to altered gene expression profiles in the hippocampus, affecting cognitive abilities (Michán et al. 2010).

11.5.2.1 Mitochondrial Function and Oxidative Stress

Mitochondrial Sirtuins Mitochondrial sirtuins like SIRT3 help maintain mitochondrial function by reducing oxidative stress. This is crucial for neuronal energy metabolism and preventing neurodegeneration (Gräff and Tsai 2013).

Neuroprotection SIRT1 protects against oxidative stress by activating antioxidant defenses and promoting mitochondrial biogenesis, which is essential for maintaining neuronal health and cognitive function (Cao et al. 2017).

11.5.2.2 Role of Other Sirtuins

SIRT2 and SIRT3 These sirtuins also play roles in synaptic plasticity. SIRT2 regulates AMPA receptor internalization, affecting synaptic strength, while SIRT3 is involved in maintaining mitochondrial function and preventing neuronal loss (Wang et al. 2017; Fagerli et al. 2022).

11.5.3 Sirtuins and Autophagy in Neurodegenerative Diseases

Neurodegenerative diseases, such as Alzheimer's disease (AD), Parkinson's disease (PD), and amyotrophic lateral sclerosis (ALS), are characterized by the progressive loss of neurons and cognitive decline. Sirtuins and autophagy play significant roles in these diseases by regulating protein homeostasis, mitochondrial function, and inflammatory pathways (Leite et al. 2022).

11.5.3.1 Role of Sirtuins in Neurodegenerative Diseases

Sirtuins, particularly SIRT1, help maintain protein homeostasis by reducing the accumulation of toxic protein aggregates. For example, SIRT1 promotes the clearance of amyloid-β plaques in AD and α-synuclein in PD (Min et al. 2013). Sirtuins enhance mitochondrial function by reducing oxidative stress, which is crucial for maintaining neuronal energy metabolism. Mitochondrial sirtuins like SIRT3 are involved in this process (Yalcin 2018; Manjula et al. 2021). SIRT1 inhibits NF-κB, a key regulator of inflammation, thereby reducing chronic inflammatory responses associated with neurodegenerative diseases (Min et al. 2013; Razick et al. 2023).

11.5.3.2 Role of Autophagy in Neurodegenerative Diseases

Autophagy is essential for removing damaged organelles and protein aggregates, such as amyloid-β and α-synuclein, which are hallmarks of neurodegenerative diseases (Min et al. 2013; Manjula et al. 2021). Autophagy, specifically mitophagy, ensures the removal of dysfunctional mitochondria, which is vital for preventing oxidative stress and maintaining neuronal health (Manjula et al. 2021).

11.5.3.3 Interplay Between Sirtuins and Autophagy

SIRT1 promotes autophagy by inhibiting mTOR signaling, a negative regulator of autophagy. This induction enhances the clearance of neurotoxic aggregates and maintains cellular homeostasis (Min et al. 2013; Razick et al. 2023). Autophagy helps maintain NAD+ levels, which are essential for sirtuin activity. Elevated NAD+ levels support sirtuin-mediated processes, including autophagy induction (Manjula et al. 2021).

11.5.4 Therapeutic Strategies Targeting Sirtuins and Autophagy for Brain Longevity

Several therapeutic strategies aim to enhance sirtuin and autophagy function to promote brain longevity:

(a) *Sirtuin Activators*: Compounds like resveratrol and SRT2104 activate SIRT1, enhancing autophagy and reducing neuroinflammation. These are being explored for their potential in treating neurodegenerative diseases (Villalba and Alcaín 2012; Gertz et al. 2012).

Sirtuin activators are compounds that enhance the activity of sirtuins, a family of proteins involved in metabolic regulation, aging, and cellular stress resistance. Here are some key points about sirtuin activators:

11.5.4.1 Types of Sirtuin Activators

Resveratrol One of the most well-known sirtuin activators, resveratrol is found in red wine and has been studied extensively for its potential health benefits, including anti-aging effects. However, its direct activation mechanism is still debated, and it may not directly activate sirtuins in vivo (Gertz et al. 2012).

Synthetic Activators Compounds like SRT-1720 and SRT-2104 are synthetic activators of SIRT1. These compounds have shown promise in preclinical studies for improving metabolic health and longevity (Bursch et al. 2024).

Natural Compounds Other natural compounds such as piceatannol, beta-lapachone, and quercetin derivatives have been identified as sirtuin activators. These compounds may have various health benefits by enhancing sirtuin activity (Bursch et al. 2024).

11.5.4.2 Mechanism of Action

Allosteric Activation Most known SIRT1 activators work through an allosteric mechanism, binding to the SIRT1 substrate-binding domain (SBD) and enhancing its enzymatic activity by lowering the Km for acetylated substrates (Bursch et al. 2024).

NAD+ Dependence Sirtuins require NAD+ for their activity, and conditions that increase NAD+ levels, such as fasting or exercise, can indirectly activate sirtuins (Sah et al. 2025).

11.5.4.3 Therapeutic Potential

Sirtuin activators are being explored for their potential to treat metabolic disorders by improving insulin sensitivity and mitochondrial function (Wu et al. 2022). Activation of sirtuins, particularly SIRT1, may offer therapeutic benefits in neurodegenerative diseases by enhancing mitochondrial function and reducing oxidative stress (Wu et al. 2022).

Despite promising preclinical data, there is currently no clinical evidence supporting the use of sirtuin activators for human longevity or disease prevention. Further research is needed to fully understand their therapeutic potential.

(b) *Autophagy Inducers*

Autophagy inducers are compounds or interventions that stimulate autophagy, a cellular process essential for removing damaged organelles and proteins, thereby maintaining cellular homeostasis. Here, we explore the mechanisms and examples of autophagy inducers.

11.5.4.4 Mechanisms of Autophagy Induction

mTOR Inhibition The mechanistic target of rapamycin (mTOR) is a key regulator of autophagy. Inhibiting mTOR, typically through nutrient deprivation or pharmacological agents like rapamycin, activates autophagy by dephosphorylating and activating ULK1/2 kinases (Yang and Klionsky 2009).

AMPK Activation AMP-activated protein kinase (AMPK) is activated during energy stress, such as fasting or exercise. AMPK inhibits mTOR signaling, thereby inducing autophagy (Yang and Klionsky 2009; Yin et al. 2016).

11.5.4.5 Examples of Autophagy Inducers

Rapamycin A well-known autophagy inducer, rapamycin inhibits mTORC1, leading to autophagy activation. It is used in research and has potential therapeutic applications (Yang and Klionsky 2009; Badadani 2012).

Spermidine This polyamine induces autophagy by inhibiting acetyltransferases, which leads to the activation of autophagy-related proteins (Badadani 2012).

Resveratrol While primarily known as a sirtuin activator, resveratrol also induces autophagy, potentially through its effects on mTOR and AMPK pathways (Levine et al. 2015).

Fasting and Caloric Restriction These dietary interventions induce autophagy by activating AMPK and inhibiting mTOR signaling, promoting cellular recycling and renewal (Shabkhizan et al. 2023).

11.5.4.6 Therapeutic Potential

Autophagy inducers are being explored for their potential in treating neurodegenerative diseases, such as Alzheimer's and Parkinson's, by enhancing the clearance of neurotoxic aggregates. Autophagy can have dual roles in cancer; it can either promote or inhibit tumor growth depending on the context. Inducers are being studied for their potential in cancer therapy.

Autophagy inducers stimulate cellular recycling processes, which are crucial for maintaining cellular health and potentially treating various diseases. Understanding these mechanisms can lead to the development of therapeutic strategies for age-related and metabolic disorders.

(c) *NAD+ Boosters* Increasing NAD+ levels using precursors like NR or NMN can enhance sirtuin activity and autophagy, offering a promising approach for age-related diseases (Lee 2019). NAD+ boosters, such as nicotinamide riboside (NR) and β-nicotinamide mononucleotide (NMN), are supplements designed to increase NAD+ levels in the body. NAD+ is crucial for various cellular processes, including energy metabolism, DNA repair, and sirtuin activity. Here, we explore the benefits and mechanisms of NAD+ boosters.

11.5.4.7 Mechanisms of Action

NR and NMN are converted into NAD+ in the body, thereby increasing its availability for cellular processes. Elevated NAD+ levels activate sirtuins, which are involved in metabolic regulation, stress resistance, and longevity. NAD+ boosters improve mitochondrial function by enhancing ATP production and promoting mitophagy, the recycling of dysfunctional mitochondria (Conlon 2022).

NAD+ is essential for mitochondrial function and ATP production. Boosting NAD+ levels can enhance energy metabolism, reducing fatigue and improving physical performance. Increasing NAD+ levels may improve cognitive function and protect against neurodegenerative diseases by enhancing DNA repair and reducing oxidative stress. NAD+ boosters are believed to have anti-aging effects by improving cellular health and potentially extending health span (Conlon 2022). Maintaining healthy NAD+ levels may help lower blood pressure and cholesterol, reducing cardiovascular risk factors.

Aging impairs both sirtuin and autophagy function, contributing to age-related cognitive decline and neurodegenerative diseases. Targeting these pathways with therapeutic strategies offers potential benefits for promoting brain longevity and mitigating neurodegenerative diseases.

11.6 Therapeutic Strategies Targeting Sirtuins and Autophagy

Sirtuins and autophagy are pivotal in maintaining cellular homeostasis and have emerged as promising therapeutic targets for various diseases, including cancer, cardiovascular diseases, and neurodegenerative disorders. This review will delve into pharmacological interventions, dietary and lifestyle modulations, and future prospects in neuroprotective therapeutics.

11.6.1 Pharmacological Interventions

Pharmacological interventions targeting sirtuins and autophagy include compounds like resveratrol, NAD+ boosters, and rapamycin.

1. *Resveratrol*: This polyphenol is known for its potential to activate SIRT1, although its role is somewhat ambiguous and requires further investigation (Sah et al. 2025). Resveratrol can modulate AMPK activity and has shown benefits in models of diabetes and inflammation by enhancing autophagy (Sah et al. 2025).
2. *NAD+ Boosters*: Nicotinamide mononucleotide (NMN) and nicotinamide riboside (NR) are NAD+ precursors that can enhance sirtuin activity by increasing NAD+ levels. NMN supplementation has been shown to mitigate age-related declines in mice by improving energy metabolism and physical activity (Sah et al. 2025).
3. *Rapamycin*: An autophagy inducer that acts by inhibiting the mTOR pathway. It has been used in studies to extend lifespan and improve healthspan in various models (Lee 2019).

Other molecules that are involved in neuroprotection by involving in sirtuin pathway are depicted in Fig. 11.3.

11.6.2 Dietary and Lifestyle Modulations

Dietary and lifestyle interventions, such as caloric restriction and intermittent fasting, have been shown to activate sirtuins and induce autophagy.

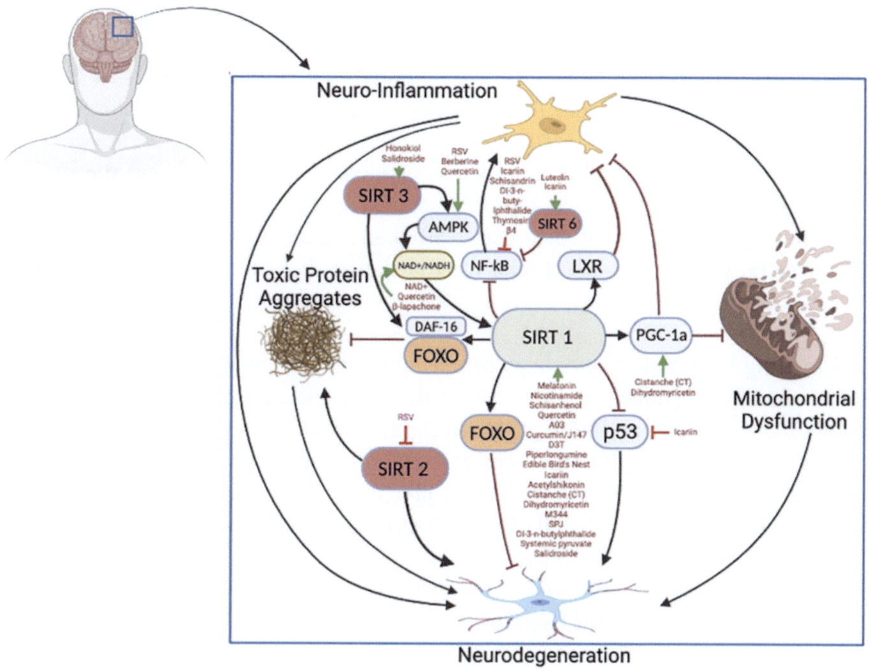

Fig. 11.3 Effect of tested novel therapeutic compounds on the sirtuin pathway in relation to biological neurocognitive effects. D3T 3 H-1,2-dithiole-3-thione, NAD+ nicotinamide adenine dinucleotide, RSV resveratrol, SPJ saponins from panax japonicas

1. *Caloric Restriction*: This dietary regimen is known to increase SIRT1 activity and induce autophagy, leading to extended lifespan in model organisms (Lee 2019; Marius 2023). Caloric restriction enhances cellular resilience against stress and promotes longevity.
2. *Intermittent Fasting*: Similar to caloric restriction, intermittent fasting activates autophagy and sirtuins, improving metabolic health and reducing the risk of age-related diseases (Michán et al. 2010). It enhances cellular cleaning processes and promotes cellular renewal.

11.6.3 Future Prospects in Neuroprotective Therapeutics

Sirtuins and autophagy play critical roles in neuroprotection, making them promising targets for neurodegenerative diseases.

1. *Neurodegenerative Diseases*: SIRT1 and SIRT3 have been implicated in protecting against neurodegeneration by promoting mitochondrial function and reducing oxidative stress (Naseer et al. 2021). Enhancing autophagy and sirtuin activity could mitigate neurodegenerative processes.

2. *Therapeutic Potential*: Activating sirtuins and inducing autophagy may offer novel therapeutic strategies for diseases like Alzheimer's and Parkinson's. NAD+ boosters, such as NMN, have shown potential in preclinical models by improving cognitive function and reducing neurodegenerative markers (Naseer et al. 2021; Sah et al. 2025).
3. *Future Directions*: Further research is needed to fully understand the mechanisms by which sirtuins and autophagy modulate neuroprotection. Clinical trials exploring the efficacy of sirtuin activators and autophagy modulators in neurodegenerative diseases are crucial for translating these findings into therapeutic applications.

Targeting sirtuins and autophagy offers a promising avenue for developing therapeutic strategies against a wide range of diseases. Pharmacological interventions, dietary modifications, and lifestyle changes can activate these pathways, potentially leading to improved health outcomes and extended lifespan. As research continues to unravel the complexities of these mechanisms, future prospects in neuroprotective therapeutics hold significant promise for combating neurodegenerative diseases.

11.7 Conclusion and Future Perspectives

Sirtuins and autophagy are emerging as promising therapeutic targets for maintaining brain health and combating neurodegenerative diseases. Their synergistic relationship offers a potential avenue for developing novel interventions that could enhance cognitive function and protect against age-related neurological disorders.

11.7.1 Sirtuins and Autophagy: Promising Targets for Brain Health

1. *Neuroprotective Roles*: Sirtuins, particularly SIRT1, have been shown to protect against neurodegenerative diseases such as Alzheimer's and Parkinson's by promoting neuronal survival and reducing oxidative stress (Lee 2019; Min et al. 2013). Autophagy, regulated by sirtuins, plays a crucial role in clearing toxic protein aggregates, further supporting brain health (Lee 2019; Marius 2023).
2. *Cognitive Function*: Sirtuins are involved in enhancing synaptic plasticity and memory, with SIRT1 and SIRT2 playing roles in regulating synaptic function and neural plasticity (Fernandez et al. 2024; Fagerli et al. 2022). Enhancing sirtuin activity could potentially improve cognitive outcomes in neurological disorders.

11.7.2 Future Directions: Unraveling the Full Potential of This Synergistic Relationship

1. *Mechanistic Insights*: Further research is needed to fully understand how sirtuins modulate autophagy and how these pathways interact in the context of neurodegenerative diseases. Elucidating these mechanisms will be crucial for developing targeted therapies (Liebl et al. 2022).
2. *Therapeutic Applications*: Developing drugs that can safely activate sirtuins or induce autophagy in humans is a promising area of research. Compounds like resveratrol and NAD+ boosters have shown potential but require more clinical validation (Lee et al. 2012; Michán et al. 2010).
3. *Combination Therapies*: Exploring combination therapies that target both sirtuins and autophagy pathways could offer synergistic benefits in treating neurodegenerative diseases. This approach might enhance efficacy and reduce side effects compared to targeting a single pathway (Lee 2019; Marius 2023).

11.7.2.1 Lifestyle Interventions to Support Sirtuin and Autophagy Function

1. *Caloric Restriction and Intermittent Fasting*: These dietary regimens have been shown to activate sirtuins and induce autophagy, potentially leading to improved brain health and reduced risk of neurodegenerative diseases (Min et al. 2013; Marius 2023).
2. *Exercise and Physical Activity*: Regular exercise can enhance autophagy and sirtuin activity, contributing to overall health and potentially reducing the risk of age-related cognitive decline (Baeken 2024).
3. *Nutritional Supplements*: Certain nutrients and supplements, such as resveratrol and omega-3 fatty acids, may support sirtuin and autophagy function, although more research is needed to confirm their efficacy in humans (Lee 2019; Marius 2023).

In conclusion, the synergistic relationship between sirtuins and autophagy presents a promising avenue for developing therapeutic strategies to enhance brain health. Future research should focus on unraveling the mechanisms underlying this relationship and exploring lifestyle interventions that can support these pathways, ultimately leading to improved outcomes in neurodegenerative diseases.

References

Alves-Fernandes DK, Jasiulionis MG (2019) The role of SIRT1 on DNA damage response and epigenetic alterations in cancer. Int J Mol Sci 20:3153

Aman Y, Schmauck-Medina T, Hansen M, Morimoto RI, Simon AK, Bjedov I, Palikaras K, Simonsen A, Johansen T, Tavernarakis N (2021) Autophagy in healthy aging and disease. Nat Aging 1:634–650

Anding AL, Baehrecke EH (2017) Cleaning house: selective autophagy of organelles. Dev Cell 41:10–22
Badadani M (2012) Autophagy mechanism, regulation, functions, and disorders. Int Sch Res Notices 2012:927064
Baeken MW (2024) Sirtuins and their influence on autophagy. J Cell Biochem 125:e30377
Bai N, LI N, Cheng R, Guan Y, Zhao X, Song Z, Xu H, Yi F, Jiang B, Li X (2022) Inhibition of SIRT2 promotes APP acetylation and ameliorates cognitive impairment in APP/PS1 transgenic mice. Cell Rep 40:111062
Bensalem J, Hattersley KJ, Hein LK, Teong XT, Carosi JM, Hassiotis S, Grose RH, Fourrier C, Heilbronn LK, Sargeant TJ (2021) Measurement of autophagic flux in humans: an optimized method for blood samples. Autophagy 17:3238–3255
Bosch-Presegué L, Vaquero A (2015) Sirtuin-dependent epigenetic regulation in the maintenance of genome integrity. FEBS J 282:1745–1767
Brito D, Esteves F, Rajado A, Silva N, Araújo I, Bragança J, Castelo-Branco P, Nóbrega C (2023) Assessing cognitive decline in the aging brain: lessons from rodent and human studies. NPJ Aging 9:23
Bursch KL, Goetz CJ, Smith BC (2024) Current trends in sirtuin activator and inhibitor development. Molecules 29:1185
Cao H, Zhou X, Xu B, Hu H, Guo J, Wang M, Li N, Zou J (2024) Advances in the study of mitophagy in osteoarthritis. J Zhejiang Univ Sci B 25:197–211
Cao Y, Yan Z, Zhou T, Wang G (2017) SIRT1 regulates cognitive performance and ability of learning and memory in diabetic and nondiabetic models. J Diabetes Res 2017:7121827
Cen X, Zhang M, Zhou M, Ye L, Xia H (2021) Mitophagy regulates neurodegenerative diseases. Cells 10:1876
Chandramowlishwaran P, Vijay A, Abraham D, Li G, Mwangi SM, Srinivasan S (2020) Role of sirtuins in modulating neurodegeneration of the enteric nervous system and central nervous system. Front Neurosci 14:614331
Chojdak-Łukasiewicz J, Bizoń A, Waliszewska-Prosół M, Piwowar A, Budrewicz S, Pokryszko-Dragan A (2022) Role of sirtuins in physiology and diseases of the central nervous system. Biomedicines 10:2434
Conlon NJ (2022) The role of NAD+ in regenerative medicine. Plast Reconstr Surg 150:41S–48S
Dharmalingam K, Prisi S, Choudhary S (2023) Mitochondria: as a protagonist in neurological disorders in brief overview: role of mitochondria in neurological disorders. J Exp Mol Biol 24:115–130
Donmez G (2013) Sirtuins as possible targets in neurodegenerative diseases. Curr Drug Targets 14:644–647
Duan W (2013) Sirtuins: from metabolic regulation to brain aging. Front Aging Neurosci 5:36
Fagerli E, Escobar I, Ferrier FJ, Jackson CW, Perez-Lao EJ, Perez-Pinzon MA (2022) Sirtuins and cognition: implications for learning and memory in neurological disorders. Front Physiol 13:908689
Fernandez F, Griffiths LR, Sutherland HG, Cole MH, Fitton JH, Winberg P, Schweitzer D, Hopkins LN, Meyer BJ (2024) Sirtuin proteins and memory: a promising target in Alzheimer's disease therapy? Nutrients 16:4088
Fernando KKM, Wijayasinghe YS (2021) Sirtuins as potential therapeutic targets for mitigating neuroinflammation associated with Alzheimer's disease. Front Cell Neurosci 15:746631
Franco AC, Martini H, Victorelli S, Lagnado AB, Wyles SP, Rowsey JL, Pirius N, Woo SH, Costa DG, Chaib S (2025) Senescent cell transplantation into the skin induces age-related peripheral dysfunction and cognitive decline. Aging Cell 24:e14340
Fujita Y, Yamashita T (2018) Sirtuins in neuroendocrine regulation and neurological diseases. Front Neurosci 12:778
Gao J, Wang W-Y, Mao Y-W, Gräff J, Guan J-S, Pan L, Mak G, Kim D, Su SC, Tsai L-H (2010) A novel pathway regulates memory and plasticity via SIRT1 and miR-134. Nature 466:1105–1109

Gertz M, Nguyen GTT, Fischer F, Suenkel B, Schlicker C, Fränzel B, Tomaschewski J, Aladini F, Becker C, Wolters D (2012) A molecular mechanism for direct sirtuin activation by resveratrol. PloS One 7:e49761

Glick D, Barth S, Macleod KF (2010) Autophagy: cellular and molecular mechanisms. J Pathol 221:3–12

Gonzalez-Rodriguez A, Mayoral R, Agra N, Valdecantos M, Pardo V, Miquilena-Colina M, Vargas-Castrillón J, Lo Iacono O, Corazzari M, Fimia GM (2014) Impaired autophagic flux is associated with increased endoplasmic reticulum stress during the development of NAFLD. Cell Death Dis 5:e1179–e1179

Gräff J, Tsai L-H (2013) Histone acetylation: molecular mnemonics on the chromatin. Nat Rev Neurosci 14:97–111

Hashemzaei M, Heravi RE, Rezaee R, Roohbakhsh A, Karimi G (2017) Regulation of autophagy by some natural products as a potential therapeutic strategy for cardiovascular disorders. Eur J Pharmacol 802:44–51

He L-Q, Lu J-H, Yue Z-Y (2013) Autophagy in ageing and ageing-associated diseases. Acta Pharmacologica Sinica 34:605–611

Herskovits AZ, Guarente L (2013) Sirtuin deacetylases in neurodegenerative diseases of aging. Cell Res 23:746–758

Huang L, Yao T, Chen J, Zhang Z, Yang W, Gao X, Dan Y, He Y (2022) Effect of Sirt3 on retinal pigment epithelial cells in high glucose through Foxo3a/PINK1-Parkin pathway mediated mitophagy. Exp Eye Res 218:109015

Iachettini S, Trisciuoglio D, Rotili D, Lucidi A, Salvati E, Zizza P, Di Leo L, Del Bufalo D, Ciriolo MR, Leonetti C (2018) Pharmacological activation of SIRT6 triggers lethal autophagy in human cancer cells. Cell Death Dis 9:996

Imai S-I, Guarente L (2014) NAD+ and sirtuins in aging and disease. Trends Cell Biol 24:464–471

Imai S-I, Guarente L (2016) It takes two to tango: NAD+ and sirtuins in aging/longevity control. NPJ Aging Mech Dis 2:1–6

Iside C, Scafuro M, Nebbioso A, Altucci L (2020) SIRT1 activation by natural phytochemicals: an overview. Front Pharmacol 11:1225

Jin SM, Youle RJ (2012) PINK1-and Parkin-mediated mitophagy at a glance. J Cell Sci 125:795–799

Johnson S, Imai SI (2018) NAD+ biosynthesis, aging, and disease. F1000Res 7:132

Kabir F, Atkinson R, Cook AL, Phipps AJ, King AE (2023) The role of altered protein acetylation in neurodegenerative disease. Front Aging Neurosci 14:1025473

Kahmini FR, Ghaleh HD, Shahgaldi S (2022) Sirtuins: subtle regulators involved in convoluted mechanisms of pregnancy. Cell Physiol Biochem 56:644–662

Kane AE, Sinclair DA (2018) Sirtuins and NAD+ in the development and treatment of metabolic and cardiovascular diseases. Circ Res 123:868–885

Ke P-Y (2024) Molecular mechanism of autophagosome–lysosome fusion in mammalian cells. Cells 13:500

Khandia R, Dadar M, Munjal A, Dhama K, Karthik K, Tiwari R, Yatoo MI, Iqbal HM, Singh KP, Joshi SK (2019) A comprehensive review of autophagy and its various roles in infectious, non-infectious, and lifestyle diseases: current knowledge and prospects for disease prevention, novel drug design, and therapy. Cells 8:674

Kim JY, Mondaca-Ruff D, Singh S, Wang Y (2022) SIRT1 and autophagy: implications in endocrine disorders. Front Endocrinol 13:930919

Kitada M, Ogura Y, Monno I, Koya D (2019) Sirtuins and type 2 diabetes: role in inflammation, oxidative stress, and mitochondrial function. Front Endocrinol 10:187

Korotkov A, Seluanov A, Gorbunova V (2021) Sirtuin 6: linking longevity with genome and epigenome stability. Trends Cell Biol 31:994–1006

Kratz EM, Sołkiewicz K, Kubis-Kubiak A, Piwowar A (2021) Sirtuins as important factors in pathological states and the role of their molecular activity modulators. Int J Mol Sci 22:630

Kroemer G, Mariño G, Levine B (2010) Autophagy and the integrated stress response. Mol Cell 40:280–293

Lahiri V, Hawkins WD, Klionsky DJ (2019) Watch what you (self-) eat: autophagic mechanisms that modulate metabolism. Cell Metab 29:803–826

Lambona C, Zwergel C, Valente S, Mai A (2024) SIRT3 Activation a promise in drug development? New insights into SIRT3 biology and its implications on the drug discovery process. J Med Chem 67:1662–1689

Lee H-J, Yoon Y-S, Lee S-J (2018) Mechanism of neuroprotection by trehalose: controversy surrounding autophagy induction. Cell Death Dis 9:712

Lee IH (2019) Mechanisms and disease implications of sirtuin-mediated autophagic regulation. Exp Mol Med 51:1–11

Lee J, Giordano S, Zhang J (2012) Autophagy, mitochondria and oxidative stress: cross-talk and redox signalling. Biochem J 441:523–540

Lee L-C, Su M-T, Bao L, Lee P-L, Tutwiler S, Yeh T-K, Chang C-Y (2025) MicroRNAs modulate CaMKIIα/SIRT1 signaling pathway as a biomarker of cognitive ability in adolescents. Brain Behav Immun Health 44:100970

Leite JA, Ghirotto B, Targhetta VP, De Lima J, Câmara NO (2022) Sirtuins as pharmacological targets in neurodegenerative and neuropsychiatric disorders. Br J Pharmacol 179:1496–1511

Levine B, Klionsky DJ (2017) Autophagy wins the 2016 nobel prize in physiology or medicine: breakthroughs in baker's yeast fuel advances in biomedical research. Proc Natl Acad Sci 114:201–205

Levine B, Packer M, Codogno P (2015) Development of autophagy inducers in clinical medicine. J Clin Invest 125:14–24

Li A, Gao M, Liu B, Qin Y, Chen L, Liu H, Wu H, Gong G (2022) Mitochondrial autophagy: molecular mechanisms and implications for cardiovascular disease. Cell Death Dis 13:444

Li H, Cai Z (2022) SIRT3 regulates mitochondrial biogenesis in aging-related diseases. J Biomed Res 37:77

Liang D, Zhuo Y, Guo Z, He L, Wang X, He Y, Li L, Dai H (2020) SIRT1/PGC-1 pathway activation triggers autophagy/mitophagy and attenuates oxidative damage in intestinal epithelial cells. Biochimie 170:10–20

Liebl MP, Meister SC, Frey L, Hendrich K, Klemmer A, Wohlfart B, Untucht C, Nuber J, Pohl C, Lakics V (2022) Robust LC3B lipidation analysis by precisely adjusting autophagic flux. Sci Rep 12:79

Lombard DB, Tishkoff DX, Bao J (2011) Mitochondrial sirtuins in the regulation of mitochondrial activity and metabolic adaptation. Handb Exp Pharmacol 206:163–188

Loos B, Toit AD, Hofmeyr J-HS (2014) Defining and measuring autophagosome flux—concept and reality. Autophagy 10:2087–2096

Maissan P, Mooij EJ, Barberis M (2021) Sirtuins-mediated system-level regulation of mammalian tissues at the interface between metabolism and cell cycle: a systematic review. Biology 10:194

Manjula R, Anuja K, Alcain FJ (2021) SIRT1 and SIRT2 activity control in neurodegenerative diseases. Front Pharmacol 11:585821

Marius W (2023) Sirtuins and their influence on autophagy. J Cell Biochem 125:e30377

Michán S, Li Y, Chou MM-H, Parrella E, Ge H, Long JM, Allard JS, Lewis K, Miller M, Xu W (2010) SIRT1 is essential for normal cognitive function and synaptic plasticity. J Neurosci 30:9695–9707

Min S-W, Sohn PD, Cho S-H, Swanson RA, Gan L (2013) Sirtuins in neurodegenerative diseases: an update on potential mechanisms. Front Aging Neurosci 5:53

Morigi M, Perico L, Benigni A (2018) Sirtuins in renal health and disease. J Am Soc Nephrol 29:1799–1809

Naseer A, Mir SS, Takacs-Vellai K, Nazir A (2021) Sirtuins and autophagy in age-associated neurodegenerative diseases: lessons from the C. elegans model. Int J Mol Sci 22:12263

Ng F, Tang BL (2013) Sirtuins' modulation of autophagy. J Cell Physiol 228:2262–2270

Orenstein SJ, Cuervo AM (2010) Chaperone-mediated autophagy: molecular mechanisms and physiological relevance. Semin Cell Dev Biol 21:719–726

Ou X, Lee MR, Huang X, Messina-Graham S, Broxmeyer HE (2014) SIRT1 positively regulates autophagy and mitochondria function in embryonic stem cells under oxidative stress. Stem Cells 32:1183–1194

Palmer RD, Vaccarezza M (2021) Nicotinamide adenine dinucleotide and the sirtuins caution: pro-cancer functions. Aging Med 4:337–344

Patel D, Soni R, Shah J (2024) Decoding the role of nuclear sirtuins in Parkinson's pathogenesis. ACS Chem Neurosci 15:3615–3625

Porter K, Nallathambi J, Lin Y, Liton PB (2013) Lysosomal basification and decreased autophagic flux in oxidatively stressed trabecular meshwork cells: implications for glaucoma pathogenesis. Autophagy 9:581–594

Rambold AS, Lippincott-Schwartz J (2011) Mechanisms of mitochondria and autophagy cross-talk. Cell Cycle 10:4032–4038

Rasti G, Becker M, Vazquez BN, Espinosa-Alcantud M, Fernández-Duran I, Gámez-García A, Ianni A, Gonzalez J, Bosch-Presegué L, Marazuela-Duque A (2023) SIRT1 regulates DNA damage signaling through the PP4 phosphatase complex. Nucleic Acids Res 51:6754–6769

Razick DI, Akhtar M, Wen J, Alam M, Dean N, Karabala M, Ansari U, Ansari Z, Tabaie E, Siddiqui S (2023) The role of sirtuin 1 (SIRT1) in neurodegeneration. Cureus 15:e40463

Roh E, Kim M-S (2020) Hypothalamic NAD+-sirtuin axis: function and regulation. Biomolecules 10:396

Sack MN, Finkel T (2012) Mitochondrial metabolism, sirtuins, and aging. Cold Spring Harb Perspect Biol 4:a013102

Sadria M, Layton AT (2021) Interactions among mTORC, AMPK and SIRT: a computational model for cell energy balance and metabolism. Cell Commun Signal 19:57

Sah P, Rai AK, Syiem D (2025) Sirtuin activators as an anti-aging intervention for longevity. Explor Drug Sci 3:100881

Saito Y, Yamamoto S, Chikenji TS (2024) Role of cellular senescence in inflammation and regeneration. Inflamm Regen 44:28

Shabkhizan R, Haiaty S, Moslehian MS, Bazmani A, Sadeghsoltani F, Bagheri HS, Rahbarghazi R, Sakhinia E (2023) The beneficial and adverse effects of autophagic response to caloric restriction and fasting. Adv Nutr 14:1211–1225

Shi H, Deng H-X, Gius D, Schumacker PT, Surmeier DJ, Ma Y-C (2017) Sirt3 protects dopaminergic neurons from mitochondrial oxidative stress. Hum Mol Genet 26:1915–1926

Smith BC, Hallows WC, Denu JM (2008) Mechanisms and molecular probes of sirtuins. Chem Biol 15:1002–1013

Son SM, Park SJ, Fernandez-Estevez M, Rubinsztein DC (2021) Autophagy regulation by acetylation—implications for neurodegenerative diseases. Exp Mol Med 53:30–41

Stampanoni Bassi M, Iezzi E, Gilio L, Centonze D, Buttari F (2019) Synaptic plasticity shapes brain connectivity: implications for network topology. Int J Mol Sci 20:6193

Sun X, Yang X, Gui W, Liu S, Gui Q (2023) Sirtuins and autophagy in lipid metabolism. Cell Biochem Funct 41:978–987

Suzuki K, Ohsumi Y (2007) Molecular machinery of autophagosome formation in yeast, Saccharomyces cerevisiae. FEBS Lett 581:2156–2161

Szenci G, Csizmadia T, Juhász G (2023) The role of crinophagy in quality control of the regulated secretory pathway. J Cell Sci 136:jcs260741

Tang Y, Ju W, Liu Y, Deng Q (2024) The role of SIRT1 in autophagy and drug resistance: unveiling new targets and potential biomarkers in cancer therapy. Front Pharmacol 15:1469830

Tchkonia T, Zhu Y, Van Deursen J, Campisi J, Kirkland JL (2013) Cellular senescence and the senescent secretory phenotype: therapeutic opportunities. J Clin Invest 123:966–972

Tong H, Yang T, Xu S, Li X, Liu L, Zhou G, Yang S, Yin S, Li X-J, Li S (2024) Huntington's disease: complex pathogenesis and therapeutic strategies. Int J Mol Sci 25:3845

Van Limbergen J, Stevens C, Nimmo E, Wilson D, Satsangi J (2009) Autophagy: from basic science to clinical application. Mucosal immunol 2:315–330

Van Meter M, Mao Z, Gorbunova V, Seluanov A (2011) Repairing split ends: SIRT6, mono-ADP ribosylation and DNA repair. Aging (Albany NY) 3:829

Vazquez BN, Thackray JK, Serrano L (2017) Sirtuins and DNA damage repair: SIRT7 comes to play. Nucleus 8:107–115
Villalba JM, Alcaín FJ (2012) Sirtuin activators and inhibitors. Biofactors 38:349–359
Wan W, Hua F, Fang P, Li C, Deng F, Chen S, Ying J, Wang X (2022) Regulation of mitophagy by sirtuin family proteins: a vital role in aging and age-related diseases. Front Aging Neurosci 14:845330
Wang C-H, Wei Y-H (2020) Roles of mitochondrial sirtuins in mitochondrial function, redox homeostasis, insulin resistance and type 2 diabetes. Int J Mol Sci 21:5266
Wang G, Li S, Gilbert J, Gritton HJ, Wang Z, Li Z, Han X, Selkoe DJ, Man H-Y (2017) Crucial roles for SIRT2 and AMPA receptor acetylation in synaptic plasticity and memory. Cell Rep 20:1335–1347
Watroba M, Szukiewicz D (2021) Sirtuins at the service of healthy longevity. Front Physiol 12:724506
Wu Q-J, Zhang T-N, Chen H-H, Yu X-F, Lv J-L, Liu Y-Y, Liu Y-S, Zheng G, Zhao J-Q, Wei Y-F (2022) The sirtuin family in health and disease. Signal transduct Target Ther 7:402
Xu H, Liu Y-Y, Li L-S, Liu Y-S (2023) Sirtuins at the crossroads between mitochondrial quality control and neurodegenerative diseases: structure, regulation, modifications, and modulators. Aging Dis 14:794
Xu Y, Wan W (2023) Acetylation in the regulation of autophagy. Autophagy 19:379–387
Yalcin G (2018) Sirtuins and neurodegeneration. J Neurol Neuromed 3:13–20
Yang Y, Liu Y, Wang Y, Chao Y, Zhang J, Jia Y, Tie J, Hu D (2022) Regulation of SIRT1 and its roles in inflammation. Front Immunol 13:831168
Yang Z, Klionsky DJ (2009) An overview of the molecular mechanism of autophagy. In: Autophagy in infection and immunity, Current Topics in Microbiology and Immunology, vol 335. Springer, Berlin, Heidelberg. https://doi.org/10.1007/978-3-642-00302-8_1
Yin Z, Pascual C, Klionsky DJ (2016) Autophagy: machinery and regulation. Microbial Cell 3:588
Youn DH, Kim Y, Kim BJ, Jeong MS, Lee J, Rhim JK, Kim HC, Jeon JP (2021) Mitochondrial dysfunction associated with autophagy and mitophagy in cerebrospinal fluid cells of patients with delayed cerebral ischemia following subarachnoid hemorrhage. Sci Rep 11:16512
Yu L, Li Y, Song S, Zhang Y, Wang Y, Wang H, Yang Z, Wang Y (2024) The dual role of sirtuins in cancer: biological functions and implications. Front Oncol 14:1384928
Zhang J, Xiang H, Liu J, Chen Y, He R-R, Liu B (2020) Mitochondrial sirtuin 3: new emerging biological function and therapeutic target. Theranostics 10:8315
Zhang L, Pitcher LE, Yousefzadeh MJ, Niedernhofer LJ, Robbins PD, Zhu Y (2022) Cellular senescence: a key therapeutic target in aging and diseases. J Clin Invest 132:e158450

Metabolism and Brain Aging: The Role of Nutrients

12

Smruti Navthare, Ruchika Rajani, Afrin Khan, Sahil Thakre, and Ashish Kumar Parashar

Abstract

The complex interaction between brain aging and metabolism has become a central theme in neuroscience studies, with important roles in cognitive well-being and longevity. This chapter delves into the complex relationships among nutritional aspects, metabolic processes, and brain health during aging. The research began by analyzing the essential association between brain aging and metabolism, emphasizing how age-related metabolic variations can directly affect cognitive performance. Major nutrients that are important for sustaining brain function and cognitive ability with aging were identified, and their mechanisms of action were explained. Exploring the negative impact of oxidative stress and inflammation on the aging brain with special reference to the possible protective action of antioxidants and anti-inflammatory nutrients in modulating cognitive deterioration. Certain vitamins, minerals, and omega-3 fatty acids have been explored for their roles in brain metabolism and overall cognition during life. Dietary habits, with special reference to the Mediterranean diet, were assessed in terms of their influence on brain health and cognition in the elderly. In addition, this chapter investigates the possible mechanisms by which nutrients modulate brain metabolism and neuroplasticity, offering insights into how nutrition interacts with complex cognitive functions. Recent evidence for and against nutritional interventions to prevent or retard cognitive decline in old age is critically

S. Navthare · R. Rajani · A. Khan
Sri Aurobindo Institute of Pharmacy, Indore, MP, India

S. Thakre
NMT Gujarati College of Pharmacy, Indore, MP, India

A. K. Parashar (✉)
Lloyd Institute of Management and Technology, Noida, UP, India

A. Kumar Singh, S. Nand Rai (eds.), *Rejuvenating the Brain: Nutraceuticals, Autophagy, and Longevity*, Nutritional Neurosciences,
https://doi.org/10.1007/978-981-95-2790-8_12

reviewed to provide valuable insights into possible therapeutic approaches. Finally, we examined the challenges and limitations inherent in studying the role of nutrition in brain metabolism and aging. These findings present a balanced view of the current state of affairs and identify areas for future research. By clarifying the complex interrelation between metabolic activities, nutritional determinants, and brain health, this research adds to the expanding evidence base for devising effective strategies to ensure healthy brain aging and prevent age-related cognitive impairment.

Keywords

Metabolism · Brain aging · Cognitive function · Nutritional factors · Oxidative stress · Inflammation · Neuroplasticity

12.1 Introduction

The human brain, a wonder about biological sophistication, experiences dramatic changes over its lifetime. As age advances, the complex interaction between brain function and metabolism becomes more evident, with diet becoming a determining factor in sustaining cognitive function and possibly preventing age-related deterioration. This chapter looks at the interaction between brain aging, metabolism, and the function of nutrients, with a view of moving from the normal adult brain to the aging brain and the function of certain nutrients in such changes. In the healthy adult brain, metabolism is a highly regulated process that guarantees optimal mental performance. Although it accounts for only approximately 2% of the body weight, it accounts for approximately 20% of the body's total energy. This high-energy requirement is chiefly served by glucose metabolism, which energizes a number of neuronal functions, such as neurotransmitter synthesis, synaptic plasticity, and ion gradient maintenance. In addition, the adult brain also depends upon a delicate system of neurotransmitters, hormones, and growth factors to ensure structural integrity and functional capability (Redolfi et al. 2016).

However, the brain experiences different physiological changes with age that influence metabolic efficiency and general function. Among these are the reductions in brain volume, especially in areas linked to cognitive activities such as memory and executive control. The density of synapses was also reduced, and there were changes in the neurotransmitter systems and cerebral blood flow. These changes contribute to the decline in cognitive processes, such as processing speed, attention, and memory.

The aging brain also experiences significant changes in metabolism. A widespread reduction in glucose utilization was attributed to mitochondrial impairment and enhanced oxidative stress. This metabolic change can drastically impact brain function, as neurons are less effective in generating and using energy. The aging brain is potentially less sensitive to insulin, a vital hormone in glucose metabolism and neuronal survival.

Nutrition has become a key component in maintaining brain health during aging. Some nutrients are especially significant in the preservation of cognitive function and possible retardation of age-related deterioration. Omega-3 fatty acids, for example, are integral parts of neuronal membranes and have been linked to enhanced cognitive function in older people. Antioxidants like vitamins C and E can potentially counteract oxidative stress, which is believed to be involved in brain aging (Igase et al. 2010) Vitamin B, specifically B12, folate, and B6, are important in neurotransmitter production, and the metabolism of the amino acid homocysteine, which is present at elevated levels, is related to cognitive deterioration. Vitamin D, which has long been associated with bone, has more recently been found to be related to cognition, and some studies indicate that a deficiency in this vitamin may contribute to a higher risk of cognitive impairment in the elderly.

Minerals are also important for brain health, but age can influence their condition. For instance, iron is necessary for producing neurotransmitters and myelination; however, its presence in some brain areas with age can also lead to oxidative stress. Zinc, important for plasticity at the synapse and learning, decreases with age and may impair cognitive performance. Magnesium, which plays a role in various enzymatic reactions and neurotransmitter release, might also decline with age, consequently influencing cognitive function (Igase et al. 2010). The degradation or perturbed metabolism of these minerals within the aging brain has wide-ranging implications. Abnormalities in calcium homeostasis have been linked to age-related neuronal impairment and neurodegenerative diseases. Changes in copper metabolism may also contribute to oxidative stress and neuroinflammation during brain aging.

A better understanding of the relationship between brain aging, metabolism, and nutrition is a promising intervention. Nutritional strategies like the Mediterranean diet or the MIND diet (Mediterranean-DASH Intervention for Neurodegenerative Delay), which focus on foods high in brain-nourishing nutrients, have been promoted in observational studies as beneficial to cognitive health in older adults (Chen et al. 2019). However, the interaction between nutrition and aging in the brain is multifactorial and interactive. Variations in genetics, lifestyle, and health status from one individual to another can mediate the way the brain ages in response to nutritional interventions. Furthermore, nutrients interact with each other and are not independent in the body. With continued research in this area, there is growing interest in individualized nutrition strategies, taking into account an individual's metabolic profile and genetics. These personalized strategies could provide better methods for maintaining brain health and cognitive function at one age. There is also beginning to be evidence from research on the gut-brain axis that the relationship between diet, gut microbiota, and brain function holds clues for understanding cognitive health in the aged (Tucker 2016).

12.2 The Intricate Architecture of the Human Brain

The human brain, a wonder about biological sophistication, is the body's command center. With a weight of about 1.4 kg, this great organ is composed of billions of neurons linked together by trillions of synapses, creating an elaborate network that allows for the processing of immense amounts of information, the regulation of body functions, and the creation of consciousness. The brain comprises several different regions, each performing a unique function. The outermost layer of the cerebral cortex is responsible for higher-order thoughts, sensory processing, and motor control. This convoluted surface was subdivided into four lobes: frontal, parietal, temporal, and occipital, each with specialized cognitive and sensory functions. For example, the frontal lobe is important for decision-making, planning, and personality, while the occipital lobe is mostly involved in visual processing (Spreng and Turner 2018).

The limbic system plays a vital role in emotion and memory consolidation in the cerebral cortex. It consists of structures such as the amygdala, which handles fear and emotional reactions, and the hippocampus, which plays a central role in transferring short-term to long-term memories. Another important part of the thalamus is a relay center for sensory and motor signals, while the hypothalamus controls hormone levels and homeostasis. The brainstem, located at the rear of the brain, links the brain to the spinal cord and controls basic functions, including breathing, heart rate, and blood pressure. The brainstem has three components: midbrain, pons, and medulla oblongata. The cerebellum, which is found at the posterior part of the brain, is responsible for motor control, balance, and coordination of movements (Spreng and Turner 2018).

Neural circuits zigzag through the brain to enable rapid communication between areas. These circuits contain white matter tracts and bundles of axons that carry signals between the neurons. The most prominent white matter structure, the corpus callosum, links the brain's two hemispheres and allows communication. The brain's capacity to change and adapt, called neuroplasticity, enables it to reorganize by creating new neural connections throughout life. This incredible ability supports learning, memory, and recovery from brain injuries. Neurotransmitters, which are chemical messengers sent out by neurons, are essential for brain function through their effects on mood, cognition, and behavior (Moya 2018).

Recent developments in neuroimaging methods, such as functional magnetic resonance imaging (fMRI) and positron emission tomography (PET), have generated unprecedented information on brain structure and function. Such technologies enable investigators to monitor brain activity in real-time, improving our understanding of cognitive functioning, neurological conditions, and the complex interactions between multiple brain regions. The complexity of the human brain extends from its physical structure to functional capabilities. It enables abstract thinking, language comprehension and production, problem-solving, and creativity. The brain also plays a crucial role in regulating emotions, forming and retrieving memories, and shaping our perceptions of the world. Although neuroscience has made tremendous progress, several brain functions remain unknown. Continued research continues to decode the nuances of this intriguing organ and promising new knowledge regarding consciousness, cognitive processes, and treatments for neurological and psychiatric diseases. With its unmatched complexity and abilities, the human brain continues to be one of the most fascinating topics of scientific examination in the known universe (Redolfi et al. 2016).

12.3 The Aging Brain: A Comprehensive Comparison with Normal Brain Function

The dynamic and intricate human brain experiences radical age-related changes that influence different cognitive tasks and neural activity. This analogy discussed the disparities between a typical healthy adult brain and an old brain, referencing the physiological differences and their resultant effects on intellectual functioning (The Aging Brain 1983). Neural networks within a healthy adult brain function incredibly effectively, enhancing quick information handling, memory, and intellectual suppleness. These networks have strong connectivity and maximal synaptic transmission, enabling unimpeded communication between various brain regions. Neurotransmitter systems operate at their highest levels, facilitating mood regulation, attention, and general cognitive functions. The plasticity of the brain, or its capacity to create new neural connections and reorganize established connections, is maintained at a high level, enabling ongoing learning and adaptation to new experiences (Marzola et al. 2023).

A healthy, mature brain also reveals superior cognitive regulation, which makes it possible to alternate among different tasks, suppress unimportant information, and persevere in goal-directed acts. Executive capabilities such as planning, choice-making, and problem-solving functions are common at their zenith in adult humans. In addition, the energetic metabolism of the brain is in good condition so that sufficient supply of blood with oxygen could occur to fulfill neural activity with higher energy (The Aging Brain 1983). However, a few important changes occur when the brain grows older. Perhaps the most prominent change is the gradual reduction in brain volume through brain atrophy. This volume loss is significant in brain parts central to learning and memory, the hippocampus and the prefrontal cortex. The shrinkage of these areas can play a role in the deterioration of episodic memory, working memory, and executive function.

Neurotransmitter synthesis and receptor sensitivity also decrease with age and may affect mood control and cognitive function. For instance, the dopaminergic system, responsible for motivation, reward, and motor control, undergoes significant alterations. Decreased levels of dopamine and dopamine receptors can lead to decreased movement, diminished motivation, and changes in reward processing. ("Brain Aging: Neuropathology & Neuropharmacology," 1983) White matter integrity, critical for effective communication between brain areas, decreases with age. This decline, known as white matter hyperintensities, can result in slower information processing and lower cognitive effectiveness. The aging brain can also

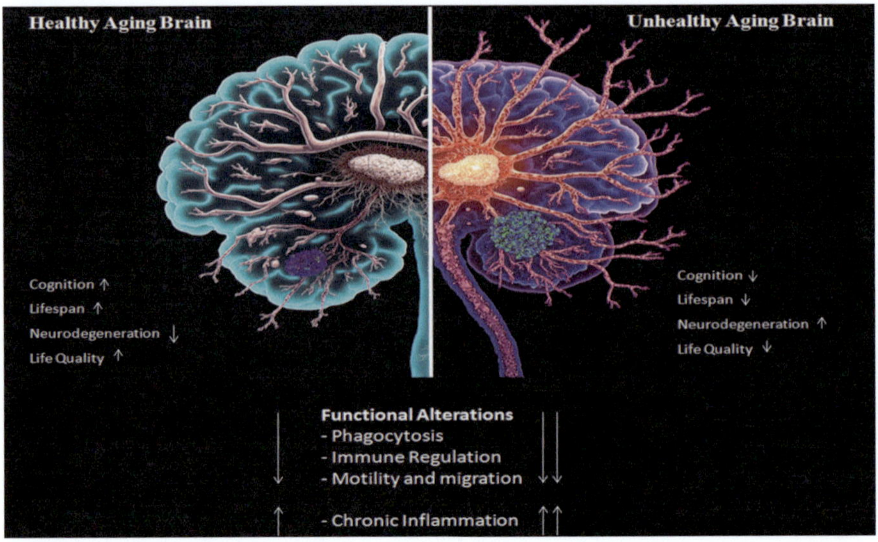

Fig. 12.1 Microglial alterations in healthy and unhealthy brain aging

experience a decrease in synaptic density, which can influence the strength and effectiveness of neural connections.

Although it still exists, neuroplasticity is diminished in the older brain (Fig. 12.1). This decrease can make creating new neural connections and learning new scenarios more difficult. Therefore, it is more difficult for older adults to acquire new skills or adapt to environmental changes. These physiological changes can manifest as mild cognitive decline, including slower processing speed, reduced working memory capacity, and difficulties in multitasking. Older adults may experience challenges in dividing attention between multiple tasks or quickly retrieving specific memories. Language abilities, particularly word-finding and verbal fluency, may also show subtle declines (Cherdak 2023). Despite these changes, aging brains tend to be highly resilient and adaptable. The brain can compensate for age-related decline by enlisting more neural networks and leveraging stored knowledge and experience through a mechanism called cognitive reserve. This compensation enables many older adults to preserve high levels of cognitive function and performance on daily tasks.

Additionally, older brains may exhibit alterations in emotional processing and regulation. Certain studies have proposed that older individuals are more likely to attend to positive information and experiences or have a positive effect. This change in emotional processing could lead to enhanced emotional well-being and life satisfaction in older age (Blumen et al. 2024; Wrigglesworth et al. 2021). Although the aging brain experiences different physiological changes that can impair cognitive ability, it also displays impressive adaptability and resilience. It is essential to understand these changes to create interventions and strategies that promote healthy cognitive aging and preserve the quality of life in older adults. It remains an active research area in the fields of neuroscience and gerontology, which continues to provide new information about the aging brain that promises to lead to novel solutions for preserving cognitive health throughout one's lifespan (Kim and Barber 2022).

12.4 Brain Aging and Metabolism

Aging brains also experience a range of structural and functional alterations, which may result in cognitive loss and neurodegenerative diseases. The Global Burden of Disease Study indicates that the risk of neurodegenerative diseases, such as Alzheimer's disease, increases markedly with age, affecting approximately 10% of adults aged 65 years and older (GBD 2019; Demographics Collaborators 2019). These alterations are linked to the loss of neuroplasticity and changes in different neurotransmitter systems (Table 12.1) (Singh et al. 2021).

The effects of these neurobiological alterations are significant, extending to cognitive functions and emotional and social performance (Fig. 12.2). Studies show that as the brain ages, there can be changes in mood control and heightened levels of anxiety or depression. This emotional turmoil can also complicate cognitive loss, making it increasingly more challenging for people to sustain social connections and participate in previously enjoyed activities (Sagar et al. 2019). Lifestyle also plays a crucial part in mitigating the impacts of brain aging. Frequent exercise, an antioxidant diet, and intellectual stimulation activities such as puzzling, reading, and staying engaged in society have all been proven to prevent some loss of cognitive ability brought about by aging. Research suggests that these lifestyle changes not only protect against brain-related illness but are also likely to promote neuroplasticity and possibly defer the onset of neurodegenerative conditions (Parrott et al. 2020).

Furthermore, recent research has highlighted the role of sleep in preserving cognitive function. Sleep disturbances are prevalent in older adults and can worsen cognitive impairment, thereby creating a cycle of deterioration. Restorative sleep is necessary for memory consolidation and removing neurotoxic waste products from the brain, such as beta-amyloid, a protein that builds up in Alzheimer's disease. Therefore, the promotion of healthy sleep habits is increasingly recognized as a critical strategy for facilitating cognitive resilience in the elderly population (Harenbrock et al. 2023). The urgency for early intervention and integrated care strategies is becoming more evident as the world population grows. Healthcare systems are increasingly concentrating on a multidisciplinary strategy of aging, uniting physical, mental, and social well-being to enable people to manage this intricate transition (Lakhman et al. 2024). By focusing on research activities, public health programs, and community resources, society can better empower itself in addressing the challenges that the aging brain presents, ultimately enhancing the quality of life among elderly individuals.

Furthermore, technologies are at the forefront of the improvements needed in the quality of care for the elderly. Wearable health devices and home automation technologies can track vital signs, monitor daily activities, and provide safety to older

Table 12.1 Age-related changes in brain-metabolism

S. no.	Metabolism change	Description	Impact on brain aging	Potential mitigation strategies
1.	Decreased glucose metabolism	Reduced ability of the brain to use glucose for energy is common in Alzheimer's disease	Cognitive decline, memory issues, and increased risk of neurodegeneration	Ketogenic diet, intermittent fasting, physical exercise, and MCT (medium-chain triglyceride) supplement
2.	Mitochondrial dysfunction	Decline in mitochondrial efficiency reduces ATP (energy) production and increases free radical damage	Impairs the contribution of neuronal communication to fatigue and accelerates brain aging	Antioxidants (Coenzyme Q (CoQ))
3.	Increased oxidative stress	Accumulation of free radicals damages neurons owing to reduced antioxidant defenses	It causes neuronal death, weakens memory, and increases the risk of neurodegenerative diseases	Antioxidant-rich diet (berries, green tea, dark chocolate), vitamins E and C, and polyphenols
4.	Altered lipid metabolism	Changes in lipid composition affect cell membrane integrity and myelin sheath function	It impairs synaptic function, slows signal transmission, and contributes to cognitive impairment	Omega-3 fatty acid (fish oil, flaxseeds), phosphatidylserine, and healthy fat intake
5.	Reduced ketone utilization	Decreased ability to use ketones as an alternative energy source when glucose metabolism declines	Lower energy availability for brain function leads to sluggish thinking and brain fogging	Ketogenic diet, exogenous ketones, and fasting to enhance metabolic flexibility
6.	Neuroinflammation	Chronic low-grade inflammation is caused by dysregulation of the immune system and the accumulation of toxic proteins	Linked to Alzheimer's, Parkinson's, and cognitive decline	Curcumin, Omega-3s, intermittent fasting, and anti-inflammatory diet (leafy greens, nuts, turmeric)
7.	Impaired insulin sensitivity	Brain cells are less responsive to insulin, which affects glucose uptake	Increases risk of neurodegeneration, brain fog, and memory loss (linked to "type 3 diabetes")	Low glycemic diet, intermittent fasting, cinnamon, berberine, and regular physical activity

(continued)

Table 12.1 (continued)

S. no.	Metabolism change	Description	Impact on brain aging	Potential mitigation strategies
8.	Decline in neurotransmitter synthesis	Reduced dopamine (motivation), serotonin (mood), and acetylcholine (memory) production	This leads to memory issues, mood disorders (depression and anxiety), and slow cognitive processing	Amino acid-rich diet (tyrosine, tryptophan), vitamin B, and regular mental stimulation
9.	Disrupted blood brain barrier	Increased permeability allows toxins and inflammatory agents to accumulate	Exposes neurons to harmful substances and accelerates neurodegeneration	Polyphenols (green tea, resveratrol), Omega-3s, and maintaining a healthy gut microbiome

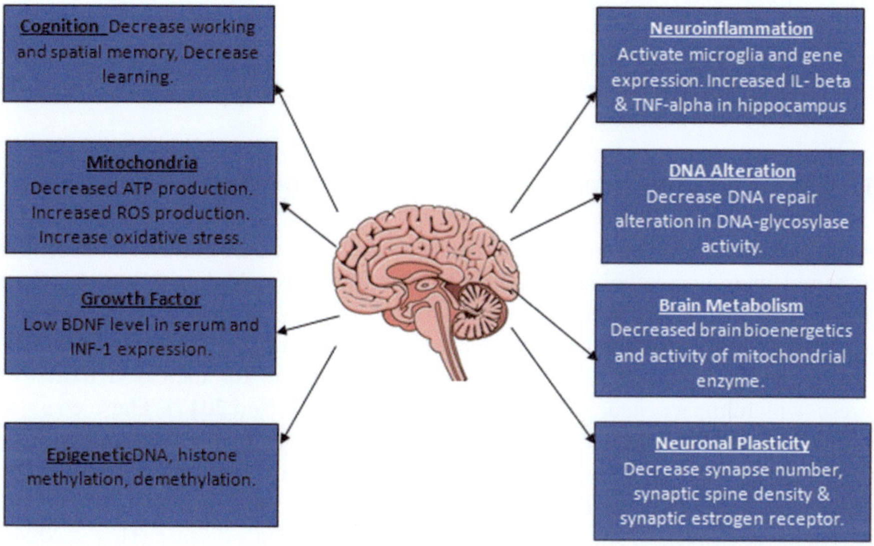

Fig. 12.2 Changes associated with brain aging

adults who live independently. These technologies encourage proactive health care and enable seniors by creating a feeling of independence and dignity. In addition, telehealth services have picked up vast momentum, enhancing the connectivity between patients and healthcare workers, most importantly in rural or underserved regions that might be unable to access geriatric specialists.

Moreover, one cannot overlook the role of professional and familial caregivers in this scenario. Providing caregivers with the necessary skills, resources, and support networks is indispensable in coping with the intricacies of caring for the elderly. Caregiving skills training, stress management, and awareness of the special needs of older persons can be significant difference-makers. Caregiver support programs, such as respite care services and counseling, must be given high priority by communities to reduce burnout and improve overall care for the elderly (Seago et al. 2025).

Incorporating mental health services into care for the elderly is a high priority. Since most older adults are lonely, depressed, and suffer from mental decline, there is a critical need to build supportive environments that foster mental well-being. Programs for social interaction, therapy sessions, and support groups can address loneliness and hopelessness. Moreover, encouraging healthy living habits such as regular exercise and a balanced diet can significantly affect physical and mental health outcomes. As attitudes toward aging change, a culture of respect and appreciation among older adults is required. Changing perceptions can result in better intergenerational relationships, where senior citizens' wisdom and experiences are appreciated and passed on to younger generations. Educational programs that enable younger people to interact with older people can help form an inclusive society to which every age group contributes. By working together across sectors, healthcare, technology, community organizations, and government—society can create a more supportive infrastructure that respects the dignity of aging individuals and nurtures their overall well-being.

12.4.1 Metabolic Changes Associated with Aging

Aging brain metabolism is characterized by decreased mitochondrial activity and impaired energy metabolism, leading to decreased cognitive function. In older adults, glucose metabolism drops by approximately 5–7% each decade, which is associated with worsening cognitive function (Kalmijn et al. 1997). Stress that disrupted insulin signaling within the brain is responsible for increased oxidative stress, which impacts neural health (Camandola and Mattson 2017). These changes in metabolic function impact cognitive function and have broader consequences for general brain health. Recent research indicates that lowering energy levels can impact the structural stability of neurons, making them more susceptible to neurodegenerative disorders, such as Alzheimer's and Parkinson's disease (Hoyer 2004). Amyloid plaque buildup and tau tangles are also linked to disrupted energy metabolism, which indicates an interplay between metabolic impairment and neurodegeneration.

Furthermore, lifestyle factors, including diet and exercise, are crucial in mitigating age-related changes. A Mediterranean diet, characterized by a high healthy fat content, antioxidants, and low refined sugar intake, has been shown to reverse inflammation and maintain mitochondrial function (Martínez-Lapiscina et al. 2013). Physical activity, particularly aerobic exercise, has also been associated with higher levels of neurogenesis and more remarkable synaptic plasticity, both essential for sustained cognitive functioning in older adults (Kirkland and Tchkonia 2017).

In addition, new evidence supports the potential of pharmacological therapies to enhance metabolic health. Medications that activate mitochondrial function or simulate caloric restriction, including metformin, have been investigated for their neuroprotective effects (Zhang et al. 2017). These results highlight the necessity of an interdisciplinary strategy to treat cognitive impairment by combining lifestyle changes with innovative therapeutic approaches to enhance brain health in older adults.

Besides pharmacological management, lifestyle parameters like diet and exercise are critical in maintaining metabolic well-being and cognition. Dietary regimens with high antioxidant content, omega-3 fatty acids, and polyphenols, such as the Mediterranean diet, have been postulated to prevent neurodegenerative disease development (Schwingshackl & Hoffmann, 2020). Regular exercise has also been associated with increased neurogenesis and synaptic plasticity, both essential for cognitive capacity (Brown et al. 2019).

Social interactions and mental stimulation are key elements of a comprehensive strategy for brain health. Social activity and lifelong learning can promote cognitive resilience and mitigate the effects of age-related decline. Interventions that encourage community participation, volunteering, and group activities not only improve emotional well-being but also lead to better cognitive outcomes (Wei et al. 2024). As our brain and aging awareness progress, collaborations across disciplines of neuroscientists, clinicians, nutritionists, and gerontologists are taking center stage. With the infusion of varied fields and expertise into the mix, robust interventional plans that combine pharmacological care, dietary intake adjustments, physical exercise, and social integration can be realized.

Finally, the aim was to develop individualized strategies considering genetic, lifestyle, and environmental differences. Advances in technology, such as wearable monitors and mobile health apps, have provided new means of tracking health metrics and promoting compliance with interventions. This personalized strategy can potentially reduce the risk of cognitive impairment significantly and improve the well-being of the aging population (Cunnane et al. 2010). Cognitive decline in the aging population can be addressed through a multifactorial approach that incorporates pharmacological investigation, life interventions, and social support networks. Prioritizing healthy brains and instilling proactive approaches to aging are steps that can be taken to ensure a future in which cognitive vigor is an attainable goal. Combining artificial intelligence and machine-learning algorithms with individualized health data may further boost the efficacy of customized interventions for improving cognitive well-being. Such sophisticated technologies can examine complicated patterns in one's health measures, life decisions, and environmental factors to deliver more precise predictions of the risk of cognitive decline and to propose highly personalized preventive actions (Blumen et al. 2024).

In addition, community-based interventions that combine social activity, cognitive stimulation, and physical exercise can promote supportive environments that are favorable for preserving cognitive function in older adults. Such programs may be carefully developed to target the cultural and socioeconomic determinants of cognitive aging so that they are both accessible and meaningful to diverse populations (Turrini et al. 2023).

12.5 Role of Macronutrients in Cognitive Function

12.5.1 Carbohydrates: Impact on Glucose Metabolism and Cognitive Performance

Carbohydrates are essential in the metabolism of energy in the brain, while high intake results in insulin resistance and compromised glucose tolerance, eventually resulting in cognitive impairment. A reduction in hippocampal glucose use among older adults with impaired glucose regulation has been previously reported (Gault and Hölscher 2008). Systematic reviews have reported improvements in cognitive functions by diets characterized by a low glycemic index, as such diets ensure blood glucose levels while preventing spikes or dips that undermine cognitive performance (Krebs-Smith et al. 2010).

Additionally, the role of antioxidants in cognitive health has garnered attention in recent years. Research has highlighted that oxidative stress can exacerbate cognitive decline, particularly in the aging population. A diet rich in antioxidants such as fruits, vegetables, nuts, and whole grains may combat this process. For example, research has shown that flavonoids, often present in berries and dark chocolates, are associated with improved memory and cognitive flexibility (Devore et al. 2012). In addition, incorporating omega-3 fatty acids, common in fish and flaxseeds, is associated with improved neuronal structure and function, thus enhancing cognitive resilience (Gómez-Pinilla 2008).

In addition, the timing of nutrient consumption is crucial. Research indicates that the consumption of breakfast, especially a mix of carbohydrates, proteins, and healthy fats, positively impacts cognition during the day. Skipping breakfast, on the other hand, has been found to result in higher levels of fatigue and reduced attention span and highlights the significance of standard meal patterns in long-term cognitive functioning (Yannakoulia et al. 2014). As we look at the larger implications of diet, the Mediterranean diet, which features a high intake of whole grains, legumes, fruits, vegetables, and healthy fats with a low intake of processed foods, has been linked to reduced risk of cognitive impairment and dementia in older adults (Yannakoulia et al. 2014). This integrated eating habits model supports both brain health and overall wellness, implying that the convergence of nutrition and cognition is critical to ongoing research and public health policy.

Based on these results, health professionals must promote dietary changes by focusing on balanced nutrition to maximize cognitive function. Incorporating educational interventions highlighting the significance of a low glycemic index diet with antioxidants and healthy fats can help prevent cognitive decline and improve the quality of life among the elderly (Gilsenan et al. 2009). Additionally, integrating health workers, nutritionists, and community organizations can increase the scope and impact of such programs. Through a program of community workshops and virtual seminars on meal preparation and planning, people can make healthier food decisions. In addition, integrating increased physical activity with dietary modifications can enable the holistic treatment of mental health. Studies indicate that exercise maintains neuroplasticity and improves brain function, thus extremely important in preventing cognitive decline.

Additionally, addressing the social determinants that affect eating habits is necessary. Most people, especially those in low-income areas, are confronted with obstacles, including the limited availability of fresh fruits and vegetables and healthy foods, which may prevent them from embracing a healthy diet. Hence, policies that enhance food availability and community gardens may enhance nutritional habits. By ensuring that all community members are provided with cheap, healthy foods, we can create an environment where everyone is cognitively healthy (Metcalfe et al. 2021).

In addition to these efforts, ongoing research on the connection between particular nutrients and brain health is crucial. Further research may reveal additional dietary factors involved in maintaining brain function. Support for health professionals' ongoing education on the latest research results will enable them to advise their patients with the best and most reliable dietary information on their diets (Tucker 2016). Finally, it is important to consider the psychological components of dietary modifications. Promoting mindfulness in eating habits and developing healthy relationships with food can assist individuals in adopting new dietary habits more easily. Counseling and support groups can make this easier by improving motivation and compliance with healthier dietary habits. Combined, these methods form a balanced strategy for maintaining cognitive function, allowing individuals to live healthy lives as they age (Lakhman et al. 2024).

12.5.2 Proteins: Amino Acids as Precursors for Neurotransmitters

Proteins are required for neurochemical synthesis. Amino acids such as tryptophan and tyrosine are precursors of serotonin and dopamine. A meta-analysis of food protein consumption among older adults indicated that increased protein intake is related to a decreased risk of cognitive impairment (Wang et al. 2014). Older adults also have a lower synthesis rate of some neurotransmitters, owing to possible amino acid deficiencies. These results highlight the significance of food habits in maintaining brain health at one age. In addition to proteins, other vitamins and minerals have supporting functions in neurobiology. For example, B vitamins, especially B6, B12, and folate, play a vital role in enzymatic reactions involved in neurotransmitter synthesis. A lack of these vitamins leads to cognitive impairment and mood disorders, further underscoring the relationship between diet and mental well-being (Coelho-Júnior et al. 2021).

Additionally, the amount of protein and its quality is important. Good sources of high-quality protein, such as lean meat, fish, eggs, and milk, provide essential amino acids that are necessary for proper neurotransmitter function. Although useful, plant proteins lack some of the amino acids needed for full neurochemical production. Therefore, a well-balanced diet comprising a combination of animal and plant proteins can provide the best results for cognitive maintenance (Phillips and Van Loon 2011). Current research has also identified the potential benefits of certain diets, such as the Mediterranean diet, which is rich in antioxidants, healthy fats, and a mix of nutrients, to improve cognitive health. Omega-3 fatty acids found in fish and flaxseeds can decrease neuroinflammation and increase synaptic plasticity, which are the key elements of cognitive resilience (Gómez-Pinilla 2008).

Furthermore, lifestyle components, including physical activity, mental challenges, and social contact, synergize with nutrition to protect cognitive abilities. Physical exercise has been established to enhance the availability of nutrients that power brain health, whereas social contact has the potential to activate neural networks and enhance mental sharpness. Thus, an integrated approach that weaves diet, physical activity, and intellectual challenges is the best way to promote cognitive longevity (Ratey and Loehr 2011). As we get older, various nutritious diets become even more crucial. Nutrition plays a vital role in the cognitive function of patient counseling and community health initiatives. Through higher awareness and knowledge of the role of dietary options, we can encourage older people to adopt better diets, thus promoting an environment that supports cognitive vitality (Lee et al. 2024).

In addition, incorporating a variety of foods that contain antioxidants, omega-3 fatty acids, and essential vitamins may greatly improve cognitive function and reduce the risk of cognitive impairment. Nutrient recommendations for older adults should emphasize berries, fatty fish, nuts, and leafy green vegetables. Such foods can also be topics of regular cooking classes or workshops on meal preparation, which may impart precious skills and information, making healthy living more convenient and enjoyable (Muth and Park 2021).

Community programs should include guest speakers such as chefs or nutritionists to provide tips and encourage older adults to try new recipes. These events encourage healthy eating and act as social gatherings that fight loneliness, a prevalent condition among the elderly. By establishing a community environment focused on nutrition, we can assist in developing peer partnerships, urging them to share their difficulties and triumphs in sustaining a healthy diet (Metcalfe et al. 2021). In addition, physical activity could not be discounted. Adding a balanced diet to regular exercise has compounded cognitive benefits. Incorporating basic exercise or mobility classes into diet education sessions integrates well-being using a whole-stake approach. Ensuring that older adults walk or participate in gentle yoga routines can build both physical and mental well-being, complementing a healthy aging model (Wrigglesworth et al. 2021).

Finally, it is necessary to harness technology to increase its reach and influence. Virtual cooking classes offered on digital platforms, nutrition apps, and online support groups can enable older adults to manage their diets from the comfort of their homes. Through both online and offline interactions, we can develop a multi-pronged nutrition education approach tailored to the special needs of older adults (Ng and Markworth 2021). Promoting a diet filled with nutrients contributes to healthy cognition, enhances community connections, fosters exercise, and accommodates an increasingly technologically driven environment. Collectively, these initiatives have the potential to shift the paradigms through which we understand aging and health, paving the way for a brighter and healthier future for older adults (Lyons 2013) (Lobuono and Milovich 2023).

12.5.3 Fats: Influence of Different Types of Fatty Acids on Brain Health

Fatty acids influence cognitive function and neural integrity. A systematic review reported that omega-3 fatty acids, specifically eicosapentaenoic acid (EPA) and docosahexaenoic acid (DHA), correlated with decreased cognitive decline rates among individuals over the age of 50 years, lowering the occurrence of Alzheimer's disease by approximately 30% ((Zhang et al. 2018). In contrast, saturated fat diet has been reported to correlate with an elevated risk of cognitive decline (Chang et al. 2009). In addition to omega-3 fatty acids, other dietary factors and nutrients may also play important roles in maintaining cognitive function. For example, antioxidants such as vitamins C and E have been shown to fight oxidative stress, which can lead to neurodegeneration (Scarmeas 2009). A diet rich in fruits, vegetables, whole grains, and healthy fats, characterized by a Mediterranean diet, maintained an environment conducive to neuronal function, lowering the risk of age-related cognitive impairment.

Furthermore, the contributions of lifestyle factors cannot be overemphasized. Regular physical exercise has been shown to increase cerebral blood flow, elevate mood, and induce neurogenesis, creating new neurons. Haag 2003 has demonstrated that moderate aerobic exercise can improve executive function and brain health. Sleep patterns significantly influence intellectual functioning. Systemic sleep deprivation is associated with disturbances in attention, memory, and decision-making abilities. Experiments indicate that stable sleep quality might increase synaptic plasticity, which is essential for learning and consolidating memories (Hawken 2020). Social interaction and mental stimulation are important for sustaining cognitive vitality. Doing group activities, learning a new skill, or even challenging puzzles can build a cognitive reserve, protecting the brain against age-related decline. A longitudinal study by Piolatto et al. 2022 showed that those with strong social networks and active lives had slower cognitive aging than their isolated peers.

12.6 Micronutrients and Brain Health

12.6.1 Vitamins: Effects of Deficiencies or Excess Intake on Neurological Function

Vitamins are important for maintaining brain function. B vitamins, especially B12 and folate, play an important role in lowering the level of homocysteine, which is a risk factor for cognitive impairment. A systematic review concluded that folate supplementation in the elderly can improve cognitive performance, as evidenced in trials that affect memory and executive function (Douaud et al. 2013). However, vitamin E supplementation has yielded inconsistent results; whereas some research has found cognitive improvement, others suggest harm when in excess (Sano et al. 1997). Additionally, the function of antioxidants such as vitamin C in maintaining cognitive well-being should not be forgotten. Studies have indicated that vitamin C can protect neurons from oxidative stress, thus contributing to healthy brain aging (Travica et al. 2017). A meta-analysis of experimental studies on vitamin C supplementation about cognitive function identified positive associations between sufficiency of vitamin C levels and enhanced cognitive outcomes, particularly among the elderly (Alavi Naeini et al. 2013).

In their role as more than just a vitamin, the omega-3 fatty acids also benefit brain function. These crucial fats, of which fish and some plant oils are the principal sources, contribute to synaptic maintenance and have been found to lower the risk of neurodegenerative disorders. A longitudinal study concluded that eating more omega-3s also diminished the rate of dementia and improved total cognitive function during aging (Yurko-Mauro et al. 2010). In addition, the significance of a balanced diet containing fruits, vegetables, and whole grains extends beyond vitamin consumption. Nutrient-rich foods contain a complex mixture of compounds that synergistically promote brain health. Flavonoid-rich foods, including berries, have been associated with enhanced cognitive function and memory preservation, indicating that lifestyle may play a significant role in determining brain health (Fig. 12.3) (Krikorian et al. 2010).

Although supplementation may be helpful, it is important to look at dietary sources as an ideal way to consume vitamins and nutrients. A whole-food diet is healthier and can contain a wider array of valuable compounds frequently lacking in extracted supplements. Tailoring nutritional strategies for unique health states and lifestyle components can lead to more efficient designs to ensure cognitive well-being across the lifespan (Klimova et al. 2020). Finally, continued studies investigate the complex relationships between aging, brain function, and nutrition,

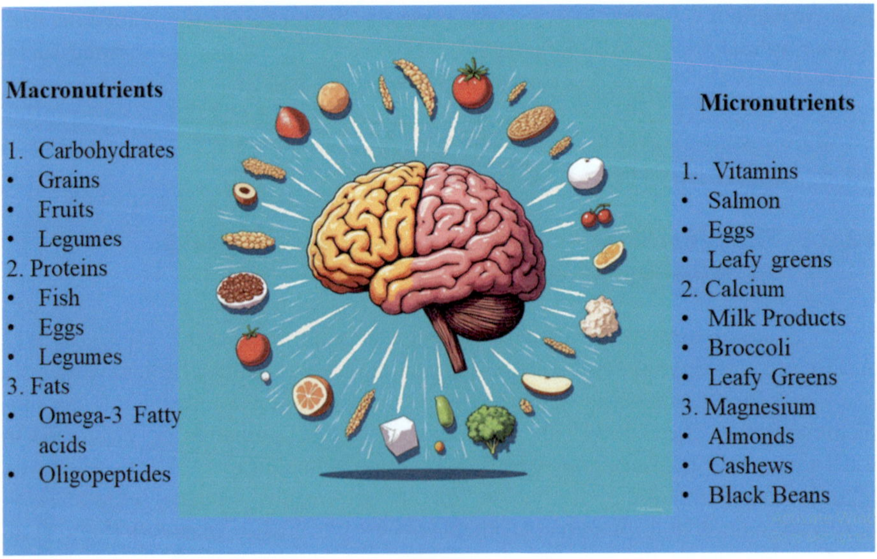

Fig. 12.3 Important macronutrients and micronutrients for healthy brain

emphasizing the necessity of a multidisciplinary strategy integrating nutritionists, clinicians, and researchers to determine optimal interventions for promoting cognitive health.

12.6.2 Minerals: Their Role in Neurochemistry and Behavior

Minerals, such as zinc and magnesium, have been recognized as central to neurochemistry. Zinc deficiency is associated with compromised cognitive ability and has been found to affect hippocampal synaptic plasticity in older individuals adversely (Koletzko 2005). Conversely, sufficient magnesium supplementation has been associated with improved cognitive performance and reduced risk of dementia (Barbagallo and Dominguez 2010).

In addition, new evidence suggests that balanced consumption of these minerals contributes to cognitive function and plays a significant role in mood control. Studies have demonstrated that zinc affects the concentration of neurotransmitters such as serotonin, which are essential for emotional well-being (Stahl et al. 2009). Zinc deficiency has been linked with higher levels of anxiety and depression, which underscores the significance of zinc in managing mental health.

In the same way, magnesium has been researched for its sedative impacts on the nervous system and has even been termed the "original chill pill." It seems to counteract the body's stress reaction by modulating the hypothalamic-pituitary-adrenal (HPA) axis, thus reducing cortisol levels, a critical component in the battle against chronic stress and its brain impairments (Gómez-Pinilla 2008). This dual function of magnesium as both a mood stabilizer and cognitive enhancer highlights its key role in a balanced dietary strategy for mental well-being (Sartori et al. 2011).

Including foods high in these essential minerals in the diet may become a cornerstone strategy for sustaining mental acuity with aging. Seeds and nuts, legumes, and dark greens are good sources of Mg, whereas oysters, red meat, and poultry are good sources of Zn. Public health efforts to raise awareness of these minerals could play an important role in averting nutritional deficiencies that could cause cognitive impairment.

Further research on the synergistic interactions of Zn and Mg in combination is warranted because, based on some investigations, the maximal utilization of these nutrients may rely on their relative availability in the body. This also indicates a very important point in food planning, as an excess of one mineral may preclude the intake of another (Chen et al. 2024). Thus, following a well-balanced diet rather than a clinically necessary supplement is preferable.

Finally, the link between micronutrients and cognition opens a hopeful path for preventive health interventions. Addressing the nutritional requirements of the older population can go a long way in preventing cognitive deterioration and encouraging mental robustness in old age (Table 12.2).

Table 12.2 Essential micronutrients and omega-3 fatty acids for neurotransmitter production and brain function

S. no.	Nutrients	Neurotransmitters supported/function in brain health	Impact on cognitive function	Food source
1.	Vitamin B6 (pyridoxine)	Supports serotonin, dopamine, GABA production	Regulates mood, reduces anxiety, and enhances cognitive function	Poultry, bananas, potatoes, fortified cereals, and nuts
2.	Vitamin B9 (folate)	Essential for dopamine and serotonin synthesis	Crucial for brain development, memory, and cognitive stability	Leafy greens, legumes, citrus fruits, eggs
3.	Vitamin B12 (cobalamin)	Involved in dopamine, serotonin, and acetylcholine production	Prevent cognitive decline and support myelin sheath integrity	Fish, meat, dairy products, eggs, and fortified cereals
4.	Vitamin C	Acts as cofactors for dopamine and serotonin synthesis	Protection against oxidative stress enhances mood and memory	Citrus fruits, bell peppers, strawberries, kiwi
5.	Magnesium	Regulates GABA and dopamine levels	Reduces stress and supports memory and cognitive stability	Nuts seeds, spinach, dark chocolate
6.	Zinc	Modulation of glutamate and GABA neurotransmitters	Enhances learning memory and neuroplasticity	Meat, shellfish, legumes, and seeds
7.	Iron	Essential for dopamine and serotonin functions	Supports oxygen transport, energy metabolism, and focus	Red meat, spinach, lentils, and fortified grains
8.	Copper	Involved in dopamine and norepinephrine metabolism	Supports neuronal communication and brain energy levels	Shellfish, nuts, seeds, and whole grains
9.	Choline	Key precursors for acetylcholine synthesis	Vital for memory, focus, and cognitive processing	Eggs, liver, soybean, and eggs
10.	DHA (docosahexaenoic acid)	Structural components of brain cells are essential for neuroplasticity	It enhances memory and learning and reduces the risk of neurodegenerative diseases	Fatty fish (salmon, sardines, mackerels), algae, fish oil
11.	EPA (Eicosapentaenoic acid)	Reduces brain inflammation and supports mood regulation	Helps combat depression anxiety, and improves focus	Fatty fish, krill oil, seaweed, and walnuts
12.	ALA (alpha linolenic acid)	The precursors of DHA and EPA support brain cell integrity	It provides mild cognitive support but is less potent than DHA/EPA	Flaxseed, chia seed, walnut, and hemp seed

12.7 Dietary Patterns and Neurodegeneration

12.7.1 Impact of High-Fat Diets on Cognitive Decline

High-fat diets, particularly those with high levels of saturated fats, are associated with higher neuroinflammation and oxidative stress, both of which have been shown to contribute to cognitive impairment. One longitudinal study determined that older adults who consumed a high-fat, low-fiber diet had cognitive impairment rates that were 50% greater than their peers on a healthy diet (Morris et al. 2015). Subsequent research has also shown that the harmful effects of fat-rich diets go beyond memory loss, since they are implicated in a greater risk of developing Alzheimer's disease and other dementias. For example, another study showed that individuals who consume high-saturated-fat diets are likely to develop brain amyloid plaques, characteristic of Alzheimer's pathology (Scarmeas 2009). These results highlight the significance of food choices in sustaining not only physical well-being but also mental sharpness. Conversely, diets high in antioxidants, omega-3 fatty acids, and fiber, including the Mediterranean diet, have been associated with better cognitive function and slower age-related cognitive decline (Fig. 12.4). A meta-analysis of studies found that adherence to a Mediterranean dietary pattern was associated with a significant decrease in the risk of developing Alzheimer's disease, which may be due to the anti-inflammatory effects of foods in the diet (Scarmeas et al. 2006).

Lifestyle itself, such as consistent physical exercise and social interaction, is highly important for reducing the adverse effects of poor eating habits. Aerobic

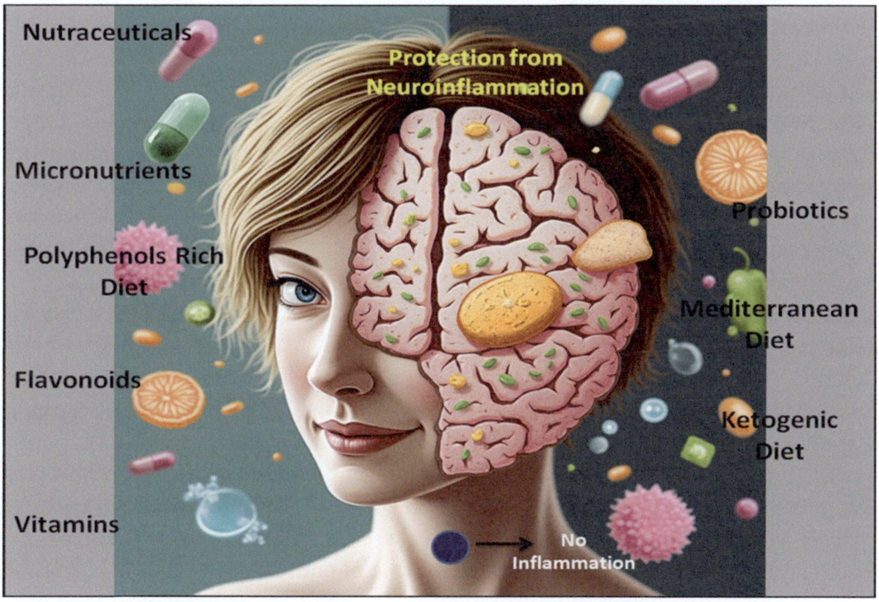

Fig. 12.4 The neuro-nutrition connection: Diets that defend against brain inflammation

exercise increases neurogenesis and the growth of new neurons within the brain, which can reverse the detrimental effects of poor diet (Kramer et al. 2006). Similarly, having good social ties has the ability to activate cognitive functions and offer emotional support, thereby developing a comprehensive means of cognitive well-being. Given this evidence, public health campaigns to promote healthier dietary habits among older adults are of utmost importance. Health education campaigns promoting the advantages of plant foods, whole grains, and unsaturated fats should be prioritized to address the increasing incidence of cognitive impairment. Ultimately, a synergy of dietary modification, exercise, and social interaction may be the key to maintaining cognitive function and improving the quality of life in aging populations.

12.7.2 Benefits of Diets Rich in Complex Carbohydrates and Fiber

Diets rich in complex carbohydrates and fibers are linked to enhanced cognitive performance in the elderly. In a previous study, increased fiber consumption was linked to superior global cognitive function, equivalent to a 15% reduction in the risk of cognitive impairment (Table 12.3) (Liu et al. 2020). The Mediterranean diet, rich in whole grains, fruits, and vegetables, has been associated with a lower incidence of neurodegenerative disorders (Martínez-Lapiscina et al. 2013). In addition, scientists have identified the antioxidant content of fruits and vegetables, especially berries and greens, which could help shield brain cells from oxidative stress (Mariani et al. 2005). These compounds not only decrease inflammation but also increase neuroplasticity and the capacity of the brain to adapt and transform in reaction to new knowledge, which is essential for maintaining function in good health as one grows older (Klimova et al. 2020).

Furthermore, lifestyle characteristics such as physical activity and socialization were also identified to influence cognition synergistically with dietary habits. Regular moderate exercise, a few times a week, has been associated with greater hippocampal volume, which is responsible for memory and learning (Erickson et al. 2011). Likewise, having active social connections leads to mental stimulation, which is crucial for maintaining the sharpness of mind. In addition, recent studies have indicated that the timing of meals can contribute to cognitive wellbeing. For example, Intermittent Fasting (IF) has been reported to cause autophagy, a mechanism that helps remove damaged cellular components, enhancing general brain function (Longo and Mattson 2014). Such a dietary regimen promotes cellular renewal and can exert neuroprotective effects, diminishing the possibility of cognitive impairment. As the aging population increases, these results can inform public health guidelines targeting dietary styles that promote brain wellbeing. A balanced diet, physical exercise, and social interaction can improve cognitive ability, and the adverse effects of neurodegenerative diseases can be minimized. Furthermore, health promotion programs that promote the role of nutrition and lifestyle habits can give individuals the power to inform their own health choices in favor of both their body and mind (Prokopidis et al. 2022).

Table 12.3 Dietary patterns and brain health

S. no.	Dietary pattern	Key characteristics	Impact on brain health	Potential risks
1.	Mediterranean diet	High in fruits, vegetables, whole grains, nuts, olive oil, and fish; moderate in dairy, poultry, and red wine; low in red meat and processed foods	It reduces the risk of cognitive decline, enhances memory and learning, and lowers Alzheimer's risk	Adjustments may be required for people with specific dietary restrictions or allergies
2.	DASH diet (dietary approaches to stop hypertension)	Rich in vegetables, fruits, whole grains, lean protein, and low-fat dairy; low in sodium, red meat, and added sugar	Improved vascular health and lowered stroke risk	Sustaining those accustomed to high-sodium diets may be challenging
3.	MIND diet (Mediterranean-DASH intervention for neurodegenerative delay)	The combination of Mediterranean and DASH diets emphasizes leafy greens, berries, nuts, whole grains, fish, and olive oil	It reduces Alzheimer's risk, slows cognitive decline, and improves brain function	Requires long-term commitment and meal planning
4.	Ketogenic diet	High levels of fat, moderate levels of protein, and very low levels of carbohydrates promote ketone production as an alternative brain fuel	It may support brain energy metabolism, reduce neuroinflammation, improve focus, and aid in neurodegenerative conditions	Potential nutrient deficiencies and the risk of the keto flu may not be sustainable
5.	Plant-based diet (vegan/vegetarian)	Emphasizes fruits, vegetables, legumes, whole grains, nuts, and seeds, avoids animal products (vegan), or limits them (vegetarian).	High levels of antioxidants and fibers may reduce neuroinflammation and slow brain aging	Risk of deficiencies in B
6.	Western diet (high-processed, high sugar diet)	High in processed foods, refined sugars, unhealthy foods, and red meats; low in fibers, antioxidants, and essential nutrients	Increased risk of cognitive decline, neuroinflammation, insulin resistance, and Alzheimer disease	Lead to metabolic disorders, obesity, and poor brain function over time

(continued)

Table 12.3 (continued)

S. no.	Dietary pattern	Key characteristics	Impact on brain health	Potential risks
7.	Intermittent fasting (time restricted eating alternate day fasting)	Periodic fasting with defined eating windows promotes metabolic flexibility during autophagy	Enhanced neuroprotection improves cognitive function and reduces the risk of neurodegeneration	It may cause initial fatigue or irritability, which is not suitable for everyone (e.g., diabetic, underweight individuals)
8.	Paleo diet	Focuses on whole foods, lean meats, fish, vegetables, fruits, nuts, and seeds, excludes grains, dairy and processed foods	Provide side anti-inflammatory benefits and stabilizes blood sugar levels, which may support brain health	It can be restrictive, leading to potential deficiencies in calcium, vitamin D, and fiber

12.8 Mechanisms Linking Nutrition to Neuroplasticity

12.8.1 Influence of Diet on Synaptic Plasticity and Neurogenesis

Diet components can significantly influence neuroplasticity and neurogenesis. A systematic review has suggested that omega-3 fatty acids increase synaptic plasticity and support neurogenesis in old rodents and humans (Dyall 2015). Additionally, polyphenols present in fruits and vegetables can enhance neurogenesis via anti-inflammatory and antioxidative mechanisms (Sarubbo et al. 2018). Furthermore, the function of the gut-brain axis in neuroplasticity has also gained greater interest. Recent studies have revealed that probiotics and dietary fiber can favorably modulate the composition of the gut microbiota, which, in turn, affects brain health. Rahmannia et al. 2024 reported that certain probiotic strains not only decrease anxiety and depressive behaviors but also enhance cognitive processes via mechanisms that involve the production of beneficial metabolites and the regulation of neuroinflammatory reactions.

In addition, lifestyle components, such as physical exercise and mental stimulation, also play important roles in promoting neuroplasticity. Exercise enhances the synthesis of brain-derived neurotrophic factor (BDNF), a protein essential for neuronal growth and survival, thereby promoting neurogenesis (Kasselman et al. 2024). A meta-analysis emphasized that even moderate exercise can profoundly improve cognitive function, primarily through neuroplasticity. Including specific dietary patterns such as the Mediterranean diet has also been linked to enhanced cognitive outcomes. High levels of omega-3 fatty acids, antioxidants, and anti-inflammatory nutrients maintain brain health and are associated with a lower risk of neurodegenerative conditions such as Alzheimer's disease (Cunnane et al. 2010). Data from longitudinal studies suggest that adherence to such dietary patterns can help maintain cognitive function in the elderly population (Mcgrattan et al. 2019).

In the future, more research is needed to understand the intricate interplay between diet, lifestyle, and neuroplasticity. Defining the precise constituents or combinations of nutrients most likely to provide positive benefits might lead to new nutritional interventions designed to support cognitive resilience and fend off decline. The more that is learned about these relationships, the more potential might be realized for creating focused interventions capable of leveraging the strength of nutrition to support brain health over the life course.

12.9 Role of Oxidative Stress and Inflammation in Brain Aging

Oxidative stress and inflammation are key components of brain aging. Studies have shown that antioxidant-rich diets can counteract these processes. A longitudinal study found that individuals with an antioxidant-rich diet significantly reduced inflammatory and oxidative stress biomarkers, leading to enhanced cognitive functioning (Saro et al. 2014). In addition, the interaction between diet and brain health extends beyond the contribution of antioxidants. Omega-3 fatty acids, especially EPA and DHA, are essential for facilitating neuroplasticity and memory preservation in the prefrontal cortex, the brain's seat of cognition. Wesołowska et al. 2025 conducted a study showing that subjects who consumed fish frequently experienced decreased rates of cognitive decline compared to those who consumed fish infrequently. The mechanisms underlying this protective effect may be attributed to the anti-inflammatory action of omega-3s, which are known to suppress neuroinflammation, a major cause of neurodegenerative diseases such as Alzheimer's and Parkinson's disease (Walters et al. 2017).

In addition to dietary lipids, the potential of flavonoids extracted from fruits and vegetables has also been investigated. Foods rich in flavonoids, such as berries, citrus fruits, and leafy green vegetables, have been found to have cognitive benefits. According to a meta-analysis by Cheng et al. (2022), participants with a higher consumption of flavonoids showed improved memory recall. They may even protect themselves against the development of dementia. These plant substances are believed to affect brain health by inhibiting oxidative damage and enhancing blood flow (Kalmijn et al. 1997). Lifestyle interventions, including physical exercise, intellectual stimulation, and social interaction, complement nutritional strategies to counteract oxidative stress and chronic inflammation. Physical exercise has been shown to increase brain-derived neurotrophic factor (BDNF) levels, with the consequent facilitation of neurogenesis and synaptic plasticity. Similarly, mentally stimulating activities, including puzzling or acquiring new skills, may support cognitive resilience. Social support through shared experiences and emotional support provides additional protection against cognitive deterioration (Mariani et al. 2005).

As the aging population expands, it is critical to understand the complex interactions between diet, lifestyle, and brain health. Future research should delineate the interaction of different components of the diet and their combined functions in maintaining cognition throughout the aging process. With the promotion of a comprehensive approach that involves dietary habits, exercise, and mental health, we can lay the groundwork for healthy brain aging and enhanced quality of life among older adults (Cunnane et al. 2010).

12.10 Interventions and Nutritional Strategies

12.10.1 Potential of Ketogenic Diets in Neurological Health

Recent studies have identified the capacity of a ketogenic diet to enhance cognitive performance among aging individuals. A high-fat, low-carbohydrate dietary regimen could help increase mitochondrial function and minimize oxidative stress, as indicated by studies in which improved memory and cognitive assessments were reported after 12 weeks of strict dietary adherence (Mazandarani et al. 2023). In addition, a ketogenic diet has been found to have neuroprotective effects, especially in diseases such as Alzheimer's disease. Researchers posit that diet can enable the production of ketone bodies, an alternative energy source for neuronal cells, thereby compensating for the compromised glucose metabolism commonly observed in patients with dementia (Zheng et al. 2021). Clinical trials providing a ketogenic diet to people with mild cognitive impairment have noted definite improvements in cognitive function, in addition to activities of daily living. In addition, the anti-inflammatory effects of a ketogenic diet may be key in preventing age-related neurodegeneration. By reducing the levels of inflammatory markers, including interleukin-6 and tumor necrosis factor-alpha, diet can provide a more conducive environment for brain health (Mazandarani et al. 2023). This action could be constructive for older populations that are more vulnerable to chronic inflammation.

As the body of research on ketogenic diets grows, scientists are exploring their long-term viability and impact. Early results suggested that a ketogenic diet can cause long-term changes in the gut microbiota, which have a direct impact on mental health and cognition (Gao and Zhou 2024). The interaction between these gut microbes and the brain has been termed the "gut-brain axis," deciphering this connection may open new doors for therapeutic interventions for cognitive decline (Mazandarani et al. 2023). Despite these encouraging results, it is essential to be careful when implementing a ketogenic diet, particularly in older patients. Nutritional imbalance, possible lack of basic vitamins, and electrolyte imbalance must be addressed carefully. Practitioners supporting nutritional therapy should first thoroughly evaluate the overall health status and provide individualized meal planning to gain maximum benefits with minimal risks. As additional information becomes available, the ketogenic diet could prove to be an important resource, not just in slowing down the cognitive decline process but also in improving the quality of life for older populations (Klimova et al. 2020).

12.10.2 Importance of Balanced Omega-3 and Omega-6 Fatty Acid Intake

A proper balance between omega-6 and omega-3 fatty acids is necessary for optimal brain function. Clinical studies have indicated that a 2:1 ratio of omega-6 to omega-3 has been shown to reduce inflammation and neurodegeneration. High levels of omega-6 are harmful to neuroinflammation and cause complex cognitive deterioration (Simopoulos 2002). In addition, including sources of omega-3s in the diet, including fatty fish, flaxseeds, and walnuts, can dramatically improve cognitive function. Studies have illustrated that individuals with greater levels of omega-3 intake also have superior memory and learning capabilities, suggesting the contribution of sources to synaptic plasticity, which is essential for efficient memory creation. Moreover, the anti-inflammatory action of omega-3 has been associated with a decreased risk of Alzheimer's disease, with great potential in the prevention and treatment of other neurodegenerative diseases (Yurko-Mauro et al. 2010).

Conversely, more characteristic Western diets with excessive intake of processed foods tend to result in an overload of omega-6 fatty acids from vegetable oils and snacks. This imbalance not only does this overbalance enhances inflammation but can also interfere with the subtle signaling processes of the brain, impacting mood and cognitive function. Correcting this imbalance with dietary modifications, where omega-3-rich foods are emphasized, and sources of omega-6 are minimized, may prove to be an intelligent strategy for maintaining brain health (Simopoulos and Cleland 2003). In addition, supplementation with omega-3 fatty acids has become an increasingly popular treatment for mental health, with the potential efficacy being proposed in studies for the control of conditions, including depression and anxiety. Mechanisms of the effects of omega-3 involve modulation of neurotransmitter function and reduction in inflammation, highlighting the significance of this role beyond simple nutrition ((Martins et al. 2008).

With further research development, general trends point toward a greater understanding of the contribution of diet to mental health and the impetus for public health initiatives that encourage populations to recognize the importance of omega fatty acids. Highlighting balanced consumption and educated food selection may eventually provide a foundation for improving cognitive resilience and enhancing general brain health in various populations (Wesołowska et al. 2025).

12.10.3 Supplement vs Whole Foods

Although supplements may allow for a concentrated intake of individual nutrients, whole foods take a balanced and synergistic approach to maintaining brain function with aging. Whole foods contain a complex mix of vitamins, minerals, antioxidants, and other plant compounds that function in combination to support the brain (Muth and Park 2021). For instance, fruits and vegetables contain fiber, vitamins, and phytochemicals with anti-inflammatory and neuroprotective properties (Mcgrattan et al. 2019). Fatty fishes provide omega-3 fatty acids, proteins, and other nutrients.

However, isolated supplements might not provide the entire range of nutrients present in food. Moreover, overconsumption of some supplements can be harmful. Targeted supplementation may be useful in patients with vitamin D deficiency (Joseph et al. 2009). Emphasizing a whole food-based diet high in nutrient-dense whole foods is usually the best way to sustain cognitive function and brain health throughout life. Whole foods also provide fiber, hydration, and increased satiety from meals. Whole foods also include complex mixtures of phytochemicals and bioactive compounds that are synergistic and collectively contribute to health effects, which might not be entirely achieved through isolated supplements (Igase et al. 2010). The natural balance and interactions between nutrients in whole foods may increase their absorption and use by the body, possibly providing better protection against age-related cognitive decline.

12.10.4 Emerging Research on Neuroprotective Compounds

New research on neuroprotective compounds for healthy brain aging has highlighted some encouraging directions. Antioxidants, such as flavonoids in berries and dark chocolate, have emerged as potential weapons against oxidative stress and inflammation contributing to cognitive impairment. Omega-3 fatty acids, especially DHA, maintain the structure and function of neurons (Sharkey et al. 2008). Curcumin from turmeric has been proven to be an anti-inflammatory and antioxidant that potentially safeguards against neurodegenerative changes. Resveratrol in grapes and red wine enhances memory and hippocampal function. Other compounds, including sulforaphane from cruciferous crops and EGCG from green tea, have been associated with neuroprotection in preclinical studies. Although more human trials are required, these natural substances hold promising potential for the facilitation of healthy brain aging through dietary intervention and strategic supplementation regimens (Vauzour et al. 2016).

Further research on the optimal dosing and bioavailability of these compounds is required to ensure their full potential benefits to brain health (Melzer et al. 2021). Combination treatment with several neuroprotective phytochemicals can provide synergistic benefits and improve overall cognitive function. Longitudinal studies across a variety of different population groups are essential to ascertain the safety and efficacy of these natural treatments in preventing age-related cognitive decline and neurodegenerative diseases (Vauzour et al. 2016).

12.11 Future Directions for Research in Nutritional Neuroscience

Longitudinal studies with explicit causative associations between certain nutrients and cognitive endpoints should be the focus of future studies. In addition, investigation of individualized nutritional interventions based on genetic and metabolic characteristics may offer personalized strategies for preventing cognitive decline in older persons (Seago et al. 2025). Such personalized nutritional strategies could revolutionize our approach to cognitive health, allowing for more effective risk assessment and management tailored to each individual's unique biochemistry. For instance, examining the impact of omega-3 fatty acids, antioxidants, and B vitamins and genetic predispositions may provide critical insights into how these nutrients influence neuroplasticity and memory. Additionally, exploring the synergistic action of different food components within a broader dietary pattern may further clarify their combined actions in maintaining cognitive function (Cross et al. 2024). Therefore, there is a need to integrate technology into this research framework. The use of wearable technology for continuous tracking of diet and exercise, combined with cognitive assessment apps, may allow real-time data acquisition that identifies developing patterns and possibly causative relationships (Linseisen et al. 2025). Thus, the social determinants cannot be ignored. Psychological well-being and community involvement are significant factors that may influence cognitive health, adding depth to the understanding of how nutrition interacts with broader environmental and social networks.

In addition, creating educational outreach programs to educate older adults and caregivers on the effects of nutrition on brain health may enable them to make informed food choices. Public health campaigns should focus on the necessity of consuming a varied diet that is full of whole foods while offering helpful recipes and advice for meal preparation (Parrott et al. 2020). Lastly, interaction among nutritionists, neurologists, and geriatricians is important for interpreting the results of these studies and converting them into effective clinical practices that value cognitive health among older adults. Through interdisciplinary collaborations, novel solutions can be created that not only alleviate nutritional deficiencies but also enhance overall brain resilience and longevity (Horne et al. 2021).

12.12 Conclusion

The aging brain is considerably affected by dietary consumption and metabolic changes. Cognitive function can be optimized, and neurodegenerative diseases can be prevented through the optimal intake of macronutrients and micronutrients, with evidence that carbohydrates, proteins, fats, vitamins, and minerals are substantial. Additionally, evidence suggests that certain dietary patterns, including a Mediterranean diet of high fruit, vegetables, whole grains, and healthy fat intake, are related to a reduced risk of cognitive decline. This diet stresses the significance of high-quality food sources and underscores the role of total diet quality. The addition of omega-3 fatty acids in fish and flaxseeds has also been vital for sustaining brain health through cellular structure support and induction of anti-inflammatory mechanisms.

Physical exercise reinforces such foods, increases blood supply to the brain, facilitates neuroplasticity, and triggers the growth of new neurons via mechanisms including neurogenesis. Exercise has a double benefit of not only aiding weight control but also mood, as well as reducing symptoms of depression and anxiety, both of which play a crucial role in influencing cognitive processes. Besides diet and exercise, the importance of hydration cannot be overlooked. Proper fluid intake is essential for peak cognitive function because dehydration can result in concentration, memory, and overall cognitive process problems. Easy habits, such as adequate daily water, can significantly affect mental acuity and energy levels. Social stimulation and mental challenges are critical for preserving cognitive function. Brain-stimulating activities, such as puzzling, reading, or skill mastery, combined with routine social interaction, can potentially preserve neural connections and delay cognitive decline. Engaging in purposeful conversations and staying active within the community provide a sense of purpose and belonging, further developing mental resilience. A multi-faceted strategy toward dietary control, physical exercise, hydration, and mental stimulation can be combined as a defense mechanism against the brain's natural aging. Through recognizing and utilizing these factors, individuals can intervene to influence their cognitive health in later life.

References

Alavi Naeini AM, Feizi A, Aghaye Ghazvini MR, Djazayery A, Elmadfa I, Djalali M, Barekatain M (2013) The effect of antioxidant vitamins E and C on cognitive performance of the elderly with mild cognitive impairment in Isfahan, Iran: a double-blind, randomized, placebo-controlled trial. Eur J Nutr 53(5):1255–1262. https://doi.org/10.1007/s00394-013-0628-1

Barbagallo M, Dominguez L (2010) Magnesium and aging. Curr Pharm Des 16(7):832–839. https://doi.org/10.2174/138161210790883679

Blumen HM, Blumen HM, Blumen HM, Blumen HM, Jayakody O, Ayers E, Barzilai N, Habeck C, Milman S, Stern Y, Weiss EF, Verghese J, Verghese J, Verghese J, Verghese J (2024) Cognitive reserve proxies are associated with age-related cognitive decline – not age-related gait speed decline. Neurobiol Aging 141:46–54. https://doi.org/10.1016/j.neurobiolaging.2024.05.012

Brown A, Corner M, Crewther S, Crewther D (2019) Age related decline in cortical multifocal flash VEP: latency increases shown to be predominately magnocellular. Front Aging Neurosci 10(Pt 7). https://doi.org/10.3389/fnagi.2018.00430

Camandola S, Mattson MP (2017) Brain metabolism in health, aging, and neurodegeneration. EMBO J 36(11):1474–1492. https://doi.org/10.15252/embj.201695810

Chang C-Y, Ke D-S, Chen J-Y (2009) Essential fatty acids and human brain. Acta Neurol Taiwanica 18(4). https://doi.org/10.29819/ant.200912.0002

Chen X, Maguire B, Brodaty H, O'Leary F (2019) Dietary patterns and cognitive health in older adults: a systematic review. J Alzheimers Dis 67(2):583–619. https://doi.org/10.3233/jad-180468

Cheng, N., Zhang, Y., Zhang, X., & Zhang, Y. (2022). Dietary flavonoids and human cognition: A meta-analysis. Frontiers in Nutrition, 9, 1008502. https://doi.org/10.3389/fnut.2022.1008502

Chen F, Wang J, Cheng Y, Li R, Wang Y, Chen Y, Scott T, Tucker KL (2024) Magnesium and cognitive health in adults: a systematic review and meta-analysis. Adv Nutr 15(8):100272. https://doi.org/10.1016/j.advnut.2024.100272

Cherdak MA (2023) Brain aging. Probl Gerosci 2:71–79. https://doi.org/10.37586/2949-4745-2-2023-71-79

Coelho-Júnior HJ, Picca A, Calvani R, Landi F, Marzetti E (2021) Protein intake and cognitive function in older adults: a systematic review and meta-analysis. Nutr Metab Insights 14:117863882110223. https://doi.org/10.1177/11786388211022373

Cross V, Stanford J, Gómez-Martín M, Collins CE, Robertson S, Clarke ED (2024) Do personalized nutrition interventions improve dietary intake and risk factors in adults with elevated cardiovascular disease risk factors? A systematic review and meta-analysis of randomized controlled trials. Nutr Rev. https://doi.org/10.1093/nutrit/nuae149

Cunnane S, Nugent S, Roy M, Courchesne-Loyer A, Croteau E, Tremblay S, Castellano A, Pifferi F, Bocti C, Paquet N, Begdouri H, Bentourkia M, Turcotte E, Allard M, Barberger-Gateau P, Fulop T, Rapoport SI (2010) Brain fuel metabolism, aging, and Alzheimer's disease. Nutrition 27(1):3–20. https://doi.org/10.1016/j.nut.2010.07.021

Demographics Collaborators. (2020). Global age-sex-specific fertility, mortality, healthy life expectancy (HALE), and population estimates in 204 countries and territories, 1950–2019: A comprehensive demographic analysis for the Global Burden of Disease Study 2019. The Lancet, 396(10258), 1160–1203. https://doi.org/10.1016/S0140-6736(20)30977-6

Devore EE, Breteler MMB, Grodstein F, Kang JH (2012) Dietary intakes of berries and flavonoids in relation to cognitive decline. Ann Neurol 72(1):135–143. https://doi.org/10.1002/ana.23594

Douaud G, Refsum H, Smith AD, Nichols TE, De Jager CA, Jacoby R, Smith SM (2013) Preventing Alzheimer's disease-related gray matter atrophy by B-vitamin treatment. Proc Natl Acad Sci 110(23):9523–9528. https://doi.org/10.1073/pnas.1301816110

Dyall SC (2015) Long-chain omega-3 fatty acids and the brain: a review of the independent and shared effects of EPA, DPA and DHA. Front Aging Neurosci 7. https://doi.org/10.3389/fnagi.2015.00052

Erickson KI, Heo S, Chaddock L, Kim JS, Vieira VJ, Mcauley E, Alves H, Martin SA, Voss MW, White SM, Mailey E, Szabo A, Basak C, Pence BD, Prakash RS, Kramer AF, Woods JA, Wojcicki TR (2011) Exercise training increases size of hippocampus and improves memory. Proc Natl Acad Sci 108(7):3017–3022. https://doi.org/10.1073/pnas.1015950108

Gao Y, Zhou T (2024) Depressive disorder prevention by DIET crosslink: diet, intestinal microbiota, enzyme and traits of metabolism. Trends Food Sci Technol 150:104596. https://doi.org/10.1016/j.tifs.2024.104596

Gault VA, Hölscher C (2008) Protease-resistant glucose-dependent insulinotropic polypeptide agonists facilitate hippocampal LTP and reverse the impairment of LTP induced by beta-amyloid. J Neurophysiol 99(4):1590–1595. https://doi.org/10.1152/jn.01161.2007

Gilsenan MB, Dye L, De Bruin EA (2009) The influence of carbohydrate on cognitive performance: a critical evaluation from the perspective of glycaemic load. Br J Nutr 101(7):941–949. https://doi.org/10.1017/s0007114508199019

Gómez-Pinilla F (2008) Brain foods: the effects of nutrients on brain function. Nat Rev Neurosci 9(7):568–578. https://doi.org/10.1038/nrn2421

Haag M (2003) Essential fatty acids and the brain. Can J Psychiatry 48(3):195–203. https://doi.org/10.1177/070674370304800308

Harenbrock J, Holling H, Reid G, Koychev I (2023) A meta-analysis of the relationship between sleep and β-Amyloid biomarkers in Alzheimer's disease. Biomark Neuropsychiatry 9:100068. https://doi.org/10.1016/j.bionps.2023.100068

Hawken J (2020) Why we sleep by Matthew Walker. Pract Neurol 20(4):345. https://doi.org/10.1136/practneurol-2019-002471

Horne JR, Nielsen DE, Madill J, Robitaille J, Vohl M-C, Mutch DM (2021) Guiding global best practice in personalized nutrition based on genetics: the development of a nutrigenomics care map. J Acad Nutr Diet 122(2):259–269. https://doi.org/10.1016/j.jand.2021.02.008

Hoyer S (2004) Causes and consequences of disturbances of cerebral glucose metabolism in sporadic Alzheimer disease: therapeutic implications. Adv Exp Med Biol 541:135–152. https://doi.org/10.1007/978-1-4419-8969-7_8

Igase M, Ogushi Y, Mizoguchi T, Ueki A, Miki T (2010) Brain aging and nutrition. Anti-Aging Med 7(14):167–173. https://doi.org/10.3793/jaam.7.167

Joseph J, Cole G, Head E, Ingram D (2009) Nutrition, brain aging, and neurodegeneration. J Neurosci 29(41):12795–12801. https://doi.org/10.1523/jneurosci.3520-09.2009

Kalmijn S, Hofman A, Witteman JCM, Ott A, Breteler MMB, Launer LJ (1997) Dietary fat intake and the risk of incident dementia in the Rotterdam study. Ann Neurol 42(5):776–782. https://doi.org/10.1002/ana.410420514

Kasselman LJ, Peltier MR, De Leon J, Reiss AB (2024) Cognitive function and the consumption of probiotic foods: a National Health and Nutrition Examination Survey Study. Nutrients 16(21):3631. https://doi.org/10.3390/nu16213631

Kim H, Barber SJ (2022) The age-related positivity effect in cognition: a review of key findings across different cognitive domains. Elsevier, pp 125–164. https://doi.org/10.1016/bs.plm.2022.08.004

Kirkland JL, Tchkonia T (2017) Cellular senescence: a translational perspective. EBioMedicine 21(132–147):21–28. https://doi.org/10.1016/j.ebiom.2017.04.013

Klimova B, Dziuba S, Cierniak-Emerych A (2020) The effect of healthy diet on cognitive performance among healthy seniors – a mini review. Front Hum Neurosci 14(155). https://doi.org/10.3389/fnhum.2020.00325

Koletzko B (2005) Developmental origins of adult disease: Barker's or Dörner's hypothesis? Am J Hum Biol 17(3):381–382. https://doi.org/10.1002/ajhb.20139

Kramer AF, Erickson KI, Colcombe SJ (2006) Exercise, cognition, and the aging brain. J Appl Physiol 101(4):1237–1242. https://doi.org/10.1152/japplphysiol.00500.2006

Krebs-Smith SM, Guenther PM, Subar AF, Kirkpatrick SI, Dodd KW (2010) Americans do not meet federal dietary recommendations. J Nutr 140(10):1832–1838. https://doi.org/10.3945/jn.110.124826

Krikorian R, Kalt W, Shukitt-Hale B, Nash TA, Vinqvist-Tymchuk MR, Joseph JA, Shidler MD (2010) Blueberry supplementation improves memory in older adults. J Agric Food Chem 58(7):3996–4000. https://doi.org/10.1021/jf9029332

Lakhman S, Gueits PG, Murzello A (2024) Physical activity and dietary interventions for mental health in ageing and longevity. Springer Nature Switzerland. https://doi.org/10.1007/978-3-031-68513-2_5

Lee JJ, Sultana N, Nishita C (2024) Participant assessment of an alternative flexible congregate nutrition meal program for older adults. J Nutr Gerontol Geriatr 43(2):1–17. https://doi.org/10.1080/21551197.2024.2367972

Linseisen J, Renner B, Gedrich K, Wirsam J, Holzapfel C, Lorkowski S, Watzl B, Daniel H, Leitzmann M, Society W (2025) Data in personalized nutrition: bridging biomedical, psychobehavioral, and food environment approaches for population-wide impact. Adv Nutr (Bethesda, Md.). https://doi.org/10.1016/j.advnut.2025.100377

Liu Y-P, Tzeng N-S, Kao Y-C, Chien W-C, Chung C-H, Chang H-A (2020) Are anticholinergic medications associated with increased risk of dementia and behavioral and psychological symptoms of dementia? A nationwide 15-year follow-up cohort study in Taiwan. Front Pharmacol 11(2). https://doi.org/10.3389/fphar.2020.00030

Lobuono DL, Milovich M (2023) A scoping review of nutrition health for older adults: does technology help? Nutrients 15(20):4402. https://doi.org/10.3390/nu15204402

Longo VD, Mattson MP (2014) Fasting: molecular mechanisms and clinical applications. Cell Metab 19(2):181–192. https://doi.org/10.1016/j.cmet.2013.12.008

Lyons BP (2013) Nutrition education intervention with community-dwelling older adults: research challenges and opportunities. J Community Health 39(4):810–818. https://doi.org/10.1007/s10900-013-9810-x

Mariani E, Polidori MC, Cherubini A, Mecocci P (2005) Oxidative stress in brain aging, neurodegenerative and vascular diseases: an overview. J Chromatogr B 827(1):65–75. https://doi.org/10.1016/j.jchromb.2005.04.023

Martínez-Lapiscina EH, Ros E, Valls-Pedret C, San Julián B, Estruch R, Martinez-Gonzalez MÁ, Toledo E, Salas-Salvadó J, Clavero P, Sanchez-Tainta A (2013) Mediterranean diet improves cognition: the PREDIMED-NAVARRA randomised trial. J Neurol Neurosurg Psychiatry 84(12):1318–1325. https://doi.org/10.1136/jnnp-2012-304792

Martins RDP, Braga HDC, Da Silva AP, Dalmarco JB, De Bem AF, Dos Santos ARS, Dafre AL, Pizzolatti MG, Latini A, Aschner M, Farina M (2008) Synergistic neurotoxicity induced by methylmercury and quercetin in mice. Food Chem Toxicol 47(3):645–649. https://doi.org/10.1016/j.fct.2008.12.020

Marzola P, Melzer T, Pavesi E, Brocardo PS, Gil-Mohapel J (2023) Exploring the role of neuroplasticity in development, aging, and neurodegeneration. Brain Sci 13(12):1610. https://doi.org/10.3390/brainsci13121610

Mazandarani M, Lashkarbolouk N, Ejtahed H-S, Qorbani M (2023) Does the ketogenic diet improve neurological disorders by influencing gut microbiota? A systematic review. Nutr J 22(1). https://doi.org/10.1186/s12937-023-00893-2

Mcgrattan AM, Passmore P, Woodside JV, Kee F, Mcevoy CT, Mckinley MC, Mcguinness B (2019) Diet and inflammation in cognitive ageing and Alzheimer's disease. Curr Nutr Rep 8(2):53–65. https://doi.org/10.1007/s13668-019-0271-4

Melzer TM, Gil-Mohapel J, Brocardo PS, Manosso LM, Yau S-Y (2021) In pursuit of healthy aging: effects of nutrition on brain function. Int J Mol Sci 22(9):5026. https://doi.org/10.3390/ijms22095026

Metcalfe JJ, Kownacki C, Mccaffrey J, Prescott MP, Schumacher M (2021) Community-based culinary and nutrition education intervention promotes fruit and vegetable consumption. Public Health Nutr 25(2):1–13. https://doi.org/10.1017/s1368980021003797

Morris MC, Tangney CC, Wang Y, Sacks FM, Bennett DA, Aggarwal NT (2015) MIND diet associated with reduced incidence of Alzheimer's disease. Alzheimers Dement 11(9):1007–1014. https://doi.org/10.1016/j.jalz.2014.11.009

Moya R (2018) P-334 – brain nutrition, aging and neuroplasticity. Free Radic Biol Med 120:S146. https://doi.org/10.1016/j.freeradbiomed.2018.04.481

Muth A-K, Park SQ (2021) The impact of dietary macronutrient intake on cognitive function and the brain. Clin Nutr 40(6):3999–4010. https://doi.org/10.1016/j.clnu.2021.04.043

Ng A, Markworth A (2021) P67 virtual nutrition education for older adults: a food smarts intervention in citywide congregate meal sites. J Nutr Educ Behav 53(7):S55. https://doi.org/10.1016/j.jneb.2021.04.126

Parrott MD, Greenwood CE, Laurin D, Carmichael P-H, Belleville S, Gaudreau P, Ferland G, Kergoat M-J, Fiocco AJ, Anderson ND, Morais JA (2020) The association between dietary pattern adherence, cognitive stimulating lifestyle, and cognitive function among older adults from the Quebec longitudinal study on nutrition and successful aging. J Gerontol B 76(3):444–450. https://doi.org/10.1093/geronb/gbaa178

Phillips SM, Van Loon LJC (2011) Dietary protein for athletes: from requirements to optimum adaptation. J Sports Sci 29(sup1):S29–S38. https://doi.org/10.1080/02640414.2011.619204

Piolatto M, Squazzoni F, Bianchi F, Rota M, Marengoni A, Akbaritabar A (2022) The effect of social relationships on cognitive decline in older adults: an updated systematic review and meta-analysis of longitudinal cohort studies. BMC Public Health 22(1). https://doi.org/10.1186/s12889-022-12567-5

Prokopidis K, Giannos P, Ispoglou T, Witard OC, Isanejad M (2022) Dietary fiber intake is associated with cognitive function in older adults: data from the National Health and Nutrition Examination Survey. Am J Med 135(8):e257–e262. https://doi.org/10.1016/j.amjmed.2022.03.022

Rahmannia M, Aalipour MA, Sadeghian M, Kheradmand A, Aslani HR, Mirzaei R, Goudarzi M, Nasiri MJ, Poudineh M, Sechi LA, Shahidi Bonjar AH (2024) Strain-specific effects of probiotics on depression and anxiety: a meta-analysis. Gut Pathogens 16(1). https://doi.org/10.1186/s13099-024-00634-8

Ratey JJ, Loehr JE (2011) The positive impact of physical activity on cognition during adulthood: a review of underlying mechanisms, evidence and recommendations. Rev Neuro 22(2):171–185. https://doi.org/10.1515/rns.2011.017

Redolfi A, Manset D, Bosco P, Frisoni G (2016) Brain investigation and brain conceptualization. Funct Neurol 28(3):175–190. https://doi.org/10.11138/fneur/2013.28.3.175

Sano M, Ernesto C, Thomas RG, Klauber MR, Schafer K, Grundman M, Thal LJ (1997) A controlled trial of selegiline, alpha-tocopherol, or both as treatment for Alzheimer's disease. New England Journal of Medicine, 336(17), 1216–1222. https://doi.org/10.1056/NEJM199704243361701

Sagar R, Agarwal V, Shukla DK, Bhatia R, Shaji KS, Odell CM, Ganguli A, Swaminathan S, Krishnankutty RP, Dandona R, Rangaswamy T, Krish V, Gururaj G, Prakash S, Murray CJL,

Santomauro D, Varghese M, Dua T, Whiteford H et al (2019) The burden of mental disorders across the states of India: the global burden of disease study 1990–2017. Lancet Psychiatry 7(2):148–161. https://doi.org/10.1016/s2215-0366(19)30475-4

Saro I, Sosa PA, González-Pérez MA, García-Verdugo C (2014) Patterns of genetic diversity in Phoenix canariensis, a widespread oceanic palm (species) endemic from the Canarian archipelago. Tree Genet Genomes 11(1). https://doi.org/10.1007/s11295-014-0815-0

Sartori S, Singewald N, Whittle N, Hetzenauer A (2011) Modulation of magnesium deficiency-induced anxiety and HPA axis dysregulation by therapeutic drug treatment. BMC Pharmacol 11(S2). https://doi.org/10.1186/1471-2210-11-s2-a40

Sarubbo F, Moranta D, Pani G (2018) Dietary polyphenols and neurogenesis: molecular interactions and implication for brain ageing and cognition. Neurosci Biobehav Rev 90:456–470. https://doi.org/10.1016/j.neubiorev.2018.05.011

Schwingshackl L, Hoffmann G (2020) Mediterranean Diet, Brain and Muscle: Olive Polyphenols and Resveratrol Protection in Neurodegenerative and Neuromuscular Disorders. Frontiers in Nutrition 7:123. https://doi.org/10.3389/fnut.2020.00123

Scarmeas N (2009) Physical activity, diet, and risk of Alzheimer disease. JAMA 302(6):627. https://doi.org/10.1001/jama.2009.1144

Scarmeas N, Luchsinger JA, Mayeux R, Stern Y, Tang M (2006) Mediterranean diet and risk for Alzheimer's disease. Ann Neurol 59(6):912–921. https://doi.org/10.1002/ana.20854

Seago ER, Rego MLM, Davy BM, Katz B (2025) Differential association of ultraprocessed food categories with risk of developing cognitive impairment in middle-aged and older adults in a longitudinal panel study. Am J Clin Nutr. https://doi.org/10.1016/j.ajcnut.2025.02.008

Sharkey JR, Liu R, Laditka S, Laditka JN, Sahyoun N (2008) Perceptions and beliefs of the role of nutrition and diet on brain health. FASEB J 22(S1). https://doi.org/10.1096/fasebj.22.1_supplement.442.1

Simopoulos AP (2002) The importance of the ratio of omega-6/omega-3 essential fatty acids. Biomed Pharmacother 56(8):365–379. https://doi.org/10.1016/s0753-3322(02)00253-6

Simopoulos AP, Cleland LG (2003) Omega-6/omega-3 essential fatty acid ratio: the scientific evidence. Karger, Basel. https://doi.org/10.1159/isbn.978-3-318-01018-3

Singh G, Steinmetz JD, Prakash S, Gupta R, Mutreja P, Bender RG, Dandona L, Tripathi M, Thomas SV, Ambekar A, Dua T, Dey S, Nichols E, Pandey A, Chowdhary N, Mathai M, Srivastava MVP, Kumar GA, Bhatia R et al (2021) The burden of neurological disorders across the states of India: the global burden of disease study 1990–2019. Lancet Glob Health 9(8):e1129–e1144. https://doi.org/10.1016/s2214-109x(21)00164-9

Spreng RN, Turner GR (2018) Structure and function of the aging brain. Center for open science. https://doi.org/10.31219/osf.io/25vbs

Stahl GK, Voigt A, Jonsen K, Maznevski ML (2009) Unraveling the effects of cultural diversity in teams: a meta-analysis of research on multicultural work groups. J Int Bus Stud 41(4):690–709. https://doi.org/10.1057/jibs.2009.85

The Aging Brain (1983) J Gerontol 38(2):245–246. https://doi.org/10.1093/geronj/38.2.245a

Travica N, Hudson I, Pipingas A, Sali A, Scholey A, Ried K (2017) Vitamin C status and cognitive function: a systematic review. Nutrients 9(9):960. https://doi.org/10.3390/nu9090960

Tucker KL (2016) Nutrient intake, nutritional status, and cognitive function with aging. Ann N Y Acad Sci 1367(1):38–49. https://doi.org/10.1111/nyas.13062

Turrini S, Wong B, Eldaief M, Press DZ, Sinclair DA, Koch G, Avenanti A, Santarnecchi E (2023) The multifactorial nature of healthy brain ageing: brain changes, functional decline and protective factors. Ageing Res Rev 88:101939. https://doi.org/10.1016/j.arr.2023.101939

Vauzour D, Sijben J, Hogervorst E, Wesnes K, Perry H, Caberlotto L, Thuret S, Perneczky R, Manach C, Kiliaan AJ, Bowman GL, Van De Rest O, Lucca U, Schuermans J, Clarke R, Mitchell ES, Williams RSB, Williams RJ, Roussel A et al (2016) Nutrition for the ageing brain: towards evidence for an optimal diet. Ageing Res Rev 35:222–240. https://doi.org/10.1016/j.arr.2016.09.010

Walters M, Isaacson R, Caesar E, Hackett K, Mosconi L (2017) Role of nutrition to promote healthy brain aging and reduce risk of Alzheimer's disease. Curr Nutr Rep 6(2):63–71. https://doi.org/10.1007/s13668-017-0199-5

Wang X, Hu FB, Ouyang Y, Liu J, Zhao G, Bao W, Zhu M (2014) Fruit and vegetable consumption and mortality from all causes, cardiovascular disease, and cancer: systematic review and dose-response meta-analysis of prospective cohort studies. BMJ 349:g4490. https://doi.org/10.1136/bmj.g4490

Wei C-C, Hsieh M-J, Chuang Y-F (2024) The effects of social interaction intervention on cognitive functions among older adults without dementia: a systematic review and meta-analysis. Innov Aging 8(10). https://doi.org/10.1093/geroni/igae084

Wesołowska W, Bachoń E, Doligalska M, Stremel A, Leszyńska A, Linke J, Bałoniak Z, Kozłowska D, Bałoniak J, Tuszyńska W (2025) Omega-3 fatty acids: key players in cognitive function and brain health. J Educ Health Sport 77:56776. https://doi.org/10.12775/jehs.2025.77.56776

Wrigglesworth J, Wu Z, Woods RL, Harding IH, Ward P, Ryan J, Nilaweera D (2021) Factors associated with brain ageing – a systematic review. BMC Neurol 21(1). https://doi.org/10.1186/s12883-021-02331-4

Yannakoulia M, Kontogianni M, Scarmeas N (2014) Cognitive health and Mediterranean diet: just diet or lifestyle pattern? Ageing Res Rev 20:74–78. https://doi.org/10.1016/j.arr.2014.10.003

Yurko-Mauro K, Mccarthy D, Rom D, Nelson EB, Ryan AS, Blackwell A, Salem N, Stedman M (2010) Beneficial effects of docosahexaenoic acid on cognition in age-related cognitive decline. Alzheimers Dement 6(6):456–464. https://doi.org/10.1016/j.jalz.2010.01.013

Zhang C, Zheng X, Huang H, Su C, Zhao H, Yang H, Guo Y, Pan X (2017) A study on the applicability of the health action process approach to the dietary behavior of university students in Shanxi, China. J Nutr Educ Behav 50(4):388–395.e1. https://doi.org/10.1016/j.jneb.2017.09.024

Zhang Y, Wei H, Cronin MJ, He N, Yan F, Liu C (2018) Longitudinal data for magnetic susceptibility of normative human brain development and aging over the lifespan. Data Brief 20(1):623–631. https://doi.org/10.1016/j.dib.2018.06.005

Zheng H, Gao H, Li X, Xu P, Xu Q, Zhang L, Yan J, Li Y, Jiang Q, Zhang X, Li C, Song W, Ning J, Zheng Y, Ji H (2021) Depletion of acetate-producing bacteria from the gut microbiota facilitates cognitive impairment through the gut-brain neural mechanism in diabetic mice. Microbiome 9(1). https://doi.org/10.1186/s40168-021-01088-9

Sleep, Circadian Rhythm, and Brain Health

13

Nandini C. Devaru, Md. Abubakar, Sana Ahmed, Sarika, Mohammad Talha, and Eswara Rao Puppala

13.1 Overview of Sleep and Circadian Rhythms

Sleep is a complex biobehavioral condition marked by shifts in brain electrical patterns, resulting in modified consciousness, diminished sensory perception, and decreased muscle activity. It is a biological process that regulates a wide array of physiological and behavioral functions in nearly all living organisms. Sleep, a reversible behavioral state characterized by reduced responsiveness to external stimuli, plays a vital role in cognitive performance, metabolic homeostasis, immune function, and overall health (Lane et al. 2023). Circadian rhythms, on the other hand, are endogenous ~24-hour cycles driven by an internal clock principally located in the suprachiasmatic nucleus (SCN) of the hypothalamus that synchronizes physiological processes to the external light–dark cycle.

These two systems, sleep, and circadian timing, are deeply interrelated, yet distinct in their regulation and expression. While sleep is homeostatically regulated and sensitive to prior wakefulness, circadian rhythms are entrained primarily by environmental signals, especially light. Disturbance in either system, whether due to lifestyle factors, neurological disorders, or shift work, has been associated with a range of adverse outcomes including mood disturbances, cognitive impairment,

Nandini C. Devaru acquired first position in contribution.

N. C. Devaru
Department of Pharmacology, JSS College of Pharmacy, JSSAHER,
Mysuru, Karnataka, India

Md. Abubakar
Department of Pharmacology and Toxicology, National Institute of Pharmaceutical Education and Research, Hajipur, Vaishali, Bihar, India

A. Kumar Singh, S. Nand Rai (eds.), *Rejuvenating the Brain: Nutraceuticals, Autophagy, and Longevity*, Nutritional Neurosciences,
https://doi.org/10.1007/978-981-95-2790-8_13

metabolic syndrome (Uthaiah et al. 2022), and increased risk for neurodegenerative diseases (Goel et al. 2013).

The current chapter provides a wide-ranging overview of the mechanisms underlying sleep and circadian rhythms, explores their interdependence, and highlights emerging evidence linking their dysregulation to disease and effect of these two systems on brain health. Understanding the intricate interplay between these systems offers promising avenues for therapeutic interventions and chronobiological strategies to enhance health and well-being.

13.1.1 Importance of Sleep on Brain Function

Sleep is a vital physiological process essential for maintaining optimal brain health and function. During sleep, the brain undergoes critical restorative activities, including synaptic remodelling, metabolic detoxification, and memory consolidation (Langille 2019). Adequate and quality sleep supports cognitive functions such as attention, learning, decision-making, and emotional regulation. Conversely, chronic sleep deprivation or disruption has been linked to impairments in neuroplasticity, increased risk of neurodegenerative disorders, mood disturbances, and reduced overall cognitive performance. Emerging research continues to highlight the role of sleep not just as a passive state, but as an active and dynamic contributor to long-term brain health across the lifespan.

13.1.2 Neurobiology of Sleep

13.1.2.1 Sleep Patterns: NREM and REM Sleep

The human sleep cycle consists of two distinct phases: rapid eye movement (REM) sleep and non-rapid eye movement (NREM) sleep. NREM sleep is further categorized into three progressive stages: N1, N2, and N3. Each phase and stage is characterized by specific variations in muscle tone, brain wave activity, and eye movement patterns. Over the course of a typical night, the body transitions through all sleep

S. Ahmed
Department of Pharmacology, Deccan College of Pharmacy Nampally, Hyderabad, Himachal Pradesh, India

Sarika
School of Biological and Environmental Sciences, Shoolini University of Biotechnology and Management Sciences, Solan, Himachal Pradesh, India

M. Talha
Department of Pharmacology & Pharmaceutical Chemistry, Anwarul Uloom College of Pharmacy, New Mallepally, Hyderabad, Andhra Pradesh, India

E. R. Puppala (✉)
Department of Surgery, School of Medicine, St. Louis, MO, USA
e-mail: eswar@wustl.edu

Table 13.1 Key metrics to quantify sleep architecture

Sleep stage	EEG characteristics	Typical adult duration	Functional interpretation (Gottesman et al. 2024)
Stage N1 (NREM)	Low-voltage, mixed-frequency theta wave signals, slow rolling eye movements, and sharp vertex waves	Less than 5% of sleep time	Characterizes the transition from wakefulness to sleep; considered the lightest sleep stage.
Stage N2 (NREM)	Sleep spindles presence (11–15 Hz, lasting 0.5–2 s) and K-complexes (sharp, biphasic delta waves)	About 45% of sleep time	Spindles are involved in memory consolidation and synaptic plasticity. K-complexes help maintain sleep and support memory-related processes.
Stage N3 (NREM/ SWS)	Dominated by high-amplitude, slow delta waves	Around 25% of sleep time	Deepest stage of sleep with high arousal threshold supports hormone regulation, energy restoration, waste clearance, and immune, cardiovascular, and brain function. Declines with age.
REM sleep	Low-amplitude, mixed-frequency activity, and rapid eye movements	Around 25% of sleep time	Characterized by intense brain activity and autonomic variability. Essential for emotional regulation and declarative memory. Reduced REM linked to depression and increased mortality.

stages approximately 4–6 times, with each complete cycle lasting an average of 90-120 minutes (Patel et al. 2025).

Key parameters used to assess sleep architecture and duration are outlined in Table 13.1, with conceptual and content credit attributed to reference article (Gottesman et al. 2024). Sleep is primarily divided into two states: non-rapid eye movement (NREM) sleep and rapid eye movement (REM) sleep. NREM sleep consists of three distinct stages—N1, N2, and N3—each characterized by unique neurophysiological patterns. Sleep progresses through cycles lasting approximately 120 minutes, with individuals experiencing 4–6 cycles per night. According to the Centers for Disease Control and Prevention (CDC), recommended sleep durations are 7–9 hours per night for adults aged 18–64 years and 7–8 hours per night for older adults (Ross et al. 2020). Internal and external stimuli that trigger arousals can disrupt the organization of sleep, commonly referred to as sleep architecture. Sleep pattern evolves across the lifespan, with aging and certain health conditions contributing to increased sleep fragmentation, greater prevalence of light sleep, and significant reductions in N3 (slow-wave sleep (SWS) and REM sleep (Sateia 2014).

Clinical polysomnography (PSG) serves as a key tool in evaluating sleep depth and duration, providing insights into the neurophysiological profiles of various sleep disorder subtypes. These profiles may help elucidate underlying pathophysiological mechanisms and influence both the presentation of symptoms and overall clinical outcomes. Currently, sleep neurophysiology is primarily assessed through metrics such as total sleep time (TST), sleep efficiency (SE), and the percentage of time spent in N1 and N3 sleep stages. Additionally, the odds ratio product (ORP) functions as a continuous measure of sleep depth and propensity, with values ranging from 0 (indicating deep sleep) to 2.5 (representing full wakefulness). Much evidence supports ORP in this regard including a high correlation ($r^2 = 0.98$) between ORP in any given epoch and probability of arousal/awakening (Eide et al. 2021).

13.1.2.2 Brain Regions and Neurotransmitters Regulating Sleep

Several interconnected brain regions control the sleep–wake cycle. The basal forebrain (BF) promotes both wakefulness and sleep through its diverse neurotransmitter systems, including cholinergic, GABAergic, and glutamatergic pathways, acting as a key relay between the reticular activating system and the cortex. The reticular activating system, extending from the medulla to the posterior hypothalamus, sustains wakefulness by relaying sensory inputs to the BF, thalamus, and hypothalamus. The thalamus modulates cortical arousal by relaying glutamatergic sensory signals, playing a central role in sleep–wake transitions. The hypothalamus contains multiple sleep- and wake-related centers: the suprachiasmatic nucleus (SCN) regulates circadian rhythms; the lateral hypothalamus houses wake-promoting hypocretin neurons; and the tuberomammillary nucleus (TMN) maintains wakefulness via histaminergic output, which is suppressed during sleep by GABAergic input from melanin-concentrating hormone neurons. Meanwhile, the preoptic area, particularly the ventrolateral (VLPO) and median preoptic nuclei (MNPO), promotes sleep through inhibitory GABA and galanin projections to arousal centers, with VLPO lesions known to significantly disrupt sleep patterns (Falup-Pecurariu et al. 2021).

Serotonin, norepinephrine (NE), histamine, hypocretins, acetylcholine (ACh), dopamine, glutamate, and GABA are primarily recognized for their wake-promoting functions (Puppala et al. 2022). The locus coeruleus serves as the principal neuronal structure responsible for NE expression, while histamine is synthesized within the tuberomammillary nucleus. Additionally, hypocretins, a class of hypothalamic neuropeptides, play a crucial role in sustaining the wake state (Oh et al. 2019).

The hypothalamus, along with the adjacent basal forebrain, plays a central role in regulating sleep duration through GABA-producing neurons. These GABAergic neurons increase activity during NREM sleep. The GABAergic neurons initiate and maintain sleep by inhibiting arousal-related neurons, including cholinergic cells in the basal forebrain and histaminergic cells in the posterior hypothalamus key regions linked to wakefulness (Steiger 2007). GABA suppresses arousal-related neurotransmitter systems such as norepinephrine (from the locus coeruleus), serotonin (from the raphe nuclei), and hypocretin. These neurotransmitters are typically active during wakefulness but become silent during REM sleep due to GABAergic inhibition. The suppression of these systems not only promotes sleep onset but also supports muscle tone regulation and REM-related phenomena. Notably, the sedative effects of antihistamines highlight the sleep-inducing impact of histaminergic inhibition. Furthermore, GABA's targeted action on serotonin and norepinephrine neurons can induce REM sleep, emphasizing its pivotal role in sleep architecture and control.

13.1.2.3 Mechanisms of Sleep Regulation: Homeostasis and Circadian Rhythms

Sleep is regulated by two primary biological processes: the homeostatic sleep drive and the circadian rhythm. The certain amount of sleep needs for a body is represented by homeostatic process, increasing with prolonged wakefulness and dissipating during sleep, largely influenced by the accumulation of sleep-promoting substances like adenosine. In contrast, the circadian process is governed by the suprachiasmatic nucleus (SCN) of the hypothalamus, which coordinates the physiological functions to the 24-hour light–dark cycle (Dijk and Archer 2009). This internal clock regulates the timing of sleep and wakefulness, independent of prior sleep duration. The interaction between these two systems ensures that sleep occurs at an optimal time and duration to maintain cognitive function, metabolic balance, and overall health. Disruptions in either system can lead to sleep disorders and impaired daytime functioning (Deboer 2018).

13.1.2.3.1 Circadian Drive

In mammals, the regulation of circadian rhythms is ruled by a biological clock located in the suprachiasmatic nucleus (SCN) of the hypothalamus. This region exhibits a daily pattern of neural activity, with increased firing rates during daylight hours and reduced activity at night. At the cellular scale, circadian rhythms are regulated by intricate transcriptional and translational feedback loops, which involve the coordinated expression of core clock genes. These molecular interactions ensure the precise timing of physiological processes and the synchronization of biological rhythms with external environmental cues. In humans, this internal clock naturally follows a cycle slightly longer than 24 hours, known as the free-running period. To align with the 24-hour day, the circadian rhythm is adjusted daily by zeitgebers' external cues that synchronize the internal clock with the environment. The most influential of these cues is the light–dark cycle. Both natural sunlight and artificial light (e.g., screens and electric lighting) act as photic synchronizers. These light signals are detected by specialized retinal cells particularly melanopsin-containing intrinsically photosensitive retinal ganglion cells (ipRGCs) and are relayed to the SCN. In addition to light, non-photic cues such as physical activity can also shift circadian timing, acting as secondary synchronizers across species, including humans (Borbély et al. 1981).

Throughout the day, the circadian system promotes alertness, peaking in the evening just before melatonin secretion begins; this period is known as the wake maintenance zone or the forbidden zone for sleep. After this peak, the alerting signal declines, allowing the drive for sleep to rise, reaching its maximum toward the end of the night when core body temperature is at its lowest. As this sleep drive decreases in the early morning, the body's readiness to awaken increases. Overall, sleep–wake timing and maintenance are strongly influenced by the circadian drive (Taillard et al. 2021).

13.1.2.3.2 Sleep Homeostatic Drive

The sleep homeostatic system generates a physiological need for sleep, often referred to as sleep pressure, which accumulates the longer one stays awake and gradually reduces during sleep typically in an exponential fashion. This mechanism ensures a balance between time spent awake and the body's need for restorative rest, thereby increasing the likelihood of sleep as wakefulness continues and reducing it as sleep progresses. It also plays a key role in regulating the intensity and duration of slow-wave activity (SWA) during deep sleep. Unlike the well-defined circadian system, the exact brain circuits and anatomical site responsible for this homeostatic regulation remain uncertain (Borbély et al. 1981).

This process may involve the build-up of a biochemical "sleep factor" during waking hours that promotes sleep by activating sleep-related neurons and inhibiting those responsible for alertness. One of the leading candidates for this factor is adenosine, a molecule whose concentration rises in the basal forebrain during extended wakefulness and falls during sleep. The influence of adenosine is also supported by the effects of substances like caffeine, which blocks adenosine receptors and delays sleep onset. Specifically, A1 and A2A receptors have been identified as key mediators of adenosine's sleep-promoting actions (Bjorness and Greene 2009).

13.1.3 Circadian Rhythms and the Brain

13.1.3.1 Role of the Suprachiasmatic Nucleus (SCN)

In mammals, the primary circadian pacemaker resides in the *suprachiasmatic nucleus (SCN)* of the hypothalamus, located just above the optic chiasm at the base of the third ventricle. Various experimental strategies have pinpointed the SCN as the seat of the internal biological clock. Initial evidence came from lesion studies damage to the SCN led to the disappearance of daily rhythms in behaviors such as locomotion, drinking, and corticosterone secretion. Similarly, disrupting the SCN eliminated the natural distribution of sleep and wake cycles across the 24-hour day (Patton and Hastings 2018).

Further support for the SCN's role came from transplantation studies. When SCNs from *tau* mutant hamsters (which have shorter circadian cycles) were grafted into wild-type animals, the recipients adopted the donor's circadian timing demonstrating that the SCN sets both rest-activity patterns and intrinsic rhythm duration (Dey et al. 2005).

Although individual SCN neurons each maintain their own circadian rhythm, they can vary widely in period length. Internally, these rhythms are regulated by a molecular feedback loop involving 5–7 key clock genes. When neurons are organized into a network such as in SCN brain slices their rhythms synchronize, and most neurons become active during the subjective daytime, indicating that intercellular communication sharpens the overall precision of the SCN's output.

The SCN in rodents contains roughly 10,000 neurons and is typically thought to generate a unified circadian rhythm. However, exceptions exist. Under certain conditions, different regions of the SCN may operate independently some being directly light-driven, while others continue to function on an intrinsic rhythm. For instance, under continuous bright light, animals may exhibit a split in their activity pattern due to the left and right sides of the SCN cycling out of phase with each other. In such cases, the split components still share the same period but occur in antiphase (Hastings et al. 2019).

The SCN's outputs regulate numerous physiological processes, including the daily sleep–wake cycle. In circadian research, the most reliable indicators of clock function are measurable biological rhythms. In animals, these typically include activity–rest cycles or fluctuations in core body temperature, while in humans, markers like body temperature rhythms or melatonin secretion patterns are commonly used to assess circadian phase.

13.1.3.2 Molecular Mechanisms of Circadian Regulation

Circadian rhythms are intrinsic, 24-hour cycles that regulate a vast array of physiological processes, including sleep–wake cycles, metabolism, hormone secretion, and body temperature. These rhythms are driven by endogenous molecular clocks present in almost every cell and are synchronized to the external environment, primarily through light–dark cycles. At the core of this system lies a transcriptional–translational feedback loop (TTFL) composed of clock genes and proteins (Anna and Kannan 2021).

Positive Limb The central activators of the mammalian circadian clock are the transcription factors CLOCK and BMAL1 (Brain and Muscle ARNT-Like 1). CLOCK and BMAL1 form heterodimers in the cytoplasm, translocate into the nucleus, and bind to E-box elements in the promoters of target genes, promoting transcription of core clock genes such as Period (Per1, Per2) and Cryptochrome (Cry1, Cry2), as well as output genes involved in various cellular processes (Trott and Menet 2018).

Negative Limb Accumulation of the PER and CRY proteins in the cytoplasm leads formation of complexes, and translocate it into the nucleus to prevent the activity of CLOCK:BMAL1, thereby repressing their own transcription in a negative feedback loop. Over time, PER and CRY are degraded by ubiquitin-proteasome pathways, relieving the inhibition on CLOCK:BMAL1 and allowing the cycle to restart. This loop takes approximately 24 hours to complete (Brown et al. 2005).

Stabilization and Fine-Tuning Several kinases, including Casein Kinase 1ε/δ (CK1ε/δ) and Glycogen Synthase Kinase 3 (GSK3), phosphorylate PER and CRY, influence their stability and nuclear entry. REV-ERBα/β and RORα/γ form an auxiliary loop that regulates Bmal1 transcription. REV-ERBs repress, while RORs activate Bmal1 expression, contributing to the robustness and precision of the circadian rhythm.

Post-translational Modifications Beyond transcriptional control, post-translational modifications (PTMs) play a critical role in determining the stability, localization, and activity of clock proteins. These include:

1. Phosphorylation: regulates degradation and nuclear localization (e.g., CK1-mediated phosphorylation of PER).
2. Ubiquitination: tags proteins for degradation (e.g., FBXL3 targets CRY).
3. Acetylation/SUMOylation: modulates transcriptional activity and stability of clock components (Albrecht 2017).

13.1.3.2.1 Peripheral Clocks and Entrainment

Although the suprachiasmatic nucleus (SCN) in the hypothalamus serves as the master pacemaker, peripheral clocks exist in organs such as the liver, heart, and kidney. These peripheral clocks are dragged by signals from the SCN and zeitgebers like feeding time, temperature, and hormones, ensuring a systemic coordination of circadian outputs (Fig. 13.1).

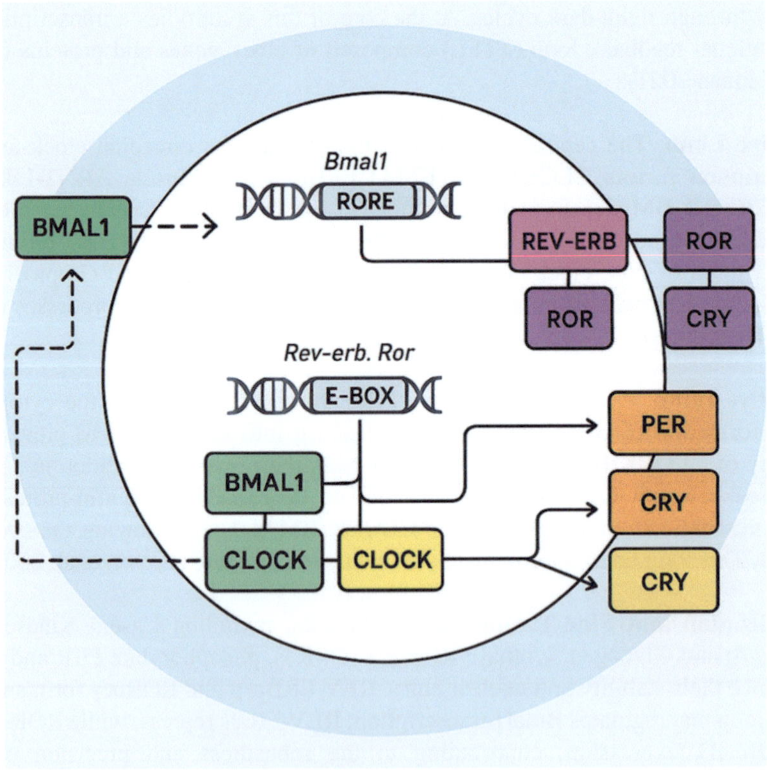

Fig. 13.1 Molecular mechanisms of circadian regulation

13.1.3.3 Synchronization of Peripheral Clocks with the Brain

Circadian clocks consist of both central and peripheral components, working together to regulate daily patterns in physiology and behavior. In mammals, nearly every cell and tissue contain a functioning circadian clock. The central pacemaker resides in the suprachiasmatic nucleus (SCN) of the hypothalamus, a key structure that orchestrates rhythmic activity. The SCN receives direct light signals from the retina via intrinsically photosensitive retinal ganglion cells (ipRGCs), which allows it to generate strong and consistent circadian rhythms. Through a variety of complex pathways some of which remain not fully elucidated, the SCN resets and synchronizes peripheral clocks throughout the body. This coordination occurs through interactions with other brain regions, secretion of signalling molecules, and regulation of behaviors like sleep–wake cycles and feeding–fasting patterns (Puppala et al. 2023).

Unlike peripheral clocks, the SCN is relatively unresponsive to non-photic signals, maintaining its rhythmic integrity even under varying environmental conditions such as temperature changes. In contrast, peripheral clocks are more sensitive to these external influences. The mechanisms by which different tissues synchronize with the SCN vary, as peripheral rhythms arise from both intrinsic molecular clocks and SCN-driven systemic signals. Notably, peripheral oscillators can, under certain conditions, operate independently of the central clock. For example, during environmental light cycle shifts or under cell culture conditions, the SCN maintains stable rhythmicity, while peripheral clocks tend to lose their oscillatory behavior over time. This has been demonstrated in transgenic rat models expressing *luciferase* under the control of the mouse Per1 promoter, where peripheral rhythmicity fades within a few days.

Research has shown that most mammalian cells possess the core molecular machinery necessary for circadian gene regulation. Research on gene expression has demonstrated that clock gene mRNAs exhibit rhythmic accumulation not only within the suprachiasmatic nucleus (SCN) but also across multiple peripheral organs, including the liver, skeletal muscle, kidney, and lungs. Additionally, circadian gene activity has been identified in diverse cell types, such as mononuclear leukocytes and vascular smooth muscle cells, highlighting the widespread influence of molecular circadian regulation. Interestingly, the timing of gene expression in peripheral tissues often exhibits a phase delay or even an opposing phase relative to the same genes in the SCN (Zhang et al. 2020).

13.1.4 Impact of Sleep on Brain Function

Sleep is a fundamental physiological process that plays a critical role in maintaining optimal brain health and function. Adequate sleep is essential for memory formation, synaptic remodelling, emotional regulation, and cognitive performance. Disruptions in sleep architecture can significantly impair these processes, increasing vulnerability to neuropsychiatric and neurodegenerative conditions.

13.1.4.1 Sleep and Memory Consolidation

Sleep facilitates the union of both declaratory (facts and events) and procedural (skills and habits) memories. The hippocampus and neocortex work in coordination during sleep to transfer newly encoded memories into long-term storage. Slow-wave sleep (SWS), which is prominent during the early part of the night, is particularly important for declarative memory consolidation, while REM (rapid eye movement) sleep supports procedural memory and emotional memory processing (Diekelmann and Born 2010). The active system consolidation hypothesis suggests that during SWS, memory traces are repeatedly reactivated in the hippocampus and gradually integrated into the neocortical networks. This reactivation strengthens synaptic connections and stabilizes memory (Rasch and Born 2013).

Sleep supports learning and memory in two key ways: it enhances the ability to acquire new information and is essential for consolidating memories after learning. While acquisition and recall happen during wakefulness, memory consolidation, the process of stabilizing and integrating memories into long-term storage primarily, occurs during sleep.

13.1.4.1.1 Memory Formation Is Typically Divided into Three Stages

- *Acquisition* (learning new information)
- *Consolidation* (stabilizing memories)
- *Recall* (retrieving stored information)

During sleep, the brain is less engaged with external stimuli, allowing it to focus on strengthening neural connections linked to recently learned material. Specific brain-wave patterns during different sleep stages, especially slow-wave and REM sleep, are believed to facilitate this process. Because encoding (learning) and consolidation may compete for shared neural resources, sleep offers an ideal window for uninterrupted consolidation. Research supports this idea interventions like protein synthesis inhibitors or electroconvulsive shocks can alter memory retention if applied soon after learning, underscoring the time-sensitive nature of consolidation. Moreover, memory consolidation may unfold in multiple phases or "waves," each governed by distinct neural plasticity mechanisms with their own timing and molecular requirements.

13.1.4.2 Sleep-Dependent Synaptic Plasticity (SHY)

Sleep is essential for the regulation of synaptic plasticity, the dynamic process by which synapses adjust their strength over time. According to the synaptic homeostasis hypothesis (SHY), wakefulness is associated with a net increase in synaptic strength due to learning and experiential input. In contrast, during sleep—particularly slow-wave sleep (SWS)—a global downscaling of synaptic strength occurs, facilitating cellular homeostasis and optimizing signal-to-noise ratios within neural circuits, thereby supporting cognitive efficiency and memory consolidation. (Tononi and Cirelli 2014).

During REM sleep, however, synaptic potentiation may occur, selectively strengthening synapses associated with salient memories or learned tasks. Thus, sleep supports both synaptic consolidation and synaptic pruning, balancing network efficiency and plasticity.

13.1.4.3 Sleep's Role in Emotional and Cognitive Processing

Sleep profoundly influences emotional regulation and cognitive performance. REM sleep plays a critical role in processing emotional experiences and integrating them into memory while reducing the emotional reactivity associated with those experiences. Sleep deprivation has been shown to impair prefrontal cortex function, top-reducing control over limbic regions such as the amygdala, thereby increasing emotional reactivity, impulsivity, and stress sensitivity.

Additionally, poor sleep negatively impacts attention, executive function, decision-making, and working memory, contributing to cognitive decline and increased risk for mood disorders and neurodegenerative diseases (Krause et al. 2017).

13.1.5 Sleep Deprivation and Brain Dysfunction

Sleep deprivation (SD) significantly impairs brain function and human performance across multiple levels. On a macroscopic level, it primarily affects executive functions, especially during novel or complex tasks. At mesoscopic and macroscopic scales, SD leads to reduced cortical responsiveness and diminished attention. Microscopically, it is associated with an accumulation of adenosine, a neuromodulator that inhibits neural activity and suppresses excitatory neurotransmitters like acetylcholine, contributing to decreased cognitive performance.

While the connections between neural disruptions at different levels remain unclear, it is well-established that SD has widespread effects, with few brain functions remaining unaffected. One key consequence is impaired memory consolidation, especially for newly learned skills or information. Beyond cognitive decline, SD also induces subjective sleepiness, negative mood, and increased stress levels. In severe cases, it may even cause hallucinations or, in animals, death after prolonged deprivation. The underlying mechanisms of SD's impact whether due solely to lack of sleep or stress-related cortisol release are still debated. Some researchers argue that chronic SD acts as a stressor, triggering allostatic load and weakening both the immune and nervous systems. However, other studies show that even without a functional stress system, the adverse effects of SD persist, suggesting stress is not the only factor. Numerous studies have documented deficits in executive function, including verbal fluency, planning, creativity, and novelty detection, along with decreased vigilance and driving performance. Tasks requiring sustained attention or non-automatic responses are particularly vulnerable to sleep loss.

Adenosine has been identified as a key neuromodulator in sleep regulation. It accumulates in the basal forebrain during wakefulness and declines with recovery sleep. Adenosine inhibits excitatory neurotransmitters such as glutamate and acetylcholine, thereby promoting sleep and reducing arousal. Pharmacological studies support its role: adenosine agonists increase sleep, while antagonists like caffeine delay it highlighting AD as a major endogenous sleep-promoting factor.

Chronic sleep deprivation has significant pathophysiological consequences, including persistent inflammation and disruptions in lipid metabolism. Research has established a correlation between self-reported sleep disturbances and the dysregulation of inflammatory markers. Neuroinflammation is characterized by the activation of microglia and astrocytes, leading to the release of pro-inflammatory molecules, which contribute to cognitive impairments that are the result of sleep loss. Furthermore, sleep deprivation influences lipid metabolism, with lipids being particularly susceptible to oxidative stress-induced degradation (Besedovsky et al. 2019).

13.1.5.1 Relationship with Psychiatric Disorders and Immunity

Chronic sleep deprivation (SD), whether due to lifestyle, occupational demands, or underlying pathology, has been strongly concerned in the onset and exacerbation of numerous psychiatric disorders. Longitudinal and cross-sectional studies reported a bidirectional relationship between sleep disturbances and psychiatric conditions such as main depressive disorder (MDD), bipolar disorder, anxiety disorders, and schizophrenia. Functional MRI studies reveal that SD disrupts top-down control from the prefrontal cortex over limbic structures like the amygdala, leading to heightened emotional reactivity and impaired affect regulation (Yoo et al. 2007). Moreover, sleep loss alters monoaminergic neurotransmission including serotonin, dopamine, and norepinephrine systems that are central to mood regulation. Disturbed circadian rhythms that often co-occur with SD have also been associated with increased risk of mood episodes and suicidal behavior in vulnerable populations.

The relationship between sleep deprivation and immune dysfunction further complicates its role in psychiatric illness. Sleep loss leads to dysregulation of both innate and adaptive immunity, evidenced by elevated levels of pro-inflammatory cytokines like, IL-1β, IL-6, and TNF-α, along with less numbers and impaired function of natural killer (NK) cells and T lymphocytes (Freeman et al. 2020). These immune alterations are particularly relevant in psychiatry, as chronic inflammation is increasingly recognized as a contributor to the pathophysiology of disorders like depression and schizophrenia (Harvey et al. 2011). For example, elevated C-reactive protein (CRP) and cytokine levels have been linked to poor antidepressant treatment outcomes, anhedonia, and neurocognitive deficits. Furthermore, inflammation-related changes in blood–brain barrier permeability and microglial activation may mediate the deleterious effects of SD on brain function and behavior (McClung 2013).

Given this complex interplay between sleep, psychiatric states, and immune regulation, addressing sleep disturbances is a promising avenue for both therapeutic intervention and prevention. Behavioral interventions such as cognitive behavioral therapy for insomnia (CBT-I) have shown efficacy in reducing both sleep-related symptoms and comorbid psychiatric symptoms (Taylor et al. 2003). Emerging treatments targeting circadian rhythms, including light therapy and melatonin agonists, offer additional tools for mood stabilization and immune modulation. Future research should aim to unravel the causal mechanisms underlying the neuroimmune consequences of SD, potentially through the identification of inflammatory and neuroendocrine biomarkers. Such insights will be crucial for designing integrative treatments that simultaneously address sleep health, psychiatric symptomatology, and immune resilience.

13.1.6 Circadian Disruptions and Brain Health

The circadian system, governed primarily by the suprachiasmatic nucleus (SCN) of the hypothalamus, orchestrates daily rhythms in sleep–wake cycles, hormonal secretion, metabolism, and cognitive function. When this internal clock is chronically misaligned with environmental cues such as through unbalanced sleep patterns or acquaintance to artificial light at night it can adversely affect brain function. Persistent circadian rhythm disturbances have been allied with impairments in memory, attention, and emotional regulation, underscoring the importance of circadian integrity for optimal neurological health (Evans and Davidson 2013). These disruptions can also alter gene expression patterns in the brain, particularly those involved in synaptic plasticity and neuroprotection, which may render neural networks more vulnerable to degeneration and dysfunction over time.

13.1.6.1 Impact of Shift Work and Jet Lag on Cognition

Occupational factors such as night shift work and frequent trans-meridian travel (jet lag) expose individuals to repeated cycles of circadian misalignment. Studies have evidenced that, shift workers often exhibit deficits in executive function, decision-making, and working memory, likely due to the desynchronization between their endogenous circadian rhythms and obligatory behavioral schedules (Benjamine). Animal models have further demonstrated that forced circadian desynchrony leads to reductions in hippocampal neurogenesis and impaired long-term potentiation both crucial for learning and memory processes. Moreover, chronic jet lag has been linked to decreased cognitive flexibility and altered stress hormone regulation, which may contribute to the long-term cognitive consequences observed in frequent travellers or individuals working in irregular schedules.

13.1.6.2 Circadian Misalignment and Neurodegenerative Diseases

Growing evidence suggests that disrupted circadian rhythms may play a role in the development and progression of neurodegenerative disorders such as Alzheimer's disease (AD), Parkinson's disease (PD), and Huntington's disease. For instance, sleep and circadian disturbances are among the earliest clinical features in AD, preceding memory impairment and amyloid plaque deposition (Musiek et al. 2015). Disrupted light–dark cycles have been shown to accelerate amyloid-beta accumulation and tau pathology in animal models, potentially through alterations in glymphatic clearance mechanisms and neuroinflammation (Xie et al. 2013). Furthermore, patients with PD often display significant circadian dysregulation, including reduced amplitude in melatonin secretion and altered rest–activity patterns, which may exacerbate motor and non-motor symptoms.

13.1.6.3 Role of Melatonin and Light Exposure

Melatonin, a hormone secreted by the pineal gland in response to darkness, plays a pivotal role in maintaining circadian synchrony by signaling night-time to the body. In addition to regulating sleep, melatonin exhibits antioxidant, anti-inflammatory, and neuroprotective properties (Nandini et al. 2021), which may be beneficial in mitigating neurodegenerative processes. Circadian disruption, particularly due to artificial light exposure at night, suppresses melatonin production and interferes with sleep architecture, thereby amplifying the risk of cognitive decline and mood disorders. Light therapy, particularly timed exposure to bright light, has shown promise in restoring circadian alignment and improving cognitive performance, especially in individuals with seasonal affective disorder or circadian rhythm sleep–wake disorders. These results underscore the promising therapeutic implications of chronobiological interventions in enhancing brain health across the lifespan (Reiter et al. 2016).

13.1.7 Sleep and Brain Aging

Sleep architecture undergoes significant alterations with aging, contributing to cognitive decline and increased vulnerability to neurodegenerative conditions. Older adults typically experience less total sleep time, longer sleep onset latency, and increased nocturnal awakenings. One of the most marked changes is the reduction in slow-wave sleep (SWS) and rapid eye movement (REM) sleep both of which are necessary for cognitive reinforcement and synaptic homeostasis. Age-related changes in the prefrontal cortex and thalamus, regions implicated in sleep regulation, have been linked to the deterioration in sleep continuity and depth (Ohayon et al. 2004). These disruptions may not merely be byproducts of aging but active contributors to cognitive impairment, given the restorative role of sleep in brain plasticity and clearance of neurotoxic waste products.

Increasing evidence supports a bidirectional relationship between sleep disturbances and neurodegenerative diseases, particularly Alzheimer's disease (AD) and Parkinson's disease (PD). In AD, sleep deprivation has been associated with accelerated accumulation of amyloid-β (Aβ) and tau proteins, which plays a main role in disease pathology. An adequate sleep facilitates the elimination of these proteins via the glymphatic system, a mechanism that is less efficient during wakefulness (Xie et al. 2013). Longitudinal studies have shown that reduced sleep efficiency and fragmented sleep patterns are predictive of higher Aβ burden and greater cognitive decline in aging individuals (Ju et al. 2014). Similarly, in PD, disrupted circadian rhythms and REM sleep behavior disorder (RBD) often precede motor symptoms and are linked to α-synuclein aggregation, suggesting that sleep alterations may be early biomarkers of neurodegeneration.

Sleep loss may also exacerbate neuroinflammation and oxidative stress that are critical mediators of age-related neurodegeneration. Chronic sleep fragmentation leads to microglial activation and pro-inflammatory cytokine release, thereby accelerating neuronal injury and synaptic dysfunction. Moreover, impaired sleep impedes hippocampal neurogenesis, diminishes memory consolidation, and increases the risk of cognitive disorders in the elderly (Lo et al. 2016). These findings underscore the potential of sleep quality as a modifiable factor in delaying the onset or progression of neurodegenerative diseases.

13.1.7.1 Therapeutic Interventions for Sleep and Circadian Disorders

Sleep and circadian rhythm disturbances are prevalent in both clinical and aging populations and have been linked to increased risk of neurodegeneration, metabolic dysfunction, and mood disorders. Pharmacological treatments remain a cornerstone in managing these disorders. Hypnotics, including benzodiazepines and non-benzodiazepine GABA-A agonists (e.g., zolpidem), are widely prescribed for short-term insomnia management but may impair sleep architecture and have dependency risks with prolonged use (Holbrook et al. 2000). Stimulants such as modafinil are used in circadian rhythm sleep–wake disorders like narcolepsy or shift work disorder, though they primarily promote wakefulness rather than address the underlying circadian misalignment. Melatonin and melatonin-receptor agonists (e.g., ramelteon) offer more physiological chrono biotic effects by reinforcing endogenous circadian signals, especially beneficial in delayed sleep phase disorder and aging individuals with decreased melatonin production (Zisapel 2018).

Given the central role of sleep in brain health, therapeutic strategies targeting sleep disturbances in aging populations have gained increasing attention. Cognitive Behavioral Therapy for Insomnia (CBT-I) has been shown to improve sleep efficiency and cognitive outcomes in older adults without the adverse effects associated with sedative hypnotics. Pharmacological interventions, including melatonin and orexin-receptor antagonists, have shown promise in restoring sleep–wake cycles and reducing nocturnal agitation in dementia patients. Light therapy and physical activity are also effective non-pharmacological approaches to enhancing circadian alignment and sleep quality in the elderly (Riemersma-van Der Lek 2008). These interventions highlight the therapeutic potential of sleep modulation as a strategy to preserve cognitive function and enhance quality of life during brain aging.

Chronotherapy, which involves strategic timing of sleep and wake schedules, can be used to realign circadian rhythms with environmental cues, particularly in delayed or advanced sleep phase disorders. One of the most effective chronotherapeutic tools is light therapy. Timed morning exposure to bright light shifts the circadian phase forward. While evening light delays it, thereby facilitating proper alignment with the desired sleep schedule (Lewy et al. 2006). Light therapy has also shown promise in treating circadian misalignment in neurodegenerative diseases such as Alzheimer's and Parkinson's, where circadian degradation often exacerbates sleep disturbances and cognitive decline (Riemersma-van Der Lek 2008). Combining light therapy with melatonin and structured behavioral routines may synergistically enhance circadian re-entrainment.

13.1.7.2 Future Directions in Sleep and Brain Health Research

Advances in technology are revolutionizing the field of sleep medicine, enabling more precise and individualized diagnostics. Emerging wearable devices and non-invasive home-based polysomnography systems now allow for continuous monitoring of sleep patterns, sleep stages, and circadian rhythms in naturalistic settings. These technologies not only improve accessibility and compliance but also provide rich longitudinal data for understanding interindividual sleep variability (Walch et al. 2016). Integration with digital platforms facilitates real-time feedback, enabling dynamic sleep coaching and behavioral intervention delivery.

Artificial intelligence (AI) is dignified to play a transformative role in custom-made sleep medicine. Machine learning algorithms can analyze vast multidimensional sleep data to identify subtle patterns predictive of sleep disorders, treatment responses, or early neurodegenerative changes. AI-powered models are being developed to optimize sleep interventions by tailoring light exposure, pharmacotherapy, or behavioral strategies based on an individual's chronotype, genetic background, and lifestyle (Verma et al. 2023). Such personalized approaches could significantly enhance adherence and efficacy, especially in populations with complex comorbidities.

Finally, novel interventions aimed at enhancing sleep quality are under exploration, including transcranial stimulation techniques (e.g., tACS, tDCS) targeting slow-wave sleep, pharmacological agents that selectively increase deep sleep (e.g., gaboxadol), and digital therapeutics that combine real-time biofeedback with cognitive training. These strategies aim not only to improve sleep outcomes but also to harness sleep's restorative potential for brain plasticity and neuroprotection (Ngo et al. 2013). As the link between sleep and brain health becomes increasingly evident, the development of multidisciplinary, personalized, and preventive interventions will be essential in mitigating cognitive decline and promoting healthy aging.

13.2 Conclusions

The intricate interplay between sleep, circadian rhythms, and brain health underscores the fundamental role of biological timing in maintaining neurological function. Sleep is not only a passive state but a dynamic process essential for synaptic plasticity, memory consolidation, and neurotoxic waste clearance via the glymphatic system. Circadian rhythms, governed by endogenous clocks, synchronize physiological processes to environmental light–dark cycles, ensuring optimal brain performance and resilience. Disruptions to these systems whether through shift work, jet lag, or chronic sleep deprivation can impair cognitive function, elevate stress responses, and increase the risk of neurodegenerative diseases. Recognizing the importance of circadian alignment and restorative sleep is therefore crucial not only for individual brain health but also for broader public health initiatives. Future research must continue to unravel the molecular mechanisms linking these domains, offering targeted strategies for prevention and intervention in neurological disorders.

Acknowledgement We thankfully acknowledge the infrastructure and fellowship support to Nandini C Devaru, JSS College of Pharmacy, Md. Abubakar, extended by Department of Pharmaceuticals, the Ministry of Chemicals and fertilizers, Government of India and NIPER-Guwahati, and Deccan School of Pharmacy, and Eswara Rao Puppala for National Institute of Health, USA.

Author Contributions Nandini C Devaru was involved in the conceptualization, writing of the original draft, mechanistic insight, input on critical points, and editing; Md. Abubakar was involved in input on critical points, editing; Sana Ahmed, Sarika, and Mohammad Talha contributed to the manuscript writing, editing; and Eswara Rao Puppala was involved in the conceptualization, manuscript writing, reviewing, conceptualization of figures, and approved the version to be published.

Availability of Data None.

Conflict of Interest None.

Funding The work received no funding support from any agency.

References

Albrecht U (2017) Molecular mechanisms in mood regulation involving the circadian clock. Front Neurol [Internet] 8. Available from: http://journal.frontiersin.org/article/10.3389/fneur.2017.00030/full. [Cited 2025 Apr 15]

Anna G, Kannan NN (2021) Post-transcriptional modulators and mediators of the circadian clock. Chronobiol Int 38(9):1244–1261

Besedovsky L, Lange T, Haack M (2019) The sleep-immune crosstalk in health and disease. Physiol Rev 99(3):1325–1380

Bjorness T, Greene R (2009) Adenosine and sleep. Curr Neuropharmacol 7(3):238–245

Borbély AA, Baumann F, Brandeis D, Strauch I, Lehmann D (1981) Sleep deprivation: effect on sleep stages and EEG power density in man. Electroencephalogr Clin Neurophysiol 51(5):483–493

Brown SA, Ripperger J, Kadener S, Fleury-Olela F, Vilbois F, Rosbash M et al (2005) PERIOD1-associated proteins modulate the negative limb of the mammalian circadian oscillator. Science 308(5722):693–696

Deboer T (2018) Sleep homeostasis and the circadian clock: do the circadian pacemaker and the sleep homeostat influence each other's functioning? Neurobiol Sleep Circadian Rhythms 5:68–77

Dey J, Carr AJF, Cagampang FRA, Semikhodskii AS, Loudon ASI, Hastings MH et al (2005) The tau mutation in the Syrian hamster differentially reprograms the circadian clock in the SCN and peripheral tissues. J Biol Rhythm 20(2):99–110

Diekelmann S, Born J (2010) The memory function of sleep. Nat Rev Neurosci 11(2):114–126

Dijk DJ, Archer SN (2009) Circadian and homeostatic regulation of human sleep and cognitive performance and its modulation by PERIOD3. Sleep Med Clin 4(2):111–125

Eide PK, Vinje V, Pripp AH, Mardal KA, Ringstad G (2021) Sleep deprivation impairs molecular clearance from the human brain. Brain 144(3):863–874

Evans JA, Davidson AJ (2013) Health consequences of circadian disruption in humans and animal models. In: Progress in molecular biology and translational science [internet]. Elsevier. p 283–323. Available from: https://linkinghub.elsevier.com/retrieve/pii/B9780123969712000105. [Cited 2025 Apr 15]

Falup-Pecurariu C, Ştefania D, Ţînţ D, Falup-Pecurariu O (2021) Neurobiology of sleep (review). Exp Ther Med 21(3):272

Freeman D, Sheaves B, Waite F, Harvey AG, Harrison PJ (2020) Sleep disturbance and psychiatric disorders. Lancet Psychiatry 7(7):628–637

Goel N, Basner M, Rao H, Dinges DF (2013) Circadian rhythms, sleep deprivation, and human performance. In: Progress in molecular biology and translational science [internet]. Elsevier. p 155–190. Available from: https://linkinghub.elsevier.com/retrieve/pii/B9780123969712000075. [Cited 2025 Apr 15]

Gottesman RF, Lutsey PL, Benveniste H, Brown DL, Full KM, Lee JM, Osorio RS, Pase MP, Redeker NS, Redline S, Spira AP (2024) Impact of sleep disorders and disturbed sleep on brain health: a scientific statement from the American Heart Association. Stroke 55(3):e61–e76

Harvey AG, Murray G, Chandler RA, Soehner A (2011) Sleep disturbance as transdiagnostic: consideration of neurobiological mechanisms. Clin Psychol Rev 31(2):225–235

Hastings M, Maywood E, Brancaccio M (2019) The mammalian circadian timing system and the suprachiasmatic nucleus as its pacemaker. Biology 8(1):13

Holbrook AM, Crowther R, Lotter A, Cheng C, King D (2000) Meta-analysis of benzodiazepine use in the treatment of insomnia. CMAJ: Canadian Medical Association Journal 162(2):225–233

Ju YES, Lucey BP, Holtzman DM (2014) Sleep and Alzheimer disease pathology—a bidirectional relationship. Nat Rev Neurol 10(2):115–119

Krause AJ, Simon EB, Mander BA, Greer SM, Saletin JM, Goldstein-Piekarski AN et al (2017) The sleep-deprived human brain. Nat Rev Neurosci 18(7):404–418

Lane JM, Qian J, Mignot E, Redline S, Scheer FAJL, Saxena R (2023) Genetics of circadian rhythms and sleep in human health and disease. Nat Rev Genet 24(1):4–20

Langille JJ (2019) Remembering to forget: a dual role for sleep oscillations in memory consolidation and forgetting. Front Cell Neurosci 13:71

Lewy AJ, Emens J, Jackman A, Yuhas K (2006) Circadian uses of melatonin in humans. Chronobiol Int 23(1–2):403–412

Lo JC, Ong JL, Leong RLF, Gooley JJ, Chee MWL (2016) Cognitive performance, sleepiness, and mood in partially sleep deprived adolescents: the need for sleep study. Sleep 39(3):687–698

McClung CA (2013) How might circadian rhythms control mood? Let me count the ways. Biol Psychiatry 74(4):242–249

Musiek ES, Xiong DD, Holtzman DM (2015) Sleep, circadian rhythms, and the pathogenesis of Alzheimer disease. Exp Mol Med 47(3):e148–e148

Nandini C, Madhunapantula SV, Bovilla VR, Ali M, Mruthunjaya K, Santhepete MN et al (2021) Platelet enhancement by Carica papaya L. leaf fractions in cyclophosphamide induced thrombocytopenic rats is due to elevated expression of CD110 receptor on megakaryocytes. J Ethnopharmacol 275:114074

Ngo HVV, Martinetz T, Born J, Mölle M (2013) Auditory closed-loop stimulation of the sleep slow oscillation enhances memory. Neuron 78(3):545–553

Oh J, Petersen C, Walsh CM, Bittencourt JC, Neylan TC, Grinberg LT (2019) The role of co-neurotransmitters in sleep and wake regulation. Mol Psychiatry 24(9):1284–1295

Ohayon MM, Carskadon MA, Guilleminault C, Vitiello MV (2004) Meta-analysis of quantitative sleep parameters from childhood to old age in healthy individuals: developing normative sleep values across the human lifespan. Sleep 27(7):1255–1273

Patel AK, Reddy V, Shumway KR, Araujo JF (2025) Physiology, sleep stages. In: StatPearls [Internet]. StatPearls Publishing, Treasure Island. Available from: http://www.ncbi.nlm.nih.gov/books/NBK526132/. [Cited 2025 Apr 15]

Patton AP, Hastings MH (2018) The suprachiasmatic nucleus. Curr Biol 28(15):R816–R822

Puppala ER, Aochenlar SL, Shantanu P, Ahmed S, Jannu AK, Jala A et al (2022) Perillyl alcohol attenuates chronic restraint stress aggravated dextran sulfate sodium-induced ulcerative colitis by modulating TLR4/NF-κB and JAK2/STAT3 signaling pathways. Phytomedicine 106:154415

Puppala ER, Yalamarthi SS, Aochenlar SL, Prasad N, Syamprasad NP, Singh M et al (2023) Mesua assamica (King&Prain) kosterm. Bark ethanolic extract attenuates chronic restraint stress aggravated DSS-induced ulcerative colitis in mice via inhibition of NF-κB/STAT3 and activation of HO-1/Nrf2/SIRT1 signaling pathways. J Ethnopharmacol 301:115765

Rasch B, Born J (2013) About sleep's role in memory. Physiol Rev 93(2):681–766

Reiter RJ, Mayo JC, Tan D, Sainz RM, Alatorre-Jimenez M, Qin L (2016) Melatonin as an antioxidant: under promises but over delivers. J Pineal Res 61(3):253–278

Riemersma-van Der Lek RF (2008) Effect of bright light and melatonin on cognitive and non-cognitive function in elderly residents of group care facilities: a randomized controlled trial. JAMA 299(22):2642

Ross R, Chaput JP, Giangregorio LM, Janssen I, Saunders TJ, Kho ME et al (2020) Canadian 24-hour movement guidelines for adults aged 18–64 years and adults aged 65 years or older: an integration of physical activity, sedentary behaviour, and sleep. Appl Physiol Nutr Metab 45(10 Suppl 2):S57–S102

Sateia MJ (2014) International classification of sleep disorders-third edition. Chest 146(5):1387–1394

Steiger A (2007) Neurochemical regulation of sleep. J Psychiatr Res 41(7):537–552

Taillard J, Gronfier C, Bioulac S, Philip P, Sagaspe P (2021) Sleep in normal aging, homeostatic and circadian regulation and vulnerability to sleep deprivation. Brain Sci 11(8):1003

Taylor DJ, Lichstein KL, Durrence HH (2003) Insomnia as a health risk factor. Behav Sleep Med 1(4):227–247

Tononi G, Cirelli C (2014) Sleep and the price of plasticity: from synaptic and cellular homeostasis to memory consolidation and integration. Neuron 81(1):12–34

Trott AJ, Menet JS (2018) Regulation of circadian clock transcriptional output by CLOCK:BMAL1. PLOS Genet 14(1):e1007156

Uthaiah CA, Devaru NC, Shivakumar NH, Rajalakshmi R, Madhunapantula SV (2022) Vitamin D mitigates hyperglycemia-induced cognition decline in Danio rerio (zebrafish) through the activation of antioxidant mechanisms. Antioxidants 11(11):2114

Verma RK, Dhillon G, Grewal H, Prasad V, Munjal RS, Sharma P et al (2023) Artificial intelligence in sleep medicine: present and future. World J Clin Cases 11(34):8106–8110

Walch OJ, Cochran A, Forger DB (2016) A global quantification of "normal" sleep schedules using smartphone data. Sci Adv 2(5):e1501705

Xie L, Kang H, Xu Q, Chen MJ, Liao Y, Thiyagarajan M et al (2013) Sleep drives metabolite clearance from the adult brain. Science 342(6156):373–377

Yoo SS, Gujar N, Hu P, Jolesz FA, Walker MP (2007) The human emotional brain without sleep—a prefrontal amygdala disconnect. Curr Biol 17(20):R877–R878

Zhang S, Dai M, Wang X, Jiang SH, Hu LP, Zhang XL et al (2020) Signalling entrains the peripheral circadian clock. Cell Signal 69:109433

Zisapel N (2018) New perspectives on the role of melatonin in human sleep, circadian rhythms and their regulation. Br J Pharmacol 175(16):3190–3199

Nutraceuticals for Brain Longevity and Autophagy Activation

14

Krishnendu Adhikary, Md. Abubakar, Krishnendu Ganguly, Riya Sarkar, Abhishek Ambawatiya, Parimala Mohanty, and Nitesh Kumar

Abstract

Brain longevity is a multifaceted concept that involves maintaining cognitive function, resilience, and neural health despite aging. Aging leads to structural and functional changes in the brain, contributing to cognitive decline and neurodegenerative diseases (NDDs) such as Alzheimer's and Parkinson's. However, emerging research highlights the role of autophagy, a natural cellular process, in mitigating these effects by removing damaged proteins and organelles, thus preserving neuronal health. Nutraceuticals such as resveratrol, curcumin, omega-3 fatty acids, and polyphenols have demonstrated potential in enhancing autophagy and promoting neuroprotection. Additionally, lifestyle modifications, including intermittent fasting (IF), exercise, stress management, and quality sleep,

Krishnendu Adhikary and Md. Abubakar contributed equally with all other contributors.

K. Adhikary · K. Ganguly
Department of Medical Laboratory Technology, Paramedical College Durgapur, Durgapur, West Bengal, India

Md. Abubakar · A. Ambawatiya · N. Kumar (✉)
Department of Pharmacology and Toxicology, National Institute of Pharmaceutical Education and Research, Hajipur, Vaishali, Bihar, India
e-mail: nitesh.kumar04@niperhajipur.ac.in

R. Sarkar
Department of Medical Lab Technology, Dr. B. C. Roy Academy of Professional Courses, Durgapur, West Bengal, India

P. Mohanty (✉)
Department of Community Medicine & Family Medicine, All India Institute of Medical Sciences (AIIMS), Bhubaneswar, India

A. Kumar Singh, S. Nand Rai (eds.), *Rejuvenating the Brain: Nutraceuticals, Autophagy, and Longevity*, Nutritional Neurosciences,
https://doi.org/10.1007/978-981-95-2790-8_14

further support cognitive longevity by modulating neuroinflammation and oxidative stress. The integration of these interventions offers a holistic strategy for sustaining brain function, delaying neurodegeneration, and improving quality of life. Future research should focus on personalized approaches, leveraging genetic and technological advancements to optimize cognitive health. By prioritizing brain longevity through diet, lifestyle, and autophagy-enhancing strategies, individuals can enhance mental clarity and overall well-being as they age.

Keywords

Autophagy · Brain longevity · Cognitive health · Neuroinflammation · Neuroprotection · Nutraceuticals · Oxidative stress.

14.1 Introduction

14.1.1 Overview of Brain Longevity

The ability of the central nervous system to sustain its optimum performance, cognitive understanding, and resilience despite age-related deterioration is referred to as brain longevity (Al Matared et al. 2024; Cammisuli et al. 2022; Sampedro-Piquero et al. 2018; Arida and Teixeira-Machado 2021; Liverman et al. 2015; Blake 2025). Memory, learning, and general cognitive ability can all be impacted by the anatomical and functional modifications that occur in the brain as people age (Glisky 2007; Fjell and Walhovd 2010; Grady et al. 2006). Preventive measures are crucial since neurodegenerative illnesses like Parkinson's disease (PD) and Alzheimer's disease (AD) present serious obstacles to brain health (Kumar and Singh 2015; Rai et al. 2024). The concept of brain longevity is not merely about preventing disease but also about enhancing cognitive potential, promoting mental clarity, and ensuring a high quality of life as individuals grow older. Understanding the biological mechanisms underlying brain aging, particularly autophagy, can offer important clues about ways to enhance neuroprotection and cognitive resilience (Glick et al. 2010).

14.1.2 Role of Autophagy in Aging and Brain Health

By eliminating damaged organelles that are present, proteins that have been improperly folded, and accumulated cellular debris, autophagy—a normal cellular process—plays a vital role in preserving cellular health (Mizushima and Levine 2020). The effectiveness of autophagy decreases with age, resulting in elevated oxidative stress, mitochondrial malfunction, and inflammation—all of which are variables that contribute to neurodegeneration and cognitive impairment (Cheon et al. 2019; Rai 2021). Research suggests that enhancing autophagy through dietary, pharmacological, and lifestyle interventions can mitigate these effects and promote neural regeneration (Hansen et al. 2018). Certain natural compounds and lifestyle practices, including fasting, physical exercise, and specific dietary components, have been shown to stimulate autophagy, thereby offering neuroprotective benefits. By understanding how autophagy influences brain health, individuals can adopt targeted strategies to slow cognitive decline and support long-term mental acuity (Navakkode and Kennedy 2024; Stacchiotti and Corsetti 2020; Ortega et al. 2024; Kocot and Wróblewska 2022; Pani 2015; Moore 2020; Mackieh et al. 2024).

14.1.3 Nutraceuticals, Autophagy, and Lifestyle: A Holistic Approach

A holistic approach to brain longevity integrates the power of nutraceuticals, autophagy-boosting strategies, and lifestyle modifications (Williams et al. 2021). Nutraceuticals that include bioactive compounds derived from food sources have been extensively studied for their neuroprotective properties (Table 14.1) (Mohd Sairazi and Sirajudeen 2020; Chib et al. 2025). Compounds such as resveratrol, curcumin, quercetin, and polyphenols found in berries and green tea have demonstrated the ability to enhance autophagy, reduce oxidative stress, and support neuronal function (Prasanth et al. 2019; Chandrasekaran et al. 2023). Furthermore, omega-3 fatty acids that are included in flaxseeds and fish oil are known to lower inflammation and support cognitive function (Singh 2020; Djuricic and Calder 2021). Beyond nutraceuticals, lifestyle factors such as intermittent fasting, calorie restriction, and high-intensity exercise can naturally stimulate autophagy and improve brain plasticity (Mattson et al. 2018). Furthermore, stress management techniques, including meditation, mindfulness, and adequate sleep, are integral to preserving brain function and supporting neurogenesis (Davidson and McEwen 2012). Chronic stress and poor sleep patterns have been linked to accelerated brain aging and cognitive decline, making mental wellness practices an essential component of brain longevity (Mukherjee et al. 2024; Addo and Khan 2024; Castruita et al. 2022). Understanding brain longevity and its association with autophagy provides a foundation for adopting a proactive approach to cognitive well-being. Through a combination of nutraceuticals, lifestyle modifications, and autophagy-enhancing strategies, individuals can protect their brains against age-related decline and promote lasting cognitive vitality (Blake 2025; Avan et al. 2022; Tenchov et al. 2024; Leong et al. 2025). By prioritizing brain health through holistic and science-backed practices, individuals can ensure mental clarity, resilience, and a higher quality of life well into their later years.

14.2 Understanding Brain Longevity

The capacity of the brain to retain its structure, function, and cognitive capacities as it matures is referred to as brain longevity. As we grow older, the brain undergoes a variety of changes that can affect memory, cognition, and overall mental health.

Table 14.1 The relationship between nutraceuticals, autophagy, and lifestyle from a holistic approach

Category	Key components	Impact on autophagy	Lifestyle practices	Examples of nutraceuticals
Diet and nutrition	Balanced, whole food diet (vegetables, fruits, lean proteins, and healthy fats)	Promotes autophagy by providing essential nutrients, antioxidants, and compounds that stimulate cellular repair.	Intermittent fasting, calorie restriction, or nutrient timing to enhance autophagy	Resveratrol (from grapes), curcumin (from turmeric), green tea extract (EGCG), and berberine
Exercise and activity	Physical activity (strength training, cardio, yoga, etc.)	Exercise induces autophagy by creating cellular stress and increasing production of autophagy-related proteins.	Regular exercise (aerobic and anaerobic), yoga, and mindfulness activities to reduce stress and promote overall health	Adaptogens (*Ashwagandha*, *Rhodiola rosea*), L-carnitine, and coenzyme Q10
Sleep and recovery	High-quality sleep, deep rest	Deep sleep is critical for autophagy activation, particularly during the REM stage of sleep.	Maintain a consistent sleep schedule, aim for 7–9 hours of sleep, and reduce stress through relaxation techniques (meditation, breathing exercises).	Melatonin, magnesium, l-theanine, and GABA (gamma-aminobutyric acid)
Stress management	Techniques like mindfulness, meditation, and breathing exercises	Chronic stress can hinder autophagy; stress reduction can help restore balance and promote autophagy.	Engage in daily stress-reducing practices like meditation, journaling, or breathing exercises.	*Rhodiola rosea, Ashwagandha, Ginseng*, and l-theanine

However, with the right lifestyle, mindset, and habits, it is possible to slow down these changes and promote brain longevity (Carmona and Michan 2016).

14.2.1 What Is Brain Longevity?

The ability of the brain to retain its structure, function, and cognitive capacities as it matures is referred to as brain longevity. As we grow older, the brain undergoes various changes that can affect memory, cognition, and overall mental health. However, with the right lifestyle, mindset, and habits, it is possible to slow down these changes and promote brain longevity.

Physical exercise Frequent exercise boosts general brain function, promotes the development of fresh brain cells, and increases blood circulation to the brain. Additionally, it can lower the chance of developing neurological illnesses like Alzheimer's. Examples include strength training, yoga, and aerobic activities including jogging, walking, and cycling (Small et al. 2006).

Healthy nutrition By lowering inflammation, enhancing blood circulation, and supplying vital nutrients including omega-3 fatty acids, essential micronutrients (vitamins B and E), and antioxidants, a nutritious, well-balanced diet promotes brain health. For instance, the Mediterranean diet, that is abundant in vegetables, whole grains, fruits, fish, and olive oil, as well as foods that are strong in antioxidants, including leafy greens, berries, and nuts (Small et al. 2006; Singh et al. 2022; Guasch-Ferré and Willett 2021).

Quality sleep Seep is essential for brain detoxification, cognitive function, and memory consolidation. The brain develops neural connections and gets rid of waste during deep sleep. For instance, adopt excellent sleep hygiene and strive for 7–9 hours of relaxation each night (Frank and Heller 2019; Desai et al. 2024; Lewis 2021; Leonidis et al. 2021).

Mental stimulation Maintaining cognitive function, enhancing memory, and promoting neuroplasticity—the brain's capacity to rearrange and create new connections—can all be achieved by partaking in mentally taxing activities. For instance, solving puzzles, picking up a new language, reading, playing an instrument, or having thought-provoking discussions.

Stress management Engaging in mentally taxing activities can maintain cognitive function, enhance memory, and promote neuroplasticity—the brain's capacity to rearrange and create new connections. You can engage in activities such as solving puzzles, learning a new language, reading, playing an instrument, or engaging in thought-provoking conversations.

Social engagement Strong social ties can boost mood, increase brain activity, and lessen loneliness and isolation—all of which are associated with risks for cognitive decline. Volunteering, participating in group activities, and upholding consistent relationships with family members, close friends, and the community are some examples.

Mental health longevity of the brain depends on addressing mental health conditions, including sadness and anxiety. Prolonged mental health issues might cause brain shrinkage and affect cognitive performance. For instance, mindfulness exercises, therapy, or counseling to preserve mental health (Syme and Hagen 2020; Logan and McClung 2019; Marin et al. 2011; Olness 2003; Costello and Walters 2022).

Brain-boosting supplement although certain supplements may promote brain health, it is crucial to use caution when using them and to speak with a healthcare provider. Examples of such supplements include vitamin D, omega-3 fatty acids, and antioxidants such as curcumin and resveratrol.

14.2.2 Aging and the Brain: A Complex Relationship

All parts of the individual's body, including the brain, are impacted by the natural process of aging. Over time, the brain undergoes various changes, some of which are normal and part of the aging process, while others can lead to cognitive decline or neurological disorders. The relationship between aging and the brain is multifaceted, involving both structural and functional changes that can impact memory, cognition, and overall brain health (Kern and Behl 2009).

14.2.3 Key Factors Contributing to Brain Health Across the Lifespan

Brain health is essential throughout the lifespan, and several factors contribute to maintaining and enhancing cognitive function, mental well-being, and overall brain health (Table 14.2). These factors can be categorized into biological, lifestyle, and environmental influences. Here are key contributors to brain health (García-García et al. 2023).

Nutrition Brain function depends on nutrients like antioxidants (found in fruits and vegetables), omega-3 fatty acids (found in fish and flaxseeds), and B vitamins (particularly folate and B12). It aids in preventing inflammation and oxidative stress that can result in cognitive deterioration.

Physical activity Frequent exercise improves blood flow to the brain, encourages the development of new brain cells, and enhances cognitive function. This is, especially true of cardiovascular exercises like swimming, jogging, or walking. Furthermore, it is associated with a decreased risk of AD (Wickramasinghe et al. 2020).

Mental stimulation Brain plasticity that aids in the formation and strengthening of synaptic connections is stimulated by mentally taxing activities like puzzles, learning new languages or abilities, reading, and problem-solving. Keeping up solid social links and taking part in worthwhile social activities will lift your spirits and stop cognitive deterioration. The social interaction can help prevent the negative consequences of stress by stimulating the brain.

Sleep Sleep is essential for brain health because it allows the brain to renew cells, eliminate waste products like beta-amyloid, and consolidate memories. Long-term

Table 14.2 Key factors contributing to brain health across the lifespan

Life stage	Key factors	Impact on brain health	Supporting practices	Examples of nutraceuticals
Childhood	Proper nutrition (essential fatty acids, protein, and vitamins)	Brain development and synaptic plasticity are highly dependent on proper nutrition during growth.	Encourage a balanced diet rich in vitamins and minerals, promote physical activity, and ensure adequate sleep.	DHA (docosahexaenoic acid), omega-3 fatty acids, vitamin D, and zinc
	Cognitive stimulation (learning, play, and social interaction)	Stimulates brain growth and development through neurogenesis and synaptogenesis.	Encourage activities like reading, creative play, and social interaction for brain stimulation.	B-complex vitamins, L-theanine (calming), probiotics (gut-brain connection)
Adolescence	Mental health (stress management, emotional regulation)	High levels of stress and poor mental health can impair brain function and emotional regulation during this period	Encourage stress management techniques (yoga, mindfulness), proper sleep hygiene, and emotional support.	*Rhodiola rosea, Aswagandha*, magnesium (for stress relief), and omega-3 s
	Physical exercise	Physical activity supports brain plasticity, cognitive function, and mood regulation.	Lifelong learning, challenging brain activities (puzzles, language learning).	*Bacopa monnieri, Ginkgo biloba*, lion's mane mushroom (neurogenesis), and phosphatidylserine
	Sleep quality (7–9 hours per night)	Sleep is crucial for memory consolidation, neural repair, and brain detoxification.	Prioritize quality sleep, maintain a consistent sleep schedule, and reduce screen time before bed.	Melatonin, magnesium, and L-theanine
	Nutrition (antioxidants, anti-inflammatory foods)	Nutrition rich in antioxidants and anti-inflammatory compounds can protect the brain from oxidative stress and aging	Incorporate anti-inflammatory foods (berries, leafy greens), limit processed foods, and stay hydrated.	Curcumin, resveratrol, omega-3 s (EPA/DHA), vitamin E, N-acetyl cysteine (NAC).

(continued)

Table 14.2 (continued)

Life stage	Key factors	Impact on brain health	Supporting practices	Examples of nutraceuticals
Older adulthood	Mental stimulation (continuous learning, social engagement)	Ongoing mental stimulation reduces cognitive decline and promotes neuroplasticity.	Stay socially active, engage in mentally challenging activities (reading, puzzles, and learning new skills).	Phosphatidylserine, *Ginkgo biloba*, b-*Bacopa monnieri*, and omega-3 s.
	Physical health (exercise, cardiovascular health)	Regular exercise, particularly aerobic exercise, improves blood flow to the brain, reduces inflammation, and supports cognition.	Maintain regular physical activity (walking, swimming, strength and training), manage chronic conditions (diabetes, hypertension).	Creatine (for cognitive energy), CoQ10, Omega-3 s

sleep deprivation raises the likelihood of mental health issues and cognitive deterioration (García-García et al. 2023).

Genetics and epigenetics Genetics play a role in brain health, influencing susceptibility to conditions like AD, mood disorders, or neurodevelopmental disorders. However, genes interact with environmental factors; so maintaining healthy habits can still influence brain health even in individuals at higher genetic risk.

Lifestyle choices and environmental exposures can influence gene expression and affect brain health throughout life. For example, avoiding toxins or practicing healthy eating can positively affect brain function through epigenetic modifications.

Environmental factors Particularly in young children and elderly individuals, exposure to environmental pollutants such as pesticides, heavy metals, and air pollution can have a detrimental effect on brain health. Preventing cognitive deterioration can be achieved by limiting exposure to such poisons.

Hormonal balance Hormonal changes across the lifespan, particularly during puberty, pregnancy, and menopause, can affect brain function. Managing hormonal imbalances through lifestyle changes or medical interventions can help maintain brain health (Wickramasinghe et al. 2020).

14.2.4 The Relationship of Genetics, the Environment, and Lifestyle

An individual's total health, including their brain health, is greatly influenced by the intersection of their genetics, environment, and lifestyle. These three factors interact in complex ways, influencing each other and contributing to various aspects of physical and mental well-being. Here is a closer look at how these three elements interconnect (Tian et al. 2024; Moffitt et al. 2006; Harris and McDade 2018; Malecki et al. 2022; Rejeski and Fanning 2019).

14.2.4.1 Genetics: Genetic Inheritance

Our genes provide the foundational blueprint for who we are, influencing a wide array of traits such as susceptibility to diseases, metabolic processes, and cognitive abilities (Timpson et al. 2018; Flint et al. 2020; Nussbaum et al. 2015; Aardema and MacGregor 2002). Regarding brain health, some genetic variations might increase a person's susceptibility to diseases including AD, PD, anxiety, and depression (Cuyvers and Sleegers 2016; Boyd et al. 2022; Oliveri et al. 2018).

14.2.4.2 Epigenetics

The genes we inherit set the scene, but environmental variables (such as stress, food, or interaction with toxins) may influence how genes are expressed. This process is known as epigenetics. These modifications do not alter the DNA sequence itself but can turn genes "on" or "off," impacting everything from memory and cognition to the development of disease. This means that even if someone has a genetic predisposition to a condition, their lifestyle and environment can influence whether it actually manifests (Abubakar et al. 2023; Sharma et al. 2024).

14.2.4.3 Environmental Toxins and Pollution

The state of our bodies and minds is greatly influenced by our surroundings. Long-term harm to brain health may result from exposure to environmental pollutants such as pesticides, heavy metals, air pollution, or chemicals in food. For instance, studies have linked pollutants like fine particulates (PM2.5) to increased risks of NDDs and cognitive impairment (Malecki et al. 2022).

14.2.4.4 Social and Physical Environment

The environment also includes the social settings we engage in—our communities, support systems, and access to resources like education, healthcare, and healthy food. Positive social interactions, access to green spaces, and a safe, nurturing environment are all associated with better mental health and cognitive function. On the flip side, exposure to chronic stress, socioeconomic hardships, or living in a highly polluted or unsafe environment can increase susceptibility to mental health issues or cognitive decline (Ugwu et al. 2025; Wiedermann et al. 2023; Hargreaves et al. 2022; Tomoh et al. 2024; Jabbar and Yusoff 2022; Modak et al. 2023; Ejiohuo et al. 2024; Nguyen et al. 2023; Tota et al. 2024; Stapelberg et al. 2022).

14.3 Lifestyle and Exercise

One of the best lifestyle decisions for brain health is to exercise regularly. The physical activity promotes neurogenesis, or the growth of new neurons, boosts blood supply to the brain, and maintains cognitive function throughout life. Additionally, it aids in preventing diseases that might impair brain function, such as diabetes, obesity, and CVDs (Dominguez et al. 2021; De Sousa et al. 2021; Chen et al. 2022; Alnawwar et al. 2023; Umegaki et al. 2021; Khalil 2024; Zubko and Kachalov 2025; Maugeri et al. 2021; Fekete et al. 2022; Izzy et al. 2023; Diab et al. 2024).

Nutrition Our diet has a direct impact on the health of our brains. The diets rich in vitamins (including B12 and D), antioxidants, and omega-3 fatty acids, among other nutrients, promote cognitive performance and guard against dementia. On the other hand, harmful dietary choices that are heavy in sugar, trans fats, and processed foods can cause oxidative stress, inflammation, and aging of the brain (Lakhman et al. 2024).

Autophagy: the body's natural repair mechanism Autophagy is a critical biological process that helps the body maintain its health by clearing out damaged cells and recycling cellular components. From immunity to longevity, autophagy is essential for many aspects of health. The Greek words "auto" (self) and "phagein" (to eat) form the term "autophagy," signifying "self-eating." This process allows the body to break down and eliminate dysfunctional or unnecessary components of cells, promoting cell survival and renewal (Dikic and Elazar 2018).

14.3.1 Defining Autophagy and Its Mechanisms

Autophagy occurs inside cells and involves the following key steps (Harris et al. 2017):

Initiation When the body detects stress (such as nutrient deprivation, infections, or damage), it triggers the autophagy process. In response to these signals, certain proteins in the cell become activated to initiate autophagy.

Formation of the autophagosome Activation of autophagy leads to the formation of a structure known as the autophagosome. The autophagosome is a double-membraned vesicle that engulfs damaged or obsolete cellular components, such as misfolded proteins, malfunctioning organelles (like mitochondria), and pathogens.

Lysosomal fusion The autophagosome then fuses with a lysosome, an organelle that contains enzymes capable of breaking down these materials. The contents are broken down into simpler molecules (such as amino acids, fatty acids, and sugars).

Recycling The breakdown products are recycled and used to fuel the cell's energy needs or to build new cellular components, ensuring that the cell remains functional and efficient (Gómez-Virgilio et al. 2022).

14.3.2 The Role of Autophagy in Cellular Maintenance

A crucial cellular mechanism that is essential to cellular maintenance, autophagy makes sure that cells continue to operate at their best and retain their integrity throughout time. This "self-eating" process allows cells to regenerate and adjust to changing circumstances by breaking down and recycling damaged, malfunctioning, or superfluous cellular components. Let us examine autophagy's function in cellular upkeep in more detail (Badadani 2012).

1. *Clearing damaged organelles:* one of the primary functions of autophagy is the removal of *damaged organelles* that can impair cellular function if left unchecked. Key examples include:

   Mitochondria: known as the "powerhouses" of the cell, mitochondria generate energy through respiration. However, over time, they can become damaged due to oxidative stress, which leads to the accumulation of defective mitochondria. If these damaged mitochondria are not cleared out, they can contribute to cell death or dysfunction. Autophagy removes these dysfunctional mitochondria in a process known as mitophagy, thereby preserving the cell's energy production capacity.

   Endoplasmic reticulum (ER): the ER plays a critical role in protein folding and trafficking. When it becomes stressed or accumulates misfolded proteins, it can trigger a condition known as ER stress. Autophagy helps remove damaged portions of the ER and misfolded proteins, thereby ensuring that the cell maintains its protein synthesis machinery (Anding and Baehrecke 2015).
2. *Protein homeostasis:* proteins are essential for nearly all cellular functions, and their proper folding, maintenance, and degradation are crucial for cell survival. Autophagy plays a key role in protein quality control (Chun and Kim 2018).

   Regulation of protein levels: autophagy helps regulate the levels of proteins within the cell by selectively degrading unnecessary or surplus proteins. The process ensures that the balance between protein synthesis and degradation is maintained, which is vital for cellular homeostasis.

   Energy regulation: autophagy is vital to restoring cellular energy balance, especially during times of nutrient scarcity. Here is how it helps with energy regulation:

   During periods of low nutrient availability, such as during fasting or calorie restriction, cells rely on autophagy to break down their internal components (e.g., damaged proteins, lipids, and even parts of organelles) and recycle these materials into basic building blocks, like amino acids and fatty acids. These recycled components are then used to generate energy or build new cellular structures (Yang et al. 2019).

3. *Cellular signaling and homeostasis:* autophagy influences a range of cellular signaling pathways that help maintain cellular homeostasis. For example, autophagy can modulate the inflammatory response by degrading inflammatory molecules or signaling complexes. This process helps prevent excessive inflammation, which, if left unchecked, can lead to chronic diseases like arthritis or autoimmune disorders. Autophagy is involved in regulating key metabolic pathways by modulating the availability of metabolites, such as glucose, lipids, and amino acids. This regulation helps cells adapt to changes in energy demand and nutrient availability (Yang et al. 2019; Monkkonen and Debnath 2018; Wu and Lu 2019; Gan et al. 2023; Yin et al. 2018; Corsetti et al. 2024).
4. *Autophagy in disease prevention and repair:* autophagy is vital in preventing various diseases by maintaining healthy cellular function and repairing cellular damage.

   Autophagy helps prevent cancer by removing damaged or dysfunctional cells that may potentially become cancerous. It also clears mutated DNA, which is essential for preventing the accumulation of mutations that could lead to tumor formation. Defective autophagy causes damaged proteins to accumulate in PD, HD, and AD. These proteins can form toxic aggregation and interfere with regular cellular activity (Meijer and Codogno 2009).
5. *Role of autophagy in immune cells:* autophagy is also crucial for immune cell function and defense against pathogens. Autophagy is involved in the defense against infections by helping immune cells degrade and eliminate intracellular pathogens (e.g., bacteria, viruses, and parasites). Autophagic vesicles can trap and degrade these pathogens, preventing them from spreading and causing harm (Puleston and Simon 2014).

14.3.3 Autophagy and the Brain: A Critical Link

Autophagy is a crucial process for maintaining brain health, playing a vital role in protecting against NDDs, supporting cognitive function, and promoting the overall longevity of neurons. The brain, being a highly active and energy-demanding organ, depends on autophagy to maintain cellular homeostasis, remove damaged proteins and organelles, and adapt to stress. Here is an in-depth look at how autophagy functions in the brain and its critical link to brain health (Lipinski et al. 2015).

14.3.4 How Autophagy Declines with Age and Its Consequences

Autophagy is a dynamic process that is continuously regulated by various factors, including nutrient availability, stress signals, and cellular signaling pathways. As we age, several factors contribute to the decline in autophagic activity.

Dysfunctional cellular signaling Changes in the cell signaling processes that control autophagy are linked to aging. Age-induced dysregulation of the mTOR process, which suppresses phagocytosis in response to poor nutrition, is one example. Even when there is cellular stress or injury, older people's mTOR pathway tends to become hyperactive, which results in decreased autophagic activity (Leidal et al. 2018).

Reduced activation of autophagic pathways Important proteins that detect cellular stress and nutrition availability, such as AMPK and ULK1, must be active in order for autophagy to begin. These proteins' activation efficiency decreases with age, which results in decreased autophagy activation and compromised cellular maintenance (Zou et al. 2022; Wang et al. 2022a; Iorio et al. 2021; Raza 2024; Kitada and Koya 2021).

Impaired lysosomal function Lysosomes are responsible for degrading the components that are engulfed by autophagosomes during autophagy. With aging, lysosomal function tends to decline, reducing the ability to break down and recycle cellular components effectively. Cellular waste, such as broken proteins and organelles, accumulates as a result (Henn et al. 2025; Patra et al. 2023; Wang et al. 2023; He et al. 2024; Malik et al. 2019; Nixon and Rubinsztein 2024).

Development of organelles and damaged proteins Over time, cells accumulate damage due to factors such as oxidative stress, environmental toxins, and metabolic waste. As autophagic activity declines with age, damaged proteins, lipids, and organelles, such as mitochondria, accumulate; cause cellular dysfunction and contribute to the aging process (Rubinsztein et al. 2011).

14.4 Nutraceuticals for Brain Longevity and Autophagy Activation

14.4.1 Important Nutraceuticals for Autophagy Promotion

In gerontology, extending the healthy life expectancy and reducing neurological disease are crucial objectives. Progressive age-related neurodegeneration results in serious illnesses that impact social life, memory, mental abilities, and mobility (González et al. 2020). Neurodegeneration and permanent neuronal loss currently have no viable therapies. In order to guarantee active aging and delay the onset of neurological disorders in senior individuals, bioactive phytochemicals may offer a natural substitute (Table 14.3) (Chiu et al. 2020). In neurons and glia, aggregate-prone organelles and protein molecules must be eliminated via autophagy, also known as macroautophagy, an evolutionarily conserved clearance mechanism. Additionally, synaptic plasticity depends on it (Fernández-Sanz et al. 2019). Aging and neurodegeneration are significantly influenced by aberrant autophagy. According to current studies, substances such as polyphenols (like resveratrol and curcumin), antioxidants (such as quercetin), and polyamines (like spermidine) prevent brain damage both in culture and in vivo (Cummings et al. 2014). By

Table 14.3 An overview of possible nutraceuticals that could affect autophagy and help prevent neurodegenerative illnesses

Nutraceuticals	Mode of action	Potential NDDs
Resveratrol	Modulates the action of SIRT1 that leads to increasing the autophagy flux.	AD, PD
Curcumin	AMPK activation, mTOR inhibition, increasing the expression of Beclin-1.	AD, PD
Epigallocatechin gallate (EGCG)	AMP-activated protein kinase (AMPK) activation, oxidative stress reduction, and increasing autophagy.	AD, PD
Spermidine	Direct interaction with the proteins of autophagy increases autophagy formation.	Age-associated NDDs
N-acetyl cysteine (NAC)	Upregulates cellular redox balance, promoting autophagy.	PD, AD
Omega-3 fatty acids	Trigger neuronal autophagy processes, break down, and recycle defective proteins as well as organelles that build up in NDDs.	AD, other forms of dementia, and PD

eliminating misfolded proteins and damaged mitochondria, their joint method of action restores efficient autophagy. The Mediterranean diet or nutritional supplementation with vegetal bioactive components shows promise in delaying the irreversible course of AD, PD, and ALS (Hornedo-Ortega et al. 2018). However, traditional medicinal herbs or vegetation with advantageous pleiotropic effects might provide a plausible preventive supplementary therapeutic opportunity, given that the "one-drug, one-target" approach to the multifactorial disease progression of NDDs is clinically unsuccessful. There is no known cure for chronic neurodegenerative illnesses, and the few approved synthetic medicines developed have only mitigated their negative effects (Vauzour et al. 2017). Additional "herbal medicine" has been proposed to treat PD and AD using secondary metabolites known as "nutraceuticals" that are produced by plants to defend themselves against infections or adverse environmental conditions (Pohl and Kong Thoo Lin 2018; Abubakar et al. 2024). Phytosterols, carotenoids, fucoidans, and polyphenols are among the neuroprotective compounds that have recently been shown to be abundant in seaweeds consumed in China and Indonesia. Nutraceuticals that alter autophagy may be helpful "functional foods," according to new research, considering the pathogenetic importance of aberrant autophagy in NDDs (Park et al. 2018).

14.4.1.1 Resveratrol (3,5,4′-trihydroxy-trans-stilbene)

It is typically present in plants and red wine because of *Vitis vinifera*, improves cognitive impairment in AD mice by activating the first phase of autophagy and modulating the AMPK/mTORC1 pathway. Nevertheless, dystrophic neuritic cells with autophagic vacuoles, higher mTOR, and reduced Beclin 1 have been found to indicate faulty enhanced autophagy as a pathogenic hallmark of AD (Castellazzi et al. 2019). All of these studies highlighted the controversial role of autophagy in AD and how it changes according to neuronal architecture and the severity of cognitive impairment. The hippocampus and para-hippocampal regions of the post-mortem brain showed reduced expression of the autophagy kinase complex gene, according to recent proteomic studies. Interestingly, the same results were seen in mice with faulty autophagy and memory deficits when the BECN1-PIK3C3 complex was absent (Lipinski et al. 2010; Liu and Li 2019).

However, activation of the nuclear receptor-binding factor 2 (NRBF2), which is associated with the PI3K complex had a significant effect on the advancement of autophagy flux and the destruction of dangerous amyloid aggregates (Lachance et al. 2019). In the rat brain, resveratrol prevented ischemic damage when glutamine-induced excitotoxicity or middle cerebral artery blockage occurred, and stilbene triggered mitophagy in these mice by activating AMPK, Beclin 1, and LC3II conversion (Reddy and Oliver 2019).

Resveratrol increased Beclin 1, LC3II, and autophagy in mice with spinal cord injury, which improved motoneuron regeneration and decreased apoptosis. Resveratrol decreased p62/SQSTM, increased Sirtuin 1/LC3II, and enhanced locomotor activity and the number of dopaminergic neurons in the substantia nigra pars compacta in the PD mouse model that was produced by methyl-4-phenyl-1, 2, 3, 6-tetrahydropyridine (MPTP) (Di Meco et al. 2020). In AD and PD, sirtuin 1 was critical to sustaining resistance to neurotoxic aggregates by inhibiting mTOR and promoting autophagy. Lastly, resveratrol restored ATG4 and stimulated LC3 II and autophagy in neuroblastoma SH-SY5Y cells expressing mutant-Huntingtin treated with dopamine (Deng and Mi 2016).

14.4.1.2 Curcumin

The major active component of curry spices, curcumin, and curcuminoid from *Curcuma longa* herbs have been shown to induce autophagy in AD mice, which inhibits mTOR/Akt signaling and preserves LC3 (Liang et al. 2019). Curcumin protected an AD neuronal cell model (N2a/APP695swe) in vitro by restoring autophagy through Beclin 1 and enhancing retrograde axonal flow. Similarly, curcumin-stimulating solid lipid nanoparticles induced CMA in human neuroblastoma cells when applied to neuronal cells from mice that were confronted with toxic beta-amyloid-inducing molecular chaperones and lysosomal activity (Perrone et al. 2019). Unfortunately, despite curcuminoids' benefits in experimental animal models, there is still disagreement about their therapeutic efficacy in AD. However, curcumin has been successfully employed as a fluorescent indicator in the retinas of both humans and animals in the early stages of AD due to its high capacity to bind β-amyloid fibrils (Jaroonwitchawan et al. 2017). Curcumin was shown to preserve neurons in hippocampal areas in mice injected with amyloid by inhibiting abnormal Beclin 1 and autophagosome synthesis. Curcumin was similarly effective in clearing α-synuclein aggregates and controlling autophagy in the PD in a vitro model of dopaminergic neurons (Li et al. 2018). Additionally, a curcumin-analog chemical, independent of mTOR, stimulated autophagy in murine neuroblastoma cells and in vivo via TFEB-lysosome signaling (Zhang et al. 2018).

14.4.1.3 Green Tea Polyphenols (EGCG) and Phenolicoleosides (OLE)

Green tea catechins, especially epigallocatechin gallate (EGCG) from *Camellia sinensis*, are diffuse phenols with several well-known antioxidant, anti-inflammatory, and neuroprotective properties. They also activate Beclin 1, autophagy, and mitophagy (Zholos et al. 2019). Researchers demonstrated that OLE may promote autophagy in AD TgRND8 mice and neuroblastoma cells in vitro. While blocking mTOR by activating AMPK, CAMKKβ (calmodulin-dependent kinase β), and releasing calcium from the sarcoplasmic reticulum. It is intriguing to note that OLE is a pleiotropic chemical that may regulate sirtuins and, therefore, when AD mice activate their Atg genes (Achour et al. 2016; Al Rihani et al. 2019).

14.4.1.4 Spermidine

Spermidine is a nontoxic polyamine that is present in many seeds and plant-based foods, including soy and wheat germ. It makes flies, yeast, and worms live longer and enhances cognitive performance in mice and the elderly (Yang et al. 2016; Schwarz et al. 2018). Furthermore, by inhibiting inhibitory acetyltransferase EP300, spermidine, an anti-inflammatory drug, decreased experimental autoimmune encephalomyelitis in animals and promoted autophagy (Muñoz-Esparza et al. 2019).

14.4.1.5 Omega-3 Fatty Acids and Their Neuroprotective Effects

Certain dietary components, like omega-3 fatty acids and docosahexaenoic acid (DHA), which are part of the Mediterranean diet, have been shown to contribute to decreasing the risk of dementia (Lu et al. 2021a). In the Framingham Progeny Research, higher levels of red blood cell DHA were linked to a lower incidence of dementia associated with AD and all-cause dementia. More DHA is also associated with fewer structural alterations on brain imaging in cognitively normal individuals homozygous for the APOE "4" gene (Lehtisalo et al. 2019; Smyth et al. 2015). Higher levels of alpha-linolenic acid, stearidonic acid, and eicosatrienoic acid among omega-3 fatty acids were linked to better achievement on fluid intelligence tasks, and in the Doetinchem Cohort Study, greater quantities of alpha-linolenic acid and related compounds were linked to slower cognitive loss in global cognition and memory in middle-aged and older individuals (Gehlich et al. 2020; Mao et al. 2019). Additionally, APOE "4" carriers show preservation of memory capacity and reduced amyloid load, and one weekly meal of fresh seafood and long-chain n-3 fatty acids was found to reduce the risk of cognitive decline in several categories (Morris et al. 2018; Li et al. 2019). Once more, however, the effects are more limited when dementia is already present. A randomized, placebo-controlled study found that high dosages of omega-3 fatty acids, including DHA, had no beneficial effects on patients with AD (Jiang et al. 2017). However, the effects are more limited when dementia is already present. A randomized, placebo-controlled study found that high dosages of omega-3 fatty acids, including DHA, had no beneficial effects on patients with AD (Yuan et al. 2019). Numerous studies have also connected a higher lifetime stress level to a higher chance of developing AD. Glucocorticoid-induced activation of the corticotrophin-releasing factor receptor increased the likelihood that mice under extended stress would accumulate potentially hazardous levels of activated tau proteins, according to animal studies (Zhang et al. 2017). Clinical investigations that assessed participants' psychological stress levels before tracking them for AD development found that those who were more likely to experience anxiety or depression had a twofold increased risk of developing AD (Wood et al. 2022).

14.4.1.6 l-Theanine and Neuroprotection

One frequent source of theanine is tea leaves (*Camellia sinensis*). It shares a similar chemical composition to glutamate; a neurotransmitter linked to memory. Theanine, which is able to pass the blood-brain barrier, has been shown to have a cerebroprotective effect and to prevent the death of neurons after brief cerebral ischemia. Theanine's antagonistic action on AMPA and kainate neurotransmitter receptor subclasses contributes to the protective impact on neurons; while its affinity is comparatively modest (Yokogoshi et al. 1995; Kakuda et al. 2000).

14.5 Lifestyle Interventions to Support Brain Longevity

14.5.1 The Role of Physical Activity in Enhancing Autophagy

Exercise training brings about multiple skeletal muscle adaptations. Resistance training is characterized by hypertrophy, or an increase in the cross-sectional area of muscle fibers, whereas endurance training is defined by metabolic adaptations. Increased muscle mass from resistance exercise training leads to improved strength and achievement, which is critical for daily bodily functions as well as sports abilities (Børsheim and Bahr 2003; Van Hooren et al. 2024). A net increase in myofilament monomers makes this possible. Research demonstrates that a single session of resistance exercise can enhance translational start and protein synthesis for up to 48 hours. The significance of a well-controlled balance between protein synthesis and degradation in the regulation of muscle growth is highlighted by the fact that resistance training also enhances protein breakdown for up to 24 hours (Hesketh et al. 2020). By raising skeletal muscle's maximum ability to oxidize and for metabolic efficiency, endurance exercise training enhances both health and endurance performance. Typically, we attribute these adaptations to the increased production of metabolic proteins (MacInnis et al. 2022).

14.5.2 Acute Exercise-Induced Autophagy

Several investigations have shown the autophagic reaction in skeletal muscle. The initial identification of vacuoles containing autophagic material in mouse fibers of muscles occurred 2–7 days following an exceptionally demanding treadmill running session (Xia et al. 2021; Mammucari et al. 2007; Rocchi and He 2017; He and Sumpter 2012; Roberts et al. 2023). Recent studies have revealed that the autophagic elimination of components of cells in human muscle is more complicated;

ultra-endurance running has been found to significantly raise the amount of LC3-II, while shorter exercise sessions have been found to decrease it. Recent research has also looked at how autophagy regulation is affected by strength and resistance training. In both young and old human muscle, it has been demonstrated that LC3-II protein and GABARAP mRNA decreased from three to twenty-four hours following a single resistance exercise session (Jamart et al. 2012; Sillanpää et al. 2020; Luk et al. 2021; Pinto et al. 2020; Sanchez et al. 2019).

We interpreted this as a reduction in autophagy activation. The postexercise consumption of essential amino acid residues was necessary for the reduction in LC3-II concentration after resistance exercise, indicating that the control of autophagy brought on by resistance exercise depends on the availability of nutrients (Pinto et al. 2020). Nevertheless, two hours after exercise, neither changes in total protein synthesis nor protein breakdown were linked to the apparent regulation of autophagy caused by resistance training combined with amino acid consumption (Holwerda et al. 2019). Since "resistance exercise" had no effect on LC3-II levels in rat skeletal muscle, rodent models of resistance training might not be relevant to human models. Nevertheless, it has been demonstrated that after high-resistance electrically driven contraction of rat skeletal muscles, the activity of VPS34, a part of the PI3K complex, increases, indicating that autophagy is triggered after recovery from resistance exercise in a manner similar to that of endurance exercise. To directly evaluate how resistance training affects autophagic flow in human muscle, more research is needed (Halling and Pilegaard 2017).

14.5.3 The Effects of Sleep on Autophagy and Brain Health

The average adult needs seven to eight hours of sleep every night for the best possible physical and mental wellness. We have noticed two significant alterations: the depth of sleep and the consistency of sleeping patterns. The health of the brain is unaffected by these changes that are a natural aspect of aging. However, there are drawbacks to getting less sleep overall (Hirshkowitz et al. 2015). The amount of time it takes older folks to fall asleep does not alter, but they frequently wake up early and have sleep disruptions during the night (Li et al. 2022; Casagrande et al. 2022; Carvalhas-Almeida et al. 2023). Furthermore, staying up late grows more challenging for older folks, who often get up sooner in the morning (Gulia and Kumar 2018; Ancoli-Israel 2005). Approximately 44% of seniors of 50 years of maturity or older ranked their sleep quality as extremely good or excellent, according to the AARP 2016 Sleep and Brain Health Survey. Of those asked, 84% said they woke up feeling largely refreshed, but just 33% said they were highly happy with the amount of sleep they had each day. According to 2016 research, more than half (56%) of those of 50 and older say they get up at around the same time every morning, but less than half (48%) go to bed at the same time, and just a third (33%) say they get seven to eight hours of sleep at night. Consistent sleep habits are important since researchers have discovered that obtaining less than eight hours of sleep can significantly damage attention, memory, and executive function (Walker 2017;

Lim and Dinges 2008). Light exposure greatly impacts the chemical balance responsible for maintaining sleep. We design treatment plans to address a variety of chronic illnesses and debilitating sleep issues that become more prevalent as people age. Fortunately, there are several lifestyle changes that can be made to counteract these detrimental changes (Brown et al. 2022; Boyce 2022; Coman et al. 2024; Medic et al. 2017; Altevogt and Colten 2006; Schuler and Hope 2020; Baranwal et al. 2023; Abe and Abe 2019). Sound sleep habits require regular sleep schedules, increasing exposure to sunshine, and learning about the deleterious effects of drugs. Additionally, proper melatonin secretion that serves as a crucial regulatory mechanism for additional chemicals in the brain to follow to maintain sleep patterns, depends on exposure to light (Czeisler and Gooley 2007; Foster 2020; Sun et al. 2019; Vitale et al. 2019; Ahmad et al. 2023; Zisapel 2018; Biggio et al. 2021). Elderly people using many medications can be quite worrisome and may change their normal sleep habits. Professionals should be aware of the negative effects that benzodiazepines, anticholinergics, hypnotics, barbiturates, antipsychotics, certain antidepressants, and prescription or over-the-counter (OTC) drugs can have on elderly patients. Additionally, patients should be urged to ask their doctors for more information regarding the possible harm that multi-drug regimens may cause to their sleep habits (Rapaport et al. 2025; Bulut and Isik 2022; Chiappini et al. 2022; Chiappini et al. 2020; Bernardo and Valentão 2024; McCuistion et al. 2021). Another frequent sleep condition in older persons is sleep apnea that is defined by the airways closing as you sleep, making it harder to breathe. According to research, with the right care, some of the adverse effects may be undone, resulting in better sleep and higher blood oxygen levels. Research has also demonstrated that when continuous positive airway pressure (CPAP) is used to treat Alzheimer's patients with sleep apnea, cognitive function may improve (Young et al. 2002; Sukhal et al. 2015).

14.5.4 Managing Chronic Stress to Protect Brain Function

Everybody has developed coping methods to deal with stress that is a constant occurrence in daily life. Prolonged stress can activate hormonal (corticosterone release) and neuroendocrine (limbic hypothalamic pituitary adrenal system) processes, among other pathophysiological consequences. Moreover, anxiety and depression can be triggered by these situations, which in turn increases oxidative stress and free radical generation. A vast range of interacting neuroendocrine and neural systems in an organism are activated by various stresses, leading to physiological and behavioral reactions (McEwen 2007; Fukuda et al. 2014; Chrousos 2009). Stressors cause arousal, which in turn alters the autonomic nervous system and behavior. However, mental and physical diseases arise when arousal rises to stress-like levels. Numerous animal models, like the chronic unpredictable stress paradigm and the olfactory bulbectomy model, have been created to study neurological problems brought on by prolonged stress. The underlying molecular mechanism in chronic stress disease is better understood, and different novel chemical entities are screened using these animal models (Ulrich-Lai and Herman 2009).

In general, stress and stressors are described as the body's reactions to demands, which can originate from any circumstance or idea that causes frustration, rage, anxiety, or nervousness. Because of its great susceptibility to NDDs brought on by stress, the brain is the target of several stressors (Sharma and Pal 2021; Bystritsky and Kronemyer 2014; Prabu 2015; Lim et al. 2023; Searle and Tuckey 2017; Migliore and Coppedè 2009; Radi et al. 2014; Rekatsina et al. 2020). A healthy stress system is necessary for the appropriate growth and maintenance of life and species. In addition to causing psychiatric, endocrine/metabolic, and/or immunological illnesses or susceptibility to them, maladaptive neuroendocrine responses, or dysregulation of the stress system, can disrupt growth and development (Lu et al. 2021b; Gunnar and Quevedo 2007; Tsigos et al. 2020; Haavik et al. 2021; Kumar et al. 2013; Noushad et al. 2021). Chronic stress grows longer and increases the level of CRF expression in the amygdala, along with additional brain regions associated with emotion and fear, and it is believed that this elevated expression of CRF causes health-related stress issues such as anxiety, depression, and infertility. Oxidative stress contributes to the destruction of neurons in the CNS as people age and acquire neurodegenerative diseases. Reactive oxygen species (ROS) production rises dramatically under several hazardous stress conditions. One component that may be responsible for the stress-induced amplification of free radicals is the rise in nitric oxide (NO) production (Kumar et al. 2013; Halliwell and Gutteridge 2015; Keifer Jr et al. 2015; Suvrathan et al. 2014; Regev et al. 2012; McEwen 2006; Houldsworth 2024; Kamboj et al. 2024; Wankhede et al. 2025; Rauf et al. 2024; Chandimali et al. 2025; Puppala et al. 2024a). Research has shown that stress depletes several enzymes responsible for detoxifying free radicals, which leads to oxidative burden; therefore, stress is connected to the pathophysiology of several disease conditions.

14.5.5 Cognitive Training and Neuroplasticity for Lifelong Learning

This section discusses five key issues that are crucial to the study of cognitive plasticity. The first inquiry explores the relationship between cognitive training and plasticity at various levels, along with the levels of complexity (behavior and brain) at which adaptation has been or needs to be studied. The second question asks about the methods or processes used to achieve plasticity; the third asks about the temporal durability of cognitive plasticity; the fourth asks about the transfer of learning impacts to abilities and methods other than those that training is intended to target; and the fifth asks about the various contexts in which cognitive adaptability needs to be studied (Pascual-Leone et al. 2005; Willis and Schaie 2009; Kachanov 2004; Appelbaum et al. 2023; Navarro and Calero 2018; Slagter et al. 2011; Keshavan et al. 2014; Finnerty et al. 2015; Nobre and Van Ede 2018; Ford et al. 2018; Tonhäuser and Büker 2016; Tan et al. 2018; Chen and Wang 2025; Demirtaş and Kara 2022; Singh et al. 2024). Research on cognitive training has concentrated on global cognition, higher-order cognitive structures, primary mental talents, and cognitive processes. It has also been investigated how noncognitive therapies affect cognition. The majority of behavioral cognitive therapies have concentrated on skills and processes linked to cognitive impairment or that show comparatively early age-related decline (Stern 2009; Jaeggi et al. 2008). The brain regions and structures of interest have been compared to the abilities that are being trained. While the prefrontal and temporal areas of the brain support executive control, fluid skills, and some memory processes, the prefrontal regions have typically been shown to have the largest normative shrinkage in cortical volume (Miller and Cohen 2001).

14.5.6 Fasting, Intermittent Fasting, and Autophagy in Aging

Age and age-related disorders are significantly influenced by macroautophagy, an early-stage cellular response to physiological and pathological stresses (Fig. 14.1). The efficacy of autophagy decreases with age in animals, resulting in the buildup of defective proteins and cellular detritus, which fuels tissue degeneration and cellular senescence. Numerous age-related illnesses, such as metabolic syndrome, CVDs, and NDDs, are linked to impaired autophagy. By improving autophagy by lifestyle interventions like exercise, CR, fasting, and calorie restriction mimetic agents, it may be able to promote healthy aging and reduce the detrimental effects of aging on cellular and systemic health (Rubinsztein et al. 2011; López-Otín et al. 2013). Intermittent fasting (IF) or CR can promote adaptive autophagy and lengthen the lifespan of eukaryotic cells; however, prolonged calorie restriction coupled with an excessive autophagic response may be detrimental and lead to type II autophagic cell death. There is still much to learn about the effects of temporal calorie restriction during diurnal intermittent fasting throughout the holy month of Ramadan on autophagic gene expressions. Approximately 1.5 billion Muslims worldwide observe Ramadan IF (RIF), during which healthy individuals must refrain from food and liquids, including water, sexual activity, and smoking from sunrise to sunset. Moreover, RIF improves indicators of oxidative stress, inflammation, and liver function, increases body weight and arrangement, and reduces the risk factors for cardiometabolic diseases. Knowledge of important genes such as LAMP2, ATG4D, ATG5, and LC3B is necessary to comprehend the mechanics of autophagy, particularly in connection with IF (Raza 2024; Shabkhizan et al. 2023; Hofer et al. 2022; Malhab et al. 2025; Hoong and Chua 2021; Visioli et al. 2022; Hoddy et al. 2020; Strilbytska et al. 2024; Elmajnoun et al. 2020; Erlangga et al. 2023; Mehrabani et al. 2020; Tang et al. 2023; Tosti et al. 2023; Brogi et al. 2024).

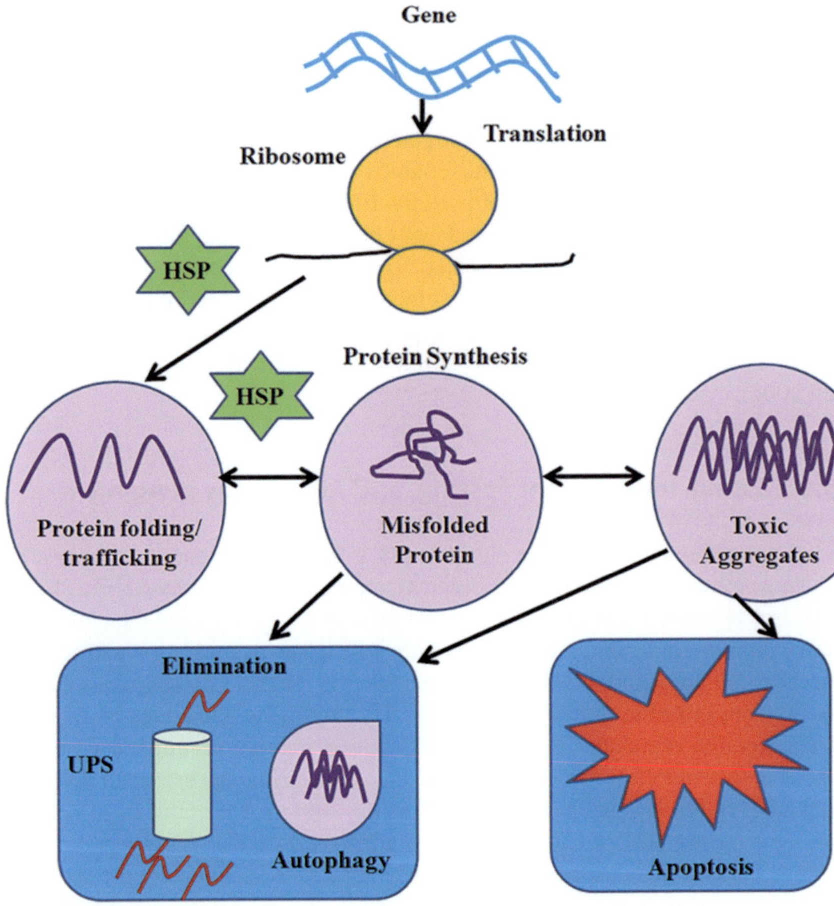

Fig. 14.1 Cell death by apoptosis and cellular proteostasis. The protein pool quality control mechanism known as the proteostasis network combines the processes involved in protein synthesis, including transformation folding of protein, protein unfolding, secretion, trafficking, and degradation or elimination. Chaperone proteins, also known as heat shock proteins or Hsps, are essential for protein folding, unfolding, refolding, and trafficking. Autophagy and ubiquitin-proteasome are two proteolytic mechanisms that break down damaged proteins during the elimination phase, which often occurs when misfolded proteins form hazardous aggregates. Cell death is prevented by the functioning of both systems. However, the cell experiences apoptosis when the proteostasis network is unable to prevent the buildup of protein aggregates. Because defective proteins and cells are eliminated via the degradation system and apoptosis, respectively, both processes are crucial for maintaining organism homeostasis

14.6 Clinical Evidence and Real-World Applications

14.6.1 Review of Clinical Trials on Nutraceuticals and Brain Health

The proteostasis network is a system for maintaining protein quality that combines the processes involved in protein synthesis, including protein folding, unfolding, secretion, trafficking, and degradation or elimination. Chaperone proteins, also known as heat shock proteins or Hsps, are essential for protein folding, unfolding, refolding, and trafficking. Autophagy and ubiquitin-proteasome are two proteolytic mechanisms that break down damaged proteins during the elimination phase, which often occurs when misfolded amino acids form hazardous aggregates. Both systems function to prevent cell death. However, the cell experiences apoptosis when the proteostasis network is unable to prevent the buildup of protein aggregates. Both the degradation system and apoptosis eliminate defective proteins and cells, making these processes crucial for maintaining organism homeostasis (Burns 2020; Jamora and Miyasaki 2017). The Mediterranean diet, characterized by a moderate consumption of wine, fish, and plant-based foods, has been shown to reduce the incidence and prevalence of some of the major clinical outcomes, such as diabetes, cancer, cardiovascular disease, and neurodegenerative disorders. Some specific nutritional food categories are rich in micronutrients and vitamins, which have been called nutraceuticals. Numerous epidemiological and laboratory studies have shown that the Mediterranean diet is rich in nutraceuticals, which has led to a lot of interest in the diet in recent decades (Sofi et al. 2010; Scarmeas et al. 2006; Puppala et al. 2024b).

14.6.2 The Impact of Autophagy-Based Interventions on Brain Aging

Most organs gradually deteriorate as a result of aging, which also throws biological systems' homeostasis off. Autophagy, an essential part of metabolic response to stress and cellular quality control, largely controls aging phenotypes (Fig. 14.2). The autophagy-lysosomal system that is extensively controlled by several signaling pathways that vary with aging, is associated '[]with complicated alterations (Kitada and Koya 2021; He et al. 2024; Moldakozhayev and Gladyshev 2023; West et al. 2019; Murrow and Debnath 2013; Ren et al. 2017; Deretic and Kroemer 2022; Palmer et al. 2025). It has long been ['=believed that lysosome function, the last stage of autophagy, declines with age. However, recent studies have shown that aged primary neurons in culture have decreased lysosomal acidification and protease activity. This is supported by the fact that older mouse brains express less v-ATPase subunit. Lack of the CCT complex can also result in age-related rises in lysosomal pH by preventing the voa1 component of v-ATPase from being trafficked to this organelle. Aging also affects cellular quality control, particularly mitophagy and selective autophagy. Research demonstrates that disruption of mitophagy

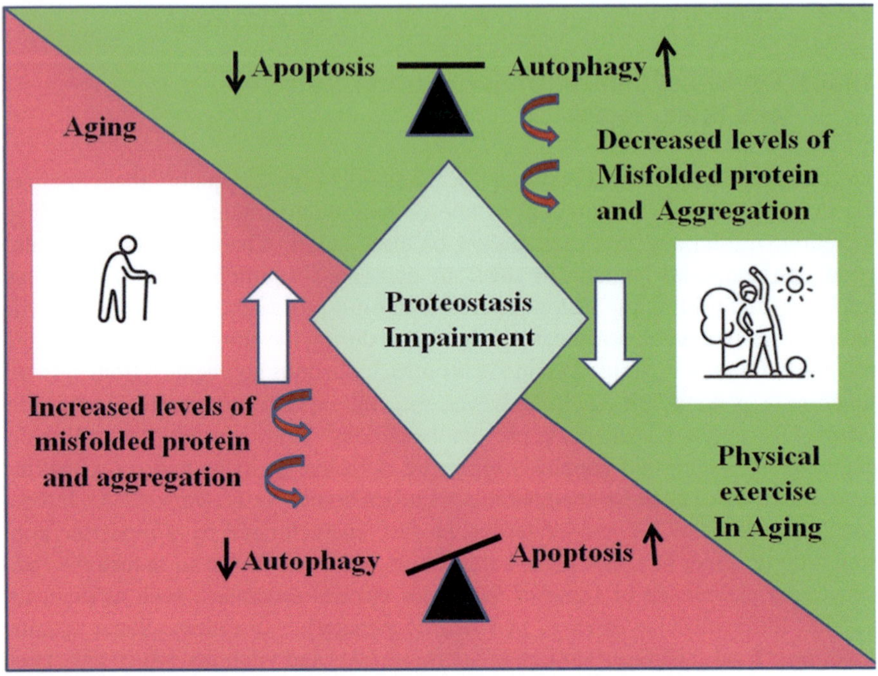

Fig. 14.2 The central nervous system exhibits greater proteostasis impairment with aging (red box), which may contribute to a rise in misfolded protein levels by reducing the autophagic process. There is an increase in programmed cell death and a loss of the equilibrium between autophagy and apoptosis. Otherwise, aging-related physical activity (green box) can partially reverse the imbalance seen in aging by increasing the autophagic process, reducing proteostasis impairment, and lowering levels of toxic protein aggregates and misfolded proteins, which reduce the activation of apoptosis

mediates aging features, including senescence, in both physiologically aged and progeria-patient-derived cells. Recent research has demonstrated that the aged mouse brain's enhanced mitophagy is a conserved mechanism that lowers neuroinflammation associated with aging. However, it is still uncertain if neuronal mitophagy increases or decreases with age. To clarify the true effect of aging on mitophagy, more research should concentrate on employing assays that measure flow through the pathway to differentiate between increased mitophagy and a block in mitophagic clearance that results in a buildup of mitophagy intermediates (Nixon 2020; Guerrero-Navarro et al. 2022; Song et al. 2020; Lee et al. 2010; Quick et al. 2023; Xie et al. 2022; Nixon and Yang 2012; Youle and Narendra 2011; Theurey and Pizzo 2018; Li et al. 2025; Fang et al. 2016; De Gaetano et al. 2021; Wang et al. 2022b; Mishra and Thakur 2023; Lautrup et al. 2019; Lu et al. 2025; Rappe et al. 2024; Fivenson et al. 2017; Williams et al. 2017; Georgakopoulos et al. 2017; Yang et al. 2018).

14.7 Conclusions and Future Perspective

In the pursuit of brain longevity, a multifaceted approach combining lifestyle modifications, dietary interventions, and autophagy-enhancing strategies proves to be a promising solution. The natural decline in cognitive function and brain health with aging can be mitigated through the activation of autophagy, which plays a crucial role in cellular maintenance and neuroprotection. Scientific evidence underscores the efficacy of nutraceuticals, including polyphenols, omega-3 fatty acids, and curcumin, in supporting brain function and reducing neurodegenerative risks. Additionally, intermittent fasting, regular exercise, and stress management techniques such as mindfulness and meditation further reinforce cognitive resilience. By integrating these scientifically backed strategies, individuals can significantly enhance cognitive function, prevent neurological disorders, and sustain mental clarity well into old age. The synergy between nutrition, lifestyle, and cellular processes like autophagy highlights the necessity of a holistic approach to brain health. Future research should continue exploring the potential of personalized interventions to optimize brain longevity further. In the end, taking proactive measures to preserve brain health promotes independence and mental clarity as people age, increases cognitive longevity, and enhances overall quality of life. Future research on brain longevity should focus on personalized interventions, leveraging advancements in genetics and neuroscience to develop targeted therapies for neurodegenerative diseases. The role of artificial intelligence in predicting cognitive decline and recommending preventive strategies holds enormous potential. The long-term impacts of autophagy-enhancing techniques and nutraceuticals on brain health also require more research. Real-time insights for early intervention can be obtained by integrating wearable technology to track lifestyle choices and cognitive performance. Extending brain health and lifespan will require a multidisciplinary strategy that incorporates lifestyle, technology, and medicinal approaches.

Acknowledgments We thankfully acknowledge the infrastructure and fellowship support to Md. Abubakar extended by the Department of Pharmaceuticals, Ministry of Chemicals and fertilizers, Government of India and NIPER HAJIPUR and Dr. Nitesh Kumar for the support and guidance in executing this work plan.

Author Contributions Krishnendu Adhikary and Md. Abubakar were involved in the conceptualization, writing of the original draft, mechanistic insight, input on critical points, and editing; Krishnendu Ganguly, Riya Sarkar, Abhishek Ambawatiya, and Riya Sarkar contributed to the manuscript writing; Parimala Mohanty contributed to the manuscript editing and critical points; and Nitesh Kumar was involved in the conceptualization, manuscript writing, reviewing, conceptualization of figures, and approved the version to be published.

Availability of Data On request

Conflict of Interest None

Funding None

References

Aardema MJ, MacGregor JT (2002) Toxicology and genetic toxicology in the new era of "toxicogenomics": impact of "-omics" technologies. Mutat Res 499(1):13–25

Abe M, Abe H (2019) Lifestyle medicine–an evidence based approach to nutrition, sleep, physical activity, and stress management on health and chronic illness. Pers Med Univ 8:3–9

Abubakar M et al (2023) Epigenetics of rhinovirus. In: Targeting epigenetics in inflammatory lung diseases. Springer, pp 185–197

Abubakar M et al (2024) GLP-1/GIP agonist as an intriguing and ultimate remedy for combating Alzheimer's disease through its supporting DPP4 inhibitors: a review. Curr Top Med Chem 24(19):1635–1664

Achour I et al (2016) Oleuropein prevents neuronal death, mitigates mitochondrial superoxide production and modulates autophagy in a dopaminergic cellular model. Int J Mol Sci 17(8):1293

Addo KM, Khan H (2024) Factors affecting healthy aging and its interconnected pathways. Turk J Healthy Aging Med 1:9–24

Ahmad SB et al (2023) Melatonin and health: insights of melatonin action, biological functions, and associated disorders. Cell Mol Neurobiol 43(6):2437–2458

Al Matared SAA et al (2024) The impact of aging on neurological function and cognitive decline. Saudi Med Horiz J 4(1):57–63

Al Rihani SB, Darakjian LI, Kaddoumi A (2019) Oleocanthal-rich extra-virgin olive oil restores the blood–brain barrier function through NLRP3 inflammasome inhibition simultaneously with autophagy induction in TgSwDI mice. ACS Chem Neurosci 10(8):3543–3554

Alnawwar MA et al (2023) The effect of physical activity on sleep quality and sleep disorder: a systematic review. Cureus 15(8)

Altevogt BM, Colten HR (2006) Sleep disorders and sleep deprivation: an unmet public health problem

Ancoli-Israel S (2005) Sleep and aging: prevalence of disturbed sleep and treatment considerations in older adults. J Clin Psychiatry 66:24

Anding AL, Baehrecke EH (2015) Autophagy in cell life and cell death. Curr Top Dev Biol 114:67–91

Appelbaum LG et al (2023) Synaptic plasticity and mental health: methods, challenges and opportunities. Neuropsychopharmacology 48(1):113–120

Arida RM, Teixeira-Machado L (2021) The contribution of physical exercise to brain resilience. Front Behav Neurosci 14:626769

Avan A et al (2022) Brain health: key to health, productivity, and well-being. Alzheimers Dement 18(7):1396–1407

Badadani M (2012) Autophagy mechanism, regulation, functions, and disorders. Int Sch Res Notices 2012(1):927064

Baranwal N, Phoebe KY, Siegel NS (2023) Sleep physiology, pathophysiology, and sleep hygiene. Prog Cardiovasc Dis 77:59–69

Bernardo J, Valentão P (2024) Herb-drug interactions: a short review on central and peripheral nervous system drugs. Phytother Res 38(4):1903–1931

Biggio G et al (2021) Melatonin: from neurobiology to treatment. Brain Sci 11(9):1121

Blake SD (2025) The end of aging: 120 Years young: the longevity blueprint. eBookIt.com

Børsheim E, Bahr R (2003) Effect of exercise intensity, duration and mode on post-exercise oxygen consumption. Sports Med 33:1037–1060

Boyce PR (2022) Light, lighting and human health. Light Res Technol 54(2):101–144

Boyd RJ et al (2022) Neuroinflammation represents a common theme amongst genetic and environmental risk factors for Alzheimer and Parkinson diseases. J Neuroinflammation 19(1):223

Brogi S et al (2024) Intermittent fasting: myths, fakes and truth on this dietary regimen approach. Foods 13(13):1960

Brown TM et al (2022) Recommendations for daytime, evening, and nighttime indoor light exposure to best support physiology, sleep, and wakefulness in healthy adults. PLoS Biol 20(3):e3001571
Bulut EA, Isik AT (2022) Abuse/misuse of prescription medications in older adults. Clin Geriatr Med 38(1):85–97
Burns A (2020) Citicoline in the treatment of acute ischaemic stroke: an international, randomised, multicentre, placebo-controlled study (ICTUS trial). Lancet 396(10248):413–446
Bystritsky A, Kronemyer D (2014) Stress and anxiety: counterpart elements of the stress/anxiety complex. Psychiatr Clin 37(4):489–518
Cammisuli DM et al (2022) What does the brain have to keep working at its best? Resilience mechanisms such as antioxidants and brain/cognitive reserve for counteracting Alzheimer's disease degeneration. Biology 11(5):650
Carmona JJ, Michan S (2016) Biology of healthy aging and longevity. Rev Investig Clin 68(1):7–16
Carvalhas-Almeida C, Cavadas C, Álvaro AR (2023) The impact of insomnia on frailty and the hallmarks of aging. Aging Clin Exp Res 35(2):253–269
Casagrande M et al (2022) Sleep quality and aging: a systematic review on healthy older people, mild cognitive impairment and Alzheimer's disease. Int J Environ Res Public Health 19(14):8457
Castellazzi M et al (2019) Autophagy and mitophagy biomarkers are reduced in sera of patients with Alzheimer's disease and mild cognitive impairment. Sci Rep 9(1):20009
Castruita PA et al (2022) Genetic, social, and lifestyle drivers of healthy aging and longevity. Curr Genet Med Rep 10(3):25–34
Chandimali N et al (2025) Free radicals and their impact on health and antioxidant defenses: a review. Cell Death Discov 11(1):19
Chandrasekaran V et al (2023) Polyphenols, autophagy and neurodegenerative diseases: a review. Biomolecules 13(8):1196
Chen F, Wang D (2025) Cross-cultural pragmatic common ground in the digital age: a synthetic analysis based on 4e cognitive framework, distributed cognition, and cognitive adaptability. In: Distributed cognition, and cognitive adaptability
Chen Y et al (2022) Defining brain health: a concept analysis. Int J Geriatr Psychiatry 37(1)
Cheon SY et al (2019) Autophagy, cellular aging and age-related human diseases. Exper Neurobiol 28(6):643
Chiappini S et al (2020) Misuse of prescription and over-the-counter drugs to obtain illicit highs: how pharmacists can prevent abuse. Pharm J 305(7943):1–27
Chiappini S et al (2022) Misuse of anticholinergic medications: a systematic review. Biomedicine 10(2):355
Chib S et al (2025) Role of flavonoids in mitigating the pathological complexities and treatment hurdles in Alzheimer's disease. Phytother Res 39(2):747–775
Chiu H-F, Venkatakrishnan K, Wang C-K (2020) The role of nutraceuticals as a complementary therapy against various neurodegenerative diseases: a mini-review. J Tradit Complement Med 10(5):434–439
Chrousos GP (2009) Stress and disorders of the stress system. Nat Rev Endocrinol 5(7):374–381
Chun Y, Kim J (2018) Autophagy: an essential degradation program for cellular homeostasis and life. Cells 7(12):278
Coman L-I et al (2024) Smart solutions for diet-related disease management: connected care, remote health monitoring systems, and integrated insights for advanced evaluation. Appl Sci 14(6):2351
Corsetti G et al (2024) Importance of energy, dietary protein sources, and amino acid composition in the regulation of metabolism: an indissoluble dynamic combination for life. Nutrients 16(15):2417
Costello CM, Walters ME (2022) Integrating resiliency into neurobiologically-focused mental health counseling through mindfulness. J Creat Ment Health 17(3):350–362
Cummings JL, Morstorf T, Zhong K (2014) Alzheimer's disease drug-development pipeline: few candidates, frequent failures. Alzheimer's Res Ther 6:1–7

Cuyvers E, Sleegers K (2016) Genetic variations underlying Alzheimer's disease: evidence from genome-wide association studies and beyond. Lancet Neurol 15(8):857–868

Czeisler CA, Gooley JJ (2007) Sleep and circadian rhythms in humans. In: Cold Spring Harbor symposia on quantitative biology, vol 72. Cold Spring Harbor Laboratory Press

Davidson RJ, McEwen BS (2012) Social influences on neuroplasticity: stress and interventions to promote Well-being. Nat Neurosci 15(5):689–695

De Gaetano A et al (2021) Mitophagy and oxidative stress: the role of aging. Antioxidants 10(5):794

De Sousa RAL et al (2021) Physical exercise effects on the brain during COVID-19 pandemic: links between mental and cardiovascular health. Neurol Sci 42:1325–1334

Demirtaş AS, Kara A (2022) Cognitive control and flexibility as predictor of career adaptability in emerging adults. Kastamonu Educ J 30(2):390–397

Deng H, Mi M-T (2016) Resveratrol attenuates Aβ 25–35 caused neurotoxicity by inducing autophagy through the TYrRS-PARP1-SIRT1 signaling pathway. Neurochem Res 41:2367–2379

Deretic V, Kroemer G (2022) Autophagy in metabolism and quality control: opposing, complementary or interlinked functions? Autophagy 18(2):283–292

Desai D et al (2024) Exploring the role of circadian rhythms in sleep and recovery: a review article. Cureus 16(6)

Di Meco A et al (2020) Autophagy dysfunction in Alzheimer's disease: mechanistic insights and new therapeutic opportunities. Biol Psychiatry 87(9):797–807

Diab R et al (2024) Intermittent fasting regulates metabolic homeostasis and improves Cardiovascular Health. Cell Biochem Biophys 82(3):1583–1597

Dikic I, Elazar Z (2018) Mechanism and medical implications of mammalian autophagy. Nat Rev Mol Cell Biol 19(6):349–364

Djuricic I, Calder PC (2021) Beneficial outcomes of omega-6 and omega-3 polyunsaturated fatty acids on human health: an update for 2021. Nutrients 13(7):2421

Dominguez LJ et al (2021) Nutrition, physical activity, and other lifestyle factors in the prevention of cognitive decline and dementia. Nutrients 13(11):4080

Ejiohuo O et al (2024) Nourishing the mind: how food security influences mental wellbeing. Nutrients 16(4):501

Elmajnoun HK et al (2020) Ramadan 2020 and beyond in the midst of the COVID-19 pandemic: challenges and scientific evidence for action. Sudan J Med Sci 15(2):85–110

Erlangga Z et al (2023) The effect of prolonged intermittent fasting on autophagy, inflammasome and senescence genes expressions: an exploratory study in healthy young males. Hum Nutr Metab 32:200189

Fang EF et al (2016) Nuclear DNA damage signalling to mitochondria in ageing. Nat Rev Mol Cell Biol 17(5):308–321

Fekete M et al (2022) Nutrition strategies promoting healthy aging: from improvement of cardiovascular and brain health to prevention of age-associated diseases. Nutrients 15(1):47

Fernández-Sanz P, Ruiz-Gabarre D, García-Escudero V (2019) Modulating effect of diet on Alzheimer's disease. Diseases 7(1):12

Finnerty GT et al (2015) Time in cortical circuits. J Neurosci 35(41):13912–13916

Fivenson EM et al (2017) Mitophagy in neurodegeneration and aging. Neurochem Int 109:202–209

Fjell AM, Walhovd KB (2010) Structural brain changes in aging: courses, causes and cognitive consequences. Rev Neurosci 21(3):187–222

Flint J, Greenspan RJ, Kendler KS (2020) How genes influence behavior. Oxford University Press

Ford JK, Baldwin TT, Prasad J (2018) Transfer of training: the known and the unknown. Annu Rev Organ Psych Organ Behav 5(1):201–225

Foster RG (2020) Sleep, circadian rhythms and health. Interface Focus 10(3):20190098

Frank MG, Heller HC (2019) The function (s) of sleep. In: Sleep-wake neurobiology and pharmacology. Springer, pp 3–34

Fukuda Y et al (2014) Structural and functional characterization of the Geobacillus copper nitrite reductase: involvement of the unique N-terminal region in the interprotein electron transfer with its redox partner. Biochim Biophys Acta 1837(3):396–405

Gan T et al (2023) Modulation of the immunity and inflammation by autophagy. Med Commun 4(4):e311
García-García I et al (2023) Maintaining brain health across the lifespan. Neurosci Biobehav Rev 153:105365
Gehlich KH et al (2020) Consumption of fruits and vegetables: improved physical health, mental health, physical functioning and cognitive health in older adults from 11 European countries. Aging Ment Health 24(4):634–641
Georgakopoulos ND, Wells G, Campanella M (2017) The pharmacological regulation of cellular mitophagy. Nat Chem Biol 13(2):136–146
Glick D, Barth S, Macleod KF (2010) Autophagy: cellular and molecular mechanisms. J Pathol 221(1):3–12
Glisky EL (2007) Changes in cognitive function in human aging. Brain aging:3–20
Gómez-Virgilio L et al (2022) Autophagy: a key regulator of homeostasis and disease: an overview of molecular mechanisms and modulators. Cells 11(15):2262
González I, Morales MA, Rojas A (2020) Polyphenols and AGEs/RAGE axis. Trends in challenges. Food Res Int 129:108843
Grady CL et al (2006) Age-related changes in brain activity across the adult lifespan. J Cogn Neurosci 18(2):227–241
Guasch-Ferré M, Willett WC (2021) The Mediterranean diet and health: a comprehensive overview. J Intern Med 290(3):549–566
Guerrero-Navarro L, Jansen-Dürr P, Cavinato M (2022) Age-related lysosomal dysfunctions. Cells 11(12):1977
Gulia KK, Kumar VM (2018) Sleep disorders in the elderly: a growing challenge. Psychogeriatrics 18(3):155–165
Gunnar M, Quevedo K (2007) The neurobiology of stress and development. Annu Rev Psychol 58(1):145–173
Haavik H et al (2021) The potential mechanisms of high-velocity, low-amplitude, controlled vertebral thrusts on neuroimmune function: a narrative review. Medicina 57(6):536
Halling JF, Pilegaard H (2017) Autophagy-dependent beneficial effects of exercise. Cold Spring Harb Perspect Med 7(8):a029777
Halliwell B, Gutteridge JM (2015) Free radicals in biology and medicine. Oxford University Press
Hansen M, Rubinsztein DC, Walker DW (2018) Autophagy as a promoter of longevity: insights from model organisms. Nat Rev Mol Cell Biol 19(9):579–593
Hargreaves D et al (2022) Strategies and interventions for healthy adolescent growth, nutrition, and development. Lancet 399(10320):198–210
Harris KM, McDade TW (2018) The biosocial approach to human development, behavior, and health across the life course. RSF: the Russell Sage Foundation. J Soc Sci 4(4):2–26
Harris J et al (2017) Autophagy and inflammasomes. Mol Immunol 86:10–15
He C, Sumpter J, Rhea, Levine B (2012) Exercise induces autophagy in peripheral tissues and in the brain. Autophagy 8(10):1548–1551
He Y et al (2024) Targeting lysosomal quality control as a therapeutic strategy against aging and diseases. Med Res Rev 44(6):2472–2509
Henn D, Yang X, Li M (2025) Lysosomal quality control review. Autophagy:1–20
Hesketh SJ et al (2020) The application of proteomics in muscle exercise physiology. Expert Rev Proteomics 17(11–12):813–825
Hirshkowitz M et al (2015) National Sleep Foundation's sleep time duration recommendations: methodology and results summary. Sleep Health 1(1):40–43
Hoddy KK et al (2020) Intermittent fasting and metabolic health: from religious fast to time-restricted feeding. Obesity 28:S29–S37
Hofer SJ et al (2022) The ups and downs of caloric restriction and fasting: from molecular effects to clinical application. EMBO Mol Med 14(1):e14418
Holwerda AM et al (2019) Leucine coingestion augments the muscle protein synthetic response to the ingestion of 15 g of protein following resistance exercise in older men. Am J Physiol Endocrinol Metab 317(3):E473–E482

Hoong CW, Chua MW (2021) SGLT2 inhibitors as calorie restriction mimetics: insights on longevity pathways and age-related diseases. Endocrinology 162(8):bqab079
Hornedo-Ortega R et al (2018) Phenolic compounds characteristic of the mediterranean diet in mitigating microglia-mediated neuroinflammation. Front Cell Neurosci 12:373
Houldsworth A (2024) Role of oxidative stress in neurodegenerative disorders: a review of reactive oxygen species and prevention by antioxidants. Brain Commun 6(1):fcad356
Iorio R, Celenza G, Petricca S (2021) Mitophagy: molecular mechanisms, new concepts on parkin activation and the emerging role of AMPK/ULK1 axis. Cells 11(1):30
Izzy S et al (2023) Long-term risk of cardiovascular disease after traumatic brain injury: screening and prevention. Lancet Neurol 22(10):959–970
Jabbar M, Yusoff MM (2022) And a. GeoJournal, Shafie, Assessing the role of urban green spaces for human well-being: a systematic review, pp 1–19
Jaeggi SM et al (2008) Improving fluid intelligence with training on working memory. Proc Natl Acad Sci 105(19):6829–6833
Jamart C et al (2012) Autophagy-related and autophagy-regulatory genes are induced in human muscle after ultraendurance exercise. Eur J Appl Physiol 112:3173–3177
Jamora RDG, Miyasaki JM (2017) Treatment gaps in Parkinson's disease care in The Philippines. Neurodegener Dis Manage 7(4):245–251
Jaroonwitchawan T et al (2017) Curcumin attenuates paraquat-induced cell death in human neuroblastoma cells through modulating oxidative stress and autophagy. Neurosci Lett 636:40–47
Jiang X et al (2017) Increased consumption of fruit and vegetables is related to a reduced risk of cognitive impairment and dementia: meta-analysis. Front Aging Neurosci 9:18
Kachanov LM (2004) Fundamentals of the theory of plasticity. Courier Corporation
Kakuda T et al (2000) Inhibiting effects of theanine on caffeine stimulation evaluated by EEG in the rat. Biosci Biotechnol Biochem 64(2):287–293
Kamboj S et al (2024) Exploring the therapeutic potential of phytoconstituents for addressing neurodegenerative disorders. Cent Nerv Syst Agents Med Chem 24(2):129–144
Keifer OP Jr et al (2015) The physiology of fear: reconceptualizing the role of the central amygdala in fear learning. Physiology 30(5):389–401
Kern A, Behl C (2009) The unsolved relationship of brain aging and late-onset Alzheimer disease. Biochim Biophys Acta 1790(10):1124–1132
Keshavan MS et al (2014) Cognitive training in mental disorders: update and future directions. Am J Psychiatry 171(5):510–522
Khalil MH (2024) The BDNF-interactive model for sustainable hippocampal neurogenesis in humans: synergistic effects of environmentally-mediated physical activity, cognitive stimulation, and mindfulness. Int J Mol Sci 25(23):12924
Kitada M, Koya D (2021) Autophagy in metabolic disease and ageing. Nat Rev Endocrinol 17(11):647–661
Kocot AM, Wróblewska B (2022) Nutritional strategies for autophagy activation and health consequences of autophagy impairment. Nutrition 103:111686
Kumar A, Singh A (2015) A review on Alzheimer's disease pathophysiology and its management: an update. Pharmacol Rep 67(2):195–203
Kumar A et al (2013) Stress: neurobiology, consequences and management. J Pharm Bioallied Sci 5(2):91–97
Lachance V et al (2019) Autophagy protein NRBF2 has reduced expression in Alzheimer's brains and modulates memory and amyloid-beta homeostasis in mice. Mol Neurodegener 14:1–13
Lakhman S, Murzello A, Gueits PG (2024) Physical activity and dietary interventions for mental health in ageing and longevity. In: Brain and mental health in ageing. Springer, pp 69–93
Lautrup S et al (2019) Microglial mitophagy mitigates neuroinflammation in Alzheimer's disease. Neurochem Int 129:104469
Lee J-H et al (2010) Lysosomal proteolysis and autophagy require presenilin 1 and are disrupted by Alzheimer-related PS1 mutations. Cell 141(7):1146–1158

Lehtisalo J et al (2019) Dietary changes and cognition over 2 years within a multidomain intervention trial—The Finnish Geriatric Intervention Study to Prevent Cognitive Impairment and Disability (FINGER). Alzheimers Dement 15(3):410–417
Leidal AM, Levine B, Debnath J (2018) Autophagy and the cell biology of age-related disease. Nat Cell Biol 20(12):1338–1348
Leong CF et al (2025) Therapeutic and pharmacological role of natural products in neurological diseases: targeting autophagy pathways. Progr Microb Mol Biol 8(1)
Leonidis A et al (2021) Improving stress management and sleep hygiene in intelligent homes. Sensors 21(7):2398
Lewis LD (2021) The interconnected causes and consequences of sleep in the brain. Science 374(6567):564–568
Li P et al (2018) Isoflavones induce BEX2-dependent autophagy to prevent ATR-induced neurotoxicity in SH-SY5Y cells. Cell Physiol Biochem 43(5):1866–1879
Li W et al (2019) The association between eating green vegetables every day and mild cognitive impairment: a community-based cross-sectional study in Shanghai. Neuropsychiatr Dis Treat Volume 15:3213–3218
Li J, Vitiello MV, Gooneratne NS (2022) Sleep in normal aging. Sleep Med Clin 17(2):161–171
Li Q et al (2025) Aging and lifestyle modifications for preventing aging-related diseases. FASEB J 39(9):e70575
Liang J et al (2019) Enhancing the retrograde axonal transport by curcumin promotes autophagic flux in N2a/APP695swe cells. Aging (Albany NY) 11(17):7036
Lim J, Dinges DF (2008) Sleep deprivation and vigilant attention. Ann N Y Acad Sci 1129(1):305–322
Lim T et al (2023) A transactional model of stress and coping applied to cyclist subjective experiences. Transport Res F 96:155–170
Lipinski MM et al (2010) Genome-wide analysis reveals mechanisms modulating autophagy in normal brain aging and in Alzheimer's disease. Proc Natl Acad Sci 107(32):14164–14169
Lipinski MM et al (2015) Function and mechanisms of autophagy in brain and spinal cord trauma. Antioxid Redox Signal 23(6):565–577
Liu J, Li L (2019) Targeting autophagy for the treatment of Alzheimer's disease: challenges and opportunities. Front Mol Neurosci 12:203
Liverman CT, Yaffe K, Blazer DG (2015) Cognitive aging: Progress in understanding and opportunities for action
Logan RW, McClung CA (2019) Rhythms of life: circadian disruption and brain disorders across the lifespan. Nat Rev Neurosci 20(1):49–65
López-Otín C et al (2013) The hallmarks of aging. Cell 153(6):1194–1217
Lu Y et al (2021a) Changes in a specific dietary pattern and incident dementia: a prospective cohort study. Clin Nutr 40(5):3495–3502
Lu S, Wei F, Li G (2021b) The evolution of the concept of stress and the framework of the stress system. Cell Stress 5(6):76
Lu Y et al (2025) Protective role of mitophagy on microglia-mediated neuroinflammatory injury through mtDNA-STING signaling in manganese-induced parkinsonism. J Neuroinflammation 22(1):55
Luk H-Y et al (2021) Differential autophagy response in men and women after muscle damage. Front Physiol 12:752347
MacInnis MJ, Egan B, Gibala MJ (2022) The effect of training on skeletal muscle and exercise metabolism. Exerc Metab:215–242
Mackieh R et al (2024) Unlocking the benefits of fasting: a review of its impact on various biological systems and human health. Curr Med Chem 31(14):1781–1803
Malecki KM et al (2022) Integrating environment and aging research: opportunities for synergy and acceleration. Front Aging Neurosci 14:824921
Malhab LJB et al (2025) Dawn-to-dusk intermittent fasting is associated with overexpression of autophagy genes: a prospective study on overweight and obese cohort. Clin Nutr ESPEN 65:209–217

Malik BR et al (2019) Autophagic and endo-lysosomal dysfunction in neurodegenerative disease. Mol Brain 12(1):100
Mammucari C et al (2007) FoxO3 controls autophagy in skeletal muscle in vivo. Cell Metab 6(6):458–471
Mao X et al (2019) Intake of vegetables and fruits through young adulthood is associated with better cognitive function in midlife in the US general population. J Nutr 149(8):1424–1433
Marin M-F et al (2011) Chronic stress, cognitive functioning and mental health. Neurobiol Learn Mem 96(4):583–595
Mattson MP et al (2018) Intermittent metabolic switching, neuroplasticity and brain health. Nat Rev Neurosci 19(2):81–94
Maugeri G et al (2021) Neuroprotective effects of physical activity via the adaptation of astrocytes. Cells 10(6):1542
McCuistion LE et al (2021) Pharmacology e-Book: a patient-centered nursing process approach. Elsevier Health Sciences
McEwen BS (2006) Protective and damaging effects of stress mediators: central role of the brain. Dialogues Clin Neurosci 8(4):367–381
McEwen BS (2007) Physiology and neurobiology of stress and adaptation: central role of the brain. Physiol Rev 87(3):873–904
Medic G, Wille M, Hemels ME (2017) Short-and long-term health consequences of sleep disruption. Nat Sci Sleep 9:151–161
Mehrabani S et al (2020) The effect of fasting or calorie restriction on mitophagy induction: a literature review. J Cachexia Sarcopenia Muscle 11(6):1447–1458
Meijer AJ, Codogno P (2009) Autophagy: regulation and role in disease. Crit Rev Clin Lab Sci 46(4):210–240
Migliore L, Coppedè F (2009) Environmental-induced oxidative stress in neurodegenerative disorders and aging. Mutat Res 674(1–2):73–84
Miller EK, Cohen JD (2001) An integrative theory of prefrontal cortex function. Annu Rev Neurosci 24(1):167–202
Mishra E, Thakur MK (2023) Mitophagy: a promising therapeutic target for neuroprotection during ageing and age-related diseases. Br J Pharmacol 180(12):1542–1561
Mizushima N, Levine B (2020) Autophagy in human diseases. N Engl J Med 383(16):1564–1576
Modak A et al (2023) The psychological benefits of breastfeeding: fostering maternal well-being and child development. Cureus 15(10)
Moffitt TE, Caspi A, Rutter M (2006) Measured gene-environment interactions in psychopathology: concepts, research strategies, and implications for research, intervention, and public understanding of genetics. Perspect Psychol Sci 1(1):5–27
Mohd Sairazi NS, Sirajudeen K (2020) Natural products and their bioactive compounds: neuroprotective potentials against neurodegenerative diseases. Evid Based Complement Alternat Med 2020(1):6565396
Moldakozhayev A, Gladyshev VN (2023) Metabolism, homeostasis, and aging. Trends Endocrinol Metab 34(3):158–169
Monkkonen T, Debnath J (2018) Inflammatory signaling cascades and autophagy in cancer. Autophagy 14(2):190–198
Moore MN (2020) Lysosomes, autophagy, and hormesis in cell physiology, pathology, and age-related disease. Dose-Response 18(3):1559325820934227
Morris MC et al (2018) Nutrients and bioactives in green leafy vegetables and cognitive decline: prospective study. Neurology 90(3):e214–e222
Mukherjee U et al (2024) Mechanisms, consequences and role of interventions for sleep deprivation: focus on mild cognitive impairment and Alzheimer's disease in elderly. Ageing Res Rev 100:102457
Muñoz-Esparza NC et al (2019) Polyamines in food. Front Nutr 6:108
Murrow L, Debnath J (2013) Autophagy as a stress-response and quality-control mechanism: implications for cell injury and human disease. Ann Rev Pathol 8(1):105–137

Navakkode S, Kennedy BK (2024) Neural ageing and synaptic plasticity: prioritizing brain health in healthy longevity. Front Aging Neurosci 16:1428244
Navarro E, Calero MD (2018) Cognitive plasticity in young-old adults and old-old adults and its relationship with successful aging. Geriatrics 3(4):76
Nguyen P-Y et al (2023) Effect of nature prescriptions on cardiometabolic and mental health, and physical activity: a systematic review. Lancet Planet Health 7(4):e313–e328
Nixon RA (2020) The aging lysosome: an essential catalyst for late-onset neurodegenerative diseases. Biochim Biophys Acta 1868(9):140443
Nixon RA, Rubinsztein DC (2024) Mechanisms of autophagy–lysosome dysfunction in neurodegenerative diseases. Nat Rev Mol Cell Biol 25(11):926–946
Nixon RA, Yang D-S (2012) Autophagy and neuronal cell death in neurological disorders. Cold Spring Harb Perspect Biol 4(10):a008839
Nobre AC, Van Ede F (2018) Anticipated moments: temporal structure in attention. Nat Rev Neurosci 19(1):34–48
Noushad S et al (2021) Physiological biomarkers of chronic stress: a systematic review. Int J Health Sci 15(5):46
Nussbaum RL, McInnes RR, Willard HF (2015) Thompson & Thompson genetics in medicine e-Book. Elsevier Health Sciences
Oliveri S et al (2018) A systematic review of the psychological implications of genetic testing: a comparative analysis among cardiovascular, neurodegenerative and cancer diseases. Front Genet 9:624
Olness K (2003) Effects on brain development leading to cognitive impairment: a worldwide epidemic. J Dev Behav Pediatr 24(2):120–130
Ortega MA et al (2024) Autophagy in its (proper) context: molecular basis, biological relevance, pharmacological modulation, and lifestyle medicine. Int J Biol Sci 20(7):2532
Palmer JE et al (2025) Autophagy, aging, and age-related neurodegeneration. Neuron 113(1):29–48
Pani G (2015) Neuroprotective effects of dietary restriction: evidence and mechanisms. In: Seminars in cell & developmental biology, vol 40. Elsevier
Park G, Kusuma IW, Kim Y-U (2018) Multiple bioactivities of traditional medicinal herbs for treatment of neurodegenerative diseases. Evid Based Complement Alternat Med 2018:3075458
Pascual-Leone A et al (2005) The plastic human brain cortex. Annu Rev Neurosci 28(1):377–401
Patra S et al (2023) Lysosome signaling in cell survival and programmed cell death for cellular homeostasis. J Cell Physiol 238(2):287–305
Perrone L et al (2019) The autophagy signaling pathway: a potential multifunctional therapeutic target of curcumin in neurological and neuromuscular diseases. Nutrients 11(8):1881
Pinto AP et al (2020) The combination of fasting, acute resistance exercise, and protein ingestion led to different responses of autophagy markers in gastrocnemius and liver samples. Nutrients 12(3):641
Pohl F, Kong Thoo Lin P (2018) The potential use of plant natural products and plant extracts with antioxidant properties for the prevention/treatment of neurodegenerative diseases: in vitro, in vivo and clinical trials. Molecules 23(12):3283
Prabu PS (2015) A study on academic stress among higher secondary students. Int J Humanit Soc Sci invent 4(10):63–68
Prasanth MI et al (2019) A review of the role of green tea (Camellia sinensis) in antiphotoaging, stress resistance, neuroprotection, and autophagy. Nutrients 11(2):474
Puleston DJ, Simon AK (2014) Autophagy in the immune system. Immunology 141(1):1–8
Puppala ER et al (2024a) Mesua assamica (King & Prain) kosterm. Bark ethanolic extract attenuates rheumatoid arthritis via down-regulating TLR4/NF-κB/COX-2/iNOS and activation of Nrf2/HO-1 pathways: a comprehensive study on in-vitro and in-vivo models. J Ethnopharmacol 335:118671
Puppala ER et al (2024b) Herbal medicines for the management of irritable Bowel syndrome and constipation problem. In: Role of herbal medicines: management of lifestyle diseases. Springer, pp 313–342

Quick JD et al (2023) Lysosomal acidification dysfunction in microglia: an emerging pathogenic mechanism of neuroinflammation and neurodegeneration. J Neuroinflammation 20(1):185
Radi E et al (2014) Apoptosis and oxidative stress in neurodegenerative diseases. J Alzheimer's Dis 42(s3):S125–S152
Rai SN (2021) Recent advances in the treatment of neurodegenerative disorders
Rai SN, Singh S, Singh SK (2024) Neurodegenerative diseases: translational models, mechanisms, and therapeutics. CRC Press
Rapaport P et al (2025) Long-term health conditions and their impact on people with sleep disturbances and dementia. BMC Geriatr 25(1):1–13
Rappe A et al (2024) Longitudinal autophagy profiling of the mammalian brain reveals sustained mitophagy throughout healthy aging. EMBO J 43(23):6199–6231
Rauf A et al (2024) Reactive oxygen species in biological systems: pathways, associated diseases, and potential inhibitors—a review. Food Sci Nutr 12(2):675–693
Raza S (2024) Autophagy and metabolic aging: current understanding and future applications. Biochim Biophys Acta:119753
Reddy PH, Oliver DM (2019) Amyloid beta and phosphorylated tau-induced defective autophagy and mitophagy in Alzheimer's disease. Cells 8(5):488
Regev L et al (2012) Site-specific genetic manipulation of amygdala corticotropin-releasing factor reveals its imperative role in mediating behavioral response to challenge. Biol Psychiatry 71(4):317–326
Rejeski WJ, Fanning J (2019) Models and theories of health behavior and clinical interventions in aging: a contemporary, integrative approach. Clin Interv Aging 14:1007–1019
Rekatsina M et al (2020) Pathophysiology and therapeutic perspectives of oxidative stress and neurodegenerative diseases: a narrative review. Adv Ther 37:113–139
Ren R et al (2017) Regulation of stem cell aging by metabolism and epigenetics. Cell Metab 26(3):460–474
Roberts MD et al (2023) Mechanisms of mechanical overload-induced skeletal muscle hypertrophy: current understanding and future directions. Physiol Rev 103(4):2679–2757
Rocchi A, He C (2017) Activating autophagy by aerobic exercise in mice. J Vis Exper 120:55099
Rubinsztein DC, Mariño G, Kroemer G (2011) Autophagy and aging. Cell 146(5):682–695
Sampedro-Piquero P, Alvarez-Suarez P, Begega A (2018) Coping with stress during aging: the importance of a resilient brain. Curr Neuropharmacol 16(3):284–296
Sanchez AM, Candau R, Bernardi H (2019) Recent data on cellular component turnover: focus on adaptations to physical exercise. Cells 8(6):542
Scarmeas N et al (2006) Mediterranean diet, Alzheimer disease, and vascular mediation. Arch Neurol 63(12):1709–1717
Schuler CB, Hope KM (2020) Circadian rhythm: light-dark cycles. In: Integrative and functional medical nutrition therapy: principles and practices, pp 577–594
Schwarz C et al (2018) Safety and tolerability of spermidine supplementation in mice and older adults with subjective cognitive decline. Aging (Albany NY) 10(1):19
Searle BJ, Tuckey MR (2017) Differentiating challenge, hindrance, and threat in the stress process. In: The Routledge companion to wellbeing at work. Routledge, pp 25–36
Shabkhizan R et al (2023) The beneficial and adverse effects of autophagic response to caloric restriction and fasting. Adv Nutr 14(5):1211–1225
Sharma SK, Pal M (2021) Stress and human body. Int J Phys Educ Sports Manage Yogic Sci 11(4):57–70
Sharma R et al (2024) Arsenic exposure and amyloid precursor protein processing: a focus on Alzheimer's disease. Curr Mol Pharmacol 17(1):e18761429272806
Sillanpää JK et al (2020) Dilution of seawater affects the Ca2+ transport in the outer mantle epithelium of Crassostrea gigas. Front Physiol 11:1
Singh JE (2020) Dietary sources of omega-3 fatty acids versus omega-3 fatty acid supplementation effects on cognition and inflammation. Curr Nutr Reports 9:264–277
Singh RB et al (2022) Why and how the Indo-Mediterranean diet may be superior to other diets: the role of antioxidants in the diet. Nutrients 14(4):898

Singh S, Rai SN, Singh SK (2024) Synaptic plasticity in neurodegenerative disorders. CRC Press
Slagter HA, Davidson RJ, Lutz A (2011) Mental training as a tool in the neuroscientific study of brain and cognitive plasticity. Front Hum Neurosci 5:17
Small GW et al (2006) Effects of a 14-day healthy longevity lifestyle program on cognition and brain function. Am J Geriatr Psychiatry 14(6):538–545
Smyth A et al (2015) Healthy eating and reduced risk of cognitive decline: a cohort from 40 countries. Neurology 84(22):2258–2265
Sofi F et al (2010) Accruing evidence on benefits of adherence to the Mediterranean diet on health: an updated systematic review and meta-analysis. Am J Clin Nutr 92(5):1189–1196
Song Q et al (2020) The emerging roles of vacuolar-type ATPase-dependent lysosomal acidification in neurodegenerative diseases. Transl Neurodegener 9:1–14
Stacchiotti A, Corsetti G (2020) Natural compounds and autophagy: allies against neurodegeneration. Front Cell Develop Biol 8:555409
Stapelberg NJ et al (2022) Environmental stressors and the PINE network: can physical environmental stressors drive long-term physical and mental health risks? Int J Environ Res Public Health 19(20):13226
Stern Y (2009) Cognitive reserve. Neuropsychologia 47(10):2015–2028
Strilbytska O et al (2024) Intermittent fasting and longevity: from animal models to implication for humans. Ageing Res Rev 96:102274
Sukhal S, Khalid M, Tulaimat A (2015) Effect of wakefulness-promoting agents on sleepiness in patients with sleep apnea treated with CPAP: a meta-analysis. J Clin Sleep Med 11(10):1179–1186
Sun Q et al (2019) Sleep problems in shift nurses: a brief review and recommendations at both individual and institutional levels. J Nurs Manag 27(1):10–18
Suvrathan A et al (2014) Stress enhances fear by forming new synapses with greater capacity for long-term potentiation in the amygdala. Philos Trans R Soc B 369(1633):20130151
Syme KL, Hagen EH (2020) Mental health is biological health: why tackling "diseases of the mind" is an imperative for biological anthropology in the 21st century. Am J Phys Anthropol 171:87–117
Tan C et al (2018) A survey on deep transfer learning. In: Artificial Neural Networks and Machine Learning–ICANN 2018: 27th International Conference on Artificial Neural Networks, Rhodes, Greece, October 4-7, 2018, Proceedings, Part III 27. Springer
Tang D et al (2023) Fasting: from physiology to pathology. Adv Sci 10(9):2204487
Tenchov R et al (2024) Antiaging strategies and remedies: a landscape of research progress and promise. ACS Chem Neurosci 15(3):408–446
Theurey P, Pizzo P (2018) The aging mitochondria. Genes 9(1):22
Tian YE et al (2024) Brain, lifestyle and environmental pathways linking physical and mental health. Nat Ment Health 2(10):1250–1261
Timpson NJ et al (2018) Genetic architecture: the shape of the genetic contribution to human traits and disease. Nat Rev Genet 19(2):110–124
Tomoh BO et al (2024) Innovative programs for community health: a model for addressing healthcare needs through collaborative relationships. Community Health 42:17
Tonhäuser C, Büker L (2016) Determinants of transfer of training: a comprehensive literature review. Int J Res Vocat Educ Train 3(2):127–165
Tosti V et al (2023) When a calorie is not a calorie: metabolic and molecular effects of intermittent fasting in humans; exploratory outcomes of a randomized clinical trial. Aging Biol 1:20230013
Tota M et al (2024) Environmental pollution and extreme weather conditions: insights into the effect on mental health. Front Psych 15:1389051
Tsigos C et al (2020) Stress: endocrine physiology and pathophysiology. Endotext [Internet]
Ugwu CN et al (2025) Sustainable development goals (SDGs) and resilient healthcare systems: addressing medicine and public health challenges in conflict zones. Medicine 104(7):e41535
Ulrich-Lai YM, Herman JP (2009) Neural regulation of endocrine and autonomic stress responses. Nat Rev Neurosci 10(6):397–409

Umegaki H, Sakurai T, Arai H (2021) Active life for brain health: a narrative review of the mechanism underlying the protective effects of physical activity on the brain. Front Aging Neurosci 13:761674

Van Hooren B, Aagaard P, Blazevich AJ (2024) Optimizing resistance training for sprint and endurance athletes: balancing positive and negative adaptations. Sports Med 54(12):3019–3050

Vauzour D et al (2017) Nutrition for the ageing brain: towards evidence for an optimal diet. Ageing Res Rev 35:222–240

Visioli F et al (2022) Traditional and medical applications of fasting. Nutrients 14(3):433

Vitale KC et al (2019) Sleep hygiene for optimizing recovery in athletes: review and recommendations. Int J Sports Med 40(08):535–543

Walker M (2017) Why we sleep: the new science of sleep and dreams. Penguin UK

Wang S et al (2022a) Role of AMPK in autophagy. Front Physiol 13:1015500

Wang Q et al (2022b) Role of mitophagy in the neurodegenerative diseases and its pharmacological advances: a review. Front Mol Neurosci 15:1014251

Wang L, Klionsky DJ, Shen H-M (2023) The emerging mechanisms and functions of microautophagy. Nat Rev Mol Cell Biol 24(3):186–203

Wankhede NL et al (2025) Etiology and treatment challenges for neurodegenerative disorders. In: Targeted therapy for the central nervous system. Elsevier, pp 1–25

West MD et al (2019) Toward a unified theory of aging and regeneration. Regen Med 14(9):867–886

Wickramasinghe K et al (2020) From lifespan to healthspan: the role of nutrition in healthy ageing. J Nutr Sci 9:e33

Wiedermann CJ et al (2023) Fortifying the foundations: a comprehensive approach to enhancing mental health support in educational policies amidst crises. In: Healthcare. MDPI

Williams JA et al (2017) New methods for monitoring mitochondrial biogenesis and mitophagy in vitro and in vivo. Exp Biol Med 242(8):781–787

Williams RJ, Mohanakumar KP, Beart PM (2021) Neuro-nutraceuticals: natural products nourish the brain but be aware of contrary effects. Elsevier, p 105159

Willis SL, Schaie KW (2009) Cognitive training and plasticity: theoretical perspective and methodological consequences. Restor Neurol Neurosci 27(5):375–389

Wood AH, Chappell HF, Zulyniak MA (2022) Dietary and supplemental long-chain omega-3 fatty acids as moderators of cognitive impairment and Alzheimer's disease. Eur J Nutr:1–16

Wu M-Y, Lu J-H (2019) Autophagy and macrophage functions: inflammatory response and phagocytosis. Cells 9(1):70

Xia Q et al (2021) The role of autophagy in skeletal muscle diseases. Front Physiol 12:638983

Xie YX et al (2022) Lysosomal exocytosis releases pathogenic α-synuclein species from neurons in synucleinopathy models. Nat Commun 13(1):4918

Yang Q et al (2016) Spermidine alleviates experimental autoimmune encephalomyelitis through inducing inhibitory macrophages. Cell Death Differ 23(11):1850–1861

Yang KC et al (2018) Evolution of tools and methods for monitoring autophagic flux in mammalian cells. Biochem Soc Trans 46(1):97–110

Yang J, Zhou R, Ma Z (2019) Autophagy and energy metabolism. In: Autophagy: Biology and diseases: basic science, pp 329–357.

Yin H et al (2018) The therapeutic and pathogenic role of autophagy in autoimmune diseases. Front Immunol 9:1512

Yokogoshi H et al (1995) Reduction effect of theanine on blood pressure and brain 5-hydroxyindoles in spontaneously hypertensive rats. Biosci Biotechnol Biochem 59(4):615–618

Youle RJ, Narendra DP (2011) Mechanisms of mitophagy. Nat Rev Mol Cell Biol 12(1):9–14

Young T, Peppard PE, Gottlieb DJ (2002) Epidemiology of obstructive sleep apnea: a population health perspective. Am J Respir Crit Care Med 165(9):1217–1239

Yuan C et al (2019) Long-term intake of vegetables and fruits and subjective cognitive function in US men. Neurology 92(1):e63–e75

Zhang S et al (2017) Citrus consumption and incident dementia in elderly Japanese: the Ohsaki Cohort 2006 Study. Br J Nutr 117(8):1174–1180

Zhang L et al (2018) The potential protective effect of curcumin on amyloid-β-42 induced cytotoxicity in HT-22 cells. Biomed Res Int 2018(1):8134902

Zholos AV, Moroz OF, Storozhuk MV (2019) Curcuminoids and novel opportunities for the treatment of Alzheimer's disease: which molecules are actually effective? Curr Mol Pharmacol 12(1):12–26

Zisapel N (2018) New perspectives on the role of melatonin in human sleep, circadian rhythms and their regulation. Br J Pharmacol 175(16):3190–3199

Zou L et al (2022) Autophagy and beyond: unraveling the complexity of UNC-51-like kinase 1 (ULK1) from biological functions to therapeutic implications. Acta Pharm Sin B 12(10):3743–3782

Zubko V, Kachalov O (2025) Physical activity as a way of influencing cognitive functions and brain activity of students during their studies at the university

Clinical Trials and Evidence-Based Nutraceutical Interventions

15

Shampa Ghosh, Krishna Kumar Singh, Manchala Raghunath, and Jitendra Kumar Sinha

Abstract

Nutraceuticals have acquired significant therapeutic relevance in the treatment and prevention of neurodegenerative illnesses, cognitive decline linked to obesity, and metabolic disorders connected with ageing. Understanding evidence-based nutraceutical therapies validated by clinical trials is essential in today's healthcare landscape. Significant bioactive compounds, including polyphenols, omega-3 fatty acids, probiotics, and neuroactive peptides, are currently being investigated for their effectiveness in modulating neuroinflammation, oxidative stress, mitochondrial function, and autophagy. Particular attention is directed towards the impact of nutraceuticals on the gut-brain axis, insulin resistance, and metabolic pathways related to cognitive function and ageing. It is essential to examine how neuroinflammation resulting from obesity accelerates brain ageing and highlights nutraceutical strategies that mitigate these effects. Furthermore, we address the methodologies of clinical trials for assessing the safety, efficacy, and translational significance of these therapies. This chapter integrates current findings to bridge the divide between clinical and preclinical studies, elucidating the potential use of nutraceuticals as adjuvant or preventative therapies for

S. Ghosh · J. K. Sinha (✉)
GloNeuro, Sector 107, Noida, Uttar Pradesh, India

K. K. Singh
Symbiosis Centre for Information Technology, Symbiosis International (Deemed University), Rajiv Gandhi InfoTech Park, Hinjawadi, Pune, Maharashtra, India
e-mail: krishnakumar@scit.edu

M. Raghunath
ICMR – National Institute of Nutrition, Tarnaka, Hyderabad, Telangana, India

A. Kumar Singh, S. Nand Rai (eds.), *Rejuvenating the Brain: Nutraceuticals, Autophagy, and Longevity*, Nutritional Neurosciences,
https://doi.org/10.1007/978-981-95-2790-8_15

neurological health and lifespan. The prospective development of personalised nutraceutical therapy and precision nutrition is considered for improving cognitive function and metabolic vitality in the elderly demographic.

Keywords

Nutraceuticals · Neuroprotection · Clinical trials · Ageing · Gut-brain axis · Obesity · Autophagy · Longevity · Metabolic health · Inflammation

Learning Objectives:

1. Identify ways in which dietary bioactive compounds are able to affect neurodegenerative processes, metabolic impairment, and age-related cognitive impairment.
2. Understand the central biological mechanisms: neuroinflammation, oxidative stress, mitochondrial impairment, and autophagy, which nutraceuticals target for therapeutic effect.
3. Distinguish between various study designs, primary outcome measures, and the primary challenges (e.g. compliance, placebo effects, bioavailability) in nutraceutical clinical trials.
4. Recognise how healthcare providers can integrate nutraceuticals into patient-focused therapies, taking into account patients' unique genetic/metabolic profiles and lifestyle.
5. Get familiar with new fields like individualised nutraceutical interventions, precision nutrition, and novel delivery systems, and the research gaps that must be filled.

15.1 Introduction

Nutraceuticals, a phrase that combines nutrition and pharmaceuticals, have garnered significant academic interest in recent decades. This increasing interest arises from their capacity to connect food-based therapies with traditional medicine. Stephen DeFelice coined the term in the late 1980s to describe "food or parts of food that provide medical or health benefits, including the prevention and treatment of disease" (DeFelice 1995); however, the overarching concept of utilising dietary components for therapeutic applications has origins in ancient medical traditions. For ages, Traditional Chinese Medicine and Ayurveda have highlighted the preventative and healing properties of specific plants and food-derived chemicals (Brower 1998). Nutraceuticals are becoming central to contemporary discourse on disease prevention, healthy ageing, and the therapy of chronic conditions, including cardiovascular disease, metabolic disorders, and neurodegenerative diseases (Rautiainen et al. 2016). The global transition to preventative healthcare, driven by the increasing prevalence of chronic diseases and an ageing demographic, has underscored the significance of lifestyle changes, particularly dietary interventions. Nutraceuticals are frequently regarded as safer alternatives or supplements to pharmaceutical treatments, although their application requires careful examination of safety, efficacy, and potential interactions with other therapies (Geller et al. 2015). This chapter emphasises the scientific and clinical factors crucial for incorporating nutraceuticals into contemporary healthcare. By emphasising evidence-based methods, it aims to harmonise the traditional dependence on anecdotal or observational data with the stringent requirements of modern clinical science.

Although the word nutraceutical lacks global standardisation, an often-used definition characterises these compounds as naturally sourced ingredients typically available in supplement or functional food formats that provide advantages beyond their fundamental nutritional worth (Shahidi and Ambigaipalan 2015). These may encompass vitamins, minerals, amino acids, fatty acids, probiotics, or phytochemicals, including polyphenols and flavonoids. The variety of chemicals categorised as nutraceuticals hinders the establishment of universal rules; in certain areas, they are regulated akin to dietary supplements, while in others, they are subject to the regulatory rigour of pharmaceutical medications (Coppens et al. 2006). Although there is no universally recognised definition, nutraceuticals uniformly exhibit bioactive potential. They can affect multiple biochemical pathways, including inflammatory cascades, metabolic processes, and mitochondrial function (Mathers 2017; Sinha et al. 2024). Certain polyphenols derived from fruits and vegetables exhibit antioxidant properties in vitro, while others demonstrate potential in modulating the gut microbiota, a burgeoning field of research associated with metabolic health and cognitive function (Gasmi et al. 2024). A crucial aspect of nutraceuticals is their marketability and consumer attraction. As populations become more cognisant of preventative health measures, markets have significantly expanded to offer functional beverages, fortified cereals, and over-the-counter nutritional supplements. This expansion poses challenges: Variations in product quality, purity, labelling accuracy, and manufacturing standards can be substantial, often leading to consumers facing inconsistent information or inferior results, especially concerning baby foods (Dickinson et al. 2015; Ghosh et al. 2019). Therefore, scientists and doctors advocate for stricter regulation and better public education to ensure that the potential benefits of nutraceuticals are realised without undue risks.

Modern healthcare increasingly adopts a preventative approach, emphasising risk reduction and the maintenance of optimal functionality rather than solely focusing on symptom management. Nutraceuticals complement this framework by enhancing conventional medical treatments or serving as part of a comprehensive strategy for chronic disease management. Their significance is derived from various factors. Chronic diseases, including type 2 diabetes, obesity, and cardiovascular disease, represent a significant portion of global morbidity and mortality (GBD 2019 Risk Factors Collaborators 2020). Epidemiological studies have linked these illnesses to lifestyle and dietary patterns, highlighting the potential for nutritional modifications and supplemental therapies to mitigate disease progression. In addition, there is increasing literature on the molecular mechanisms underlying the efficacy of nutraceuticals. Omega-3 fatty acids have been intensively investigated for their cardioprotective and anti-inflammatory activities. There is evidence to suggest that they affect eicosanoid production and cytokine network modulation to justify their role in cardiovascular risk control. The research study also examined the micronutrients' functions such as vitamin D and certain probiotics in immune modulation and communication within the gut-brain axis. Basically, the therapeutic potential of nutraceuticals is enhancing systemic and neurological well-being (Jacquier et al. 2024). The onset of nutrigenomics and personalised medicine is an opportunity for supplementing nutraceutical therapies. Higher numbers of genetic tests will motivate patients to adopt individually designed nutrition programmes in the form of taking specific supplements or functional foods for fighting genetic weakness or metabolic deficiencies (Mehta et al. 2024). These uses are major leaps, showing the growing relevance of nutraceuticals within a more tailored healthcare landscape. Much remains to be overcome. The regulatory uncertainty of nutraceuticals could lead to irregular product quality and significant differences in labelling claims. Together, this dual legislation can overwhelm both practitioners and consumers, who are forced to navigate an ever more complicated economy with no easy, standardised guidelines. In addition, although numerous nutraceuticals have been investigated in vitro or in small human trials, broad long-term clinical data are absent for most agents (Schultz and Garbe 2023). The absence of standardised procedures for evaluating outcomes like cognitive function, metabolic indicators, or quality of life complicates the ability to make definitive conclusions regarding efficacy. A burgeoning consensus advocates for stringent clinical validation to enhance confidence in nutraceutical application and to guide optimal patient care practices.

The definition of evidence-based medicine, as articulated by Sackett and associates, amalgamates the most reliable research outcomes with clinical acumen and patient preferences (Sackett et al. 1996). The implementation of this approach is especially relevant in the realm of nutraceuticals. Although historical and anecdotal evidence may guide researchers to potential candidates, it is the methodologically rigorous, peer-reviewed clinical trials that ascertain whether a certain nutraceutical yields clinically significant advantages. These trials often utilise randomised controlled designs, including suitable controls (including placebo groups), and assess validated outcomes that signify tangible health gains (Batailler et al. 2020). Evidence-based methodologies facilitate the systematic evaluation of safety. A widespread fallacy asserts that nutraceuticals, due to their "natural" status, pose negligible danger. While some nutraceuticals possess substantial safety margins, specific substances may elicit undesirable effects, especially when ingested at elevated amounts or in conjunction with prescription drugs (Geller et al. 2015). The bioavailability of nutraceuticals depends on formulation strategies and specific patient characteristics for instance, gut microbiota diversity (Ali Redha et al. 2024). The evidence-based research enhances clinical applicability as well. It is also vital in public health policy, as results from strictly conducted studies can influence recommendations for dietary supplements and functional foods. Regulatory agencies often rely on clinical evidence to differentiate between general health claims and those specifically related to the prevention or treatment of particular illnesses (Santini et al. 2018). Nutraceuticals can gain legitimacy through evidence-based approaches, which facilitates their adoption by healthcare practitioners and their integration into established treatment protocols.

15.2 Nutraceuticals in Neurodegenerative Disorders

Neurodegenerative diseases, such as Alzheimer's disease (AD) and Parkinson's disease (PD), are defined by the progressive loss of neuronal structure and function. Notwithstanding the differential clinical features of each disease, the pathophysiology often overlaps with key molecular mechanisms, including protein misfolding, chronic neuroinflammation, and oxidative stress (Sinha et al. 2014). Nutraceuticals have received academic interest in recent years for their potential to modulate these pathogenic mechanisms. In discussing the overview of AD and PD pathophysiology, it is critical to discuss four prominent nutraceuticals: curcumin, resveratrol, omega-3 fatty acids, and coenzyme Q10, and to provide an overview of the main clinical trials that have explored their therapeutic effectiveness.

15.2.1 Pathophysiology of Neurodegenerative Disorders

Neurodegenerative diseases represent a heterogeneous group of conditions, but they share certain common pathophysiological features. One important topic is the formation of misfolded proteins that have harmful effects on neurons. In AD, extracellular deposition of β-amyloid (Aβ) peptides, which are derived from the amyloid precursor protein, causes plaque formation in the brain, while intracellular accumulation of hyperphosphorylated tau protein gives rise to the formation of neurofibrillary tangles (Serrano-Pozo et al. 2011; Singh et al. 2025). Misfolded proteins inhibit synaptic contact, influence neuronal metabolism, and can stimulate microglia, initiating an inflammatory cascade that amplifies neuronal damage. Mitochondrial dysfunction is another key feature of AD, increasing oxidative stress and further eroding neuronal survival.

Parkinson's disease is marked by the specific degeneration of dopaminergic neurones in the substantia nigra pars compacta. A defining characteristic is the aggregation of α-synuclein (Lewy bodies) within neurones. This pathogenic process disrupts cellular homeostasis and hinders multiple physiological activities, including autophagy and proteasomal pathways (Dexter and Jenner 2013). Mitochondrial abnormalities are crucial, as numerous cases, especially those with genetic origins, are linked to alterations in genes that govern mitochondrial quality control (Kaur and Aran 2025). The resultant energy shortfalls and persistent oxidative stress establish a detrimental cycle, exacerbating neuronal death. In both AD and PD, persistent inflammation mediated by activated microglia and astrocytes exacerbates tissue damage (Tansey et al. 2022). The brain, previously seen as an immune-privileged region, is now recognised as being dynamically susceptible to both peripheral and central inflammatory signals. This pro-inflammatory environment can induce further oxidative stress, resulting in a harmful feedback loop. Considering this intricate pathophysiology, therapies that diminish inflammation, enhance mitochondrial function, and safeguard against oxidative damage are particularly promising in decelerating or altering the illness trajectory.

15.2.2 Essential Nutraceuticals and Their Mechanisms

15.2.2.1 Curcumin

Curcumin is a polyphenolic substance derived from the rhizome of Curcuma longa (turmeric). Curcumin, historically utilised in Asian food and traditional medicine, has attracted significant scientific interest due to its anti-inflammatory and antioxidant characteristics. Its chemical configuration enables it to eliminate reactive oxygen species (ROS) and reactive nitrogen species, hence alleviating oxidative stress (Hewlings and Kalman 2017). In the realm of neurodegeneration, curcumin may inhibit the nuclear factor-κB (NF-κB) pathway, thereby diminishing the synthesis of pro-inflammatory cytokines such as TNF-α and IL-1β. Curcumin attenuates the local inflammatory response by regulating the activity of enzymes such as cyclooxygenase-2 (COX-2) and lipoxygenase.

Curcumin seems to disrupt amyloid aggregation, specifically the generation of β-amyloid fibrils in AD models (Ringman et al. 2005). Certain in vitro and animal investigations suggest that curcumin can interact with Aβ peptides, diminishing their tendency to aggregate. Moreover, curcumin's ability to increase autophagy may facilitate the removal of misfolded proteins. A significant restriction is its limited bioavailability resulting from fast metabolism and inadequate intestinal absorption (Sabet et al. 2021). There is an urgent need for innovative formulations, e.g. liposomal encapsulations, nanoparticles to be explored to address this limitation and enhance medication delivery.

15.2.2.2 Resveratrol

Resveratrol is a natural polyphenolic compound mostly located in grapes, red wine, and berries. The neuroprotective benefits are frequently ascribed to the activation of sirtuins, especially SIRT1, a category of NAD^+-dependent deacetylases associated with ageing and metabolic control (Bastianetto and Quirion 2002). Resveratrol can boost mitochondrial biogenesis and function through the activation of SIRT1, potentially mitigating the energy deficiencies seen in AD and PD (Rakshe et al. 2024).

In AD models, resveratrol has been demonstrated to reduce β-amyloid formation and deposition via altering secretase enzymes and affecting amyloid precursor protein processing (Sawda et al. 2017). It may also block oligomer formation and minimise neuroinflammation via inhibition of pro-inflammatory cytokines. Resveratrol has been reported to have protective effects in models of PD on dopamine neurones, potentially through α-synuclein aggregation inhibition and inflammation. However, as with curcumin, bioavailability is the key issue; resveratrol undergoes extensive first-pass metabolism in the gut and liver and is thus required to be delivered by alternative methods or derivatives to reach appropriate therapeutic brain levels.

15.2.2.3 Omega-3 Fatty Acids

Omega-3 polyunsaturated fatty acids (PUFAs), particularly eicosapentaenoic acid (EPA) and docosahexaenoic acid (DHA), are essential constituents of neuronal membranes, affecting fluidity, signal transduction, and synaptic plasticity (Bazan 2018). Their anti-inflammatory effects are partially facilitated by the synthesis of resolvins and protectins, which attenuate inflammation and enhance neuronal viability. A low dietary intake of omega-3 fatty acids has been epidemiologically associated with increased risks of cognitive decline and dementia, leading to enquiries into their therapeutic potential.

DHA supplementation in AD models has been linked to less amyloid deposition and enhanced synaptic integrity, potentially by altering lipid raft composition and reducing the synthesis of pro-inflammatory mediators. Omega-3 fatty acids seem to safeguard dopaminergic neurones in PD models by reducing oxidative damage and maintaining mitochondrial function (Yang et al. 2024). The extent of these effects in humans varies significantly, likely because of changes in dosage, treatment duration, and baseline nutritional intakes among study populations. Concerns about the sustainability and quality of fish oil supplies have prompted interest in algal-based or krill oil supplements, highlighting the practical aspects of omega-3 supplementation.

15.2.2.4 Coenzyme Q10

Coenzyme Q10 (CoQ10), or ubiquinone, is an essential element of the mitochondrial electron transport chain, enabling ATP synthesis. It serves as a powerful lipophilic antioxidant, neutralising free radicals in cellular membranes (Spindler et al. 2009). Neurodegenerative disorders like PD frequently demonstrate mitochondrial impairment and energy deficiencies, rendering CoQ10 supplementation a compelling approach. Preclinical investigations indicate that CoQ10 may stabilise mitochondrial membranes, diminish lipid peroxidation, and augment ATP generation. Furthermore, it may have protective effects against α-synuclein–induced cellular damage in PD models.

In AD, CoQ10 has been investigated for its ability to alleviate oxidative stress in neurones by restoring diminished antioxidant defences (Gutierrez-Mariscal et al. 2020). Nonetheless, similar to numerous nutraceuticals, oral CoQ10 encounters bioavailability challenges. Elevated dosages and lipid-soluble formulations are frequently necessary to attain clinically significant concentrations. Current research seeks to enhance delivery techniques, such as nanocarriers and co-administration with dietary lipids, to maximise absorption into the central nervous system.

15.2.3 Clinical Trials and Evidence

A critical inquiry for neurologists, geriatricians, and other healthcare professionals is the extent to which preclinical findings on nutraceuticals correlate with clinical significance. In the past two decades, several pilot studies, randomised controlled trials, and meta-analyses have been undertaken on curcumin, resveratrol, omega-3 fatty acids, and CoQ10 in individuals with neurodegenerative illnesses. Preliminary trials examining curcumin in cases of mild cognitive impairment or mild-to-moderate AD have produced inconsistent outcomes. Some trials indicated slight enhancements in cognitive metrics and reduced plaque accumulation on imaging (Ringman et al. 2005), whereas others observed no meaningful advantages, potentially attributable to insufficient plasma levels of curcumin. Nevertheless, limited sample quantities, brief durations, and formulation challenges hinder conclusive determinations. A recent double-blind, placebo-controlled trial of a bioavailable curcumin formulation showed enhancements in cognitive performance and decreased amyloid/tau signals on positron emission tomography scans (Cox et al. 2015). The findings indicate that the advancement of innovative delivery systems may be crucial for utilising curcumin's neuroprotective characteristics.

Resveratrol trials in individuals with mild to severe AD have concentrated on biomarkers indicative of neurodegeneration and inflammation. A randomised controlled trial indicated that high-dose resveratrol (up to 1 g/day) was safe and well-tolerated, suggesting a reduction in cerebrospinal fluid Aβ40 levels, although clinical outcomes related to cognition exhibited negligible change (Turner et al. 2015). Subsequent analyses suggested possible stabilisation of critical neurodegenerative biomarkers, underscoring the necessity for more extensive, prolonged trials. In PD, resveratrol has been mostly assessed in preclinical investigations; however, preliminary human trials are now in progress. Clinical research investigating omega-3 fatty acid supplementation in AD has yielded inconclusive results. Certain research, especially those focusing on mild cognitive impairment or early AD, indicates slight cognitive enhancements, improved mood, or a deceleration of decline in persons with low initial omega-3 levels (Freeman et al. 2006). In contrast, trials including advanced AD populations have typically not shown significant results, perhaps due to the restricted ability to impede established neuropathology. Meta-analyses underscore the significance of timing, dosage, and initial nutritional state in influencing the results of omega-3 therapies. Limited clinical trials have particularly investigated PD; nevertheless, initial findings indicate that omega-3 supplementation may enhance non-motor symptoms, including depression and anxiety, through the modulation of neuroinflammation.

Coenzyme Q10 (CoQ10) has undergone extensive evaluation in PD, prompted by preliminary evidence suggesting its potential to decelerate the functional deterioration of dopaminergic neurones. A large investigation referred to as the QE3 trial assessed high-dose CoQ10 (up to 2400 mg/day) in PD patients but failed to demonstrate a statistically significant advantage on clinical outcomes (Beal 1999). Although this was disheartening, subsequent analyses have suggested that patient variability, disease progression, and CoQ10 formulation may have impacted the outcomes. In AD, CoQ10 trials are scarce; however, tiny pilot studies indicate possible enhancements in cognitive metrics (Gutierrez-Mariscal et al. 2020). Bioavailability continues to be a significant challenge, and future research aims to improve CNS penetration. These clinical studies collectively highlight the complex hurdles in assessing nutraceuticals for neurodegeneration. The variability in trial design, participant selection, disease staging, dose regimens, and formulations hinders direct comparisons among studies. Moreover, the extended natural progression of illnesses such as AD and PD requires protracted intervention durations to obtain significant therapeutic results. Nevertheless, the prevailing opinion indicates that although none of these nutraceuticals should be regarded as a cure, specific compounds, particularly when administered in optimised formulations and commenced early, may demonstrate a protective or disease-modifying effect (Table 15.1).

Table 15.1 Essential nutraceuticals, mechanisms of action, and neurodegenerative targets

Nutraceutical	Primary mechanisms of action	Neurodegenerative targets	References
Curcumin	*Anti-inflammatory* via downregulation of NF-κB and COX-2	Alzheimer's disease (cognitive impairment)	Hewlings and Kalman (2017)
	Antioxidant properties scavenging ROS and RNS	Possible role in other neurodegenerative conditions where protein aggregation is relevant	Ringman et al. (2005)
	Inhibition of amyloid aggregation (Aβ fibrils)	Alzheimer's disease	Cuccioloni et al. (2022)
	Possible enhancement of autophagy	Alzheimer's disease	Mukherjee et al. (2021)
Resveratrol	*SIRT1 activation* improving mitochondrial function and longevity pathways	Alzheimer's disease	Bastianetto and Quirion (2002)
	Reduction of β-amyloid (Aβ) production and deposition	Preliminary indications in Parkinson's disease research	Turner et al. (2015)
	Anti-inflammatory by inhibiting pro-inflammatory cytokines	Alzheimer's disease, Parkinson's disease	Meng et al. (2021)
	Antioxidant properties	Alzheimer's disease, Parkinson's disease	Meng et al. (2021)
Omega-3 Fatty Acids (e.g. DHA, EPA)	*Membrane fluidity* support, crucial for synaptic function and plasticity	Alzheimer's disease (particularly in early stages)	Bazan (2018)
	Anti-inflammatory by generating resolvins and protectins	Possible protective effects in Parkinson's disease	Freeman et al. (2006)
	Reduced oxidative stress and improved neuronal survival	General cognitive decline prevention	Yang et al. (2024)
	Potential *modulation* of amyloid/tau pathologies	Alzheimer's disease	Green et al. (2007)
Coenzyme Q10	*Mitochondrial support* through enhanced electron transport and ATP production	Parkinson's disease (most studied; e.g. QE3 trial)	Beal (1999)
	Antioxidant protection within lipid membranes	Early-stage research in Alzheimer's disease	Gutierrez-Mariscal et al. (2020)
	Possible stabilisation of dopaminergic neurons	Parkinson's disease	Spindler et al. (2009)
	Potential reduction of oxidative stress associated with α-synuclein	Parkinson's disease	Xu et al. (2022)

15.3 Nutraceuticals in Cognitive Decline Associated with Obesity

Obesity, a multifaceted metabolic disorder frequently linked to excessive fat accumulation and systemic inflammation, has become a notable risk factor for cognitive deterioration and neurodegenerative diseases. Although bodyweight and body mass index (BMI) have traditionally been utilised to evaluate health status, it is becoming increasingly evident that the relationship between metabolic dysregulation and cerebral health is complex. In this context, specific nutraceuticals, such as substances sourced from foods or dietary supplements with potential medicinal or health advantages, may provide viable approaches for alleviating obesity-related cognitive impairments. This section investigates the pathways connecting obesity and cognitive function, analyses essential nutraceuticals that influence metabolic health, and evaluates clinical evidence about their impact on obese or overweight individuals.

15.3.1 Association Between Obesity and Cognitive Function

Obesity is frequently linked with metabolic syndrome, characterised by a constellation of risk factors including insulin resistance, dyslipidemia, hypertension, and central adiposity (Alberti et al. 2009). Metabolic abnormalities can impair brain structure and function (Ghosh et al. 2025) (Fig. 15.1). One postulated mechanism involves chronic low-grade inflammation activated by expanded adipose tissue,

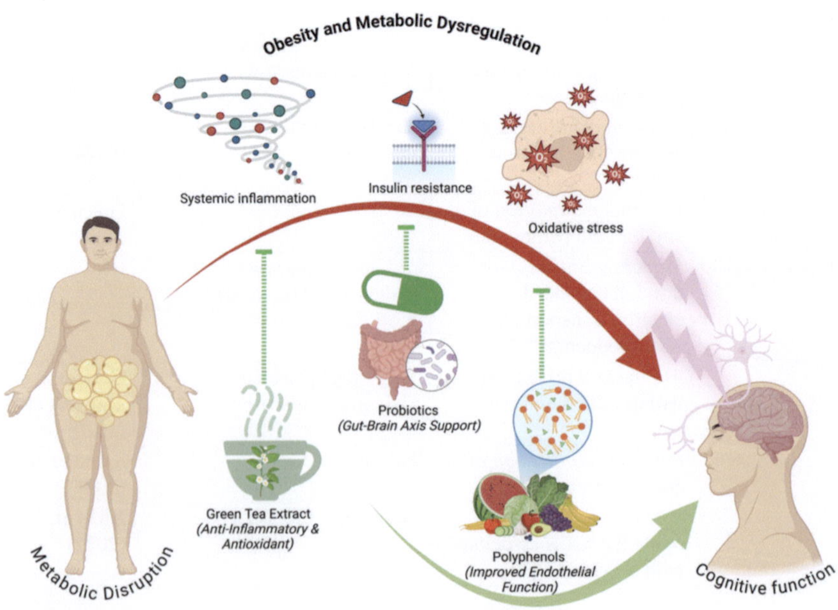

Fig. 15.1 Mechanisms of nutraceuticals in cognitive decline associated with obesity

releasing pro-inflammatory cytokines like interleukin-6 (IL-6) and tumour necrosis factor-alpha (TNF-α). This inflammatory milieu increasingly leads to endothelial dysfunction, impaired glucose uptake in the brain, and oxidative stress, potentially accelerating neurodegenerative pathways (Cao et al. 2018).

Insulin resistance, a central component of metabolic syndrome, also inhibits neuronal insulin signalling, thereby modifying pivotal processes like synaptic plasticity and memory consolidation (Kullmann et al. 2015). In addition, hormonal disturbances that accompany obesity, such as elevated leptin and decreased adiponectin, have been linked to hippocampus dysfunction. The hippocampus is responsible for the formation of memories, and metabolic disturbances within this region can lead to serious cognitive impairments. Together, our findings underscore the urgent need to explore treatments that could target both metabolic and neurological pathways at the same time.

15.3.2 Metabolic Health-Targeting Nutraceuticals

15.3.2.1 Green Tea Extract

Green tea extract, which contains polyphenolic compounds known as catechins (epigallocatechin-3-gallate, EGCG), has been of particular interest due to its potential for anti-obesity and neuroprotective effects. Green tea's thermogenic contribution by catechins and caffeine may increase fat oxidation and energy expenditure, supporting weight regulation (Hursel and Westerterp-Plantenga 2013). Additionally, EGCG was shown to have the capacity for neutralising free radicals and downregulating pro-inflammatory mediators, resulting in a holistic mitigation of oxidative stress and inflammation (Chu et al. 2017). Cognitively, green tea catechins can also protect neurons by modulating signalling pathways involved in synaptic plasticity. Animal studies show that EGCG might enhance hippocampal neurogenesis and enhance tests of learning and memory but are inconsistent for humans. Some clinical studies have noted small improvements in cognitive functioning among overweight individuals with green tea supplement intake, although the effects vary and might depend on dose, baseline metabolic state, and treatment duration.

15.3.2.2 Probiotics

The gut microbiota, acknowledged as a crucial factor in metabolic health, can be utilised by probiotic supplementation to affect both peripheral and central physiology. Dysbiosis, characterised by an imbalance in the gut microbiota, frequently coexists with obesity and may exacerbate metabolic syndrome through heightened intestinal permeability, systemic inflammation, and insulin resistance (Carding et al. 2015). Probiotics are live bacteria that provide health advantages when ingested in sufficient quantities. They can influence the gut-brain axis by fostering a beneficial microbial equilibrium, generating short-chain fatty acids, and improving mucosal barrier integrity. Probiotics may mitigate neuroinflammation by downregulating inflammatory cytokines that traverse the blood-brain barrier or indirectly affect neuronal circuits (Ghosh et al. 2024). Early studies have linked probiotic supplementation with improvement in mood, stress resistance, and cognitive function in individuals presenting with metabolic risk factors (Fang et al. 2025). Such results open up avenues for further larger-scale clinical trials to determine whether probiotics may be used as a viable adjunct to lifestyle therapy for cognitive impairment due to obesity.

15.3.2.3 Polyphenols

Polyphenols are a miscellaneous group of phytochemicals found in plant foods, chocolates, tea, and wine. They include flavonoids (e.g. quercetin, catechins), phenolic acids (e.g. chlorogenic acid), and stilbenes (e.g. resveratrol). Polyphenols exert strong antioxidant and anti-inflammatory actions, primarily due to the neutralisation of reactive oxygen species and the modulation of signalling pathways such as nuclear factor-κB (NF-κB) (Scalbert et al. 2011). In obese patients, polyphenols may protect against endothelial dysfunction and decreased insulin sensitivity and, in turn, counter two vital features of metabolic syndrome detrimental to brain integrity. Animal and human studies confirm that high polyphenol dietary intakes ameliorate brain function by keeping vascular integrity intact and lessening neuronal destruction. Blueberries, rich in anthocyanins, have been linked to enhanced memory and executive function in older individuals (Miller et al. 2018). While further study is necessary to clarify the effects of particular polyphenols, these preliminary findings underscore their potential as an accessible, dietary approach to mitigate obesity-related cognitive impairments.

15.3.3 Clinical Evidence

Empirical research into nutraceutical treatments for obesity-related cognitive decline is in its nascent phase but is progressively expanding. Numerous small-scale clinical studies have assessed the effects of green tea extract on cognitive function in overweight or obese persons. Findings indicate slight enhancements in executive function and working memory, as well as decreases in body weight and waist circumference (Boschmann and Thielecke 2007). Nonetheless, divergent results among research underscore the necessity for standardised dosing and extended intervention durations. Randomised controlled trials for probiotics frequently emphasise measures of metabolic health, such as insulin resistance and lipid profile, while increasingly integrating cognitive evaluations. Several pilot investigations suggest that probiotic administration may diminish systemic inflammation and enhance insulin sensitivity, hence indirectly enhancing cognitive performance (Fang et al. 2025). Despite the encouraging results, extensive trials with stringent cognitive objectives are essential to validate these initial findings.

Although polyphenol research is considerable in vitro and in animal models, it necessitates more substantial human evidence. Research indicates that flavonoid-rich foods, including berries and cocoa, may enhance attention, memory, and executive functioning in obese or metabolically vulnerable people (Bell et al. 2015). However, the diversity of polyphenols and their differing bioavailability provide difficulties in establishing effective dosages and formulations. Additionally, confounding dietary and lifestyle variables must be meticulously regulated in clinical environments to isolate the distinct effects of polyphenols on cognitive function. Although the data is still developing, recent research indicates that nutraceuticals may serve as beneficial supplements for alleviating cognitive impairment associated with obesity. Mechanistic processes including the attenuation of inflammation, enhancement of insulin sensitivity, and antioxidant activities seem to underpin these effects. Subsequent research would be enhanced by increased sample numbers, extended durations, and standardised procedures to produce conclusive findings about efficacy, safety, and applicability across many groups.

15.4 Nutraceuticals and Metabolic Disorders Associated with Ageing

The worldwide demographic transition towards an ageing population has heightened focus on age-related metabolic disorders, such as type 2 diabetes, dyslipidaemia, and cardiovascular diseases. These disorders not only diminish longevity but also impair health span (the duration of life devoid of substantial sickness or impairment). Nutraceuticals, a diverse category of dietary substances, have surfaced as promising solutions for alleviating metabolic dysfunctions associated with ageing. It is essential to investigate the effects of ageing on metabolic pathways and evaluate the efficacy of key nutraceuticals (nicotinamide riboside, berberine, and quercetin) in older persons.

15.4.1 Senescence and Metabolic Dysregulation

Ageing is characterised by progressive impairment of physiological function, governed by molecular, cellular, and systemic changes. One of the most important features of ageing is the accumulation of cellular damage caused by oxidative stress and impaired autophagic function (López-Otín et al. 2013). Mitochondrial dysfunction plays a central role, since mitochondria produce the majority of cellular energy (ATP) and regulate apoptotic pathways. With time, ROS generated by non-functional mitochondria can damage lipids, proteins, and DNA, thus contributing to worsening metabolic decline. Further, ageing is closely linked to insulin resistance that impairs glucose uptake and causes hyperglycemia, inflammation, and dyslipidaemia. This disease is fuelled by the gradual impairment of insulin receptor sensitivity and intracellular signalling pathways (Fabbri et al. 2016). Chronic inflammation, which is otherwise termed "inflammaging", is responsible for a self-perpetuating mechanism typified by enhanced levels of pro-inflammatory cytokines, such as TNF-α and interleukins like IL-6, IL-1β (Franceschi and Campisi 2014). The cumulative impact of these contributes to making the older individuals vulnerable to metabolic abnormalities like insulin resistance, non-alcoholic fatty liver disease, and atherogenic dyslipidaemia, thus placing them at a higher risk for cardiovascular disease and other diseases of ageing. In light of these interconnected pathways, therapies that enhance mitochondrial function, reduce inflammatory load, and increase insulin sensitivity have gained significant interest. Nutraceuticals constitute an adjuvant approach to conventional therapy through the provision of bioactive compounds capable of modifying different aspects of metabolic derangement.

15.4.2 Nutraceutical Interventions

15.4.2.1 Nicotinamide Riboside

Nicotinamide riboside (NR) is a vitamin B3 analogue that acts as a precursor to nicotinamide adenine dinucleotide (NAD^+), a coenzyme that is important for many metabolic processes. The level of NAD^+ reduces with age, which contributes to the reduction in mitochondrial efficiency and dysfunctional sirtuin action, a family of enzymes important for DNA repair and cell homeostasis (Canto et al. 2015). NR supplementation has been shown to enhance mitochondrial performance, enhance energy metabolism, and reduce oxidative stress through NAD^+ supplementation. Preclinical models have reproducibly indicated that NR could enhance health span, protecting against obesity, insulin resistance, and neurodegeneration (Zhang et al. 2016). Human trials, however, are still preliminary and have indicated that NR is capable of increasing NAD^+ levels modestly and potentially enhancing some cardiometabolic biomarkers. NR has been of interest as it is capable of enhancing muscle function, brain function, and overall metabolic health in older persons. More extensive studies are needed to optimise dosing, long-term safety, and efficacy for age-related metabolic disease.

15.4.2.2 Berberine

Berberine is an isoquinoline alkaloid mostly obtained from plants like Berberis aristata and Coptis chinensis. Berberine, a drug in Chinese and Ayurvedic medicine, has shown diverse pharmacological actions, such as enhanced insulin sensitivity, lipid metabolism regulation, and anti-inflammatory activity (Ding et al. 2025). Berberine activates adenosine monophosphate-activated protein kinase (AMPK), an important energy sensor, which enhances fatty acid oxidation and glucose uptake and suppresses lipogenesis. Berberine, in several clinical trials, has been compared with other classical hypoglycemic drugs such as metformin and has been found to be equally effective in reducing fasting blood sugar and HbA1c concentrations in patients with type 2 diabetes (Zhang et al. 2008). Berberine is also capable of reducing total and low-density lipoprotein (LDL) cholesterol and thereby benefits elderly persons predisposed to dyslipidaemia. The poor oral bioavailability of berberine has prevented its widespread application; however, new drug delivery systems such as liposomal and nanoparticle-encapsulated berberine aim to address this shortcoming and maximise its therapeutic effects.

15.4.2.3 Quercetin

Quercetin is a flavonoid commonly found in onions, apples, berries, and various other fruits and vegetables. It is recognised for its antioxidant and anti-inflammatory properties, which can reduce cellular damage caused by reactive oxygen species and inhibit pro-inflammatory signalling pathways (Li et al. 2016). Quercetin has also attracted significant attention because of its senolytic activity, i.e. its capacity to trigger apoptosis in senescent cells (Zhu et al. 2015). Senescent cells accumulate over time, releasing inflammatory mediators that impair tissue function and general health. Quercetin, in addition to its senolytic activity, can enhance insulin sensitivity and lower dyslipidaemia by modulating genes associated with glucose and lipid metabolism. Several animal studies validate these benefits; however, human trials indicate promise but yield conflicting results based on variations in dosage and formulations. Bioavailability of quercetin is influenced by the composition of gut microbiota and the co-consumption of other polyphenols, and therefore, individualised strategies may enhance its efficacy in the elderly.

15.4.3 Clinical Trials and Results

Preclinical reports on NR, berberine, and quercetin are plentiful, while human trials are scarce. Early evidence indicates some benefits for aged people with metabolic disease. Anyways, preliminary findings indicate some advantages for elderly individuals with metabolic disorders. Table 15.2 reports a list of clinical trials based on these nutraceuticals in ageing models.

Now, it is clear that these findings collectively highlight the potential of nutraceuticals as adjunct treatments for age-related metabolic disease. Nicotinamide riboside supplementation consistently increases NAD^+ levels, a critical biomarker

Table 15.2 Selected clinical trials of key nutraceuticals in ageing populations

Nutraceutical	Study design & population	Key findings	Reference/ Clinical Trial #
Nicotinamide Riboside	Randomised, double-blind, placebo-controlled; older adults (ages 55–79) with overweight/obesity, 8-week intervention	Mild but significant increase in NAD^+ levels; trends towards improved insulin sensitivity and reduced blood pressure (no significant difference vs. placebo in body composition)	Martens et al. (2018), NCT02921659
Berberine	Randomised controlled trial; middle-aged and older adults with type 2 diabetes, 12-week intervention	Significant reductions in fasting glucose, HbA1c, and LDL cholesterol; comparable efficacy to metformin in glucose control	Zhang et al. (2008), NCT03267846
Quercetin	Randomized, double-blind, placebo-controlled crossover; overweight/obese adults 25–65 y with metabolicsyndrome traits; 150 mg/day quercetin for 6 weeks with a 5-week washout	Lowered systolic blood pressure (about 2.6 mmHg overall) and plasma oxidized LDL; no effect on CRP or TNF-α; HDL decreased, other lipids and insulin sensitivity not improved.	Egert et al. (2009),

of cellular health. However, its translation into determinative therapeutic effects such as dramatic improvements in insulin sensitivity or body composition has yet to be widely established. Berberine's efficacy in suppressing hyperglycemia and dyslipidaemia in elderly subjects with type 2 diabetes places it among the most promising of the phytochemicals to be used in metabolic control. Finally, quercetin's action in inhibiting inflammatory and senescent cell burdens could suppress some systemic factors that hasten ageing; however, further studies are needed to standardise dosage and test long-term safety. With advancements in the discipline, wider and more rigorous clinical trials, using standardised criteria, stratified patient groups, and long follow-ups will be required to unravel the full scope of such nutraceuticals. Synergy between these approaches can also come from combining the administration of NR with berberine or quercetin through combination therapy. In addition, individually tailored strategies taking into consideration genetic predisposition, epigenetic status, and microbiome composition may boost nutraceutical therapies and yield better outcomes in ageing populations.

15.5 Mechanisms of Action

Nutraceuticals are bioactive substances included in foods and dietary supplements that can have diverse impacts on the central nervous system. Advancements in neuroscience and gerontology reveal that numerous pathogenic processes contributing to cognitive decline and neurodegeneration are dependent on shared biological mechanisms. These encompass persistent neuroinflammation, oxidative stress, mitochondrial malfunction, and compromised autophagic activity. Nutraceuticals may enhance neuronal health, decelerate disease development, and elevate cognitive outcomes by targeting one or more of these processes. This section explores four principal pathways (modulation of neuroinflammation, decrease of oxidative stress, mitochondrial support, and autophagy induction) to demonstrate how nutraceuticals provide neuroprotective advantages.

15.5.1 Regulation of Neuroinflammation

Chronic neuroinflammation has emerged as a key factor in the aetiology of diseases like AD, PD, and other forms of cognitive impairment (Heneka et al. 2015). Neuroinflammation is largely caused by activated microglia and astrocytes, which secrete pro-inflammatory cytokines (e.g. interleukin-1β, tumour necrosis factor-alpha) and chemokines that can damage nearby neurones and synapses. While acute inflammation is protective by removing pathogens or debris, chronic inflammatory responses may be detrimental in nature. Nutraceuticals have been shown to modulate neuroinflammation through different mechanisms. Polyphenols like resveratrol and curcumin suppress inflammatory processes by suppressing the activity of transcription factors like nuclear factor kappa B (NF-κB). This inhibition prevents gene activation for inflammatory mediators, hence decreasing oxidative and nitrosative stress in the brain (Reyes-Corral et al. 2024). Likewise, omega-3 fatty acids improve the production of special pro-resolving mediators such as resolvins and protectins, which actively resolve inflammation instead of passively suppressing it (Calder 2017). These nutraceuticals enhance neuronal integrity and synaptic function by shifting the balance from pro-inflammatory to anti-inflammatory signalling.

A particularly interesting aspect of neuroinflammation is the interrelation between systemic metabolic dysfunctions and brain immune responses. Both obesity and type 2 diabetes, characterised by increased pro-inflammatory markers, have the potential to exacerbate microglial activation by increased circulating cytokines (Kullmann et al. 2015). Under these conditions, nutraceuticals (with the capacity to increase insulin sensitivity or reduce systemic inflammation) can potentially counteract various neuroinflammatory mechanisms. In a recent study, berberine has been shown to regulate AMP-activated protein kinase (AMPK) and suppress the production of inflammatory cytokines. This suggests a dual role in metabolic regulation and neuroprotection (Ding et al. 2025). The potential of nutraceuticals to counteract neuroinflammation provides a solid rationale for their use as adjuvant treatments in neurodegenerative disease. The effectiveness of such anti-inflammatory effects can vary, influenced by such parameters as bioavailability, dose, and the specific pathological phase of administration.

15.5.2 Mitigation of Oxidative Stress

Oxidative stress, defined by an imbalance between the production of reactive oxygen species (ROS) and the antioxidant protective mechanisms within the body, is intimately connected with neuronal damage and death. The brain is particularly vulnerable to oxidative damage from its high oxygen usage and relatively low levels of some endogenous antioxidants (Sies 2015). Build-up of reactive oxygen species (ROS) can damage lipids within cell membranes (lipid peroxidation), proteins, and DNA, impacting critical neural functions and amplifying inflammatory signalling. A range of nutraceuticals exhibits both direct and indirect antioxidant effects. Curcumin possesses the ability to bind metal ions that facilitate oxidative processes and directly neutralise free radicals (Hewlings and Kalman 2017). Furthermore, curcumin may regulate essential antioxidant enzymes, including superoxide dismutase (SOD) and catalase, via affecting transcription factors such as nuclear factor erythroid 2-related factor 2 (Nrf2). Quercetin, a flavonoid present in various fruits and vegetables, serves as a powerful free radical scavenger and can enhance endogenous antioxidant defences, so bolstering neural resilience against oxidative stress (Li et al. 2016). Omega-3 fatty acids and coenzyme Q10 play significant roles in combating oxidative stress. By maintaining the functional integrity of mitochondrial membranes, these chemicals stabilise electron transport pathways, therefore decreasing ROS leakage. Coenzyme Q10 (ubiquinone) is essential to the electron transport chain; supplementation may replenish reduced endogenous CoQ10 levels and improve cellular antioxidant capacity, although clinical outcomes have been inconsistent, partly due to bioavailability issues (Gutierrez-Mariscal et al. 2020).

The antioxidative characteristics of nutraceuticals are essential for avoiding or decelerating the progression of neurodegenerative diseases. By protecting cellular structures from ROS-induced damage, these chemicals create an environment conducive to neuronal survival and plasticity.

15.5.3 Mitochondrial Assistance

Mitochondria function as the energy generators of cells, synthesising adenosine triphosphate (ATP) via oxidative phosphorylation. Neurones, due to their substantial energy requirements, depend significantly on optimal mitochondrial activity to sustain electrochemical gradients, neurotransmitter production, and axonal transport. Mitochondrial malfunction, frequently caused by oxidative damage, compromised biogenesis, or changes in mitochondrial DNA, has been associated with disorders such as AD and PD (Lin and Beal 2006).

Nutraceuticals can provide mitochondrial assistance in several ways. Nicotinamide riboside (NR) restores nicotinamide adenine dinucleotide (NAD^+) levels, essential for mitochondrial integrity and the function of sirtuins, a group of enzymes that govern ageing and metabolism (Canto et al. 2015). Augmented sirtuin activity can enhance mitochondrial biogenesis, facilitate DNA repair, and regulate inflammatory pathways—thereby providing neuroprotection. Resveratrol, often linked to red wine, stimulates sirtuins and has been demonstrated to stimulate the peroxisome proliferator-activated receptor gamma coactivator-1 alpha (PGC-1α), a crucial regulator of mitochondrial biogenesis (Bastianetto and Quirion 2002). Animal research underscores resveratrol's capacity to safeguard dopaminergic neurones, sustain synaptic plasticity, and uphold cognitive function among metabolic and oxidative stressors. Berberine enhances mitochondrial health by activating AMPK, which increases fatty acid oxidation and improves mitochondrial quality control (Ding et al. 2025). Coenzyme Q10 is particularly important for both its function in ATP synthesis and the scavenging of free radicals in the mitochondrial membrane (Spindler et al. 2009). CoQ10 enhances electron transport and neutralises free radicals, minimising mitochondrial damage and maximising neuronal energetics. Further research is required to establish optimal form and dosages, but the capacity of some nutraceuticals to enhance mitochondrial function does indicate a rational approach to brain health maintenance.

15.5.4 Induction of Autophagy

Autophagy is a cellular housekeeping mechanism that breaks down and recycles damaged organelles, protein aggregates, and other potentially toxic components. Dysregulated autophagy is also involved in the pathophysiology of several neurodegenerative diseases, where the occurrence of misfolded proteins (β-amyloid and α-synuclein) can disrupt neuronal function (Menzies et al. 2017). Enhancing autophagic function may be a therapeutic approach to eliminate aberrant aggregates and preserve neuronal homeostasis. Nutraceuticals such as curcumin and resveratrol can induce autophagy by various molecular mechanisms. Curcumin stimulates the beclin-1 and microtubule-associated protein 1A/1B-light chain 3 (LC3) signalling pathway, thereby facilitating autophagosome production (Ghaeini Hesarooeyeh et al. 2024). Resveratrol stimulates autophagy via the sirtuin-FoxO pathway, enhancing the expression of genes related to the autophagic process (Pallauf et al. 2017). Polyphenols have the capacity to inhibit the accumulation of detrimental protein aggregates and enhance cellular longevity. Other nutraceuticals may enhance autophagy indirectly by addressing metabolic and oxidative conditions that hinder autophagic flux. Berberine inhibits endoplasmic reticulum stress and fosters an intracellular environment conducive to effective proteostasis (Powell et al. 2021). Omega-3 fatty acids and other anti-inflammatory compounds may enhance autophagy by diminishing pro-inflammatory signals that could impede lysosomal activity. Nutraceutical-induced autophagy remains unexamined in clinical contexts; nonetheless, preclinical evidence underscores its potential importance in the prevention or mitigation of neurodegenerative diseases.

15.6 Methodologies for Clinical Trials

Research into nutraceuticals has significantly increased over the last decades, which has required strong clinical study designs to accurately determine efficacy and safety parameters. Nutraceuticals are controlled differently from conventional drugs; nevertheless, they are still under the same fundamental principles of clinical research. Now we shall describe and discuss the major study types and settings most frequently utilised, the principal outcome measures most commonly used to evaluate the efficacy of nutraceuticals, and the major challenges in carrying out the studies.

15.6.1 Nutraceutical Research Study Designs

The randomised controlled trial (RCT) serves as the gold standard for evaluating clinical efficacy. The random assignment of participants to intervention or control groups in randomised controlled trials reduces selection bias and confounding variables (Friedman et al. 2010). In nutraceutical research, randomised controlled trials commonly employ parallel-group or crossover designs. Parallel-group designs assess outcomes in separate groups receiving either a nutraceutical or a placebo, while crossover designs allow each participant to serve as their own control, thereby reducing inter-individual variability (Sibbald and Roland 1998).

Cohort studies are one of the more common methodologies involving the longitudinal follow-up of a group of individuals to determine the relationship between health outcomes and nutraceutical consumption (Song and Chung 2010). While observational study designs do not define causality as strongly as randomised controlled trials, they can define relationships and suggest hypotheses for investigation.

Case-control studies, which compare individuals with a specific outcome (e.g. cognitive deterioration) to those without, are rarely employed in nutraceutical research due to challenges in accurately assessing long-term supplement intake.

15.6.2 Performance Indicators

Critical endpoints in nutraceutical trials depend on the specific condition being addressed. In neurodegenerative and cognitive research, metrics typically include standardised neuropsychological assessments, such as the Mini-Mental State Examination, and biomarker evaluations, including β-amyloid concentrations in Alzheimer's studies (Cummings 2018). In metabolic diseases, key outcomes may concentrate on fasting blood glucose, insulin sensitivity, and lipid profiles (Donovan et al. 2025). Inflammatory markers, such as C-reactive protein and interleukin-6, function as general indicators of systemic inflammation and are commonly monitored in nutraceutical trials designed to mitigate inflammation-related diseases. Trials consistently integrate quality-of-life evaluations to obtain patient-reported results wherever feasible.

15.6.3 Obstacles and Factors

Nutraceutical trials encounter distinct challenges compared to pharmaceutical investigations. A prevalent concern is the placebo effect, where participants' faith in a supplement's efficacy may result in subjective enhancements regardless of the active ingredient (Dunn et al. 2018). Carefully structured experiments employ placebos that match in taste, look, and packaging to reduce biases. Compliance is an important issue in this area. Nutraceuticals often require long-term bioavailability at large dosages, posing adherence challenges (Rautiainen et al. 2016). Researchers can use pill counts, electronic monitoring, or biochemical measures (e.g. plasma levels of the active drug) to measure compliance accurately, but other challenges can change the suitability of the results. Finally, it is important to understand that bioavailability plays a major role in determining trial outcomes. Many nutraceuticals like curcumin have inherently low bioavailability and rapid metabolism (due to processing variations). This clearly necessitates new formulations or delivery systems (Abuhassira-Cohen and Livney 2022). Omitting to take into account these factors may conceal true effectiveness and create conflicting study results.

15.7 Discussion

Nutraceuticals have become increasingly significant for their functions in alleviating chronic diseases, cognitive deterioration, and age-related ailments. However, as research progresses, it is becoming clear that individuals do not uniformly respond to identical interventions. Genetic variations, metabolic profiles, and lifestyle factors can substantially influence the efficacy and safety of a specific nutraceutical. It is important to address the implications of personalised nutraceutical therapy, the notion of precision nutrition, and newly emerging research aims to enhance both efficacy and delivery. The premise of personalisation of nutraceutical therapy arose from the identification that individuals have varying genetic proclivities and metabolic reactions towards food components (Zeisel 2020). Nutrigenomics as a field concerned with the nature of how components of diet can affect gene function offers a gateway to personalising nutraceuticals.

Single-nucleotide polymorphisms (SNPs) of genes implicated in inflammation, oxidative stress, or metabolism may affect an individual's reaction to some substances such as curcumin, resveratrol, or omega-3 fatty acids (Singar et al. 2024). There are specific polymorphisms in the APOE gene that are related to different responses towards lipid-lowering nutraceuticals; therefore, it is advisable to take genotype-based advice that may improve benefits and decrease excess intake. Besides genetics, nutri-metabolomics (the comprehensive analysis of metabolites following the consumption of food or nutraceuticals) offers insights into a person's metabolic status (de la O et al. 2024). Clinicians and researchers are able to establish in real time whether a nutraceutical therapy is effective through the monitoring of biomarkers such as lipid profiles, inflammatory cytokines, or markers of oxidative stress. Such individualised treatment is particularly critical in the treatment of metabolic disorders like obesity, metabolic syndrome, or insulin resistance because the uniform standard treatments tend to be less uniform in their efficacy for different individuals. Individualised intervention emphasises the necessity of considering factors such as age, sex, gut microbiome composition, and circadian rhythms. Older adults with reduced gastric acid secretion can fail to absorb certain polyphenols effectively, requiring reformulated dosages or specific preparations (Chopra et al. 2022). A disturbed gut microbiome can inhibit the biotransformation of polyphenols or probiotics, thus wasting their bioactivity. Personalised nutraceutical therapies help in a sound understanding of an individual's genetic profile, metabolic state, and environment shaping. Precision nutrition further helps by extending the principles of personalisation by including a balanced diet, exercise, and nutraceuticals in an all-encompassing, data-led approach.

Tailored nutraceutical therapies have concentrated on the use of a single or short list of supplements, whereas precision nutrition seeks to equilibrate the overall makeup of the diet—both macro- and micronutrients—to maximise the efficiency of targeted applications of nutraceuticals (Bray et al. 2016). Precision nutrition becomes achievable through applying digital health platforms and biometric monitoring. Sleep, activity, and heart rate variability wearables provide clinicians and consumers with the means to correlate lifestyle factors with variation in metabolic indicators or cognitive function. These real-world data points provide dynamic compensation to one's diet regimen or nutraceutical consumption, evolving beyond fixed protocols to a reactive, individualised system (Donovan et al. 2025). Also, machine learning software is used in more amounts for genetic markers, food intake, and clinical outcomes. These programmes or software can determine which patients are most likely to benefit from a given intervention by recognising patterns that traditional statistical methods may not catch. The integration of nutrigenomic and real-time lifestyle information can lead to more accurate dietary advice, which could minimise the risk of chronic diseases and maximise overall health span. Despite the promise of precision nutrition, there are still challenges. Genetic and metabolic testing may be limited by cost or healthcare infrastructure, and data privacy and ethical use concerns are still widespread. In addition, translating genetic results into actionable dietary advice requires reproducible evidence from robust clinical trials, an area that is underdeveloped for a number of nutraceuticals. When these challenges are overcome, precision nutrition can transform the use of nutraceuticals by physicians and consumers in pursuit of maximal health. Advancement in nutraceutical science is driven by the ongoing identification of new chemicals and the development of novel delivery systems.

Polyphenols of lesser crops have some anti-inflammatory or antioxidant functions that could be superior to those of known nutraceuticals such as curcumin. Psychoactive plants such as some adaptogens and fungi are being researched with more interest into their impact on stress adaptation and cognition (Mehla et al. 2020). Delivery system advances aim to avoid problems of bioavailability, of particular concern for compounds such as curcumin and resveratrol. Processes like nanoparticle encapsulation, liposomal carriers, and the development of microemulsions can enhance absorption, prolong circulation time, and facilitate targeted delivery to tissues like the brain (Ali Redha et al. 2024). Latest strategies can even use probiotic carriers or biopolymer-based hydrogel systems having a dual action of stimulating gut health and releasing active ingredients at enhanced levels. The interface between nutraceuticals and immunometabolism is coming under greater examination. Immunometabolism, an area of research that appreciates the interrelation between immune activation and metabolic pathways and vice versa, can provide novel approaches to the development of nutraceuticals that appropriately modulate inflammatory responses (Hotamisligil 2017). Understanding these cross-talk mechanisms can explain the best combination of nutraceuticals with drugs or diet modifications to maximise synergistic effects. Finally, longitudinal, large-scale, and multi-omic approaches are going to revolutionise nutraceutical research. Studies that integrate genomic, transcriptomic, proteomic, and metabolomic information over extended periods of time can reveal the dynamic interactions between nutraceutical treatments and the biological processes underlying ageing, cognition, and metabolic disorders. Through this, they provide evidence-based recommendations that make nutraceuticals part and parcel of healthcare, not ancillary supplements.

15.8 Conclusions

Nutraceuticals are promising supplementary treatments for several health conditions, particularly in the case of neurodegenerative diseases and metabolic syndrome linked with ageing. Evidence-based research shows that bioactive chemicals, whether of spice origin such as turmeric (curcumin) or marine origin (omega-3 fatty acids). This may affect interrelated systems such as neuroinflammation, oxidative stress, mitochondrial dysfunction, and also impaired autophagy. Clinical trial methodology has tightened up, using randomised controlled designs and robust outcome measurement to elucidate efficacy and safety. But still issues such as placebo effect, patient compliance, and variable bioavailability underscore the complexities of emerging nutraceutical findings into mainstream clinical practice. Healthcare practitioners can incorporate nutraceuticals by first examining individual patient profiles, considering genetic makeups, baseline metabolic state, and any interactions with existing drugs. Dietitian-chemist-doctor cooperation is needed to maximise dose regimens, formulations, and supplementation duration for therapeutic impact. Additionally, personalised and precision nutrition paradigms suggest that tailoring nutraceutical therapies can generate more consistent outcomes, leveraging advances in nutrigenomics and ongoing monitoring technologies. Substantial gaps remain in our understanding of the proper utilisation of nutraceuticals for long-term therapy and prevention of disease. Longitudinal, large-scale studies that include multi-omics data (genomics, metabolomics, proteomics) will be critical to identifying biomarkers of nutraceutical response. Additionally, follow-up studies will need to advance new delivery mechanisms and define standardised dosing regimens. Addressing these gaps will solidify the position of nutraceuticals in evidence-based practice, ultimately allowing healthcare professionals to guide patients towards holistic approaches that improve cognitive function, metabolic health, and overall lifespan.

15.9 Summary

This chapter discusses the scientific and clinical evidence of nutraceuticals in relation to brain health and metabolic well-being. The discussion explores significant research areas, including neurodegenerative diseases such as Alzheimer's and Parkinson's, cognitive impairment associated with obesity, and the metabolic dysregulation that occurs with ageing.

Nutraceuticals work through multiple mechanisms, including modulation of neuroinflammation, attenuation of oxidative stress, enhancement of mitochondrial function, and activation of autophagy. We focused on compounds like curcumin, resveratrol, omega-3 fatty acids, and coenzyme Q10 have been thoroughly investigated using in vitro, in vivo models, and clinical trials. Although several studies have reported positive results in the research literature, there are several issues that call for the application of well-planned and rigorous RCTs in future studies. For example, limited bioavailability, placebo response, and heterogeneity of patient populations. This chapter also addresses the incorporation of nutraceuticals into practice with a focus on the importance of personalised and precision nutrition. We have also elaborated that genomic and metabolic profiling allows for personalisation of intervention based on the individual's requirements. This improves outcomes and reduces some risks. Recent advances in formulation and delivery technologies, including nanoparticle encapsulation and liposomal carriers, are optimising the therapeutic action of nutraceuticals through enhanced stability, absorption, and targeted delivery. We have also identified some gaps in existing knowledge, necessitating large-scale longitudinal studies and multi-omic approaches to optimise dosing regimens, clarify pharmaceutical-nutraceutical interactions, and develop comprehensive safety profiles.

References

Abuhassira-Cohen Y, Livney YD (2022) Enhancing the bioavailability of encapsulated hydrophobic nutraceuticals: insights from in vitro, in vivo, and clinical studies. Curr Opin Food Sci 45:100832. https://doi.org/10.1016/j.cofs.2022.100832

Alberti KGMM, Eckel RH, Grundy SM, Zimmet PZ, Cleeman JI, Donato KA et al (2009) Harmonizing the metabolic syndrome. Circulation 120(16):1640–1645. https://doi.org/10.1161/CIRCULATIONAHA.109.192644

Ali Redha A, Kodikara C, Cozzolino D (2024) Does encapsulation improve the bioavailability of polyphenols in humans? A concise review based on in vivo human studies. Nutrients 16(21):3625. https://doi.org/10.3390/nu16213625

Bastianetto S, Quirion R (2002) Natural extracts as possible protective agents of brain aging. Neurobiol Aging 23(5):891–897. https://doi.org/10.1016/s0197-4580(02)00024-6

Batailler C, Swan J, Sappey Marinier E, Servien E, Lustig S (2020) New technologies in knee arthroplasty: current concepts. J Clin Med 10(1):47. https://doi.org/10.3390/jcm10010047

Bazan NG (2018) Docosanoids and elovanoids from omega-3 fatty acids are pro-homeostatic modulators of inflammatory responses, cell damage and neuroprotection. Mol Asp Med 64:18–33. https://doi.org/10.1016/j.mam.2018.09.003

Beal MF (1999) Coenzyme Q10 administration and its potential for treatment of neurodegenerative diseases. BioFactors (Oxford, England) 9(2–4):261–266. https://doi.org/10.1002/biof.5520090222

Bell L, Lamport DJ, Butler LT, Williams CM (2015) A review of the cognitive effects observed in humans following acute supplementation with flavonoids, and their associated mechanisms of action. Nutrients 7(12):10290–10306. https://doi.org/10.3390/nu7125538

Boschmann M, Thielecke F (2007) The effects of epigallocatechin-3-gallate on thermogenesis and fat oxidation in obese men: a pilot study. J Am Coll Nutr 26(4):389S–395S. https://doi.org/10.1080/07315724.2007.10719627

Bray MS, Loos RJ, McCaffery JM, Ling C, Franks PW, Weinstock GM et al (2016) NIH working group report-using genomic information to guide weight management: from universal to precision treatment. Obesity 24(1):14–22. https://doi.org/10.1002/oby.21381

Brower V (1998) Nutraceuticals: poised for a healthy slice of the healthcare market? Nat Biotechnol 16(8):728–731. https://doi.org/10.1038/nbt0898-728

Calder PC (2017) Omega-3 fatty acids and inflammatory processes: from molecules to man. Biochem Soc Trans 45(5):1105–1115. https://doi.org/10.1042/BST20160474

Canto C, Menzies KJ, Auwerx J (2015) NAD^+ metabolism and the control of energy homeostasis: a balancing act between mitochondria and the nucleus. Cell Metab 22(1):31–53. https://doi.org/10.1016/j.cmet.2015.05.023

Cao T, Zhou X, Zheng X, Cui Y, Tsien JZ, Li C, Wang H (2018) Histone deacetylase inhibitor alleviates the neurodegenerative phenotypes and histone dysregulation in presenilins-deficient mice. Front Aging Neurosci 10:137. https://doi.org/10.3389/fnagi.2018.00137

Carding S, Verbeke L, Vipond DT, Corfe BM, Owen LJ (2015) Dysbiosis of the gut microbiota in disease. Microb Ecol Health Dis 26:26191. https://doi.org/10.3402/mehd.v26.26191

Chopra AS, Lordan R, Horbańczuk OK, Atanasov AG, Chopra I, Horbańczuk JO, Jóźwik A, Huang L, Pirgozliev V, Banach M, Battino M, Arkells N (2022) The current use and evolving landscape of nutraceuticals. Pharmacol Res 175:106001. https://doi.org/10.1016/j.phrs.2021.106001

Chu C, Deng J, Man Y, Qu Y (2017) Green tea extracts Epigallocatechin-3-gallate for different treatments. Biomed Res Int 2017:5615647. https://doi.org/10.1155/2017/5615647

Coppens P, da Silva MF, Pettman S (2006) European regulations on nutraceuticals, dietary supplements and functional foods: a framework based on safety. Toxicology 221(1):59–74. https://doi.org/10.1016/j.tox.2005.12.022

Cox KH, Pipingas A, Scholey AB (2015) Investigation of the effects of solid lipid curcumin on cognition and mood in a healthy older population. J Psychopharmacol (Oxford, England) 29(5):642–651. https://doi.org/10.1177/0269881114552744

Cuccioloni M, Cecarini V, Bonfili L, Pettinari R, Tombesi A, Pagliaricci N, Petetta L, Angeletti M, Eleuteri AM (2022) Enhancing the amyloid-β anti-aggregation properties of curcumin via Arene-Ruthenium(II) derivatization. Int J Mol Sci 23(15):8710. https://doi.org/10.3390/ijms23158710

Cummings J (2018) Lessons learned from Alzheimer disease: clinical trials with negative outcomes. Clin Transl Sci 11(2):147–152. https://doi.org/10.1111/cts.12491

de la O V, Fernández-Cruz E, Valdés A, Cifuentes A, Walton J, Martínez JA (2024) Exhaustive search of dietary intake biomarkers as objective tools for personalized nutrimetabolomics and precision nutrition implementation. Nutr Rev:nuae133. Advance online publication. https://doi.org/10.1093/nutrit/nuae133

DeFelice SL (1995) The nutraceutical revolution: its impact on food industry R&D. Trends Food Sci Technol 6(2):59–61. https://doi.org/10.1016/S0924-2244(00)88944-X

Dexter DT, Jenner P (2013) Parkinson disease: from pathology to molecular disease mechanisms. Free Radic Biol Med 62:132–144. https://doi.org/10.1016/j.freeradbiomed.2013.01.018

Dickinson A, MacKay D, Wong A (2015) Consumer attitudes about the role of multivitamins and other dietary supplements: report of a survey. Nutr J 14:66. https://doi.org/10.1186/s12937-015-0053-9

Ding L, Xia J, Wang H, Qian J, Jin X, Yang Y, Xia J, Shang W, Chen M (2025) Berberine improves glucose and lipid metabolism in obese mice through the reduction of IRE1/GSK-3β Axis-mediated inflammation. Endocr Metab Immune Disord Drug Targets. Advance online publication. https://doi.org/10.2174/0118715303319434241113161606

Donovan SM, Abrahams M, Anthony JC, Bao Y, Barragan M, Bermingham KM, Blander G, Keck AS, Lee BY, Nieman KM, Ordovas JM, Penev V, Reinders MJ, Sollid K, Thosar S, Winters BL (2025) Personalized nutrition: perspectives on challenges, opportunities, and guiding principles for data use and fusion. Crit Rev Food Sci Nutr:1–18. Advance online publication. https://doi.org/10.1080/10408398.2025.2461237

Dunn AG, Coiera E, Mandl KD (2018) Social media interventions for precision public health: promises and risks. NPJ Digit Med 1:47. https://doi.org/10.1038/s41746-018-0054-0

Fabbri E, An Y, Zoli M, Tanaka T, Simonsick EM, Kitner-Triolo MH, Studenski SA, Resnick SM, Ferrucci L (2016) Association between accelerated multimorbidity and age-related cognitive decline in older Baltimore longitudinal study of aging participants without dementia. J Am Geriatr Soc 64(5):965–972. https://doi.org/10.1111/jgs.14092

Fang RY, Pan XR, Zeng XX, Li ZZ, Chen BF, Zeng HM, Peng J (2025) Gut-brain axis as a bridge in obesity and depression: mechanistic exploration and therapeutic prospects. World J Psychiatry 15(1):101134. https://doi.org/10.5498/wjp.v15.i1.101134

Franceschi C, Campisi J (2014) Chronic inflammation (inflammaging) and its potential contribution to age-associated diseases. J Gerontol Ser A Biol Med Sci 69(S1):S4–S9. https://doi.org/10.1093/gerona/glu057

Freeman MP, Hibbeln JR, Wisner KL, Davis JM, Mischoulon D, Peet M, Keck PE Jr, Marangell LB, Richardson AJ, Lake J, Stoll AL (2006) Omega-3 fatty acids: evidence basis for treatment and future research in psychiatry. J Clin Psychiatry 67(12):1954–1967. https://doi.org/10.4088/jcp.v67n1217

Friedman LM, Furberg C, DeMets DL (2010) Fundamentals of clinical trials, 4th edn. Springer. https://doi.org/10.1007/978-1-4419-1586-3

Gasmi A, Noor S, Mujawdiya PK, Akram M, Manzoor A, Bjørklund G (2024) Herbal treatments for obesity: a review. Curr Med Chem, Advance online publication. https://doi.org/10.2174/0109298673287491240315055726

GBD 2019 Risk Factors Collaborators (2020) Global burden of 87 risk factors in 204 countries and territories, 1990–2019: a systematic analysis for the Global Burden of Disease Study 2019. Lancet 396(10258):1223–1249. https://doi.org/10.1016/S0140-6736(20)30752-2

Geller AI, Shehab N, Weidle NJ, Lovegrove MC, Wolpert BJ, Timbo BB et al (2015) Emergency department visits for adverse events related to dietary supplements. N Engl J Med 373(16):1531–1540. https://doi.org/10.1056/NEJMsa1504267

Ghaeini Hesarooeyeh Z, Basham A, Sheybani-Arani M, Abbaszadeh M, Salimi Asl A, Moghbeli M, Saburi E (2024) Effect of resveratrol and curcumin and the potential synergism on hypertension: a mini-review of human and animal model studies. Phytother Res 38(1):42–58. https://doi.org/10.1002/ptr.8023

Ghosh S, Raghunath M, Das BC, Sinha JK (2019) High sugar content in baby food: an Indian perspective. Lancet Diabetes Endocrinol 7(10):748–749. https://doi.org/10.1016/S2213-8587(19)30291-8

Ghosh S, Dhungel S, Shaikh MF, Sinha JK (2024) Editorial: world digestive health day: investigating the link between neurodegenerative disease and gut microbiota. Front Aging Neurosci 15:1351855. https://doi.org/10.3389/fnagi.2023.1351855

Ghosh S, Singh KK, Jha S, Raghunath M, Sinha JK (2025) Interconnected dynamics of 'Obesageing': intricate relationship between obesity and ageing processes. Alzheimers Dement 20(Suppl 1):e085889. https://doi.org/10.1002/alz.085889

Green KN, Martinez-Coria H, Khashwji H, Hall EB, Yurko-Mauro KA, Ellis L, LaFerla FM (2007) Dietary docosahexaenoic acid and docosapentaenoic acid ameliorate amyloid-beta and tau pathology via a mechanism involving presenilin 1 levels. J Neurosci 27(16):4385–4395. https://doi.org/10.1523/JNEUROSCI.0055-07.2007

Gutierrez-Mariscal FM, Arenas-de Larriva AP, Limia-Perez L, Romero-Cabrera JL, Yubero-Serrano EM, López-Miranda J (2020) Coenzyme Q_{10} supplementation for the reduction of oxidative stress: clinical implications in the treatment of chronic diseases. Int J Mol Sci 21(21):7870. https://doi.org/10.3390/ijms21217870

Heneka MT, Golenbock DT, Latz E (2015) Innate immunity in Alzheimer's disease. Nat Immunol 16(3):229–236. https://doi.org/10.1038/ni.3102

Hewlings SJ, Kalman DS (2017) Curcumin: A review of its' effects on human health. Foods 6(10):92. https://doi.org/10.3390/foods6100092

Hotamisligil GS (2017) Inflammation, metaflammation and immunometabolic disorders. Nature 542(7640):177–185. https://doi.org/10.1038/nature21363

Huang Y, Guo X, Wu Y, Chen X, Feng L, Xie N, Shen G (2024) Nanotechnology's frontier in combatting infectious and inflammatory diseases: prevention and treatment. Signal Transduct Target Ther 9(1):34. https://doi.org/10.1038/s41392-024-01745-z

Hursel R, Westerterp-Plantenga MS (2013) Catechin- and caffeine-rich teas for control of body weight in humans. Am J Clin Nutr 98(6 Suppl):1682S–1693S. https://doi.org/10.3945/ajcn.113.058396

Jacquier EF, van de Wouw M, Nekrasov E, Contractor N, Kassis A, Marcu D (2024) Local and systemic effects of bioactive food ingredients: is there a role for functional foods to prime the gut for resilience? Foods (Basel, Switzerland) 13(5):739. https://doi.org/10.3390/foods13050739

Kaur M, Aran KR (2025) Unraveling the role of Nrf2 in dopaminergic neurons: a review of oxidative stress and mitochondrial dysfunction in Parkinson's disease. Metab Brain Dis 40(2):123. https://doi.org/10.1007/s11011-025-01552-7

Kullmann S, Heni M, Fritsche A, Preissl H (2015) Insulin action in the human brain: evidence from neuroimaging studies. J Neuroendocrinol 27(6):419–423. https://doi.org/10.1111/jne.12254

Li Y, Yao J, Han C, Yang J, Chaudhry MT, Wang S et al (2016) Quercetin, inflammation and immunity. Nutrients 8(3):167. https://doi.org/10.3390/nu8030167

Lin MT, Beal MF (2006) Mitochondrial dysfunction and oxidative stress in neurodegenerative diseases. Nature 443(7113):787–795. https://doi.org/10.1038/nature05292

López-Otín C, Blasco MA, Partridge L, Serrano M, Kroemer G (2013) The hallmarks of aging. Cell 153(6):1194–1217. https://doi.org/10.1016/j.cell.2013.05.039

Martens CR, Denman BA, Mazzo MR, Armstrong ML, Reisdorph N, McQueen MB, Chonchol M, Seals DR (2018) Chronic nicotinamide riboside supplementation is well-tolerated and elevates NAD+ in healthy middle-aged and older adults. Nat Commun 9(1):1286. https://doi.org/10.1038/s41467-018-03421-7

Mathers JC (2017) Nutrigenomics in the modern era. Proc Nutr Soc 76(3):265–275. https://doi.org/10.1017/S002966511600080X

Mehla J, Gupta P, Pahuja M, Diwan D, Diksha D (2020) Indian medicinal herbs and formulations for Alzheimer's disease, from traditional knowledge to scientific assessment. Brain Sci 10(12):964. https://doi.org/10.3390/brainsci10120964

Mehta NH, Huey SL, Kuriyan R, Peña-Rosas JP, Finkelstein JL, Kashyap S, Mehta S (2024) Potential mechanisms of precision nutrition-based interventions for managing obesity. Adv Nutr (Bethesda, Md.) 15(3):100186. https://doi.org/10.1016/j.advnut.2024.100186

Meng T, Xiao D, Muhammed A, Deng J, Chen L, He J (2021) Anti-inflammatory action and mechanisms of resveratrol. Molecules (Basel, Switzerland) 26(1):229. https://doi.org/10.3390/molecules26010229

Menzies FM, Fleming A, Caricasole A, Bento CF, Andrews SP, Ashkenazi A et al (2017) Autophagy and neurodegeneration: pathogenic mechanisms and therapeutic opportunities. Neuron 93(5):1015–1034. https://doi.org/10.1016/j.neuron.2017.01.022

Miller MG, Hamilton DA, Joseph JA, Shukitt-Hale B (2018) Dietary blueberry improves cognition among older adults in a randomized, double-blind, placebo-controlled trial. Eur J Nutr 57(3):1169–1180. https://doi.org/10.1007/s00394-017-1400-8

Mukherjee S, Mishra AK, Peer GDG, Bagabir SA, Haque S, Pandey RP, Raj VS, Jain N, Pandey A, Kar SK (2021) The interplay of the unfolded protein response in neurodegenerative diseases: a therapeutic role of curcumin. Front Aging Neurosci 13:767493. https://doi.org/10.3389/fnagi.2021.767493

Noshadi N, Bonyadian A, Hojati A, Abbasalizad-Farhangi M, Heidari M, Darzi M, Seyedhosseini-Ghaheh H, Khajeh M, Pourteymour Fard Tabrizi F, Vajdi M, Askari G (2024) The effect of quercetin supplementation on the components of metabolic syndrome in adults: A systematic review and dose–response meta-analysis of randomized controlled trials. J Funct Foods 116:106175. https://doi.org/10.1016/j.jff.2024.106175

Powell M, Blaskovich MAT, Hansford KA (2021) Targeted protein degradation: the new frontier of antimicrobial discovery? ACS Infect Dis 7(8):2050–2067. https://doi.org/10.1021/acsinfecdis.1c00203

Rakshe PS, Dutta BJ, Chib S, Maurya N, Singh S (2024) Unveiling the interplay of AMPK/SIRT1/PGC-1α axis in brain health: promising targets against aging and NDDs. Ageing Res Rev 96:102255. https://doi.org/10.1016/j.arr.2024.102255

Rautiainen S, Manson JE, Lichtenstein AH (2016) Dietary supplements and disease prevention—a global overview. Nat Rev Endocrinol 12(7):407–420. https://doi.org/10.1038/nrendo.2016.54

Reyes-Corral M, Gil-González L, González-Díaz Á, Tovar-Luzón J, Ayuso MI, Lao-Pérez M, Montaner J, de la Puerta R, Fernández-Torres R, Ybot-González P (2024) Pretreatment with oleuropein protects the neonatal brain from hypoxia-ischemia by inhibiting apoptosis and neuroinflammation. J Cerebr Blood Flow Metab, 271678X241270237. Advance online publication. https://doi.org/10.1177/0271678X241270237

Ringman JM, Frautschy SA, Cole GM, Masterman DL, Cummings JL (2005) A potential role of the curry spice curcumin in Alzheimer's disease. Curr Alzheimer Res 2(2):131–136. https://doi.org/10.2174/1567205053585882

Sabet S, Rashidinejad A, Melton LD, McGillivray DJ (2021) Recent advances to improve curcumin oral bioavailability. Trends Food Sci Technol 110:253–266. https://doi.org/10.1016/j.tifs.2021.02.006

Sackett DL, Rosenberg WM, Gray JAM, Haynes RB, Richardson WS (1996) Evidence based medicine: what it is and what it isn't. BMJ 312(7023):71–72. https://doi.org/10.1136/bmj.312.7023.71

Santini A, Cammarata SM, Capone G, Ianaro A, Tenore GC, Pani L, Novellino E (2018) Nutraceuticals: opening the debate for a regulatory framework. Br J Clin Pharmacol 84(4):659–672. https://doi.org/10.1111/bcp.13496

Sawda C, Moussa C, Turner RS (2017) Resveratrol for Alzheimer's disease. Ann N Y Acad Sci 1403(1):142–149. https://doi.org/10.1111/nyas.13431

Scalbert A, Andres-Lacueva C, Arita M, Kroon PA, Manach C, Urpi-Sarda M, Wishart DS (2011) Databases on food phytochemicals and their health-promoting effects. J Agric Food Chem 59(9):4331–4348. https://doi.org/10.1021/jf200591d

Schultz F, Garbe LA (2023) How to approach a study in ethnopharmacology? Providing an example of the different research stages for newcomers to the field today. Pharmacol Res Perspect 11(4):e01109. https://doi.org/10.1002/prp2.1109

Serrano-Pozo A, Frosch MP, Masliah E, Hyman BT (2011) Neuropathological alterations in Alzheimer disease. Cold Spring Harb Perspect Med 1(1):a006189. https://doi.org/10.1101/cshperspect.a006189

Shahidi F, Ambigaipalan P (2015) Phenolics and polyphenolics in foods, beverages and spices: antioxidant activity and health effects—a review. J Funct Foods 18:820–897. https://doi.org/10.1016/j.jff.2015.06.018

Sibbald B, Roland M (1998) Understanding controlled trials: why are randomised controlled trials important? BMJ 316(7126):201. https://doi.org/10.1136/bmj.316.7126.201

Sies H (2015) Oxidative stress: a concept in redox biology and medicine. Redox Biol 4:180–183. https://doi.org/10.1016/j.redox.2015.01.002

Singar S, Nagpal R, Arjmandi BH, Akhavan NS (2024) Personalized nutrition: tailoring dietary recommendations through genetic insights. Nutrients 16(16):2673. https://doi.org/10.3390/nu16162673

Singh A, Sethi P, Ghosh S, Singh KK, Rohatgi S, Singh P, Sinha JK (2025) Current status of prevention strategies adapted for Alzheimer's disease. In: A new era in Alzheimer's research: pathogenesis, prevention, and treatment. Elsevier, pp 91–108

Sinha JK, Ghosh S, Swain U, Giridharan NV, Raghunath M (2014) Increased macromolecular damage due to oxidative stress in the neocortex and hippocampus of WNIN/Ob, a novel rat model of premature aging. Neuroscience 269:256–264. https://doi.org/10.1016/j.neuroscience.2014.03.040

Sinha JK, Jorwal K, Singh KK, Han SS, Bhaskar R, Ghosh S (2024) The potential of mitochondrial therapeutics in the treatment of oxidative stress and inflammation in aging. Mol Neurobiol, Advance online publication. https://doi.org/10.1007/s12035-024-04474-0

Song JW, Chung KC (2010) Observational studies: cohort and case-control studies. Plast Reconstr Surg 126(6):2234–2242. https://doi.org/10.1097/PRS.0b013e3181f44abc

Spindler M, Beal MF, Henchcliffe C (2009) Coenzyme Q10 effects in neurodegenerative disease. Neuropsychiatr Dis Treat 5:597–610. https://doi.org/10.2147/ndt.s5212

Tansey MG, Wallings RL, Houser MC, Herrick MK, Keating CE, Joers V (2022) Inflammation and immune dysfunction in Parkinson disease. Nat Rev Immunol 22(11):657–673. https://doi.org/10.1038/s41577-022-00684-6

Turner RS, Thomas RG, Craft S, van Dyck CH, Mintzer J, Reynolds BA et al (2015) A randomized, double-blind, placebo-controlled trial of resveratrol for Alzheimer disease. Neurology 85(16):1383–1391. https://doi.org/10.1212/WNL.0000000000002035

World Health Organization (WHO) (2013) Promoting health through the life-course. WHO. Retrieved from https://iris.who.int/handle/10665/205670

Xu J, Jin X, Ye Z, Wang D, Zhao H, Tong Z (2022) Opposite roles of co-enzyme Q10 and formaldehyde in neurodegenerative diseases. Am J Alzheimers Dis Other Dement 37:15333175221143274. https://doi.org/10.1177/15333175221143274

Yang L, Zhao F, Sun Y, Wang Z, Li Q, Wang H, Lu Y (2024) N-3 polyunsaturated fatty acids in elderly with mild cognitive impairment: a systemic review and meta-analysis. J Alzheimer's Dis 99(s1):S81–S95. https://doi.org/10.3233/JAD-220863

Zeisel SH (2020) Precision (Personalized) nutrition: understanding metabolic heterogeneity. Annu Rev Food Sci Technol 11:71–92. https://doi.org/10.1146/annurev-food-032519-051736

Zhang Y, Li X, Zou D, Liu W, Yang J, Zhu N et al (2008) Treatment of type 2 diabetes and dyslipidemia with the natural plant alkaloid berberine. J Clin Endocrinol Metabol 93(7):2559–2565. https://doi.org/10.1210/jc.2007-2404

Zhang X, Zhao Y, Xu J, Xue Z, Zhang M, Pang X et al (2016) Modulation of gut microbiota by berberine and metformin during the treatment of high-fat diet-induced obesity in rats. Sci Rep 5:14405. https://doi.org/10.1038/srep14405

Zhu Y, Tchkonia T, Pirtskhalava T, Gower AC, Ding H, Giorgadze N, Palmer AK, Ikeno Y, Hubbard GB, Lenburg M, O'Hara SP, LaRusso NF, Miller JD, Roos CM, Verzosa GC, LeBrasseur NK, Wren JD, Farr JN, Khosla S, Stout MB et al (2015) The Achilles' heel of senescent cells: from transcriptome to senolytic drugs. Aging Cell 14(4):644–658. https://doi.org/10.1111/acel.12344

MIX
Papier aus verantwortungsvollen Quellen
Paper from responsible sources
FSC® C105338

If you have any concerns about our products,
you can contact us on
ProductSafety@springernature.com

In case Publisher is established outside the EU,
the EU authorized representative is:
Springer Nature Customer Service Center GmbH
Europaplatz 3, 69115 Heidelberg, Germany

Printed by Libri Plureos GmbH
in Hamburg, Germany